IV International Conference on Safety Problems of Civil Engineering Critical Infrastructures (Safety 2018)

IOP Conference Series: Materials Science and Engineering
Volume 481

Ekaterinburg, Russia
4 – 5 October 2018

ISBN: 978-1-5108-8328-4
ISSN: 1757-8981

Printed from e-media with permission by:

Curran Associates, Inc.
57 Morehouse Lane
Red Hook, NY 12571

Some format issues inherent in the e-media version may also appear in this print version.

Printed by Curran Associates, Inc. (2019)

For permission requests, please contact the Institute of Physics at the address below.

Institute of Physics
Dirac House, Temple Back
Bristol BS1 6BE UK

Phone: 44 1 17 929 7481
Fax: 44 1 17 920 0979

techtracking@iop.org

Additional copies of this publication are available from:

Curran Associates, Inc.
57 Morehouse Lane
Red Hook, NY 12571 USA
Phone: 845-758-0400
Fax: 845-758-2633
Email: curran@proceedings.com
Web: www.proceedings.com

Table of contents

Volume 481

IV International Conference on Safety Problems of Civil Engineering Critical Infrastructures

4–5 October 2018, Russian Federation

Accepted papers received: 3 January 2019 Published online: 11 March 2019

Preface

IV International Conference «Safety Problems of Civil Engineering Critical Infrastructures

Organizing Committee

Acknowledgements

Peer review statement

Papers

Assessment of human factor in critical infrastructures

L V Poluyan and M G Malukova.....1

Assessment of large-scale catastrophes in complex engineering systems

N A Makhutov and D O Reznikov.....7

Cyberattack intensity forecasting on informatization objects of critical infrastructures

Y M Krakovsky, A N Luzgin and Y M Ivanyo.....15

Verification of recovery of distribution density function of the random variable by methods of nonparametric statistics

V N Syzrantsev and K V Syzrantseva.....22

Advantages of the rain-flow method at the post-processing stage in comparison with the spectral approach

I Gadolina and R Zaynetdinov.....34

The reduction the load on the MSW landfill by optimizing the waste management system of residential territories

K V Afonin, T S Zhilina and A A Zagorskaya.....43

Structural Design Development of a Float Type Wave Micro Power Plant

I Hossain, V I Velkin, S E Shcheklein and A V Eliseev.....49

Analysis of Interdependence of Arctic Critical Infrastructures as Transportation Networks

S A Timashev and A V Bushinskaya.....55

Cyber Reliability, Resilience, and Safety of Physical Infrastructures

S A Timashev.....66

Assessing risk and resilience of critical infrastructures in the fuel and energy economy involving failure cascading effects

LV Poluyan.....81

Influence of oil spills on tundra degradation

S I Chelombitko.....90

Statistical modelling of impacts on a pile-raft foundation

V O Efimov, A I Osokin, L N Kondrat'eva and N R Stepanova.....96

Structural safety of buildings in excess values of differential settlements

Y A Pronozin, L R Epifantseva and M D Kajgorodov.....102

Constitutive relations for isotropic materials allowing quasi-linear approximation of the deformation law

A A Treschev, A A Bobrishev and L N Shafigullin.....109

An indirect method for determining the dynamic properties of soils at modeling the vibrating effects of technological equipment on the building structures

I Sh Alirzaev, A Sh Rakhimov and N S Sova.....116

Mathematical modeling technology of a foundation slab concreting on a frozen base

V V Molodin, M S Gorlatov and S M Zaika.....123

Geotechnical monitoring of a residential building being erected on strip-shell foundations in "Aquarel" housing complex in Tyumen

Y A Pronozin, L R Epifantseva and M A Stepanov.....131

Empirical Period-Height Relationship for reinforced Concrete Moment Resisting Buildings in India

S P Kewate and M M Murudi.....142

Natural frequencies of a vertical pipeline element

A A Sekacheva, L G Pastukhova, V N Alekhin and A S Noskov.....152

Concept of Organic Architecture in the Second Half of the XXth Century in the Context of Sustainable Development

T Y Bystrova.....162

Efficiency of Installation of Acoustic Structures in Urban Environment

A V Kolmakov and V O Ignatyeva.....169

Dependency of noise level indicators on the heating power of rooftop boiler room

A Plotnikov and T Zhilina.....175

Infranetics: The New MAICS-convergent Technology Science

S A Timashev.....180

Principles of Renovation the Territory of Residential Buildings Dating from 1960s to 1970s: Coping with Modern Housing Crisis

O Ryzhova and A Khrichenkov.....192

Reindustrialization, International Practices, and the Opportunities of Ekaterinburg (the case of the Uralmash district)

A V Alekseeva, S V Pridvizhkin and V G Veniaminov.....200

Big data in urban planning and territory management

E A Pentsev and O A Makarova.....205

Modern Models of Public-Communication spaces of General Education Establishments within the Process of Social Adaptation of Students

I N Maltceva, N N Kaganovich, A I Redinkina and K A Tkachuk.....214

Urban Planning Principals of Sustainable Neo-Industrial Territory Development

M V Pastukhova and L G Pastukhova.....222

Improvement the scheduling based on the reducing of the dimensionality of the system

M G Dobrosotskikh, V Ya Mishchenko and M A Preobrazhenskii.....232

Numerical simulation of thermal bridges in enclosure structures as an express-method for the expert assessment of thermal comfort in domestic premises

N D Korsun, D A Prostakishina and A S Novikova.....241

Increase of energy efficiency during overhaul of housing stock in Russian Federation

V Ya Mishchenko, S G Sheina and E P Gorbaneva.....247

Technical indexes - building energy in tropical climate

Tran Ngoc Chan and Bui Quang Trung.....253

Investigation of heat transfer coefficient of spherical element using infrared thermography (IR) and gas - water droplets (mist) as working medium

A H Abed, S E Shcheklein and V M Pakhaluev.....258

The analysis of the possibility of use of nanofluid as the heat carrier for increase in efficiency of heat supply systems

V V Girfanova, A G Gevorgyan and V I Velkin.....265

Optimization of Work of Ventilation System in the Building with "Warm Attic"

A D Sukhov and M G Ushakov.....269

Resource-saving in the operation of a building with a correctly executed waterproofing

A V Vasilev and V I Yamov.....273

Review of effective tools for monitoring the performance of membrane bioreactors

A S Ilinykh, V A Ilinykh and E V Migalatiy.....278

Climatized module as the basic unit of high-tech medicine. Problems of development and application

V A Tomilin, M Yu Ananin and N D Perfilieva.....283

Reducing the risks of water and territory pollution by toxic metal-containing effluents of machine-building industry

G B Brayalovsky, O B Naschetnikova and E V Migalatiy.....290

Modern Tendencies in Design of Public Spaces in term of Sustainable Development

O S Burukhina, S A Serbin, S V Vartanyan and I N Maltceva.....296

Numerical and experimental studies of gas cleaning in multi-cyclone elements with filter inserts

G I Belyaeva, A T Zamalieva and M G Ziganshin.....304

BIM technologies in customer service business processes. Commercial pricing classifier

S V Pridvizhkin, N S Davydov, A V Belkevich and M K Kozhevnikova.....310

Drying and shrinkage of cement paste for 3D printable concrete

G S Slavcheva.....316

Estimates of the influence of the solar irradiation spectrum energy on the intensity of heat treatment of solid waste with hexavalent chromium

M I Bilalov and M G Ziganshin.....325

Application of the laser cutting of wood-containing materials in construction

A T Gabdrakhmanov, A A Bobrishev and L N Shafigullin.....332

Testing steelfibroconcrete centrifuged specimens of pipes used in drainage and water disposal systems at places of automobile and railway transport heavy traffic

M S Kuznetsov, A S Kuznetsov, A S Noskov and V G Dubinina.....337

Experience of using the ultrasonic low-frequency tomograph for inspection of reinforced concrete structures

M S Kuznetsov, O V Maltseva, A S Noskov and A S Kuznetsov.....343

Modified fine-grained concretes based on highly filled self-compacting mixtures

T A Nizina, A S Balykov, D I Korovkin and V V Volodin.....353

Fly Ash-based Geopolymer Binders: Perspective Materials for Sustainable Building

I K Domanskaya and A Yekimovskaya.....360

Composition and Properties of Non-Fired Fly Ash Gravel and Concrete on Its Basis

F L Kapustin, D V Kokorina and I V Fomina.....366

The analysis of technological parameters of precast-monolithic system with permanent formwork walls

S A Serbin, P O Dedyukhin and N I Fomin.....371

New Design and Technological Solutions for Ensuring Technological Reliability of Erecting Pre-Cast with Cast-in-Place Slabs in Civil Buildings

N I Fomin, K V Berngardt and A V Vorobiyov.....379

Structure of a lesson on the development of the students' creative personality in the CFCT-TRIZ pedagogical system

Vyacheslav Kim, Marina Kim and Michael Kim.....385

Training Business Simulation Game «Commercial Property Management in a Competitive Environment»

A S Salamatina, A A Andronova and I E Alekseeva.....391

The Assortment Approach to the Selection of Building Materials for the Construction of Real Estate

D N Krivogina, N I Safonov and V A Kharitonov.....402

Enhancement of residential construction of urban agglomerations based on the modern financial mechanisms

V Ya Mishchenko, O K Meschcherjakova and M A Meschcherjakova.....410

Ways to Improve the Safety of Built-Up and Adjacent Areas

A E Sheveleva, N R Stepanova and P V Bochkov.....415

Sustainability analysis of the management strategy in a shopping centers

A O Alekseev, V S Spirina and A A Andronova.....423

Technologies of Development Decisions Making in Residential Civil Engineering

A O Alekseev, K S Koskova and E R Galiaskarov.....434

Challenges of Entering International Markets for Russian Construction Companies

A Platonov, V Larionova and Yu Davy.....440

PREFACE

This volume contains conference proceedings of the IV International Conference on safety problems of civil engineering critical infrastructures «Safety2018» held on 4th – 5th October 2018 at Ural Federal University in Ekaterinburg, Russia

The main conference purpose is integration and coordination of efforts of scientists, experts and specialists in the field of integrated risk analysis related to building interdependent critical infrastructures throughout their life cycle at the international level. Design, fracture mechanics problems, diagnostics, monitoring, maintenance, survivability and optimal management of infrastructure systems issues is considered in the context of the creation and operation of safe, smart and durable infrastructure systems. Question of the creation of real conditions and a practical mechanism for an innovative scientific, educational and organizational breakthrough is one of the most important problems of the twenty-first century – safety and sustainable development of cities, regions and worldwide taking into account the human factor.

The objectives of Safety2018 are to collect and disseminate state-of-the-art research and technology for design, fracture mechanics problems, diagnostics, monitoring, maintenance, survivability and optimal management of infrastructure systems issues is considered in the context of the creation and operation of safe, smart and durable infrastructure systems.

The conference proceedings contain 60 papers, which are divided into 9 sections:

Safety technologies for building critical infrastructure and territories; Vitality of interdependent critical infrastructures; Statistical modeling of loads and impacts on building structures and buildings; Problems of viable and smart cities and sustainable development of territories; Energy efficiency and resource saving in the civil engineering; Information modeling technology in the construction industry; Building technology and materials of innovative type; Education in architecture and civil engineering; Expertise and management in construction.

We thank all authors of submitted papers for their participation. They contributed a great deal of effort and creativity to produce this work. We appreciate that they have chosen the IV International Conference on safety problems of civil engineering critical infrastructures to present their work. Many thanks also to all editors and reviewers, who donated substantial time to carefully read and evaluate the submitted papers.

At last but not at least, we would like to appreciate to all supporting organizations for their support and encouragement and making the event a success.

In addition, we would like to thank also to Organizing Committee for their great preparing work.

We look forward to the next 5th International Conference on Safety Problems of Civil Engineering Critical Infrastructures «Safety2019» that will be held next year on May at the same location. We hope that it will as be interesting and enjoying as all four predecessors.

Ekaterinburg, October 2018

Professor Vladimir Alekhin	Professor Lawrence Boswell	Professor Svyatoslav Timashev

IOP Conf. Series: Materials Science and Engineering **481** (2019) 011002 doi:10.1088/1757-899X/481/1/011002

Organizing Committee

Chairman

V.A. Koksharov – UrFU, Rector

Vice-chairmen:

A. V. Krupkin – Institute of Civil Engineering and Architecture UrFU, Director

V. N. Alekhin – Institute of Civil Engineering and Architecture UrFU, Deputy Director for Strategic Development and Infrastructure Projects – First Deputy Director, Head of the Department «CAD Systems in Civil Engineering », Professor; Member of Bureau UTB RAACS; Adviser RAACS;

S. A. Timashev – Science and Engineering Center "Reliability and Safety of Large Systems and Machines" SEC UB RAS, Academic Leader and Principal Scientist; Institute of Civil Engineering and Architecture UrFU, Department «CAD Systems in Civil Engineering», Professor;

G.V. Mazaev – UTB RAACS, Chairman, Academician RAACS;

A. Musaio – University of Genoa, Head of Support Service for International Strategies;

D. V. Ulrikh – SUSU, Institute of Architecture and Civil Engineering, Director;

V.Z.Marukhyan – NPUA, Rector

Committee members:

V. Bianco – University of Genoa, As. Professor;

L. Boswell– City University, London, Professor, the President of the Association European Civil Engineering Education and Training – EUCEET;

G.Capannelli – President of TICASS Consortium, Professor, Italy;

I. Fontanals– OptiCits, Barselona, Spain;

A. Gheorghe – Old Dominion, USA, Professor;

D. Lo Presti – University of Pisa, Professor; the Vice-President of the Association European Civil Engineering Education and Training – EUCEET;

R. Aghgashyan – NPUA, Vice-Rector;

P. Akimov – RAACS, Executive Scientific Secretary, Academician RAACS, Professor;

A. Amirkhanian – American University of Armenia, "Acopian Center for the Environment", Director;

A. A. Antipin – Institute of Civil Engineering and Architecture UrFU, the Department «CAD Systems in Civil Engineering», As. Professor;

Z. V. Beliaeva – Institute of Civil Engineering and Architecture UrFU, Head of the Department «Building structures and soil mechanics»;

E. S. Guryev – Science and Engineering Center "Reliability and Safety of Large Systems and Machines" SEC UB RAS, Scientific Secretary of the Academic Council; Institute of Civil Engineering and Architecture UrFU, the Department «CAD Systems in Civil Engineering», As. Professor;

I. N. Maltseva – Institute of Civil Engineering and Architecture UrFU, Deputy Director for International Relations, the Department "Architecture", As. Professor;

O. V. Mashkin – Institute of Civil Engineering and Architecture UrFU, the Department «Industrial, Civil Engineering and real estate expertise», Senior Lecturer;

O. B. Naschetnikova – Institute of Civil Engineering and Architecture UrFU, the Department «Water management and water technology»;

A. S. Noskov – Institute of Civil Engineering and Architecture UrFU, Head of the Department «Hydraulics», Professor; Adviser RAACS;

L. G. Pastukhova – Institute of Civil Engineering and Architecture UrFU, Deputy Director for Science and Innovation, the Department «Hydraulics», As. Professor;

L. V. Poluyan – Science and Engineering Center "Reliability and Safety of Large Systems and Machines" SEC UB RAS, Director; Institute of Civil Engineering and Architecture UrFU, the Department « CAD Systems in Civil Engineering», As. Professor;

S. V. Pridvizhkin – Institute of Civil Engineering and Architecture UrFU, Head of the Department « Pricing in Construction and Industry », Professor; Adviser RAACS;

V. I. Travush – Vice-president RAACS, Academician RAACS, Professor;

A. V. Khrichenkov A.B. – Institute of Civil Engineering and Architecture UrFU, Head of the Department «City Planning and Municipal Engineering»;

N. P. Shiriaeva – Institute of Civil Engineering and Architecture UrFU, Deputy Director for Education, Head of the Department « Heat-, Gas- Supply and ventilation»;

V. V. Shikhirin – Institute of Civil Engineering and Architecture UrFU, Head of the Department «Industrial, Civil Engineering and real estate expertise».

ACKNOWLEDGEMENTS

The conference organizers acknowledge the support towards the organization of the IV International Conference «Safety Problems of Civil Engineering Critical Infrastructures» (Safety2018) to the following organizations:

Ural Federal University named after the first President of Russia B.N.Yeltsin (UrFU)

Ural Branch of the Russian Academy of Sciences, Science and Engineering Center "Reliability and Safety of Large Systems and Machines" (SEC UB RAS)

Ural Territorial Branch of the Russian Academy of Architecture and Construction Sciences (UTB RAACS)

University of Genoa, Italy

South-Ural State University (SUSU)

National Polytechnic University of Armenia (NPUA)

The European Civil Engineering Education and Training EUCEET – Association

Peer review statement

All papers published in this volume of *IOP Conference Series: Materials Science and Engineering* have been peer reviewed through processes administered by the proceedings Editors. Reviews were conducted by expert referees to the professional and scientific standards expected of a proceedings journal published by IOP Publishing.

IOP Conf. Series: Materials Science and Engineering **481** (2019) 012001 doi:10.1088/1757-899X/481/1/012001

Assessment of human factor in critical infrastructures

LV Poluyan[1,2], M G Malukova[1]

[1]Science and Engineering Centre «Reliability and Safety of Large Systems and Machines», Ural Branch, Russian Academy of Sciences, Studencheskaya, 54-A,Yekaterinburg, Russia, 620049
[2]Ural Federal University, Mira street, 19Yekaterinburg, Russia, 620002

E-mail: sec@wekt.ru

Abstract. Methodology for assessing human error probability under emergency conditions, taking into account internal and external factors, the degree of uncertainty, which can influence the reliability of a "man-machine" system, is presented. A model based on the HRA (Human reliability analysis) methodology is used. Probability of a human error is determined, which combines the nominal error and the influence of environmental conditions. The error is compared to the rated (design) permissible critical value dependent on the system state. It enables steadily monitoring the technical state of the system. In case of an unsatisfactory state, correcting actions are performed and probability for a human error is again assessed as well as the extent of the system improvement is tested. The methodology uses reliability criteria taking into account the degree of human involvement into the process of controlling critical infrastructures. The methodology includes efficiency analysis and assessment of the safeguarding system with regard to the human factor. Currently, there is no special documentation setting the procedure of human behavior under emergency conditions by means of quantitative methods. The methodology can serve the purpose of developing a practical manual for people having to make decisions; it can be used in training the personnel for acting in emergency situations. It is useful for assessing the influence of human factor on operation of critical infrastructures used for various purposes.

1. Introduction

The standard of the SSSRIEC 62508-2014 (the Russian State Standard Specification, the International Electrical Commission) [1] cites the first manual on system reliability in the part concerning human activities. The standard can be applied in any field of man-machine interaction. It is designed to be used by engineering personnel and heads of organizations. Consideration is given to ensuring reliability related to the Human Factor (HF) and aimed at methods of a human-centered approach to design-engineering and reliability improvement. These techniques can be used at all stages of a system life cycle. The document allows taking into account the human effect on system reliability, involving application of human engineering principles.

Analyzing the overall reliability of a technical system, it is necessary to take into consideration all aspects of human activity, including its positive features, constraints, opportunities and areas for improvement [2–7]. If operator is part of a complex technical system with safety ensured, all possible unfavourable consequences of outages caused by HF (errors, violations, omissions or malicious acts) are critical.

The standard presents only qualitative methods and a brief review of foreign quantitative techniques. To date, the remaining challenge is developing manuals on quantitative assessment of HF.

2. Methodology for human factor assessment in critical infrastructures

The system the operation of which necessarily involves man-machine interaction includes man (operators/process controlling men), machine (equipment, mechanisms) as well as social and physical environment where the interaction takes place. Man can be involved in various stages of the technical system life cycle, he can influence its operational reliability by virtue of his activity and making decisions in accordance with the functional responsibilities (as a designer, a manual developer, a functional manager and observer, an operator, an instructor, a maintenance and repair force) [2, 8-13,17 - 19].

Human error probability (HEP) is a measure of the likelihood that plant personnel will fail to initiate the correct, required, or specified action or response in a given situation, or by commission will perform a wrong action. The HEP is the probability of the human failure event.

Performance shaping factor (PSF) is a factor that influences human performance and human error probabilities. In SPAR-H, this includes: time available, stress/stressors, complexity, experience/training, procedures, ergonomics/human-machine interface, fitness for duty, and work processes [2].

Negative PSFs in SPAR-H, negative performance shaping factors (PSFs) are those PSF values that increase the nominal value rate, i.e., the PSF values which are greater than 1 are referred to as negative PSFs and figure in conjunction with positive PSFs in the overall HEP calculation [2, 3]. When the number of negative PSFs is three or greater, then the HEP adjustment factor is applied.

2.1. Probability assessment of human error under emergency conditions

Initial data:

- The number of people (operators) involved into the system operation – n.
- Working time of the n-th operator – t_n.
- Parameter characterizing reliability of the n-th operator – k_n.
- Characteristics of the environmental influence on the human reliability – PSFs factors: PSF_1, $PSF_2, \ldots, PSF_m$.
- The number of external factors – m.
- Parameter describing the environmental conditions- PSF_{comp}.
- Permissible critical value of human error probability - HEP_{lim}.

2.2. Main methodology stages

In accordance with the research in [8] (included into the book [8]), it is possible to define:

1. Setting the number of n individuals (operators) who can be involved into the failure (the failure can be caused by a chance, system outage, a human error, etc.).

2. Assessing probability of a human error with regard to internal and external factors which can influence human reliability. The methodology (algorithm) makes use of the hybrid model based on HRA methodology and which takes into account: 1) internal human factors (based on the application of the Weibull Distribution function); 2) external factors related to the environmental conditions (using PSFs factors specifying the system performance).

In case 1, the Weibull Distribution function enables determining a human error, taking into consideration the difference between the first hour of the operator's work and other time periods:

$$HEP_{nom}(t_n) = 1 - k_n \cdot e^{-\alpha(1-t_n)^\beta}, \forall\ t_n \in [0; 1], \quad (1)$$

$$HEP_{nom}(t_n) = 1 - k_n \cdot e^{-\alpha(t_n-1)^\beta}, \forall\ t_n \in]1; \infty[, \quad (2)$$

where HEP_{nom} is the rated probability of the human error dependent on human behavior only; t_n is a moment of the operator' work; k_n is the parameter specifying reliability of the n-th operator; β is the parameter defining the reliability curve shape (it is noted in [8] that human behaviour is best characterized using the Weibull Distribution function with β parameter equal to 1.5); α is the parameter dependent on the reliability of the n-th operator (k_n values):

$$\alpha_n = \frac{-\ln[k_n]}{(t_n - 1)^{\beta}} \tag{3}$$

Figure 1 a, b demonstrate 6 graphs of combining two Weibull distributions with various combinations of k_n and α parameters; this distribution functions' combination is used to determine probability of a human error in the intervals of 0 to 1 and 1 to 8 hours of the operator's work. Only internal factors are taken into account in the calculated rated probability of the human error.

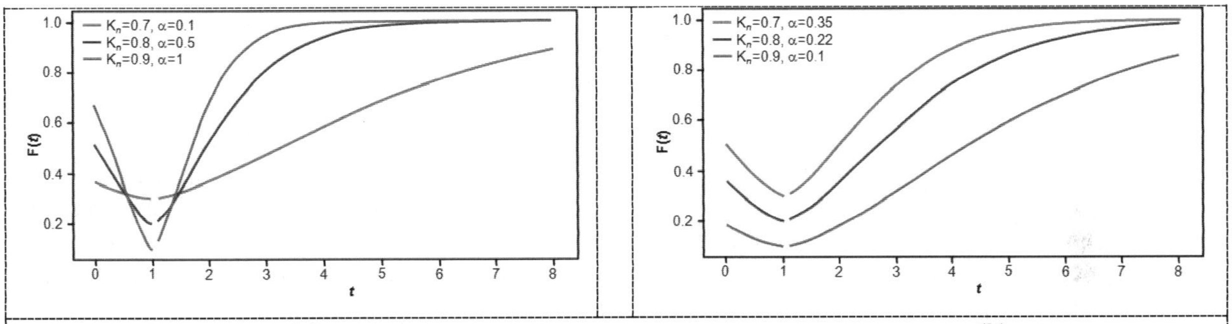

(a) **(b)**

Figure 1. Combination of two Weibull functions used to determine probability of a human error

In case 2, the effect external factors of environmental conditions have on operators' reliability is taken into account in the model by introducing PSFs factors (PSF_1, PSF_2,…, PSF_m) [2, 8, 15].

The product of all PSFs determines the total effect environmental conditions have on operators:

$$PSF_{comp} = PSF_1 \cdot PSF_2 \cdot \ldots \cdot PSF_m \tag{4}$$

Contextual probability of a human error, combining the human rated error and the influence of the environment is calculated by [2, 4, 8]:

$$HEP_{cont} = \frac{HEP_{nom} \cdot PSF_{comp}}{HEP_{nom} \cdot (PSF_{comp} - 1) + 1}, \tag{5}$$

where HEP_{cont} is the contextual probability of a human error dependent on human behavior and environmental conditions; HEP_{nom} is the human rated probability of a human error dependent on human behavior only; PSF_{comp} is the parameter determining the total effect environmental conditions have on operators.

3. Comparing the contextual *probability* value *of a human error* HEP_{cont} and the rated *permissible critical value* HEP_{lim} dependent on the system state. If the values HEP_{cont} and HEP_{lim} are close to each other, the system doesn't need any improvement. Otherwise improvement is necessary to reduce the probability of a human error.

4. Determining a set of KPIs (key performance indicators) for the system analysis (in particular, KPIs are to deal with external factors influencing probability of a human error. If the influence of PSFs factors representing environmental conditions is measurable, then they can be used as KPIs (otherwise they should be transformed into measurable elements).

5. System improvement is performed in order to reduce probability of a human error. In doing so, KPIs indicators are monitored using, for example, an instrument panel allowing keeping track of KPIs values.

6. At this stage, a *new probability of a human error* is assessed and the extent of the system improvement is examined. If probability of a human mistake HEP_{cont} in this case is lower than the rated critical value HEP_{lim}, then this probability is considered to be permissible and the improvement measures taken to be successful.

7. Reliability assessment of the "man-machine" system [14]. Man is represented as a system component. Cases are specified for reliability assessment of the system with the interaction of technical facilities and a human operator, assuming that equipment failures and operator's errors are rare, simple, separate events. Occurrence of more than one failure is of the same type; failures in the course of the system operation from t_0 till $t_0 + t$ are unfeasible, the operator's ability to compensate for errors and ensure the faultless operation is an independent operator's feature. If it is impossible to compensate for the errors made by a human operator and equipment failures, then probability of the system failure – free operation is $P_1(t_0,t) = P_T(t_0,t)P_0(t)$, where $P_T(t_0,t)$ is the probability of failure-free operation of technical facilities for a time (t_0,t_0+1); P_0 *(t)* is the probability of the operator's errorless work in the course of time *t* provided the equipment operation is trouble-free; t_0 is the total system operation time; *t* is the time interval under consideration. With the immediate compensation for the operator's errors and *p* probability, the probability for the failure-free system operation is $P_2(t_0,t) = P_T(t_0,t)\{P_0(t)+[1-P_0(t)]\}$. In case of only technical facilities' failure, the probability for the failure – free system operation is $P_3(t_0,t) = P_0(t)[P_T(t_0,t)+P_{COND.}(t_0,t,\tau)]$, where $P_{COND}(t_0,t,\tau)$ is the conditional probability of the failure-free system operation for a time $(t_0 + t)$, with failure consequences being compensated for, provided that a failure occurs at the moment of τ, where $t_0 < \tau < t_0 + t$. The probability for the failure-free system operation, including compensation for the operator's errors and facilities' failure, is $P_4(t_0,t) = \{P_0(t)+[1-P_0(t)]p\}[P_T(t_0,t)+P_{COND}(t_0,t,\tau)]$. A considerable gain in reliability by the probability G_p for the failure-free operation through errors and failures compensation is characterized by the relationship: $G_p = [P_4(t_0,t)]/[P_1(t_0,t)]$. Reliability gain increases as *p* and $P_{COND}(t_0,t,\tau)$ increase, i.e. when the operator's level of training (expertise) to compensate for the errors and failures has been improved. If systems are investigated with regard to the extent to which human involvement into the system control is uninterrupted, there can be distinguished corresponding reliability criteria for each of these "man-machine" system (MMS) types. Probability for a failure-free, errorless and appropriate progress of the process under control for a *t* time given is such a criterion for systems of the first type. This progress of the process is possible in the following cases: 1) there are no outages in the equipment operation; 2) there happened to be an outage but in the course the operator accurately and timely took the necessary measures to eliminate the emergency situation; 3) the operator made erroneous actions but corrected them just-in-time. In accordance with the designation assumed earlier, the reliability of the MMS will be written as follows: $P_1 = P_T(t) + [1-P_T(t)]\cdot K_{rf}[P_0 P_{toa} + (1-P_0)P_{rec}]$, where $K_{rf} = 1-(T_n/T)$ is readiness factor characterizes the probability of including a human operator in the work at any arbitrary time, T_n is time during which the human operator cannot accept information received to him, T is total operating time of the human operator; $P_{toa} = (N - N_{\text{unt.n.t.}})/N$ is timeliness of the operator's actions is assessed by the probability of the task within a given time; *N* и $N_{\text{unt.n.t.}}$ are the total and untimely number of tasks; $P_{rec} = P_{\text{scs}} \cdot P_{\text{ds}} \cdot P_{\text{corr.err}}$ is the recoverability of the operator estimating by the probability of correction of the error, P_{scs} is the probability of a signal from the control system, P_{ds} is the probability that the operator will detect the signal, $P_{corr.err}$ is the probability of correcting erroneous actions when the entire operation is again performed. For MMSs of the second type, the reliability criterion is probability for a failure - free, errorless and appropriate problem solution. The system can perform the task in the event that at the required moment the operator is ready to receive the incoming information and, in addition: 1) there was no equipment outages during the temporary stop and the time of performing the task, the operator took accurate and correct measures or 2) there was an equipment outage but the operator timely eliminated it and made no errors in solving the problem, or 3) in the case of failure-free operation of the equipment the operator made an error but timely and accurately compensated for it. Reliability calculations will look like $P_2 = K_{rf}[P_T P_0 P_{toa} + (1-P_T)P_{ER}P_0P_{toa} + (1-P_0)P_T P_{rec}]$, where P_{ER} is the equipment reconditioning (ER) probability. For the systems of the third type, the reliability criterion is similar to that for the second type. The task can be considered to be performed by the system if 1) at the moment required, the equipment was in operable condition, there was no equipment outage during the time of performing the task, the operator's actions were timely and errorless, or 2) the equipment which wasn't ready or was malfunctioning was timely reconditioned while the operators made an error; 3) there was no equipment outage but the operator made an error and timely compensated for it. In this case, reliability can be calculated by $P_3 = K_r P_T P_0 P_{toa} + (1-P_T K_r)P_{ER}P_0P_{toa} + (1-P_0)P_T P_{rec}$, where K_r is the equipment availability. The correct and valid account of HF at each of these stages contributes to ensuring maximal efficiency and reliability.

8. Efficiency analysis of the safe guarding system (SGS). There is an uncertainty in determining probability of a human error. An interesting approach to allowing for HF in the SGS efficiency analysis is considered in [15, 16]. Risk assessments of safety disturbance caused by human errors enable determining the value of residual entropy of the measures taken to eliminate HF, using:

$$Э_{\text{ост}}^{\text{чф}} = -\log P^{\text{hf}}, \tag{6}$$

where P^{hf} is the efficiency of safety hazard countermeasures, related to HF.

The risk of violating the security of critical infrastructure with probability of a human error will be written as:

$$R = 1 - C^{-\log P^{\text{hf}}}, \tag{7}$$

where C – the base of the logarithm.

3. Future work

Further investigations in modeling human behavior in stress situations for various fields of industry are necessary. There is an urgent need in developing an algorithm based on this methodology and its software for critical infrastructures used for various purposes. Further investigation and a complete implementation of the methodology proposed in a critical infrastructure of the fuel and energy sector is planned.

4. Conclusion

A comprehensive procedure for assessing probability of a human error under emergency conditions has been developed, which is based on the human reliability analysis methodology taking into account human factors using a combination of the Weibull distributions functions) and external factors related to the environmental conditions (using a table of factors contributing to the system efficiency (serviceability). The Weibull function enables determining probability of a human error, taking into consideration the difference between the first hour of the operator's work and other time periods. The procedure assesses reliability of a "man-machine" system, efficiency of the safeguarding system in a critical infrastructure. Risk assessments of safety disturbance caused by human errors and the safeguarding system enable determining the value of residual entropy of the measures taken to eliminate HF. As nowadays, there is no special documentation setting the procedure of human behavior under emergency conditions by means of quantitative methods, the procedure proposed can be used to fill in the gap. It enables developing adequate and efficient techniques for mitigating harmful influence of human errors on safety of the critical infrastructures.

References

[1] The standard of the SSSRIEC 62508-2014 (the Russian State Standard Specification, the International Electrical Commission) Risk management. Impact analysis to dependability of human aspects 2015

[2] NUREG/CR-6883 INL/EXT-05-00509. 2005 The SPAR-H Human Reliability Analysis Method. Prepared for Division of Risk Analysis and Applications Office of Nuclear Regulatory Research U.S. Nuclear Regulatory Commission Washington, DC 20555-0001 NRC Job Code W6355 230 pp.

[3] NUREG-1921 2009 EPRI/NRC-RES Fire Human Reliability Analysis Guidelines 322 pp.

[4] Akhmedzhanov F M, Krymsky V G 2017 Interval-probabilistic algorithm for the evaluation of human reliability: a new modification of the method SPAR-H. J *Electrical and informational complexes and systems* №4 **1**(**13**) pp 81-91

[5] Barseghyan P 2017 Elements of the mathematical theory of human systems (Part 1: Assessment of the Results of Human Actions and Activities Based on the Method of State Equations *PM World Journal* **VI** Issue XII [https://pmworldjournal.net/wp-content/uploads/2017/12/pmwj65-Dec2017-Barsegyan-mathematical-equation-of-human-systems-part1-featured-paper.pdf]

[6] IEC 60300-2 Dependability management - Part 2:Guidelines for dependability management)

[7] IEC 60300-3-15 Dependability management - Part 3-15: Application guide – Engineering of system dependability).

[8] Fabio De Felice, Antonella Petrillo, and Federico Zomparelli Fabio 2018 *Human Factors and Reliability Engineering for Safety and Security in Critical Infrastructures Decision Making, Theory, and Practice* 258 p

[9] Stephen J. Guastello 2014 *Human Factors Engineering and Ergonomics: A Systems Approach* (CRC Press, Taylor&Francis Group) 417 p

[10] Dhillon B S 2017 *27th European Safety and Reliability Conference* (CRC Press) ISBN 9781138629370

[11] Marko Cepin, Radim Bris 2017 *27th European Safety and Reliability Conference* (CRC Press) ISBN 9781138629370

[12] Moura R, Patelli E, Levis J, Morais C, Beer M 2017 *27th European Safety and Reliability Conference* (CRC Press) ISBN 9781138629370

[13] Jang S, Jae M 2017 A development of a methodology to calculate human error probability of severe accident mitigating strategy using dynamic HRA method *27th European Safety and Reliability Conference* (CRC Press) ISBN 9781138629370

[14] Lapin V L, Popov V M, Ryzhkov F N and Tomac V I 1995 *Safe human interaction with technical systems* (Kursk: KSTU) 254 pp.

[15] Tsygichko V N 2016 Estimation risks of a security breach for critical object and critical infrastructure *Scientific and Practical Journal Issues of Risk Analysis* **13** No. 5 pp 6-11

[16] Tsygichko V N 2010 *Guide to decision-making* (Moskow: KASAND) 352 p

[17] Fabio De Felice, Antonella Petrillo, and Federico Zomparelli 2017 Human Factors Challenges in Disaster Management Scenario *27th European Safety and Reliability Conference* (CRC Press) ISBN 9781138629370 pp 171-187

[18] Timashev S A, Poluian L V, Yurchuk Z V 2004 Role of human factor in pipeline safety *International Pipeline Conference* Volumes 1,2 and 3 pp 2471-2478 doi:10.1115/ipc2002-0297

[19] Timashev S A, Poluian L V, Yurchuk Z V 2005 The human dimension of pipeline integrity and safety. *J. Pipeline Integrity* (Houston, USA) **4** No 1 pp 27-53

Assessment of large-scale catastrophes in complex engineering systems

N A Makhutov [1], D O Reznikov[2]

[1] Mechnical Engineering Research Institute of the Russian Academy of Sciences. 4 Maly Kharitonievsky lane, Moscow, 101990 Russia
[2] Mechnical Engineering Research Institute of the Russian Academy of Sciences. 4 Maly Kharitonievsky lane, Moscow, 101990 Russia

E-mail: mibsts@mail.ru

Abstract. The paper considers a traditional approach to ensuring safety of complex technical system that is based on development of protection barriers against the bounding set of credible (design-basis) impacts and presents a modern approach. The approach is based on combined efforts to improve the system protection from design basis impacts and to increases the system resilience towards beyond design basis events allowing one to include low probability –high consequences (beyond design-basis) events into the scope of risk assessment and management framework

1. Introduction

Modern society cannot exist without stable and reliable operation of complex engineering systems (*CES*) whose functioning is vital for any national economy and constitute the back-bone of modern industry and society. Such systems include electric power transmission and distribution; rail and road transportation systems; oil, gas, and water processing, storage and delivery; telecommunication and information networks. Their performance is commensurate with storing, processing and transmission of huge amounts of hazardous substances, energy and information. The unauthorized release of these substances, energy and information may cause disastrous consequences and trigger cascading failures in interrelated infrastructures.

An extensive bank of knowledge has been created by scientists of different countries for analysis and classification of natural and man-made emergencies, and in-depth study of the processes of their initiation and propagation in order to reduce vulnerability of *CES* in case of natural or manmade catastrophes. Currently available methodologies of risk assessment and *CES* design codes have however been developed without accounting for risks of beyond design-basis accidents. It means that these methodologies and codes should be revised to include into consideration risks of beyond design-basis impacts on complex engineering systems.

2. Fundamentals of safety analysis for complex systems

Complex engineering systems are referred to the group of high risk facilities. The failure to provide for basic characteristics of strength, reliability, resilience and robustness with regard to a range of criteria leads to the possibility of accidents and catastrophic situations arising and developing at all stages of the life cycle of high risk facilities. Over the past decade, institutes of the Russian Academy of Sciences, the Russian Ministry of Emergency Situations, the Russian Ministry of Education and Science, have synthesized a substantial volume of fundamental information on accidents and catastrophes of an industrial and natural character as part of the State Scientific-Technical Program for *Safety for Population and Economic Facilities Considering the Risk of Natural and Manmade Disas*ters ("SSTP Safety"). These efforts were continued in the frame of the Federal Research Program *Reduction of Risks and Mitigation of Consequences of Natural and Manmade Emergencies in the Russian Federation'***.** In carrying out these programs, participants analyzed and generalized information on the basic characteristics, conditions, and scenarios of accidents and catastrophes in the natural and industrial spheres engendered by complex dangerous phenomena and processes in various regions of the world. Hazardous components of high risk facilities (atomic power stations, petroleum chemical plants, hydropower engineering facilities, etc.) might create catastrophes in the following classes (from 7 to 1): planetary, global, national, regional, local, facility-level, and localized (Figure 1). The potential losses and periodicity of occurrence were evaluated depending on the class of accidents and catastrophes (beginning from the global and ending with the localized) [1].

Legislation and regulatory framework in the Russian Federation uses six classes of catastrophes (from 6 to 1): global, international, national, regional, local, facility-level.

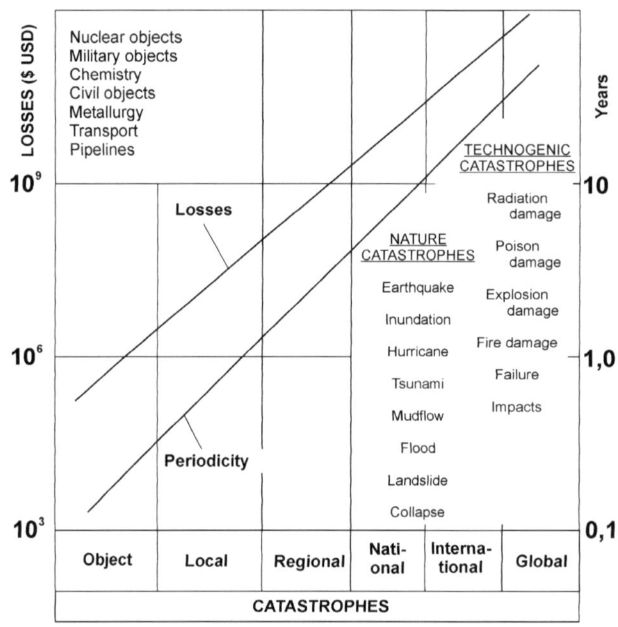

Figure 1. Losses and periodicity of natural and manmade catastrophes

Based on the results of this summary analysis, a classification of catastrophes was constructed, taking into account the losses U and the periodicity ΔT of their occurrence (see Table 1). Here losses U for each catastrophe decreases from $10^{10}\div10^{11}$ to $10^3\div10^4$ dollars, while the periodicity of their occurrence declines from $(3\div5)\cdot10^1$ to 10^{-1} years. Thus, the variation in losses (dollars per catastrophe) for various types of disasters could reach seven orders of magnitude, while that of the probability of occurrence $P = 1/\Delta T$ (1/year) could reach three orders of magnitude.

The concept of risk is the key one in resolving problems related to ensuring safety. Risk is defined by means of the functional F_R:

$$R = F_R\{P,U\} = \sum_{i=1}^{n} R_i = \sum_{i=1}^{n} P_i \cdot U_i = \int U(P)\cdot dP = \int P(U)\cdot dU \qquad (1)$$

where R represents the risk associated with a natural or manmade catastrophe, P is its probability, and U is loss.

Table 1. Risks of accidents and catastrophes

No.	Class of accidents and catastrophes	P (1/year)	U (dollars)	R (dollars/year)
1	Object	$5.0\cdot 10^{0}$	$5.0\cdot 10^{3}$	$2.5\cdot 10^{4}$
2	Facility-level	$1.2\cdot 10^{0}$	$4.0\cdot 10^{5}$	$4.8\cdot 10^{5}$
3	Local	$5.0\cdot 10^{-1}$	$7.0\cdot 10^{6}$	$3.5\cdot 10^{6}$
4	Regional	$1.6\cdot 10^{-1}$	$1.0\cdot 10^{8}$	$1.6\cdot 10^{7}$
5	National	$1.2\cdot 10^{-1}$	$1.5\cdot 10^{9}$	$1.8\cdot 10^{8}$
6	Global	$8.0\cdot 10^{-2}$	$1.0\cdot 10^{10}$	$8.0\cdot 10^{8}$

Here risks vary within the bounds of four orders of magnitude. For Russia, the probability of the occurrence of national and regional natural-manmade emergencies differ by 1.4 times and are approximately on the order of magnitude lower than the risk for local situations; the likelihood of local and facility-level accidents differ by 5 times.

The assessment of the probability P, losses U, and risks R of accidents and catastrophes involves a group of risk assessment methods, including various methods for analyzing statistical information on natural and manmade catastrophes of a particular type in the region being studied, as well as methods for analyzing the reliability of equipment and technological processes and the effectiveness of management and control. Methods for calculating the magnitude of loss substantially differ for various technical facilities and natural systems. Therefore, specialists in Russia and other countries have developed a group of special methods aimed at analyzing natural-manmade processes capable of leading to accidents and catastrophic situations at engineering infrastructures.

These methods are based on special graph models called scenario trees. The system is designed to fulfill a so called success scenario S_0 (i.e. a transition from its initial state IS to the designed end state ES_0). Since any failure scenario S_i (Figure 2) presents a deviation from the success scenario S_0 that corresponds to the successful functioning of the CES, the scenario S_i must have a disturbance point at which an extreme event (loading regime) A_i $(i = 1, 2, ..., m)$, occurs (Fig. 2). Each attack gives rise to a branch of a scenario tree that has a corresponding scenario S_j $(j = 1, 2, ..., n)$ that ends with an end state ES_i [2, 3].

In this case one can get a risk index using the following matrix expression:

$$R = \underbrace{\{P(A_1); P(A_2); \ldots; P(A_n)\}}_{Threat\ \mathbf{T}} \times \underbrace{\begin{bmatrix} P[ES_1 \mid A_1] & P[ES_2 \mid A_1] & \cdots & P[ES_m \mid A_1] \\ P[ES_1 \mid A_2] & P[ES_2 \mid A_2] & \cdots & P[ES_m \mid A_2] \\ & \cdots & & \\ P[ES_1 \mid A_n] & P[ES_2 \mid A_n] & \cdots & P[ES_m \mid A_n] \end{bmatrix}}_{Vulnerability\ \mathbf{V}} \times \underbrace{\begin{Bmatrix} U_{ES_1} \\ U_{ES_2} \\ \cdots \\ U_{ES_m} \end{Bmatrix}}_{Consequences\ \mathbf{C}} \quad (2)$$

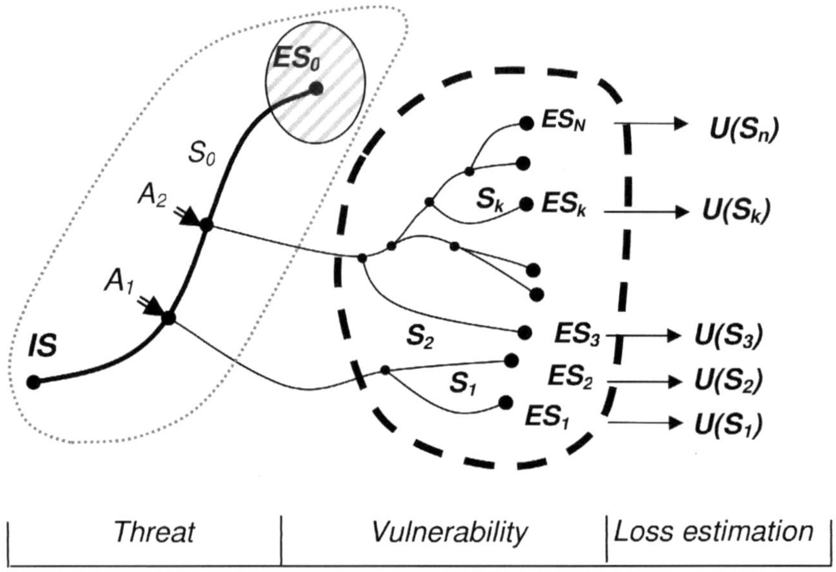

Figure 2. General risk assessment framework.

3. New approach to ensuring safety of complex systems: moving from "Protection" to "Protection + Resilience"

The complexity of modern engineering systems and their interdependence with other systems makes them vulnerable to emergency situations triggered by natural and manmade catastrophes or terrorist attacks. These complexities largely stem from the vast functional and spatial dependencies and nonlinear interactions between the components of *CES* as well as from interdependencies that exist among the *CESs*, which enable failures to cascade within one system and from a system to other system.

Different historical, economical, political, social as well as cultural traditions have formed different approaches to ensuring safety of *CES*. Contemporary *CES*s, i.e. power, transport, and telecommunication networks, are becoming transbordery. Their significant spatial extension makes their functioning dependent on many factors and events in different parts of the world. The ensuring of complex engineering systems' safety is a complex interdisciplinary problem. It is impossible to solve this problem without joining efforts of experts in different fields and taking into account technical, social, psychological, and cultural-historical aspects.

Analysis of major disasters at *CES* in different countries shows that high-risk systems in many cases are being designed and constructed according to traditional design codes and norms that are based on common and quite simple linear 'sequential' risk assessment models and employ traditional design, diagnostics and protection methods and procedures. This is being done in the assumption that a bounding set of credible (design-basis) impacts and subsequent failure scenarios could be determined for the *CES* thus allowing one to create a system of protection barriers and safeguards that could secure the *CES* from the identified impacts with required substantially high probability. This bounding set of impacts referred to as design-basis impacts includes both normal and abnormal operation events (components failures, human errors, extreme environmental loads, intentional unauthorized impacts on *CES*) that are expected to occur or might occur at least once during the lifetime of the *CES*.

In this approach a number of low probability impacts of extreme intensity are neglected as being practically incredible. Other impacts or impact combinations are not identified and, consequently, not analyzed. Such impacts and impact combinations of different types are classified as beyond design-basis impacts. Thus the issue of protection of CES from beyond design-basis impacts has not been addressed in a proper manner. These impacts however can cause large-scale disasters of extreme

IOP Conf. Series: Materials Science and Engineering **481** (2019) 012002 doi:10.1088/1757-899X/481/1/012002

severity and induce tremendous property losses and a great number of victims. Such disasters are referred to as beyond design-bases accidents.

A typical example of a beyond design-basis accident is the recent accident at the Fukushima NPP where the height of the Tsunami wave exceeded the designed level. Another example is the Chernobil catastrophe that was caused by a combination of technical failures, operator's errors and organizational violations.

It should be noted that the efforts on protection of *CES* has been traditionally focused on technical issues. All this allows one to substantially increase reliability of the technical components of *CESs*. This approach however is close to its exhaustion. This is due to the fact that *CESs* are no longer predominantly technical but socio-technical systems. The performance of *CES* is becoming more and more dependable on human and organizations factors. The statistics of accidents at *CES* reads that from 70 to 90% of accidents at complex engineering systems are caused by human errors made at the stage of design, construction, maintenance and operation of complex engineering systems. It means that these accidents can not be prevented by technical measures only. Some of the emergencies are typical and predictable but others are not. The latter are classified as beyond design-basis events. Some of human errors should also be classified as beyond design-basis impacts on *CES*.

The traditional approach to modeling socio-technical systems implies decomposition the system into technical, human and organizational subsystems (levels) that are then studied within different disciplines. The interdependencies between subsystems and their influences on the system's safety are not considered properly.

It should be noted that design, construction and operation of complex engineering systems are being done in competitive environment. There is always a temptation to increases effectiveness by means of reducing investments in safety related activity. In order to justify such reduction organizations operating *CESs* may intentionally underestimate safety and security risks. The easiest way to do it is to unfoundly reduce the set of design-basis impacts and design-basis accidents or in other words to extend the list of beyond design-basis impacts in order not to consider them. The question "How to assess risks resulted from beyond design-basis events?" requires a special detailed investigation.

Currently available methodologies of risk assessment and *CESs* design codes have been developed without accounting for risks of beyond design-basis accidents. It means that these methodologies and codes should be revised to include into consideration risks of beyond design-basis impacts on complex engineering systems.

Complex engineering systems are becoming global networks. Protection of complex engineering systems requires pulling the efforts of specialist from different countries. Traditional risk assessment methodologies were developed for complicated technological systems with fixed boundaries, well-specified targets of hazards and for systems for which exists historical and/or actuarial data on accidents initiation events, components failure rates and accidents' consequences which allow one to quantify and verify models taking into account uncertainties deriving from both natural variations of the systems parameters (and performance conditions) as well as from lack of knowledge about the system itself.

The current accident models and risk assessment techniques such as fault and event tree analysis are not adequate to account for the complexity of modern engineering systems. The performance of complex engineering systems (as socio-technical systems) depends on the interaction between technical, human and organizational factors. Due to rapid technological and societal developments of the recent decades complex engineering systems are becoming steadily more complex. It means that (1) in safety assessments for *CES* there are too many details to be considered and (2) some modes of *CES's* operation may be incompletely known due to complex nonlinear interactions between components of *CES*, tight couplings among different systems and because *CES* and its environment may change faster than they can be described. As a result complex engineering systems are becoming underspecified and therefore intractable [7]. Thus for *CES* it is not possible to describe its performance in every detail, and their performance must therefore be flexible and adaptive rather than rigid. The distinction between tractable and intractable systems is very important in development safety management systems for *CESs*. Intractable systems cannot be completely described or specified. In other words for complex engineering systems it is practically impossible to define a bounding set of design-basis impacts that are expected to occur or might occur at least once during the lifetime of the *CES*.

Most current safety methods however have been developed on the assumption that systems are tractable. As this assumption is not valid for *CESs*, there is a need to develop methods to deal with intractable systems. This could be done based on Resilience and Robustness Engineering. The proposed approach should not be considered as a substitute rather a supplement to the traditional one. Adopting this view creates a need to move beyond traditional "threat-vulnerability-consequence" models that are limited to analyzing design-basis events and deal with beyond design-basis impacts (and impact combinations). This new approach will be based on such concepts as robustness and resilience to provide more comprehensive explanations of accidents as wells as identify ways to reduce risks caused by beyond design-basis impacts. In other words the new safety paradigm for complex engineering systems and other complex socio technical systems should focus the efforts not only on development of protection barriers and safeguards against design-basis accidents but also on increasing the *CES's* robustness and resilience towards beyond design-basis impacts.

Traditionally, accidents at *CES* have been viewed as resulting from a chain of failure events, each related to its "causal" event or events. It means that currently available risk assessment techniques are based on this linear notion of causality, which have severe limitations in the modeling and analysis of modern complex socio-technical systems that include nonlinear interactions between components, feedback loops, multiple source of failure, etc. Thus traditional accident modeling approaches are not adequate to analyze accidents that occur in complex engineering systems where accidents do not occur due to a single technical failure or human error; rather they arise from the interconnection of several causal factors originating at many levels in a system: hardware failures, human errors (mistakes and/or procedural violations committed by *CES* operators), latent conditions (arising from such aspects as management decision practices, or cultural influences), local triggering events (such as extreme weather conditions).

Thus, the study of complex engineering systems (being socio-technical systems) requires an understanding of the interactions and interrelationships between the technical, human, social and organizational aspects of systems. These interactions and interrelationships are complex and non-linear, and traditional modeling approaches cannot fully analyze the behaviors and failure modes of such systems. Thus each complex engineering systems should be treated as an integrated whole, and the emphasis will be made on the simultaneous consideration of technical and social aspects of systems, including engineering design and maintenance strategies of *CESs* as well as social structures and cultures, social interaction processes, and individual factors such as capability and motivation of operators. Interdisciplinary research should be applied to capture the complexity of complex engineering systems from a broad systemic view for understanding the multi-dimensional aspects of safety and modeling accidents in socio-technical systems.

As previously stated due to the complexity of complex engineering systems and their potential to large-scale catastrophes in order to ensure safety of such systems one needs to move beyond traditional design-basis risk assessment framework. Here a new paradigm is needed that is focused on increasing *CES's* robustness and resilience (Figure.3). That means that if the beyond design-basis accidents are to be considered the scope of the analysis should be widened. Safety related effots should be focused not only on development of protection barriers and safeguards from predetermined (postulated) set of maximal credible accidents but also on increasing complex engineering system robustness and resilience that would prevent catastrophic failure and long-term disfunctioning of *CESs* in case of beyond design-basis accident.

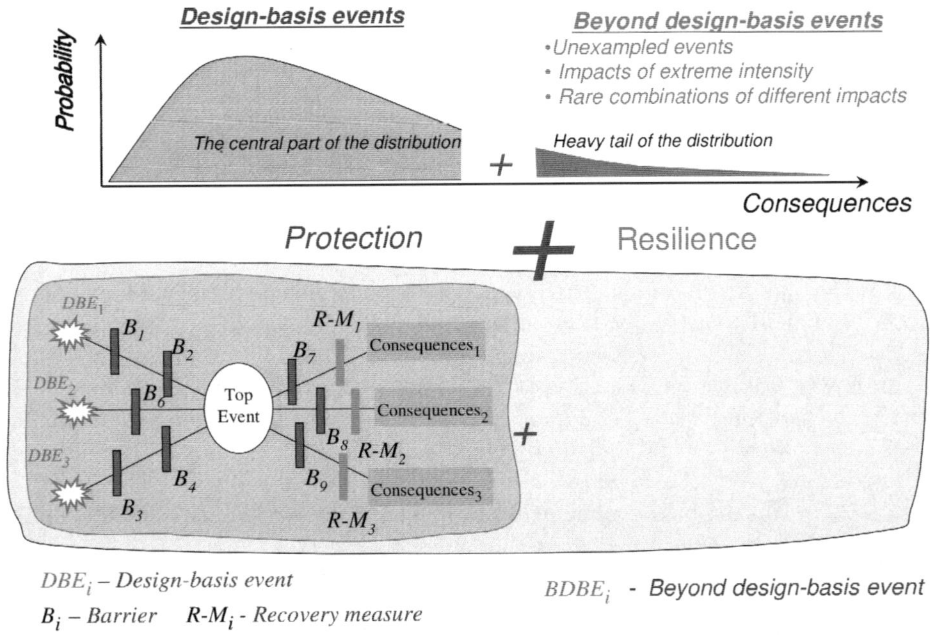

Figure 3. A new paradigm of CES safety: transition from "Protection" to "Protection + Resilience"

Quantitative measures of robustness (based on energy absorption, the ratio between direct and indirect risks and others) and resilience (based on shape of the *CES's* recovery curve) should be developed as well as a set of measures to increase *CES's* robustness and resilience will be identified.

4. Conclusion

Complex engineering systems should be considered as *techno-social* systems whose performance depends on tightly coupled technical, human and organizational functions. A new approach to ensuring *CES's* safety will be developed to address the new safety concerns that arise from the use of increasingly complex socio-technical systems. This approach should be based on the following premises:

1. Complex engineering systems should be considered as socio-technical systems whose performance is determined by interaction of technical, human and organizational factors.

2. Performance conditions for *CES* are always underspecified because they are too complex to be described in every detail and because they are not stable and fixed. Thus it is impossible to characterize the *CES's* performance by analyzing its response to a bounding set of design-basis impacts.

3. Low probability –high consequences (beyond design-basis) events should be included into the scope of *CES* risk assessment and management framework. This requires development of additional measures aimed at increasing *CES's* robustness and resilience in case of beyond design-basis impacts.

4. Many adverse events in *CES* can not be attributed to a single failure or malfunctioning of components. They result from unexpected combinations of various adverse factors. Occurrences of accidents in *CES* may be due to a combination of performance variability, malfunctions, human errors, adverse unauthorized impacts. Safety management must therefore be not only reactive but also proactive as well.

5. Risk assessment and management models for complex engineering systems should include several levels (legislators, managers, system operators, hardware components ets).

6. Impacts of technological and intelligent terrorism should be included in the framework of *CES* risk assessment. This requires that a number of specific features of terrorist impacts (feedbacks between vulnerability and consequences) be accounted for.

Acknowledgments

This work was financially supported by the Russian Foundation for Basic Research (grant no. 14-19-00776-P).

References

[1] Makhutov N and Reznikov D 2010 Engineering Infrastructures: Problems of Safety and Security in the Russian Federation *European Perspectives on Security Research* pp 93-106

[2] Makhutov N, Khaziakhmetov R and Reznikov D 2012 Basic Scenarios of Terrorist Attacks at Hydropower Engineering Facilities *Proceedings of the 3rd International Forum on Risk Analysis, Dam Safety Dam Security and Critical Infrastructure Management* pp 389-394

[3] Makhutov N, Reznikov D and Petrov V 2014 Specific Futures of Ensuring Critical Infrastructures Safety *Safety in technosphere* **3** No **1**(**46**) pp 3-14

[4] Epstein S 2006 Unexampled events, resilience and PRA *Proceedings of the Second Resilience Engineering Symposium* pp 105-116

[5] Schintler L A, Gorman S, Kulkarni R and Stough R 2007 Moving from Protection to Resiliency:A Path to Securing Critical Infrastructure *Critical Infrastructure* pp 291-307

[6] Lundberg J and Johansson B 2006 Resilience, stability and requisite interpretation in accident investigations *Proceedings of the Second Resilience Engineering Symposium* pp 191-198

[7] Cacciabue P & Hjälmdahl Magnus & Lüdtke Andreas & Riccioli Costanza 2011 *Human Modeling in Assisted Transportation: Models, Tools and Risk Methods* 10.1007/978-88-470-1821-1

IOP Conf. Series: Materials Science and Engineering **481** (2019) 012003 doi:10.1088/1757-899X/481/1/012003

Cyberattack intensity forecasting on informatization objects of critical infrastructures

Y M Krakovsky[1], A N Luzgin[2], Y M Ivanyo[3]

[1]Irkutsk State Transport University, 15 Chernyshevskogo Street, Irkutsk 664074, Russia
[2]Administration of Irkutsk city, 14 Lenina, Irkutsk 664025, Russia
[3]Irkutsk State Agrarian University named after A.A. Ezhevsky, Irkutsk 664038, Russia

E-mail: alexln@mail.ru

Abstract. In regulatory documents of recent years in the field of information security, much attention is paid to information systems of critical infrastructures. This, in turn, justifies the need for scientific research on the development of new methods of protection against cyberattacks on such information systems. For this task, interval forecasting is recommended based on a probabilistic neural network with dynamic updating of the smoothing parameter. As benchmarks for comparing the interval forecasting results, the naive Bayesian model and the probabilistic cluster model were chosen.

1. Introduction

In last years, in Russia and World, much attention has been paid to the security of critical infrastructures. In accordance with the federal law «About security of a critical information infrastructure of Russian Federation» adopted in 2017 [1], the information systems (IS) are the important objects of protection. These objects fall under the Decree of the President of the Russian Federation 15.01.2013 N31s «On creation of the state system of detection, prevention and liquidation of consequences of computer attacks to information resources of Russian Federation». In a development of this Decree, in December 2014, the President of the Russian Federation approved a Concept of a state system for Russia's information resources for detecting and preventing computer attacks, and mitigating their consequences. In accordance with this Concept, the main functions of the system are: to identify signs of computer attacks, to determine their sources and other related information, to forecast a situation in the field of information security of the Russian Federation, to collect and analyze information about computer attacks on information resources of the Russian Federation, and to react to attacks and eliminate their consequences [2].

In 2016, the «Information security doctrine of the Russian Federation» was adopted, where it is noted that «the state of information security in the field of state and public security is characterized by a constant increase in complexity, an increase in a scale and growth of cyberattacks to objects of critical infrastructures» [3]. The federal law «About security of a critical information infrastructure of Russian Federation» [1] notes that there is a mandatory requirement for an implementation of a state system for detecting and preventing cyberattacks on IS of critical infrastructures, and mitigating their consequences. This once again confirms the importance and relevance of issues of cybersecurity of these facilities for the Russian Federation. Thus, scientific research on the development of new methods of protection against cyberattacks on IS of critical infrastructures is important and necessary.

One of the most promising research directions for solving the task of protecting against cyberattacks on IS of critical infrastructures is to create cyberattack intensity forecasting methods based on machine learning [4, 5]. Note that the cyberattack intensity is the total number of these attacks per unit time. If a forecast is received that the cyberattack intensity on IS exceeds a predetermined value, additional protection measures can be taken; for example, the more detailed analysis of traffic. It should be noted that in the federal law «About security of a critical information infrastructure of Russian Federation», as well as in «Concept of the state system for detecting, preventing and eliminating of consequences of computer attacks on the information resources of Russian Federation», the need for using forecasts in the cybersecurity field is underlined. Thus, in research related to protection against cyberattacks, in addition to assessing different risks and using traditional protection systems, attention should be paid to cyberattack intensity forecasting [6].

In the past few years, there has been an increasing interest of researchers in probabilistic forecasting [7, 8]. This can be explained by the fact that probabilistic forecasts make it possible to obtain not only forecasts of future events, but also probabilistic estimates of these events. One type of probabilistic forecasting is interval forecasting (IF) [9-11]: this involves forecasting of an interval (from two predetermined intervals) in which a future value of an indicator will be located. Probability estimates are used for this purpose. A dividing bound of these intervals is determined by a calculation method based on statistical characteristics of the indicator.

In this paper, for forecasting cyberattack intensity on IS of critical infrastructures it is recommended carry out IF based on a probabilistic neural network with dynamic updating of the smoothing parameter value (PNN) [10,11]. As a standard for comparing the results of IF, a naive Bayesian model (NBM) and probabilistic cluster model (PCM) were selected [12].

2. Description and formalization of a cyberattack intensity indicator

Given that information about the cyberattack intensity on IS is confidential, we have used another public indicator of cyberattack intensity. This indicator is the cyberattack number per day that occurred from 1998 to 2015 in South Korea [14]. This indicator has a large volume of values, suitable for constructing various machine learning models for IF. On the other hand, the chosen indicator is non-stationary with respect to the location and the scale parameters, which underlines its dynamic statistical characteristics [10]. Thus, if this indicator shows good results of cyberattack intensity IF, then we can more confidently draw similar conclusions with regard to IS.

This indicator was formalized as the time series:

$$\mathbf{z} = \{z_t : t \in \boldsymbol{t}\}, \tag{1}$$

z_t is the value of the indicator at the discrete moment of time t; $t \in \mathbf{t}$; $\mathbf{t} = \{1, \dots, n\}$; and n is the number of values. For the chosen indicator, $n = 1552$.

Let $[z_{min}; z_{max}]$ be the conditional range of possible values of the indicator (1), then z_α is the threshold of the cyberattack intensity ($z_{\min} \le z_\alpha \le z_{\max}$). The threshold of the cyberattack intensity z_α is a value for which the probability that $z_t \le z_\alpha$ equals α. Thus, z_α is the quantile of a probability distribution function of (1) for a given probability α.

Note that in a future scenario, with respect to the selected indicator, it is sufficient to take only the integer part of z_α, since the cyberattack intensity is always an integer value.

Further, it is proposed to perform the following completely reversible transformation of the original indicator (1):

$$\boldsymbol{q} = log(\mathbf{z} + 1) - log(z_\alpha + 1) = \{q_t : t \in \boldsymbol{t}\}. \tag{2}$$

Here q_t is the value of the indicator at the discrete moment of time t; $t \in \mathbf{t}$; $\mathbf{t} = \{1, \dots, n\}$; n is the number of values; and z_α is the threshold of the cyberattack intensity.

This conversion is useful for several reasons:

1) The values of the initial indicator (1) are in a very wide range, and some (extreme) values significantly exceed the others. Logarithmic transformation helps to improve visual work with such data and their associated graphs;

2) The indicator (2) contains both positive and negative values, in contrast to the indicator (1). Some forecasting models (including PNN) are sensitive to the sign of predictors and demonstrate better IF accuracy after such transformations;

3) The equivalent of z_α for the indicator (2) is always 0. That is, the distribution of positive and negative values relative to z_α for indicator (1) and relative to 0 for the obtained indicator (2) is identical, and this slightly simplifies the formalization of IF for the indicator (2) without distorting the essence and interpretation of the obtained results.

Figure 1 shows the graph of the obtained indicator (2).

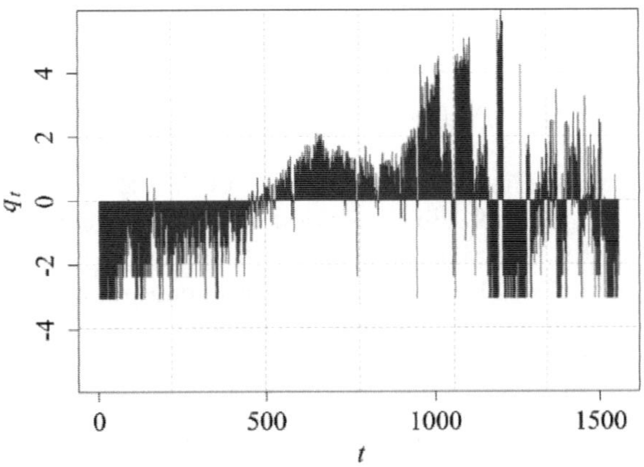

Figure 1. The graph of **q** (2) for $\alpha = 0.5$.

Thus, it should be noted that the transformation of the indicator (1) into the indicator (2) is an integral part of the implementation of IF.

For obtaining some statistical characteristics of this indicator, its class was determined by the method described in [10]. This indicator is an indicator of the first class, non-stationary in terms of the location and scale parameters, which indicates its distinct statistical nature among indicators of other classes [9,10].

3. Formalization of cyberattack intensity interval forecasting

Let $[q_{min}; q_{max}]$ be the conditional range of possible values of the indicator (2). Construct two intervals:

$$I^- = [q_{min}; 0],\ I^+ = (0; q_{max}]. \quad (3)$$

At time $t = n - 1$ it is necessary to identify the interval (3) in which the future (unknown) value q_{t+p} will be located. The following estimates of probabilities are required: ρ_{t+p}^+ and ρ_{t+p}^- where $p = 1, \ldots, r$ is the look-ahead period, ρ_{t+p}^+ is the probability that the indicator future value $q_{t+p} \in I^+$; and ρ_{t+p}^- is the probability that the indicator future value $q_{t+p} \in I^-$; $\rho_{t+p}^+ + \rho_{t+p}^- = 1$.

Let $\tilde{\rho}_{t+p}^+$ and $\tilde{\rho}_{t+p}^-$ be probability estimates of ρ_{t+p}^+ and ρ_{t+p}^-. The interval forecasting is carried out according to the following rules: the future value $q_{t+p} \in I^+$ if $\tilde{\rho}_{t+p}^+ > \tilde{\rho}_{t+p}^-$; and the future value $q_{t+p} \in I^-$ if $\tilde{\rho}_{t+p}^+ \le \tilde{\rho}_{t+p}^-$.

4. Formalization of training set for learning of probabilistic models

It is necessary to consider some features of the formation of a training set for the implementation of IF.

IOP Conf. Series: Materials Science and Engineering **481** (2019) 012003 doi:10.1088/1757-899X/481/1/012003

Suppose $t = n$ and there is a sequence of values $q_{t-f+1}, \ldots, q_t$ in a number of f. Create a vector $\mathbf{h} = (q_{t-f+1}, \ldots, q_t)$. Let y_{t+p} be a dependent variable (or a response) the true value of which is unknown and that can take only two possible values: $y_{t+p} = 1$ if $q_{t+p} \in I^+$ and $y_{t+p} = -1$ if $q_{t+p} \in I^-$.

Performing IF using $\mathbf{h}$ requires making a forecast of y_{t+p} based on probability estimates that $q_{t+p} \in I^+$ and $q_{t+p} \in I^-$. Recall that if $\tilde{p}^+_{t+p} > \tilde{p}^-_{t+p}$, then $y_{t+p} = 1$, else $y_{t+p} = -1$.

Next, create a training set based on the values of $\mathbf{q}$ (1) for $t = 1, \ldots, m$, where $m = n - f - p + 1$ (this value is chosen so that the responses' values can be calculated based on pre-history values of the indicator):

$$\mathbf{x} = \begin{pmatrix} q_1 & \cdots & q_{1+f-1} \\ & \cdots & \\ q_m & \cdots & q_{m+f-1} \end{pmatrix}, \quad \mathbf{y} = (y_1, \ldots, y_m). \tag{4}$$

Here $\mathbf{x}$ is the matrix of dimensions $m \times f$ (a training set); $\mathbf{y}$ is a vector of responses of size m (these responses are calculated based on the pre-history values of the indicator); and m is the number of training samples.

Each row i of the matrix $\mathbf{x}$ corresponds to the response of $\mathbf{y}$ (4): $\mathbf{x}_i \to y_i$.

Using the training set (4), it is possible to build and train some forecasting model, and also implement the IF.

Often the matrix of predictors $\mathbf{x}$ is used not in pure form, but in the transformed one. For example, for PNN each row of $\mathbf{x}$ is transformed so that the sum of squares of each row of values is equal to 1. For NBM and PCM, this is not necessary.

5. General algorithm of interval forecasting of cyberattack intensity

The IF algorithm in its general form consists of the following stages:

- Prepare initial data: $\mathbf{z}$ (1);
- Set the parameter: α;
- Construct a piecewise linear probability distribution function of $\mathbf{z}$ (1) and estimate z_α for selected value of α;
- Transform $\mathbf{z}$ (1) to $\mathbf{q}$ (2);
- Set the parameters: p, f;
- Create the training set (4);
- Select a forecasting model and set its parameters (parameter values can be optimized based on the training set; for example, by cross-validation methods [15]);
- Carry out IF.

Thus, this algorithm has three parameters: α is the probability with which the cyberattack intensity will be below the threshold cyberattack intensity z_α; p is the ahead time; and f is the dimension of a training set.

6. Interval forecasting results and prospects of their practical application

For an analysis of IF results of cyberattack intensity, several scores were used. They are considered in more detail, and reasons are provided for choosing each of them.

First of all, we are interested in an accuracy with which the forecasting of events is carried out, $q_{t+p} \in I^+$. In fact, when obtaining such a forecast, it is necessary to take additional measures to protect against increasing cyberattacks. The more accurate such forecasts, the fewer mistaken additional measures will be taken to protect against cyberattacks (i.e. false positives). The fewer false positives, the more effective the system of protection against cyberattacks will be. For estimating the accuracy of such forecasts, it is proposed to use the score:

$$pr^+ = l^+/u^+, \tag{5}$$

where pr^+ is the estimation of forecasting accuracy of events $q_{t+p} \in I^+$, l^+ is the number of justified forecasts that $q_{t+p} \in I^+$, and u^+ is the total number of forecasts that $q_{t+p} \in I^+$, $0 \leq pr^+ \leq 1$.

Also, we are interested in the accuracy with which the forecasting of events $q_{t+p} \in I^-$ is carried out. Here, when we get a forecast that $q_{t+p} \in I^-$, the system of protection against cyberattacks continues to work in its regular mode. The more accurate such forecasts, the less likely situations will arise when, in fact, additional measures of protection from cyberattacks were required, but this was not done (i.e. false negative). This also affects the effectiveness of protection systems against cyberattacks. To estimate the corresponding accuracy of such forecasts, the following score was used:

$$pr^- = l^-/u^-, \tag{6}$$

where pr^- is the estimation of forecasting accuracy of events $q_{t+p} \in I^-$, l^- is the number of justified forecasts that $q_{t+p} \in I^-$, and u^- is the total number of forecasts that $q_{t+p} \in I^-$, $0 \leq pr^- \leq 1$.

Thus, the larger the both values pr^+ and pr^-, the better. It should be noted that the forecasting model should forecast both variants of events: $q_{t+p} \in I^+$ and $q_{t+p} \in I^-$. For example, the model that gives the result $pr^+ = 0.75$ and $pr^- = 0.80$ is preferable to the one that gives the result $pr^+ = 0.55$ and $pr^- = 0.95$. This allows to determine the final score characterizing the accuracy of the IF based on any selected model:

$$pr = min(pr^+, pr^-). \tag{7}$$

Here pr^+ is the estimation of forecasting accuracy of events $q_{t+p} \in I^+$ (5), and pr^- is the estimation of forecasting accuracy of events $q_{t+p} \in I^-$ (6). The larger the value of pr (7), the more accurate IF.

The testing of the selected models was carried out as follows. The training set (4) was divided into two parts. The first part included the even rows of **x** and elements of **y**. This part was used for training and optimizing models. The second part with odd rows of **x** and elements of **y** is used for obtaining forecasts. Subsequently, the second part was used for training and optimizing selected models, and the first part for obtaining forecasts. Next, the values (5-7) were estimated by the cross-validation method for two blocks [15].

The estimates of the scores (5-7) were carried out for different values of α from 0.20 to 0.80 with the step 0.1. In all cases, the parameter p was fixed and equal to 1. At a fixed valueα a sequential search of the parameter f from 1 to 10 values was carried out (this parameter is common for PNN, CPM, and NBM). For NBM, for each new value of f, the values of the smoothing parameter of a nonparametric density function of predictors are changed from 0.1 to 1 with the step 0.1. Among all the estimates obtained (7), such a model was chosen in its class, for which the value of (7) was maximal. All algorithms were implemented using the R language [16-18]. Table 1 shows the results obtained.

Table 1. Interval forecasting results.

Parameter, α	Threshold, z_α	PNN, pr	NBM, pr	PCM, pr
0.2	3	0.75	0.60	0.66
0.3	6	0.76	0.70	0.75
0.4	11	0.84	0.81	0.83
0.5	20	0.88	0.88	0.88
0.6	35	0.88	0.83	0.86
0.7	55	0.81	0.75	0.80
0.8	79	0.79	0.77	0.74

As follows from the above results, PNN is more accurate and acceptable in all cases. The highest accuracy is observed in the middle of the parameter values of α. In practice, the choice of the value of

α can be done experts. It should be noted that the range of α from 0.20 to 0.80 is quite sufficient for solving practical problems. It is not advisable to specify larger or smaller values of α, as this will lead to a serious «imbalance» in the training set, and the results of IF can consequently be unstable and inadequate.

It is possible that the additional measures to protect against cyberattacks should not be applied at the first hit of the future value in the interval I^+, but after several of hits. More research is needed in this direction.

7. Conclusion

As follows from the results of this work, interval forecasting of cyberattacks intensity on IS of critical infrastructures is a necessary and important practical task. Experiments showed that interval forecasting of cyberattack intensity based on a probabilistic neural network for the selected indicator is more accurate than other models.

Given that information about cyberattack intensity on IS of critical structures is confidential, the number of cyberattacks per day that occurred from 1998 to 2015 in South Korea was considered as an alternative indicator in this work [14]. This indicator was chosen because it is publicly available and it has a large volume of data, suitable for constructing various models of machine learning for the purpose of IF. On the other hand, the selected indicator is non-stationary in terms of the location and scale parameters, which underlines its dynamic statistical nature. Since this indicator showed good results of interval forecasting of cyberattack intensity, similar conclusions can be drawn with respect to the IS of critical infrastructures.

8. References

[1] Federal act of security of a critical information infrastructure of Russian Federation on 26–07–2017 189–FZ

[2] Extract from Concept of the state system for detection, prevention and liquidation of the consequences of computer attacks on information resources of Russian Federation, approved by the President of the Russian Federation on 12–12–2014 K1274.

[3] Presidential Decree of the approval of the Doctrine of information security of the Russian Federation.

[4] Petrenko S A and Petrenko A S 2016 The concept of early detection and prevention of computer attacks *Materials of Russian Scientific and Practical Conference «Information Systems and Technologies in Modeling and Control»* p 82–6

[5] Petrenko S A and Stupin D D 2017 National early warning system of computer attacks (Publishing house: «Athena»)

[6] Gandotra E, Bansal D and Sofat S 2015 Computational Techniques for Predicting Cyber Threats *Proceedings of Intelligent Computing, Communication and Devices* p 247–253

[7] Zhan Z, Xu M, and Xu S 2015 Predicting cyberattack rates with extreme values *IEEE Transactions on Information Forensics and Security* **10** (**8**) p 1666–1677

[8] Werner G, Yang S and McConky K 2017 Time series forecasting of cyberattack intensity *Proceedings of the 12th Annual Conference on Cyber and Information Security* p 224–240

[9] Krakovsky Y M and Luzgin A N 2017 *Applied aspects of application of interval forecasting in system analysis* Modern technology. System Analysis. Modeling **2**(**54**) p 115–121

[10] Kargapoltsev S K, Krakovsky Y M and Luzgin A N 2017 Nonparametric classification of technical condition parameters based on shift and scale tests *International Conference on Industrial Engineering, Applications and Manufacturing* p 1–5

[11] Kargapoltsev S K, Krakovsky Y M, Lukanov A V and Luzgin A.N 2017 A dynamic updating algorithm of smoothing parameter values of probabilistic neural networks *Far East Journal of Electronics and Communications* **17**(**4**) p 909–914

[12] Ethem A 2014 Introduction to Machine Learning: Massachusetts Institute of Technology *The MIT Press Cambridge* p 616

[13] Krakovsky Y M and Luzgin A N 2018 Robust interval forecasting algorithm based on a probabilistic cluster model *Journal of statistical computation and Simulation* **88**(**12**) p 2309–2324

[14] Han L, Han Ch, Kang R, Kwak I, Mohaisen A and Kim H 2016 WHAP: Web-hacking profiling using Case-Based Reasoning *IEEE Conference on Communications and Network Security* p 344–5

[15] Browne M 2000 Cross-validation methods *Journal of Mathematical Psychology* vol 44 p 108–132

[16] The R Project for Statistical Computing https://www.r-project.org

[17] Krakovsky Y M, Kurchinsky B V and Luzgin A N 2018 Cyber-attack intensity interval forecasting on objects of critical information infrastructure *Proceedings of Tomsk State University of Control Systems and Radioelectronics* **21** no 1 p 71–9

[18] Krakovsky Y M and Luzgin A N 2015 Software for interval prediction of non-stationary dynamic indicators *Proceedings of Irkutsk State Technical University* **1** no 4 p 12–16

IOP Conf. Series: Materials Science and Engineering **481** (2019) 012004 doi:10.1088/1757-899X/481/1/012004

Verification of recovery of distribution density function of the random variable by methods of nonparametric statistics

V N Syzrantsev[1], K V Syzrantseva[1]

[1]Department of Machines and equipment of oil and gas industry, Industrial University of Tyumen, 38, Volodarskogo str., Tyumen, 625000, Russia
[2]Department of Cybernetic systems, Industrial University of Tyumen, 38, Volodarskogo str., Tyumen, 625000, Russia

E-mail: kv.syzr@gmail.com

Abstract. The paper considers an algorithm of recovery of distribution density function of a random variable by methods of nonparametric statistics. For samples of various length (from 20 to 2000) of normal distribution the quantile estimates of a random variable are calculated by methods of parametrical and nonparametric statistics. It is established that errors of the quantile estimates obtained by parametrical and nonparametric methods are values of the same order.

1. Introduction

When assessing strength reliability of elements of different machines, stresses and their limit values obtained during an experiment or calculation in the form of random variable $x_i, i = \overline{1,N}$ samples are used. Consequently, determining failure-free operation probability of the elements and metal construction of machines requires knowledge of the functions of the above mentioned random variables distribution density.

In the practice of processing experimental data for solving strength reliability problems it is traditionally accepted [1, 2] that required function $P(X)$ is known up to the finite number of the parameters. In this case, the problem of density recovery is well-posed and its solution uses methods of parametric statistics [2, 3, 4, 5]. However, in general case, the class of functions to which $P(X)$ can belong is substantially wider than those discussed in papers [1, 2, 3, 4, 5] and many others.

To date, in the theories of probability and mathematical statistics over a hundred of random variables distribution laws have been studied. At the same time, the experience of processing experimental data on machine parts and metal constructions loading and durability during operation indicates that the distribution of random variables of operating and limit stresses, the number of cycles prior to parts and carrying systems destruction do not follow the standard laws described in the theory of parametric statistics.

To study the processes containing random variables not described by standard distributions nonparametric methods have been developed in the theory of mathematical statistics [6, 7, 8, 9, 10, 11, 12, 13]. When processing experimental data in problems of recovery of unknown function of random variables distribution density the methods of nonparametric statistics [14, 15, 16, 17, 18] based on Parzen-Rosenblatt estimates are mostly used [11, 12], or the decomposition method with respect to basic functions [6].

IOP Conf. Series: Materials Science and Engineering **481** (2019) 012004 doi:10.1088/1757-899X/481/1/012004

2. Nonparametric Estimates of the Distribution Density Function

Based on the physical meaning of X, we present the only requirement – its continuity and consider the problem of $P(X)$ recovery based on the existing sample $x_i, i=\overline{1,N}$ by means of nonparametric statistics [7, 8].

It is known [2, 4, 5, 8], that the probability distribution density is related to the probability distribution function by the ratio:

$$\int_{-\infty}^{y} P(x)dx = F(y), \tag{1}$$

presented in the form of:

$$\int_{-\infty}^{\infty} \theta(y-x)\cdot P(x)dx = F(y), \ \theta(s)=\begin{cases}1, \ if \ s\geq 0\\ 0, \ if \ s<0\end{cases}, \tag{2}$$

where $\theta(s)$ is a function of the unit step (Heaviside function).

The optimal nonparametric estimate at each point y for theoretical distribution function $F(y)$ is [8] empirical cumulative probability distribution function $F_N(y)$. If value y exceeds k of the sampling elements $x_i, i=\overline{1,N}$ by volume N function $F_N(y)$ looks like:

$$F_N(y) = \frac{1}{N}\sum_{i=1}^{N}\theta\cdot(y-x_i). \tag{3}$$

With the growth of sample volume N function $F_N(y)$ with probability of one uniformly approaches $F(y)$:

$$\Pr\left\{\lim_{N\to\infty} \sup|F_N(y)-F(y)|=0\right\}=1. \tag{4}$$

In statistical processing the data of experimental stress studies the right part of equation (2) – distribution function $F(y)$ is replaced by empirical distribution function $F_N(y)$ obtained on the basis of a sample of limited volume stresses. Therefore, the solution of the equation (2) will always be approximate. To recover the distribution density function by solving the equation (2) within the theory of nonparametric statistics, special procedures have been developed [6, 7, 8], providing equation solutions sequence convergence (2) to the required probability density $P(X)$ with increasing N and taking into account the incorrect formulation of the problem (2), associated with the requirement to differentiate the inexact right part of the equation (2). Based on the physical meaning of the distribution density function, it has only positive values on the entire axis of change X – from $-\infty$ to $+\infty$ – and fulfills the following condition:

$$\int_{-\infty}^{+\infty} P(x)dx = 1. \tag{5}$$

Under the condition of continuity of function $P(X)$ the solution (2) is the only one.

3. Algorithms for recovery of unknown distribution density function

The methods of distribution density function estimation proposed by Parzen and Rosenblatt [11, 12], use a smoothed empirical distribution function in the form of:

IOP Conf. Series: Materials Science and Engineering **481** (2019) 012004 doi:10.1088/1757-899X/481/1/012004

$$F_N(y) = \frac{1}{N}\sum_{i=1}^{N} G\left(\frac{y - x_i}{h_N}\right), \tag{6}$$

where $G(t)$ is a steadily non-decreasing function from 0 to 1 of its argument; in this casc $G(t) = 1 - G(-t)$, i.e. $G(t)$ is a symmetric function relative to zero; h_N is a fuzziness parameter (bandwidth).

After differentiation (6) we have:

$$P_N(y) = F'_N(y) = \frac{1}{N \cdot h_N}\sum_{i=1}^{N} G'\left(\frac{y - x_i}{h_N}\right) = \frac{1}{N \cdot h_N}\sum_{i=1}^{N} K\left(\frac{y - x_i}{h}\right), \tag{7}$$

where $K(t) = G'(t)$ is distribution density $G(t)$ or kernel function (kernel).

The theoretical studies [8] of the function (6) show that displacement and variation of the estimate (7) depend on kernel type $K(t)$ and the value of fuzziness parameter h_N. Various dependencies are used as kernel functions [8, 9, 17]. The most common of them, used in processing experiments data by assessing loading and durability of mechanical engineering products, are presented in Table 1.

Recovery of distribution density function by means of the Parzen–Rosenblatt method based on (7) is performed in two stages. At the first stage, kernel function $K(t)$ is selected from among the known ones (Table 1 or others). At the second stage, the problem of determining an optimal value of fuzzing parameter h_N is solved.

In physical meaning, use of function (7) allows describing a random variable, the discrete values of which are recorded in the course of experimental studies, with continuous function which provides a possibility of its application to solve numerous practical problems of differential and integral methods.

To select the best one from among a finite number of functions $K(t)$, it is required to have a selection criterion. Information functional of the form [8, 13, 15]:

$$J = \int \ln\left[K(t)\right]P(t)dt = \int \ln\left[K(t)dF(t)\right], \tag{8}$$

the maximum value of which fulfills condition $K(t) = P(t)$ can be adopted as such a criterion.

Then the search for optimal h_N^* and $K(t) \in K = \{K_1(t),...K_{12}(t)\}$ comes down to solving the following problem [8, 16, 17]:

$$(h_N^*, K^*(t)) = \arg\max_{h_N, K(t)} J_N(h_N, K(t)) = \arg\max_{h_N, K(t)} \left\{\frac{1}{N}\sum_{i=1}^{N} \ln\left[\frac{1}{(N-1)h_N}\sum_{j \neq i}^{N-1} K\left(\frac{x_i - x_j}{h_N}\right)\right]\right\}. \tag{9}$$

Determination of optimal value h_N^*, as shown in papers [6, 8], is a more difficult task than a recovery of distribution density, since optimal value h_N^* depends on unknown distribution density, especially, its unknown derivatives. In practical applications density estimates are often required in particular areas; for example, in prediction of resource and reliability, first of all, density estimate in the distribution tail area is very important. Therefore, in solving this problem, the algorithms providing determination of optimal values of parameter h_N^* on the basis of only available sample x_i, $i = \overline{1, N}$ of random variable X are required.

For some functions $K(t)$ in Table 1, the dependences for calculating estimates of optimal value h_N^* based on different estimates of sample x_i, $i = \overline{1, N}$ have been obtained. For example, when using kernel function as normal distribution (Table 1):

$$K\left(\frac{y-x_i}{h_N}\right)=\frac{1}{\sqrt{2\cdot\pi}}\exp\left[-0.5\left(\frac{y-x_i}{h_N}\right)^2\right]. \tag{10}$$

Table 1. Kernel functions [8, 9, 17].

Uniform kernel	Triangle kernel No.1	Triangle kernel No.2
$K_1(t)=\begin{cases}0,5;\lvert t\rvert\le 1\\ 0;\lvert t\rvert>1\end{cases}$	$K_2(t)=\begin{cases}1-\lvert t\rvert;\lvert t\rvert\le 1\\ 0;\lvert t\rvert>1\end{cases}$	$K_3(t)=\begin{cases}\frac{1}{\sqrt{6}}-\frac{\lvert t\rvert}{6};\lvert t\rvert\le\sqrt{6}\\ 0;\lvert t\rvert>\sqrt{6}\end{cases}$
Quadratic kernel	Parabolic kernel	Epanechnikov kernel
$K_4(t)=\frac{3\cdot(1-t^2)}{4};\lvert t\rvert\le 1$	$K_5(t)=\begin{cases}\frac{9}{8}\left(1-\frac{5}{3}t^2\right);\lvert t\rvert\le 1\\ 0;\lvert t\rvert>1\end{cases}$	$K_6(t)=\begin{cases}\frac{3}{4\sqrt{5}}\left(1-\frac{t^2}{5}\right);\lvert t\rvert\le\sqrt{5}\\ 0;\lvert t\rvert>\sqrt{5}\end{cases}$
Gaussian kernel	Laplace kernel	Fisher kernel
$K_7(t)=\frac{1}{\sqrt{2\cdot\pi}}\exp\left(-\frac{t^2}{2}\right)$	$K_8(t)=\frac{1}{2}\exp(-\lvert t\rvert)$	$K_9(t)=\frac{1}{2\cdot\pi}\left(\frac{2}{t}\sin\left(\frac{t}{2}\right)\right)$
Logistics kernel	Trigonometric kernel	Cauchy kernel
$K_{10}(t)=\frac{\exp(-t)}{[1+\exp(-t)]^2}$	$K_{11}(t)=\begin{cases}0,54+0,46\cos(\pi\cdot t);\lvert t\rvert\le 1\\ 0;\lvert t\rvert>1\end{cases}$	$K_{12}(t)=\frac{1}{\pi}\left(\frac{1}{1+t^2}\right)$
Vallee-Poussin	Jacobi kernel	Chebyshev-Hermite kernel
$K_{13}(t)=\frac{1}{2\cdot\pi}\left(\frac{2}{t}\sin\left(\frac{t}{2}\right)\right)^2$	$K_{14}(t)=\begin{cases}\frac{15}{2^5}(3-10t^2+7t^4);\lvert t\rvert\le 1\\ 0;\lvert t\rvert>1\end{cases}$	$K_{15}(t)=\frac{3}{2}\left(1-\frac{t^2}{3}\right)\frac{1}{\sqrt{2\pi}}\exp\left(-\frac{t^2}{2}\right)$

Optimal value of parameter h_N^* («bandwidth»), is defined by formula [6]:

$$h_N^*=D_N\cdot N^{-0.2}, \tag{11}$$

where D_N is sampling variance, calculated based on the available values sample $x_i, i=\overline{1,N}$:

$$D_N^2=\frac{1}{N-1}\sum_{i=1}^{N}\left(x_i-\frac{1}{N}\sum_{i=1}^{N}x_i\right)^2. \tag{12}$$

As a result, to estimate the density with kernel (10) and fuzziness parameter (11) based on (7) we have the following expression:

$$P_N(y)=\frac{1}{\sqrt{2\cdot\pi}\cdot N\cdot h_N^*}\sum_{i=1}^{N}\exp\left[-0.5\left(\frac{y-x_i}{h_N^*}\right)\right]. \tag{13}$$

IOP Conf. Series: Materials Science and Engineering **481** (2019) 012004 doi:10.1088/1757-899X/481/1/012004

Analyzing (13), it is easy to see that the implementation of the Parzen-Rosenblatt method involves two steps of calculation. First, rough characterization of the sample, sampling variance (12), is calculated, which is further used to correct the estimation of distribution density (13) via fuzziness parameter h_N^*. Since value D_N is sensitive to emissions and does not reflect the nature of the density function change (unimodal, multimodal) the information about the distribution density obtained with D_N may not be sufficient to solve the problem correctly with the method under consideration.

To solve the problem of recovery of distribution density unknown function using (9) on the basis of the kernel functions presented in Table 1, a set of programs in the Mathcad system has been developed. The search for function maximum (9) for each kernel function $K(t)$ from Table 1 by value h_N is carried out with the numerical method taking into account the initial value h_N calculated by expression (11).

4. Task of Verifying the Recovery of Unknown Random-Variable Distribution Density Function by Means of Nonparametric Statistical Methods

To date, the apparatus of nonparametric statistics has been widely used in processing experimental data to solve the image (target) identification tasks, as well as in medicine, cryptography, macroeconomics, reliability screening for parts and metal structures in various machines and equipment. At the same time, nonparametric statistical methods are reluctantly applied for processing experimental data, which is explained by wider confidence intervals of a random variable, provided by nonparametric statistics in comparison with traditional methods. In this paper, using the samples of a random variable of a different length (from 20 to 2000) from the normal distribution, we calculated its quantile estimators through parametric and nonparametric statistical methods and analyzed the error of these estimators to verify the recovery of the density function by means of nonparametric statistical methods.

It is known that the normal distribution density function looks like:

$$P(x)=\frac{1}{s\sqrt{2\cdot\pi}}\exp\left[-0,5\left(\frac{x-m}{s}\right)^2\right], \qquad (14)$$

where m and s constitute an average deviation and a mean square deviation of the random variable x, respectively.

Let us assume that x changes within the limits of $x_{\min}=10$ and $x_{\max}=2000$, and the distribution parameters (14) have the values of $m=800$; $s=80$.

The quantile values (x_α) of the random variable x are determined by solving the following equation through the calculus of approximations:

$$\int_{x_{\min}}^{x_\alpha} P(x)dx=\frac{1}{s\sqrt{2\cdot\pi}}\int_{x_{\min}}^{x_\alpha}\exp\left[-0,5\left(\frac{x-m}{s}\right)^2\right]dx=\alpha . \qquad (15)$$

The calculation results x_α based on the equation (15) with $m=800$, $s=80$ and $x_{\min}=10$ for the range of values α, which correspond to the «tails» of the normal distribution, are given at the top of Tables 2 and 3.

Table 2. The results of calculation of x_α (left tail of distribution).

α	0.005	0.010	0.050	0.100
	Normal distribution low (14)			
Value x_α	593.933	613.901	668.410	697.475
	$n = 20$			

Value $\hat{x}_\alpha$	642.191	659.440	706.509	731.613
Error (%) $\delta\hat{x}_\alpha$	+8.12	+7.42	+5.70	+4.89
Value $\tilde{x}_\alpha$	626.205	642.490	690.990	718.037
Error (%) $\delta\tilde{x}_\alpha$	+5.43	+4.66	+3.38	+2.95
		$n = 50$		
Value $\hat{x}_\alpha$	594.826	615.220	670.905	700.593
Error (%) $\delta\hat{x}_\alpha$	+0.15	+0.21	+0.37	+0.45
Value $\tilde{x}_\alpha$	533.821	560.640	643.170	683.530
Error (%) $\delta\tilde{x}_\alpha$	-10.12	-8.67	-3.77	-1.99
		$n = 100$		
Value $\hat{x}_\alpha$	587.842	609.090	667.100	698.025
Error (%) $\delta\hat{x}_\alpha$	-1.03	-0.78	-0.19	+0.08
Value $\tilde{x}_\alpha$	543.635	564.880	639.260	681.840
Error (%) $\delta\tilde{x}_\alpha$	-8.47	-7.98	-4.36	-2.24
		$n = 200$		
Value $\hat{x}_\alpha$	602.783	622.790	677.410	706.530
Error (%) $\delta\hat{x}_\alpha$	+1.48	+1.44	+1.35	+1.30
Value $\tilde{x}_\alpha$	568.535	590.500	659.545	697.330
Error (%) $\delta\tilde{x}_\alpha$	-4.28	-3.81	-1.33	-0.02
		$n = 500$		
Value $\hat{x}_\alpha$	597.553	617.380	671.530	700.400
Error (%) $\delta\hat{x}_\alpha$	+0.61	+0.57	+0.47	+0.42
Value $\tilde{x}_\alpha$	577.823	594.110	660.650	695.519
Error (%) $\delta\tilde{x}_\alpha$	-2.71	-3.22	-1.16	-0.28
		$n = 2000$		
Value $\hat{x}_\alpha$	564.738	618.160	672.185	700.990
Error (%) $\delta\hat{x}_\alpha$	-4.91	-0.69	+0.56	+0.50
Value $\tilde{x}_\alpha$	588.884	607.320	666.440	697.980
Error (%) $\delta\tilde{x}_\alpha$	-0.85	-1.07	-0.29	+0.07

Table 3. The results of calculation of x_α (right tail of distribution).

α	0.900	0.950	0.990	0.995
	Normal distribution low (14)			
Value x_α	902.525	931.590	986.100	1006.09
		$n = 20$		
Value $\hat{x}_\alpha$	908.696	933.800	980.883	998.140
Error (%) $\delta\hat{x}_\alpha$	+0.68	+0.24	-0.53	-0.79
Value $\tilde{x}_\alpha$	927.680	952.457	993.856	1007.95
Error (%) $\delta\tilde{x}_\alpha$	+2.78	+2.23	+0.78	+0.18
		$n = 50$		

Value $\hat{x}_\alpha$	910.051	939.740	995.440	1015.80
Error (%) $\delta\hat{x}_\alpha$	+0.83	+0.87	+0.95	+0.96
Value $\tilde{x}_\alpha$	925.666	955.045	1003.05	1019.03
Error (%) $\delta\tilde{x}_\alpha$	+2.56	+2.51	+1.71	+1.29
		$n = 100$		
Value $\hat{x}_\alpha$	916.224	947.150	1005.16	1026.43
Error (%) $\delta\hat{x}_\alpha$	+1.67	+1.67	+1.93	+2.02
Value $\tilde{x}_\alpha$	922.315	949.64	1003.55	1026.00
Error (%) $\delta\tilde{x}_\alpha$	+2.19	+1.93	+1.77	+1.97
		$n = 200$		
Value $\hat{x}_\alpha$	911.992	941.110	995.747	1015.76
Error (%) $\delta\hat{x}_\alpha$	+1.02	+1.02	+0.99	+0.96
Value $\tilde{x}_\alpha$	917.378	944.730	995.670	1014.70
Error (%) $\delta\tilde{x}_\alpha$	+1.64	+1.41	+0.97	+0.86
		$n = 500$		
Value $\hat{x}_\alpha$	904.069	932.940	987.100	1006.90
Error (%) $\delta\hat{x}_\alpha$	+0.14	+0.14	+0.10	+0.08
Value $\tilde{x}_\alpha$	905.201	935.705	993.090	1013.85
Error (%) $\delta\tilde{x}_\alpha$	+0.30	+0.44	+0.71	+0.77
		$n = 2000$		
Value $\hat{x}_\alpha$	904.210	933.015	987.050	1040.46
Error (%) $\delta\hat{x}_\alpha$	+0.15	+0.15	+0.09	+3.41
Value $\tilde{x}_\alpha$	907.718	938.180	989.690	1008.56
Error (%) $\delta\tilde{x}_\alpha$	+0.57	+0.71	+0.36	+0.24

Using the random number generator (the algorithm and the program to realize it are presented in the work [16]), according to the law (14), we generate a sample with the length of $n = 2000$ for the random variable x_i, $i = \overline{1,n}$, which becomes the basis for determining the parameters of the law (14), with the sample volume sequentially increasing:

$$\hat{m} = \frac{1}{n}\sum_{i=1}^{n} x_i \quad \hat{s} = \sqrt{\frac{1}{n-1}\sum_{i=1}^{n}(x_i - \hat{m})^2} \quad n = 20, 50, 100, 200, 500, 2000\,. \qquad (16)$$

In this case, the normal distribution density function looks like the following:

$$\hat{P}(x) = \frac{1}{\hat{s}\sqrt{2\cdot\pi}} \exp\left[-0,5\left(\frac{x-\hat{m}}{\hat{s}}\right)^2\right], \qquad (17)$$

and the quantile estimators $\hat{x}_\alpha$ of the random variable x, calculated on the basis of the sample, are determined by solving an equation similar to (15):

$$\int_{x_{\min}}^{\hat{x}_\alpha} \hat{P}(x)dx = \frac{1}{\hat{s}\sqrt{2\cdot\pi}} \int_{x_{\min}}^{\hat{x}_\alpha} \exp\left[-0,5\left(\frac{x-\hat{m}}{\hat{s}}\right)^2\right] dx = \alpha\,. \qquad (18)$$

As the truth values x_α are known, we are able to analyze the errors $\delta\hat{x}_\alpha$ of the obtained estimators $\hat{x}_\alpha$:

$$\delta\hat{x}_\alpha = 100\left(\hat{x}_\alpha - x_\alpha\right)/\,x_\alpha\,. \quad (19)$$

The values $\hat{x}_\alpha$, $\delta\hat{x}_\alpha$ calculated with different sample volumes of the random variable x ($n = 20, 50, 100, 200, 500, 2000$)) are presented in Tables 2 and 3.

In Figures 1–6, the dashed line shows the density function $P(x)$ and histograms of the random variable x, which correspond to different sample volumes.

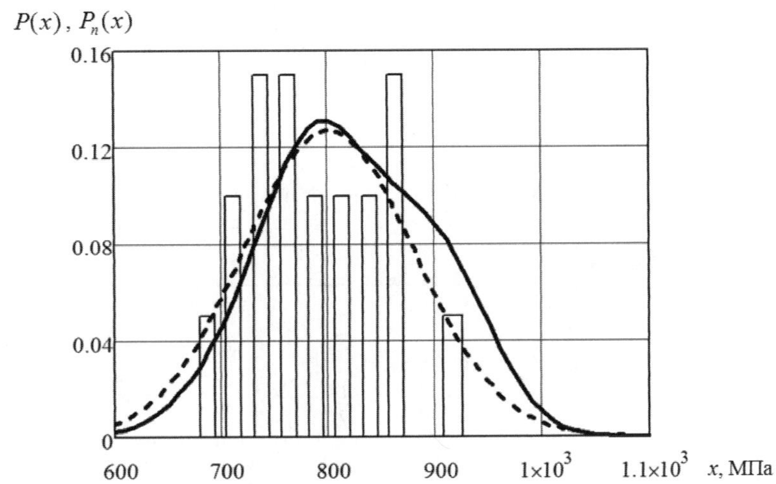

Figure 1. Distribution density functions $P(x)$, $P_n(x)$ and the histogram of distribution of a random variable x at $n = 20$

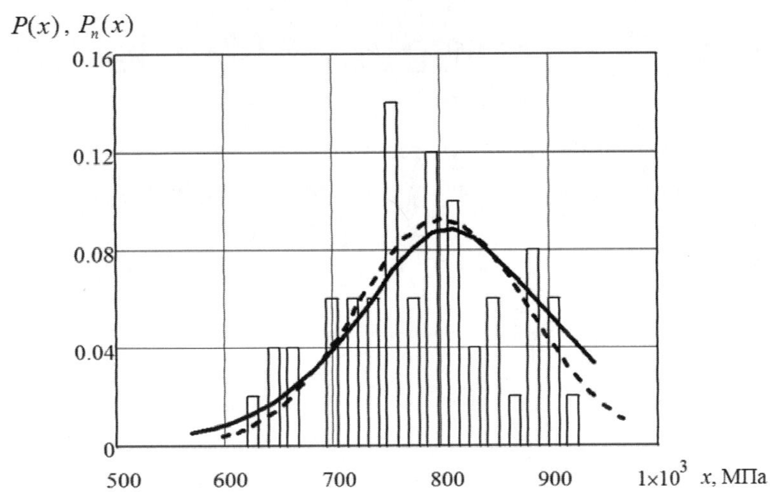

Figure 2. Distribution density functions $P(x)$, $P_n(x)$ and the histogram of distribution of a random variable x at $n = 50$

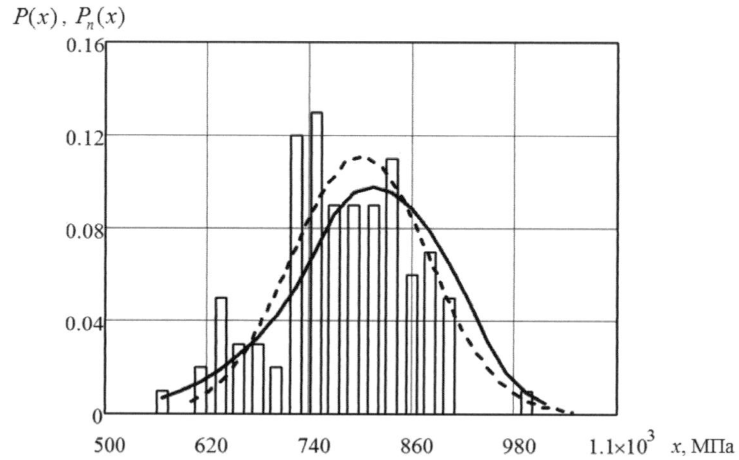

Figure 3. Distribution density functions $P(x)$, $P_n(x)$ and the histogram of distribution of a random variable x at $n = 100$

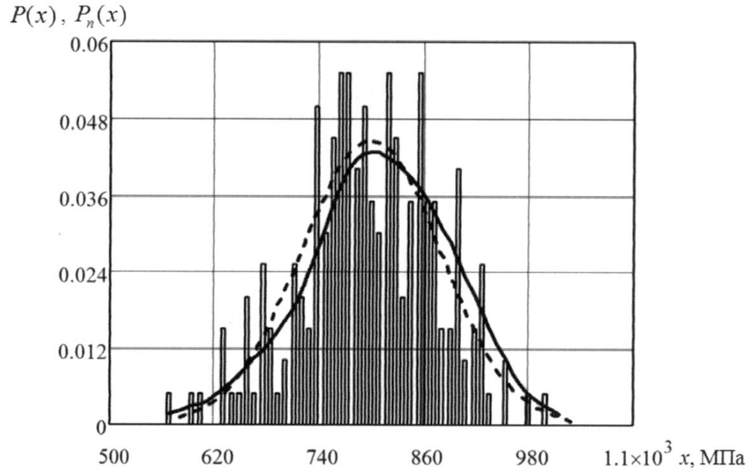

Figure 4. Distribution density functions $P(x)$, $P_n(x)$ and the histogram of distribution of a random variable x at $n = 200$

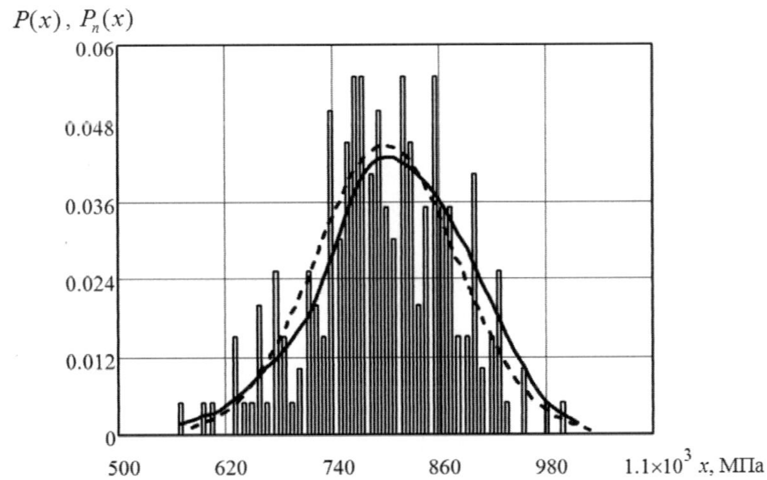

Figure 5. Distribution density functions $P(x)$, $P_n(x)$ and the histogram of distribution of a random variable x at $n = 500$

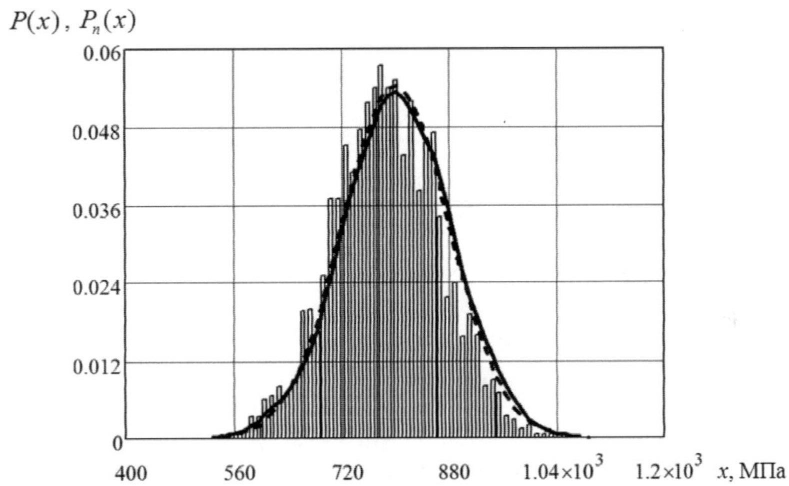

Figure 6. Distribution density functions $P(x)$, $P_n(x)$ and the histogram of distribution of a random variable x at $n = 2000$

Next, having the samples of the random variable x, we recover the unknown distribution density function $P_n(x)$ by means of nonparametric statistical methods. Using the Gaussian as the kernel function (Table 1), we present the desired function $P_n(x)$ as follows:

$$P_n(x) = \frac{1}{\sqrt{2 \cdot \pi} \cdot n \cdot h_n^*} \sum_{i=1}^{n} \exp\left[-0,5 \left(\frac{x - x_i}{h_n^*} \right)^2 \right], \tag{20}$$

where the optimal value of fuzziness parameter h_n^* corresponds to the functional maximum:

$$\max_{h_n}\left\{\frac{1}{n}\sum_{i=1}^{n}\ln\left[\frac{1}{(n-1)h_n}\sum_{j\neq i}^{n-1}\frac{1}{\sqrt{2\pi}}\exp\left[-0,5\left(\frac{x_i-x_j}{h_n}\right)^2\right]\right]\right\}. \tag{21}$$

When solving the problem (21), the initial estimate h_n is calculated according to the dependence (11), taking into account (12).

Using the distribution density function $P_n(x)$, the quantile estimators $\tilde{x}_\alpha$ of the random variable x are determined by solving the following equation through the calculus of approximations:

$$\int_{x_{\min}}^{\tilde{x}_\alpha} P_n(x)dx = \frac{1}{n\cdot h_n^*\sqrt{2\cdot\pi}}\int_{x_{\min}}^{\tilde{x}_\alpha}\sum_{i=1}^{n}\exp\left[-0,5\left(\frac{x-x_i}{h_n^*}\right)^2\right]dx = \alpha\,. \tag{22}$$

With the known truth values x_α, the errors of the estimators $\tilde{x}_\alpha$ are calculated according to the similar formula (19) of dependence:

$$\delta\tilde{x}_\alpha = 100\left(\tilde{x}_\alpha - x_\alpha\right)/\,x_\alpha\,. \tag{23}$$

The values $\tilde{x}_\alpha$, $\delta\tilde{x}_\alpha$ obtained with different sample volumes of the random variable x $(n = 20, 50, 100, 200, 500, 2000)$ are given in Tables 2 and 3, and in Figures 1–6 the function $P_n(x)$ recovered by means of nonparametric statistical methods with different n is indicated with a solid line.

The analysis of the results presented in Tables 2 and 3 shows that for the left distribution «tail» the maximum error $|\delta\hat{x}_\alpha| = 8.12\%$ (for the sample $n = 20$ and $\alpha = 0.005$), and for $|\delta\tilde{x}_\alpha| = 10.12\%$ (for the sample $n = 50$ and $\alpha = 0.005$). For the right distribution «tail» the maximum error $|\delta\hat{x}_\alpha| = 3.41\%$ (for the sample $n = 2000$ and $\alpha = 0.995$), and for $|\delta\tilde{x}_\alpha| = 2.78\%$ (for the sample $n = 20$ and $\alpha = 0.900$). In other words, the errors of the quantile estimators obtained by means of parametric and nonparametric statistical methods constitute equal-order values.

5. Conclusion

In connection to the task of verifying the recovery of random-variable distribution density function having the normal distribution law by means of nonparametric statistical methods through correlating the values of the random variable quantile estimators in the left and right distribution «tails», the results presented in this work prove that the errors of these estimators calculated by means of parametric and nonparametric statistical methods with the sample length range from 20 to 2000 constitute equal-order values.

This work was supported by grant (project N9.6355.2017/BCh) of government order of Ministry of education and science of Russian Federation for the period 2017–2019 in Tyumen Industrial University.

References

[1] Kapur K C and Lamberson L R 1977 Reliability in Engineering Design (New Jork: John Wiley & Sons) 586 p
[2] Gnedenko B V 1962 The Theory of Probability *Physics Today* **15** iss. 10 pp 62–64
[3] Gadolina I and Zaynetdinov R 2018 The estimation of the sufficient random loading realization length in the problem of machine parts longevity Proc. IEEE 9th Int. Conf. on Dependable Systems, Services and Technologies (Kiev) pp 159–162
[4] Collins J A 1981 Failure of Materials in Mechanical Design: Analysis, Prediction, Prevention (New York: John Wiley & Sons) 630 p

[5] Timashev S A, Bushinskaya A V, Malukova M G and Poluyan L V 2013 Integrity and Safety of Pipeline Systems ed S A Timashev (Yekaterinburg: UB RAS) 590 p
[6] Vapnik V 1999 An overview of statistical learning theory *IEEE Transactions on Neural Networks* **10** iss. 5 pp 988–999
[7] Orlov A I 2006 Applied statistics (Moscow: Exam) 671 p (in Russian)
[8] Simakhin V A 2006 Nonparametric robust regression estimate Proc. SPIE 13th Joint Int. Symp. on Atmospheric and Ocean Optics vol.6522 (Tomsk) pp 130–139
[9] Koshkin G M 2016 Smooth nonparametric estimation of the failure rate function and its first two derivatives *Russian Physics Journal* **59** iss.6 pp 833–844
[10] Porshnev A S and Koposov A S 2014 Quantitave estimation peculiarities of densities blur parameter of random sequences distribution in Parzen-Rosenblatt approximation *In the World of Scientific Discoveries, Series B* **2** №1 2014 pp 120–129
[11] Parzen E 1962 On estimation of a probability density function and mode *Ann. Math. Statist.* **33** No 3 pp 1065-1076
[12] Rozenblatt M 1956 Remarks on some nonparametric estimates of a density functions *Ann. Math. Statist.* **27** No 3 pp 832–837
[13] Timashev S A and Tyrsin A N 2011 Entropy approach to risk-analysis of critical infrastructures systems Proc. First Int. Conf. on Vulnerability and Risk Analysis and Management and Fourth Int. Symp (Hyattsville) pp 147–154
[14] Syzrantseva K and Syzrantsev V 2017 Determination of parameters of endurance limit distribution law of material by the methods of nonparametric statistics and kinetic theory of high-cycle fatigue *Key Engineering Materials* **736** pp 52–57
[15] Syzrantseva K V 2009 Predicting the failure of gear transmissions by nonparametric statistics *Russian Engineering Research* **29** No 2 pp 1206–1208
[16] Syzrantsev V N, Golofast S L and Nevelev Ya P Calculation of equipment durability based on the methods of the distribution-free statistics (Novosibirsk: Nauka) 216 p (in Russian)
[17] Syzrantsev V N, Novoselov V V, Sozonov P M and Golofast S L Estimation of safety and durability of technological pipelines by methods of nonparametric statistics (Novosibirsk: Nauka) 172 p (in Russian)
[18] Syzrantseva K and Syzrantsev V 2016 Reliability estimation of machine parts with complicated geometry on a base of methods of nonparametric statistics *Journal of Engineering and Applied Sciences* **11** iss. 2 pp 204–209

IOP Conf. Series: Materials Science and Engineering **481** (2019) 012005 doi:10.1088/1757-899X/481/1/012005

Advantages of the rain-flow method at the post-processing stage in comparison with the spectral approach

I Gadolina[1], R Zaynetdinov [2]

[1]IMASH RAS, Maly Kcharitonievsky, 4, Moscow, 101990, Russia
[2] Independent researcher, Moscow, Russia

E-mail: gadolina@mail.ru

Abstract. In some massive constructions, like off-shore platforms for example, it is sometimes difficult or impossible to register the loading process during many years of exploitation. Knowing the input random processes characteristics, it sometimes seems promising to estimate the output characteristics using the structural dynamic equation. Following this approach, the spectral densities are employed into longevity estimation. Many methods for estimation of the fatigue damage thought the spectral densities were developed. The question arises: is it practical to use the spectral approach in case of post-processing when the process realization already exists? The aim of this paper is to show some limitations of the spectral approach. First, the short review with a brief analysis of the accuracy of the spectral method is presented. It is worth mentioning, that the ultimate goal of the authors of those methods is to achieve the results, similar to those, which were obtained from the rain-flow method. Next, the real loading process is analyzed. To show the fact, that some of the assumptions of spectral methods are doubtful, the method of the digitally modelled process is introduced. Pro and contras of the spectral methods are discussed.

1. Introduction

Nowadays many works are dedicated to machine longevity estimation in the frequency domain, i.e. though operating with spectral densities of the random loading signal. Indeed, there are some situations, when the problem is already described somehow in the frequency domain. That means that the spectral densities of the element loading are known. Spectral densities could be obtained, for example, through the decision of the structural dynamic task. In these situations, spectral methods for estimating longevity represent the natural way to solve the problem [1]. The power spectra of wave loading, for example, can serve as an input in a structural dynamic analysis for estimating of the off-shore longevity by spectral methods at the design stage [2].

On the other hand, during the post-processing stage, when the researchers could register the random loading realization during the exploitation, there is no need to take that controversial way of random loading proceeding. Following the well-known and experimentally proved rain-flow method [3], engineers could estimate longevity for almost every type of the random processes, including non-stationary and non-gaussian types without any difficulty. The rain-flow method is among the best methods of cycle counting – the procedure of substitution of the random amplitudes by the set of

IOP Conf. Series: Materials Science and Engineering **481** (2019) 012005 doi:10.1088/1757-899X/481/1/012005

harmonics in a special way. Most of the standards recommend the rain flow method [4-6]. Together with the Miner's summation rule [7] the rain-flow method gives good agreement with the experiment.

In this paper 1) the short review of the spectral methods is presented; 2) the analysis of some real loading data are shown; 3) special method of process simulation was proposed with the purpose to reveal some contradictions in spectral approach.

2. Short review of the spectral methods

Since the introduction of the famous Rice's formula [8] in 1944, describing the distribution of peaks of narrowband random processes in a closed form, a lot of effort has been made to find the ways to use the spectral representation of random processes of loading in the evaluation of durability.

2.1. Narrow-band approximation

The Rayleigh approximation, also referred to as the narrowband approximation, assumes that the stress ranges are distributed as the maximums of the stationary Gaussian process . In the narrow-band process for one peak the mean level crossing takes place. In 1965 the formula for accessing the fatigue damage in case on narrow -band process was proposed [9]. The fatigue damage evaluated by the narrow-band Rayleigh approximation (*NB*) can be defined as D_{NB}.

2.2. Wide-band approximation by correction factor

Most processes are not narrow band. The more complex the random process, the more distant the distribution of peaks is from the distributions of amplitudes of the rain flow amplitudes. They might be called as the wide-band process or as the processes with the complex structure.

The damage due to a wideband stress process (*D*) in some cases is proposed to be estimated through the correction of the narrowband estimated damage as:

$$D = \lambda D_{NB} \tag{1}$$

In (1) λ is the correction factor. Some authors [10], [11] proposed varied methods for estimation of λ. In Table 1 and Fig.1 these results will be referred to as WL for [10] and AL for [11].

2.3. Wide-band approximation by composing a few distributions

Some authors tried to compose two or more varied distributions to get the rain-flow amplitudes distributions. For a long time considered as the best method – Dirlik method [12] will be referred to as DK. The other method of this kind is [13] (ZB).

2.4. Methods, which employ spectral moments

The geometrical characteristics of spectral density W(f), which represents the frequency spectrum of the random stress, or the moments m_j, for longevity estimation are calculated by [1]:

$$m_j = \int_0^{\infty} f^j W(f) df \tag{2}$$

2.5. Brief comparison

Detailed review with the thorough description of the methods is presented in [11]. Some extracts from [11] in chronological order are shown in Tabl.1.

IOP Conf. Series: Materials Science and Engineering **481** (2019) 012005 doi:10.1088/1757-899X/481/1/012005

Table 1. List of some popular spectral methods (from [17] with the permission of the author).

Definition	Method	Year	Reference
NB	narrow-band method	1965	9
WL	Wirsching-Light method	1980	10
DK	Dirlik method	1985	12
ZB	Zhao-Baker method	1992	13
AL	the α0.75 method	2004	11
PZ	Petrucci-Zuccarello method	2004	15
TB	Tovo-Benasciutti method	2006	14

In [17&18] the accuracy of fatigue-life estimates of the frequency-domain methods was compared to the life estimate in the time domain using a combination of the rain-flow count [4,5] and the Palmgren-Miner [7] hypothesis. This value was denoted as TRF and was assumed to be an exact reference value in the studies. The relative error is calculated by:

$$ERR = (T^{XX} - T_{RF})/T_{RF} \quad (3)$$

where T^{XX} –the estimations based on the methods shown in Table 1.

Several random processes were modelled and the longevity estimations for varied fatigue exponent *m* values were performed. Here, among many other processes, the process MM (multi-modal, [18]) has been chosen, as the most interesting example, because of its wide-band nature. In Figure 1 according to works [17] and [18] the accuracy of some spectral methods is shown.

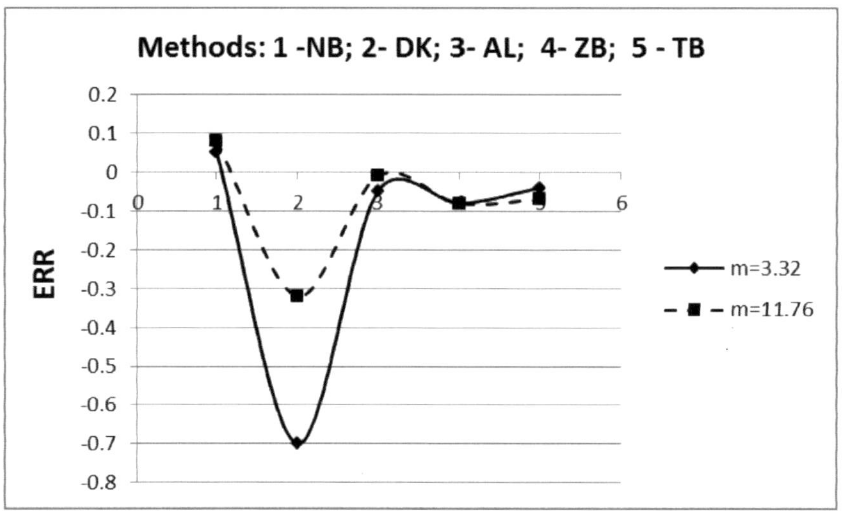

Figure 1. The comparison of varied methods accuracy in relation to rain-flow. (Notations according to the Table 1).

Note, that the results in Fig.1 were taken just for example. It is impossible to judge about the accuracy of the methods looking at this Figure. More thorough investigations usually conclude about advantages of DK and TB methods above the other methods.

The fact, that we have learned from this analysis, is the following. The accuracy of the estimations by the spectral approach methods is satisfactory, no matter which of the methods was used.

3. Material

As an example of the random loading, the real loading processes in the torsion shaft of the cantilever machine during exploitation were investigated [19]. Figure 2 shows one realization registered during 1,5 min. move. It represents a typical loading of the machine part.

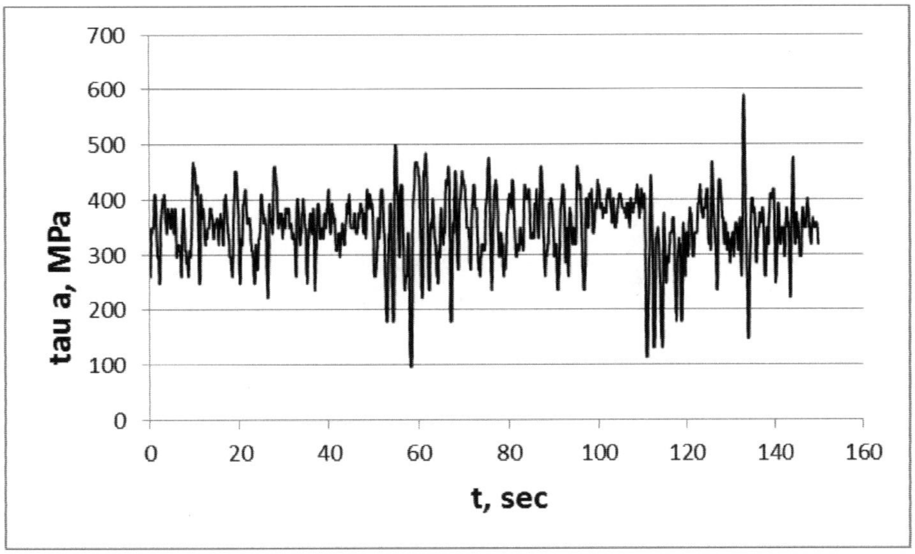

Figure 2. The realization #1 of random loading in the part of cantilever machine.

The main statistical characteristics of the process are:

- Mean value = 348.9 MPa;
- Std = 60.84 MPa.

Two studies of the random processes were performed.

3.1. Stationarity test

To calculate moments m_j (2) the spectral densities W(f) should be evaluated. By the definition they exist only for stationary processes. In this study Augmented Dickey-Fuller Test was employed. The analysis was performed in R statistical package [20]. Since the p-values for the sample processes were small, then with the significance level of 0.01 those series could be considered stationary (the hypothesis of the presence of a single root is rejected). It allows estimating the spectral densities (Fig.5, the solid line).

3.2. Analysis the of irregularity factor

The important characteristic, which shows the degree of complexity of the process is irregularity factor, IF. Knowing $W(f)$ and the moments m_j the irregularity factor might be estimated [8]:

$$IF = m_2 \sqrt{m_0 m_4}$$

In other way it might be defined as in [4]:

$$IF = N_O/N_E \quad (4)$$

where N_O and N_E are the numbers of mean level crossing and the number of extremums accordingly.

The study of dependability of the *IF* indicator was performed by the investigation of several realizations of stresses in the torsion shaft, like what one which is shown in the Fig. 2. The initial process was divided by the classes by the ordinate [4]. Classes number in this study was taken in the interval $M = 12 \ldots 100$. This procedure of dividing by classes might be used as a first step of filtration. The dependence of *IF* on *M* for some realizations is shown in the Fig. 3 .

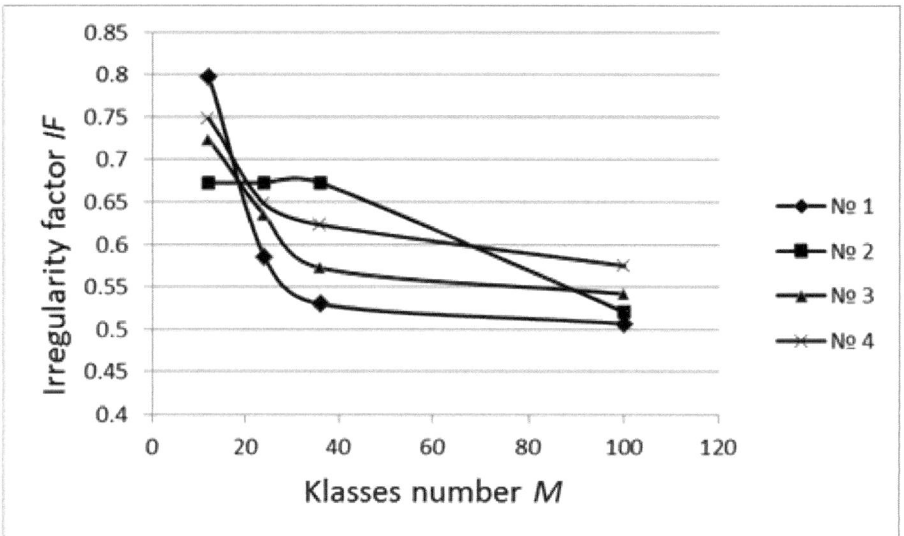

Figure 3. Dependence the estimated *IF* on the number of classes

It can be seen, that with the increase in the number of classes, *IF* decreases by up to 35%. Hence the conclusion: the value *IF* cannot be used as an objective characteristic of the random process. If the value M is taken $M = 36$, (following the recommendations [4]), the estimated for realization #1 by equation (4) is $IF = 0.525$, that means, that the process is complex enough.

4. Method

As a method to confirm the validity of the transition from a continuous process to a sequence of extrema, the “saw”, the numerical simulation was carried out. That question is of great importance while speaking of spectral densities because during this transition the information of time is lost. Fig. 4 schematically shows the three stages of the process transformation. As an example, the initial part of the random process #1 was taken (Fig.2). The first transition from "initial" (I) to the "saw" (II) is recommended by the majority of researchers [15, 16] and is included in all standards [4, 5]. This transition is a necessary stage for all the cycle counting methods, including rain-flow. What happens with the spectral density during this transformation? This question is important while speculating about the spectral approach.

There is a desire to compare spectral densities of those processes. However, it is impossible to judge of the spectral density of the “saw”: there is no time factor, the process is not differentiable, because there are singularities. To show the fact, that the spectra do not influence much upon the longevity, a method of the peaks approximating of the process II by the half-waves was developed. It was proposed to connect the peaks of the "saw" by the half- cosines. The equation of the smoothing process corresponds for each half-cosine wave would be:

$$X = A\cos(wt + \varphi) \tag{5}$$

where x is the resulting extrapolating process. For each half-wave (5) the parameters A, w and φ are unique. The amplitude A is defined as the module of the half difference of successive extremums $e[I-1]$ and $e[i]$:

$$A_i = \mathrm{abs}(e[i] - e[i-1])/2 \tag{6}$$

The frequency w_i is chosen in a such way, that the half-period would have enough points ($k = 5 \ldots 40$), depending on the A_i value. The initial cosine phase φ depends on the type of half-wave. If the downward sweep is modelled, $\varphi = 0$. For the rising half-wave $\varphi = \pi/2$. Of course, the choice of k will influence the frequency characteristics of the modelled process (Fig. 4, III).

This modelling was performed in R language [20].

The resulting process, obtained as outcome of this approximation, is shown in Fig. 4, III. After this transformation it becomes possible to estimate the its spectral density, since process III represents the discretization of a continuous function. Spectral densities of the processes I and III are shown in Fig. 5. The dashed line, laying below in the high- frequency domain, corresponds to the process III.

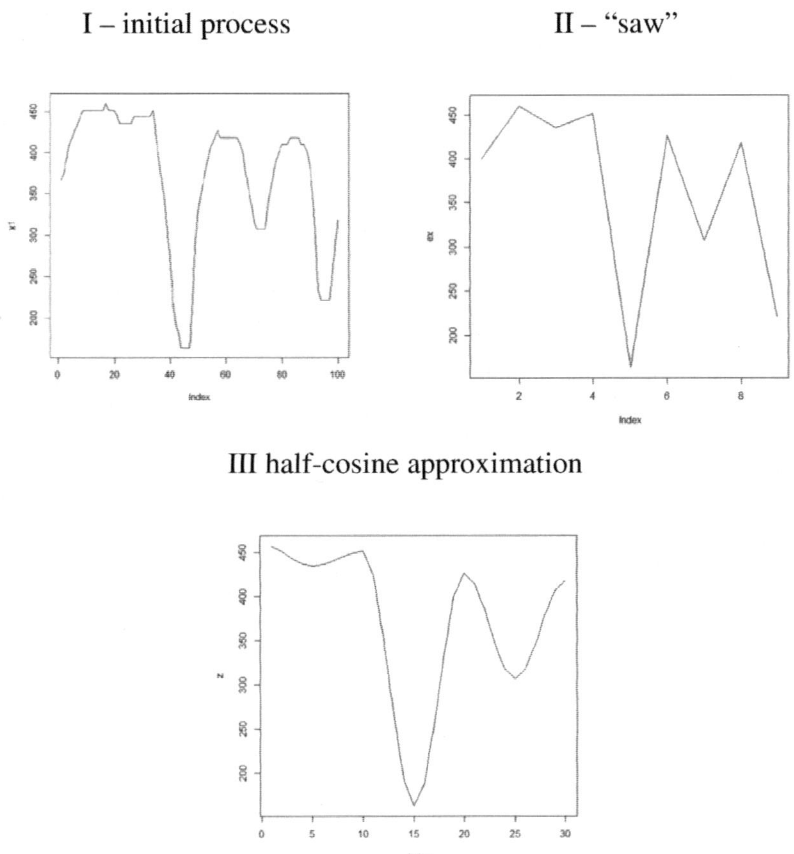

Figure 4. Three processes

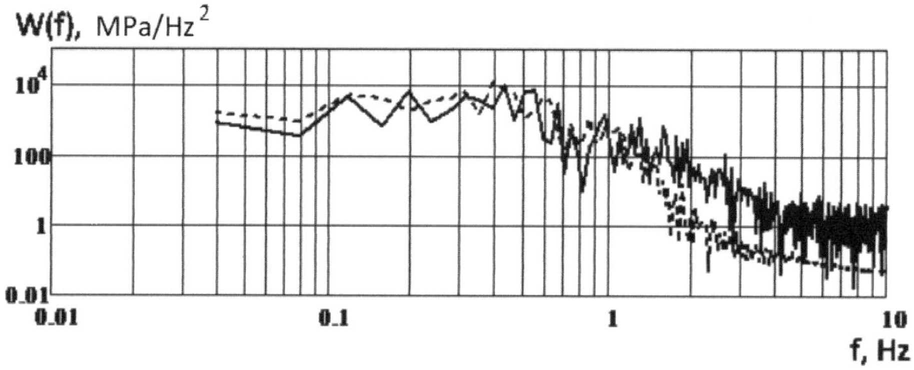

Figure 5. Comparing the spectral densities of two similar by damage level processes

Figures 4 and 5 show that the processes I and III are similar at some extent, the frequency characteristics are close. On the other hand, the natures of the spectra are different. In the region of high frequencies (above 1.5 Hz), process III has significantly lower level than I: the simulated process has less "high-frequency" noise for the power spectrum by two orders of magnitude. This difference cannot but affect the calculated moments of m_j, according to the formula (2), and hence the durability estimates, produced by the processes I and III using moments (3) will be different. Research practice shows, on the contrary, the legitimacy of the transition from I to III and the similarity of these processes by the damaging impact.

5. Discussion

In this paper, only the stage of post-processing is for uniaxial loading was considered. The question of multiaxial loading was not covered. Multiaxial loading is addressed in [1, 21].

Table 2. Arguments and counterarguments concerning rain-flow and spectral approach.

Person	Argument	Counterargument
D. Benashutti, private exchange of views in reseatchgate	"If you have, say, only one sample value from a Gaussian distribution, what can you say about the parameters (true 'mean' and 'std') of that distribution? Very little... …nothing can be said about the true mean value of the damage statistical distribution. In addition, nothing can be said about cycles not directly observed in that time-history. By contrast, if you have a PSD, you can estimate the amplitude PSD and the expected damage from PSD data. The main difference between time- and frequency-domain approach basically relies in the reliability of statistical estimates"	Estimating the spectral density by only ONE realization also restricts the horizons. Some extra steps concerning the loading variability should be taken in the both approaches
F.Cianetti [16]	Rain-flow has a heavy computational cost…	While choosing the proper classes number [4] it is possible to perform the rain-flow treating in a real-time mode

Currently, many researchers are working to expand spectral methods towards non-Gaussian and non-stationary processes [1, 22]. Using rain-flow method the researchers do not experience special difficulties because of these peculiarities.

As for the discussion about the use of spectral methods in the post-processing phase, table 2 attempts to collect some arguments and respond to them.

6. Conclusions

In some situations (especially at the design stage) methods for estimating longevity by the spectral density might be helpful.

Many methods for estimating the longevity by the spectral methods exist and it is tough to choose the best one for all the situations concerning the varied processes types and material properties.

The irregularity factor seems not to be the robust indicator of the random processes.

All types of random loading processes, including non-Gaussian and non-stationary, could be easily treated by rain-flow method.

Developed intendedly for the analysis of possibility of transmission from continuous process to the extremum discontinuous sequence method proved the fact, that there is no strong correlation between the spectral density and longevity.

Following the recommendation [23], which was given many years ago, the spectral methods would be recommended at the design stage only.

References

[1] Benasciutti D, Tovo R 2018 Frequency-based analysis of random fatigue loads: Models, hypotheses, reality. *Mat. Wiss. Werkstof.* 49 345–367

[2] Larsen C, Lutes L 1991 Predicting the Fatigue Life of Offshore Structures by the Single-Moment Spectral Method. In: Elishakoff I., Lin Y.K. *Stochastic Structural Dynamics* 2. Springer, Berlin, Heidelberg

[3] Matsuishi M and Endo T 1968 Fatigue of metals subject to varying stress, Presented to the Japan Society of Mechanical Engineers, Fukuoka, Japan

[4] GOST 25.101 Calculations and strength tests. Methods of cycle counting of random loading processes of machine elements and structures and statistical representation of the results. Approved 21.04.2018 (in Russian).

[5] ASTM E 1049-85 (Reapproved 2005). Standard practices for cycle counting in fatigue analysis. ASTM International.

[6] Tutorial for WAFO version 2017. by the WAFO group. Lund University 195 p

[7] Miner MA 1945 Cumulative damage in fatigue. *J Appl Mech* **12** pp A159–A164

[8] Rice SO. 1944 Mathematical analysis of random noise. *J Bell Syst. Tech* **23**(3)

[9] Miles JW 1965 On structural fatigue under random loading *J. Aeronaut. Soc.*, **21** pp 753-762

[10] Wirsching PH and Light MC 1980 Fatigue under wide band random stresses *J. Struct. Div. ASCE*, **106**(7) pp 1593–1607

[11] Benasciutti D 2004 Fatigue analysis of random loadings. PhD thesis, University of Ferrara, Italy

[12] Dirlik T 1985 Application of Computers in Fatigue Analysis. PhD thesis, Univ Warwick, UK

[13] Zhao W and Baker MJ 1992 On the probability density function of rainflow stress range for statonary gaussian processes *Int. J. Fatigue*, **14**(2) pp 121–135

[14] Benasciutti D and Tovo R. 2005 Spectral methods for lifetime prediction under wide-band stationary random processes *Int. J. Fatigue*, **27**(8) pp 867–877

[15] Petrucci G and Zuccarello B. 2004 Fatigue life prediction under wide band random loading. *Fatigue & Fract. Eng. Mater. & Struct.*, **27**(12) pp 1183–95

[16] Braccesi C, Cianetti F, Tomassini L 2015 Random fatigue. A new frequency domain criterion for the damage evaluation of mechanical components *Int Journal of Fatigue* **70** pp 417-427

[17] Mrsnik M, Slavic J and M. Bolteza M 2013 Frequency-domain methods for a vibration-fatigue-life estimation -Application to real data//Intern J of Fatigue **47** pp 8-17

[18] Larsen C, Irvin T. 2015 A rewiew of spectral methods for variable amplitude fatigue prediction and new results *Procedia Engin* **101** pp 243-250

[19] Dubin D A, Nakaznoi O A, Smirnov A A, Shleev A N 2016 Experimental determination of kinematic and power parameters of loading of elements of the system of high-speed tracked vehicle suspension *Trudy NAMI* . **266** pp 45-53 (in Russian)

[20] R Core Team 2014 R: A language and environment for statistical computing. R Foundation for Statistical Computing, Vienna, Austria. URL http://www.R-project.org/

[21] Niesłony A, Macha E 2007 *Spectral method in multiaxial random fatigue, Springer*, Berlin, Germany.

[22] Syzrantseva, K. 2017 Determination of parameters of endurance limit distribution law of material by the methods of nonparametric statistics and kinetic theory of high-cycle *Key Engineering Materials* **736** 52-57

[23] Kogaev V P 1993 *Strength Calculations at stresses variable in time*. M. Mashinostr (in Russian) 364 p

The reduction the load on the MSW landfill by optimizing the waste management system of residential territories

K V Afonin[1], T S Zhilina[1], A A Zagorskaya[1]

[1]Tyumen Industrial University, Lunacharskogo street, Tyumen 625001, Russia

E-mail: zagorskajaaa@tyuiu.ru

Abstract. The article considers the possibility of reducing the load on municipal solid waste landfills in the real conditions of Tyumen. Analysis of the content of bio fermented components in wastes in Tyumen was carried out. A decentralized scheme for collecting, accumulating and disposing of municipal solid waste (MSW) of multi-apartment buildings for the purpose of generating and using methane is given. It is established that the amount of methane produced is sufficient to compensate for the communal needs of apartment buildings. Reduction of anthropogenic impact of landfills of solid domestic waste is possible with the introduction of utilization and use of biogas at the place of waste generation. A project has been developed for the placement of decentralized MSW processing facilities in the area of the existing residential district. Potential options for utilization of methanogenesis products were selected.

1. Introduction

According to [1], almost all waste of the housing and communal sector in the south of the Tyumen Region is located in the dumps and landfills of municipal solid waste (MSW), the total area of which in the region is 3 880 hectares. The distribution of waste storage facilities by type is shown in Figure 1.

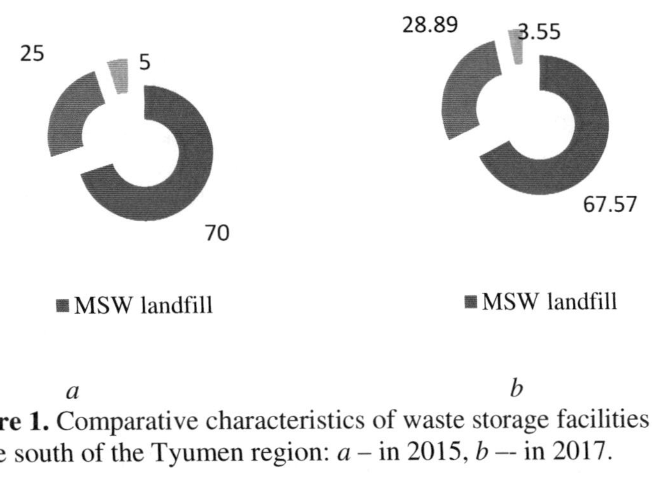

Figure 1. Comparative characteristics of waste storage facilities in the south of the Tyumen region: *a* – in 2015, *b* –- in 2017.

Analyzing the situation with the placement of waste, we can say with confidence that over the past two years the number of unauthorized objects has increased, and the number of officially authorized ones has decreased. This indicates a low environmental and economic efficiency of the existing waste management scheme in the region [2, 3]. The purpose of this research is to consider an alternative scheme of waste management on the example of the city of Tyumen.

2. Materials and methods

The largest MSW landfill in the Tyumen region - Velizhansky, has been in operation since 2010 (Figure 2). Occupies more than 30 hectares and takes about 230 tons of waste [4]. It is located in the agricultural zone of the city of Tyumen, near the large water body - Lake Lipovoye. According to studies [5], the Velizhansky landfill is a serious source of atmospheric pollution and the cause of regular fires.

a *b*

Figure 2. Waste Disposal at the Velizhansky landfill: *a* – according to Google earth, b – waste at the landfill photo of the authors.

The landfill accepts MSW and waste products that are close to them in composition. This object is used only for disposal of waste, does not have a centralized sorting and processing of waste. According to the administrations of municipal districts and urban districts, sorting for the purpose of extracting secondary material resources from MSW throughout the South of the Tyumen region was not carried out [1, 5]. Only in Tyumen the technology of separate collection of municipal solid waste is partially implemented. At the same time, this practice in 2017 had an episodic character with a low level of efficiency. The introduction of separate collection of municipal solid waste in Tyumen in 2017 had no significant impact on the waste management system.

Solid domestic waste has a significant resource value. The Wastes of the Velizhansky polygon [6, 7] contain 75% recycled components, including biologically active (Figure 3).

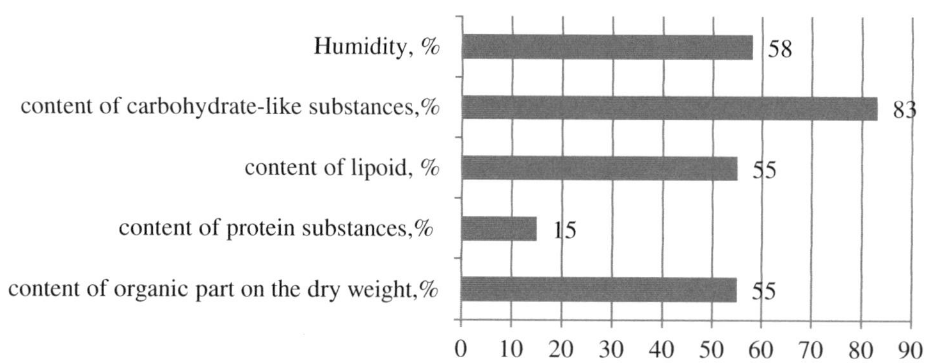

Figure 3. The content of fermented components in MSW of the Velizhansky polygon.

This confirms the possibility of methane fermentation of solid waste in order to obtain biogas. However, significant areas occupied by landfills, as well as the lack of control and technological methods of intensification of methanogenesis in the landfill do not allow doing it in natural conditions. The construction of reactors for fermentation requires significant investment and increased requirements for the safety of the facility. All this contributes to the gradual abandonment of centralized waste collection, storage and disposal [8, 10].

3. Results

The concept of processing and use of solid waste at the place of formation is shown in Figure 4.

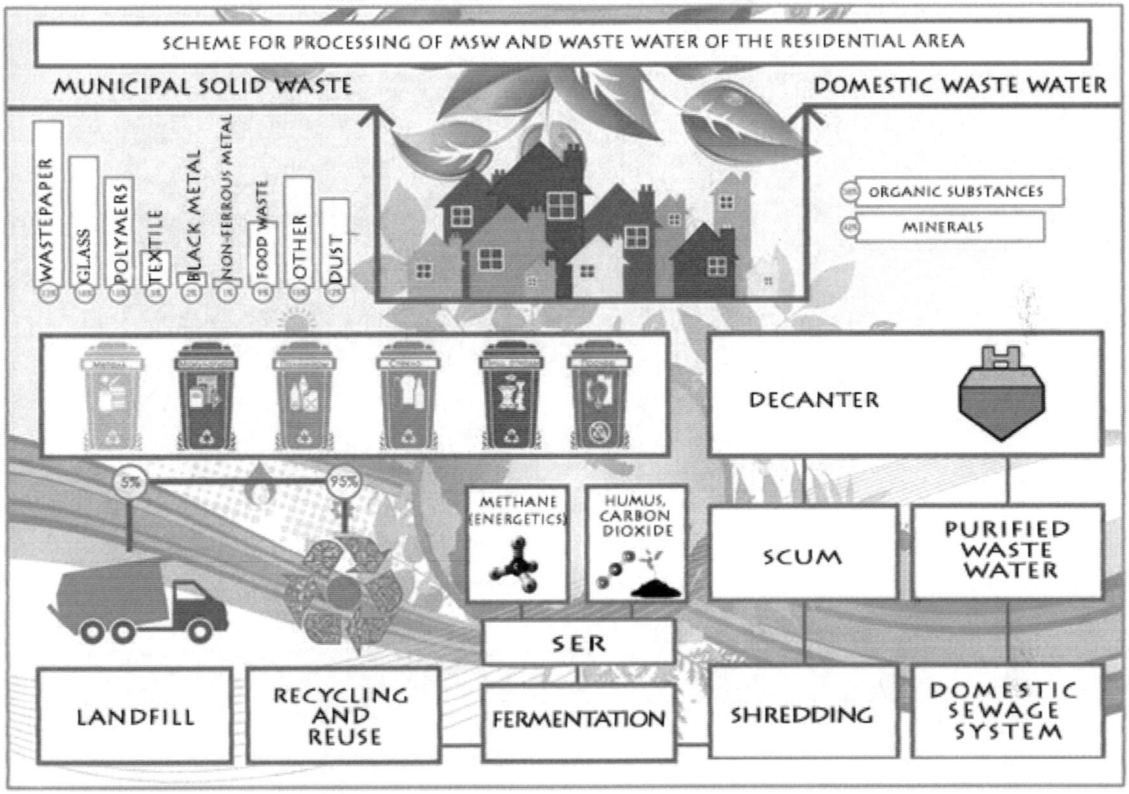

Figure 4. The concept of co-utilization of municipal waste of apartment buildings.

The presented scheme of waste management involves:

1. Separate collection of municipal solid waste from apartment buildings on specialized place
2. Primary wastewater treatment with the removal of large objects that prevent the transportation of wastewater
3. Grinding and dehydration of a mixture of MSW suitable for fermentation and sewage sludge
4. Production of methane, carbon dioxide and humus-like product as a result of methane fermentation
5. Use of the received products for needs of housing and communal services of the residential district

The introduction of a decentralized scheme solves several pressing problems in the field of housing and communal services:

1. Reduce the amount of waste to be accumulated to 5% of the total mass and, as a result, reduce the load on landfills, thereby minimizing their area and increasing their service life.

2. It will contribute to the elimination of blockages of intra-house and intra-municipal sewerage networks, reducing accidents and ensuring the stable operation of urban networks and sewage facilities.
3. It will produce methane up to 1.5 m^3 per day per inhabitant suitable for use in roof boilers [11].

All volume of engineering constructions according to the presented scheme can be placed within the residential district. The architectural concept is presented in Figures 5 and 6.

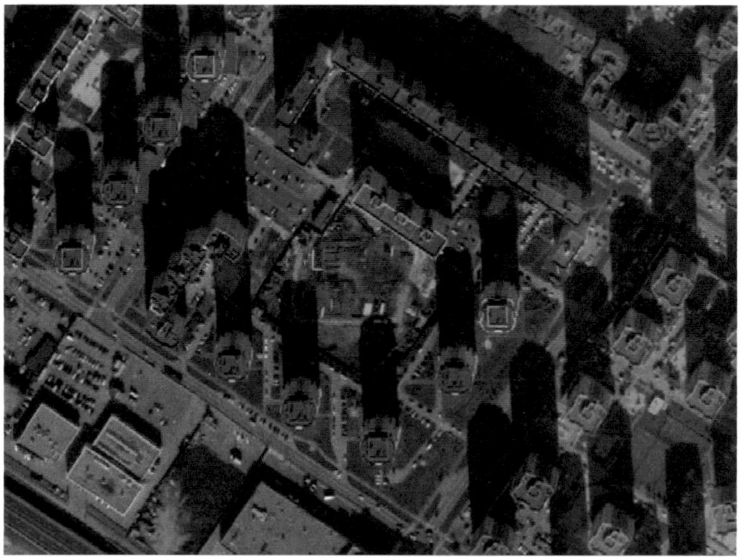

Figure 5. Locations of the installation of decentralized use of municipal solid waste in the quarter (according to Google earth).

Figure 6. The project of construction of methanogenesis plants in the existing residential area.

The modeling of methane formation conditions and the calculation of the main characteristics of the process equipment were carried out for the quarter shown in Fig. 5. As a result of preliminary calculations, the overall dimensions of the blocks were determined (Fig. 7), which is included in the technological scheme of co-digestion.

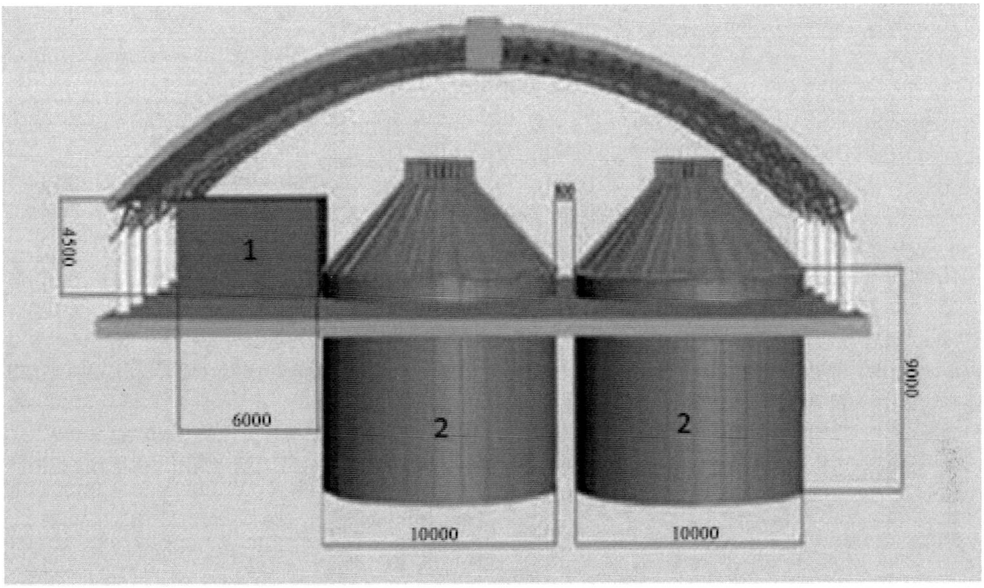

Figure 7. Prototype of methanogenesis installation
1 – Grinding and dewatering unit; *2* – Methane reactors.

4. Conclusion

The accumulation of waste in the environment dramatically affects all elements of the biosphere: changes in relief, evaporation of moisture and groundwater, the nature of vegetation, surface runoff regimes, accumulation of toxic products in biota and transmission along the food chain [12, 13]. Storage of waste is associated with the alienation of large areas, often not subject to restoration and reclamation after the end of the service life [14]. The total impact of waste causes significant environmental and economic damage [15, 16, 17, 18].

Recycling of municipal solid waste by anaerobic digestion reduces the load on municipal solid waste landfills, which will lead to positive changes in the biosphere as a whole.

Co-processing of municipal waste in apartment houses at the place of education allows us to improve the hydraulic modes and conditions of operation of the system of collection of sewage water due to the extraction of floating large impurities (waste).

In addition, biomethanogenesis products are suitable for use in the national economy, as a source of thermal, mechanical and electrical energy, the catalyst for the growth of plants or soil [19, 20].

Reference

[1] *Report "On the environmental situation in the Tyumen region in 2017"* 2018 (Department of subsoil use and ecology of the Tyumen region)

[2] Lebedeva A A, Lebedev D A 2015 Indicators of the condition of the system of managing waste. *Innovative science* **5** pp 229-231

[3] The regional cadaster of waste in Tyumen region 2015 (Tyumen)

[4] Hygienic aspects of waste production and consumption, and protecting the health of the population of the region. *Labor medicine and human ecology*, 2015 - №3

[5] Zhilina T, Afonin K, Zagorskaya A 2017 Reduction of environmental impact of solid domestic landfills of residential area due to their recycling. *MATEC Web of Conferences. "International Science Conference SPBWOSCE-2016 "SMART City""* p 07004.
[6] Afonin K V, Zhilina T S, Zagorskaya A A 2017 Processing of waste of residential areas in order to obtain secondary energy and material resources *Bulletin of the engineering school of the far Eastern Federal University* **4** (**33**). pp 65-73.
[7] Marakulina S P, Choi V E and Emel'yanova D A 2018The use of the system of internal drainage to remove organic waste from the buildings *System technology* **27**
[8] Chudakova O G, Bloodless D V, Methanetank as a device for obtaining fuel from industrial waste. *Bulletin of Kazan technological University* **19(1**8)
[9] SP 32.13330.2012 Sewerage. Pipelines and wastewater treatment plants
[10] Azarova T B, Buslaeva D G, Zagorskaya A A, Zhilina T S, Afonin K V 2017 The Use of biogas for heating residential houses *Engineering science in Russia and abroad* **2**
[11] Skipin L and others 2016 Biogeochemistry of heavy metals in trophic chain in terms of the south of Tumen region *Procedia Engineering* **165** pp 860-868.
[12] Kalyuzhina E A, Samarskaya N S, 2014 Environmental features of the effects of polygons of solid domestic wastes on the condition of the environment in the regions of their location. *Engineering Bulletin of the Don* **3**
[13] SP 2.1.7.1038-01 "Hygienic requirements to the device and the maintenance of ranges for municipal solid waste" (approved. resolution of the Chief state sanitary doctor of the Russian Federation of may 30, 2001 N 16)
[14] Sharova O A 2014 Influence of solid domestic waste landfill on the state of groundwater (for example, landfill R. P. Krasny Yar, Astrakhan Region) *Bulletin of the Tambov University.* Series: Natural and Technical Sciences **19** (5)
[15] Ernst Ulrich von Weizsäcker, Anders Wijkman 2018 *A Report to the Club of Rome.*
[16] Zhukova N S, Samarskaya N S 2014 Environmental and economic peculiarities of the system for hardening waste consumption *Engineering Bulletin of the Don* **3**
[17] Dzhamalova G A 2013 Biotesting of quality of disturbed geotechnogenic ecosystems of the MSW landfill *News of the Russian State Pedagogical University A.I. Herzen* pp55-62
[18] Bashirov V D and others 2014 Modern technologies for the security and processing of solid domestic wastes *News of the Orenburg State Agrarian University* pp 77-80.
[19] Baramzin K N and others 2017 Efficiency of modernization of MSW landfills *Investment valuation* **3**(**17**) pp 9-31.
[20] Mochalova L A 2017 System of handling municipal waste: foreign and domestic experience. *News of the Ural State Mining University* **3**(**47**) pp 97-101.

Structural Design Development of a Float Type Wave Micro Power Plant

I Hossain[*1, 2], V I Velkin[1], S E Shcheklein[1] and A V Eliseev[3]

[1]Ural Federal University, Yekaterinburg, Russia
[2]University Of Dhaka, Bangladesh
[3]Ocean Rus Energy, Russia

*E-mail: hossain.ismail44@yahoo.com

Abstract. The purpose of this study was to improve the design construction of a float type wave micro energy converter. Especially we emphasized on the modelling of a metal capsule float type converter in order for better performance of its kind. At the same time we discussed some parameters on the basis of experimental study. Data concerning on the status of wave energetics and the process of development were discussed. The operating principle, advantages of the wave plant and its benefits as compared to the existing technology were reported. The project of a test installation of wave simulator and the research results were presented. The authors also investigated the electro technical characteristics of the generating unit. The developed micro unit of WMPP showed the possibility of energy generation at wave amplitude from 20cm and more when the wave oscillation period T within 1-5sec. The field research with the wave oscillation simulator confirmed the generation values of one micro unit within the capacity limits from 20 to 60W. It will broaden the possible opportunities and increase the WMPP capacity by a certain level if the micro units are constructed in cluster systems.

Keywords: micro generator, wave power unit, wave simulator, cluster system, energy converter

1. Introduction

In Russia the wave energetics stands apart from the various types of renewable energy resources (RER). Meanwhile, in recent times, it arouses a greater interest among the experts of all over the world [1]. The application for a wave power plant patent was introduced in Paris for the first time in 1799. In the year of 1890 the attempt of the practical use of wave energy comes in real for the first time. The first industrial wave power plant with the capacity 850KW had been operating in Norway since 1985 [2]. In the year of 2008 next wave power plant with the capacity 2.25MW was brought into commercial operation 5km offshore in Agucadoura, Portugal. The power plant project belongs to a Scottish company Pelamis Wave Power which in 2005 had concluded a wave power plant building contract with the Portuguese energy producer Enersis. In 2009 wave power plant was brought into operation on Orkney Islands. The installation plan of a new wave power plant with the capacity of 20MW has been taken into account in the United Kingdom [3].

2. Aspects of wave energy efficiency

Wave energy potential depends on the height, width and wave period. Most of the characteristics of wave energy mostly depend on meteorology and geographical location of the surface of heaving. Experts says that in the Northern hemisphere the medium specific capacity of wind waves is about 25 kW/m. In Atlantic Ocean, 1 linear meter of wave surface is capable of producing up to 30000kWh when the wave height is 2.7 m and wave period is 7.1 sec. per 60 h [4]. Regions with the highest specific energy of waves in different parts of the world and regions of the world and seas with the highest specific energy of waves are presented in Table 1 and 2.

Table 1. Regions with the highest specific energy of waves.

Region	million KWh/ linear meter of wave surface per year
North-West of Ireland	0.535
South-East of Canada	0.465
West of France	0.420
West of USA	0.415
Kuril Islands	0.370

Table 2. Regions of the world and seas with the highest specific energy of waves.

Region	KWh per linear meter of wave surface
North Atlantic	40-80
India	50-90
West of Canada	up to 60
New Zealand	up to 35
Barents Sea	22-29
Norway	up to 20
Sea of Okhotsk	12-20

The practicability of wave energy use is determined by its high specific capacity [5]. At sea the specific capacity can be up to 2MW/m for the wave of more than 10m height. Technically, the wave energy can be used only in coastal zones, where the specific capacity does not exceed 80 KW/m. The specific capacity of wind-induced waves in different seas are presented in Table 3 [6].

Table 3. The specific capacity of wind-induced waves in seas.

Region	kW/m
Caspian sea	7-11
Barents sea	22-29
Baltic sea	7-8
Sea of Okhotsk	12-20

Cockerel Raft, Salter's nodding «duck», oscillating water column, Masuda's pulsating water column are usually considered as the most common and well known hydro-electrical plant of float type [7]. The best use of wave plants can be very useful for its high level of efficiency coefficient (80%) and possibility of full extraction of wave energy. In the year of 1980 on trial basis Japan installed "oscillating water columns" with quite low level of efficiency coefficient. Several of oscillating water columns were constructed and experimented through the period of 1980s and 1990s in Japan, India, China, Norway, Portugal and Britain. The largest of those built in Japan was the port of Sakata [8]. However the development of wave energy plants of such a type is still continued. The installations of "Rubber tube" and "Cockerel Raft" type plants are not much profitable at present for its massive construction cost and its low efficiency.

3. Modelling and design basis of a micro power plant (WMPP)

A closed metal capsule wave energy converter of negative buoyancy has been taken into account for construction development. The body frame of WMPP is made of steel rolled tubular products with the diameter of 180mm consisting a spring-loaded pendulum inside.

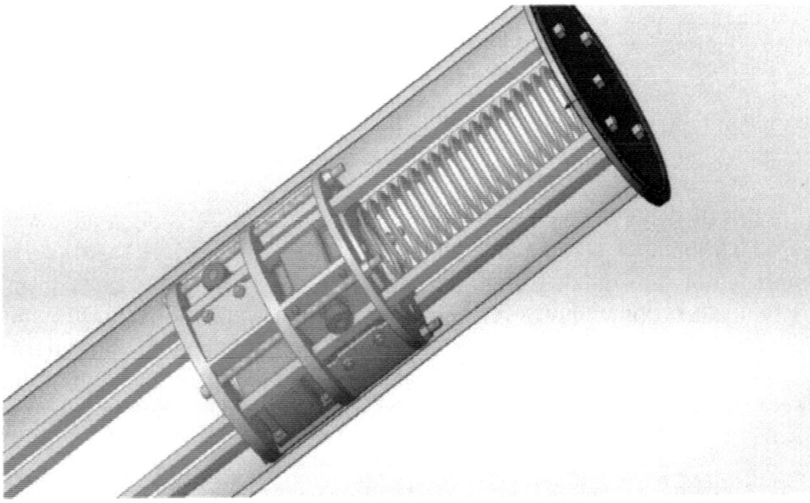

Figure 1. Unit arrangements of structural settings of WMPP.

A small triple-wound generator has been fixed inside the pendulum case. The cog wheel is located on the input drive of the generator. The cog wheel moves in engagement with a cogged rail, fastened longitudinally inside the wave generator frame. Wave electricity generating unit consists of vertical cylindrical case with a mechanical converter of sea wave energy.

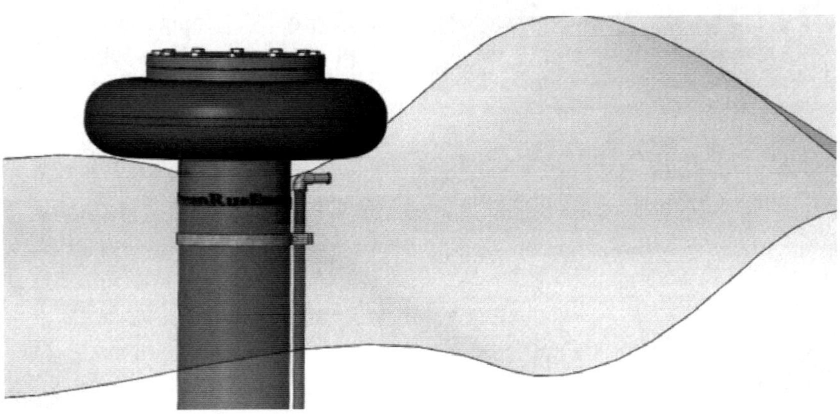

Figure 2. Diagram of wave action on the float micro unit of WMPP.

The rest of it comprises a spring with a plumb, screw pair, clutches, adaptor, idle gear and generator. The construction has embedded speed up gear this is why the rotation to generator is transmitted from the screw pair through the clutches, gear wheels and speed up gear. The clutches are pinion overrunning. A figure of wave motion simulator was presented in the (Fig. 3.).

IOP Conf. Series: Materials Science and Engineering **481** (2019) 012007 doi:10.1088/1757-899X/481/1/012007

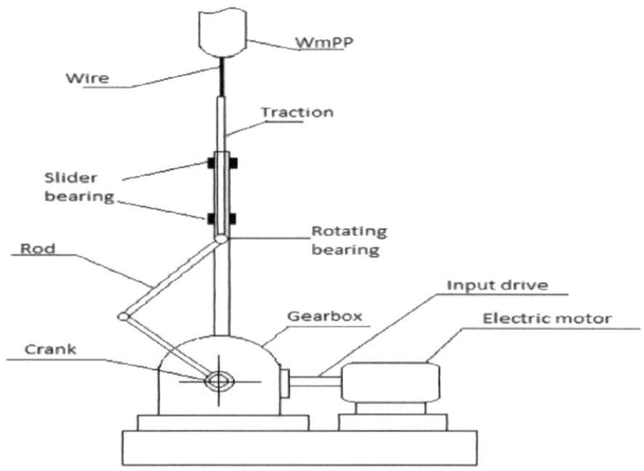

Figure 3. Simulator of wave-generation for the WMPP.

The unique feature of the cylindrical case of WMPP has two units separated by a baffle-one is accelerating part which includes screw pair and spring with plumb (Fig.1.) and the other generating part which includes pinion overrunning clutches, adaptor, idle gear, speed up gear and generator [9]. When a wave motion in the upper and lower points appears, the pendulum shuttles and accumulates a potential energy in the spring. When input drive of the generator rotates, the alternating-current is produced. Small rectifier units are provided for creation of a direct-current according to the Larionov's scheme, making it possible to charge the battery. The diagram of wave action on the float micro unit of WMPP is illustrated in (Fig. 2).

Figure 4. Image of WMPP exploratory prototype in the laboratory at the Ural Federal University.

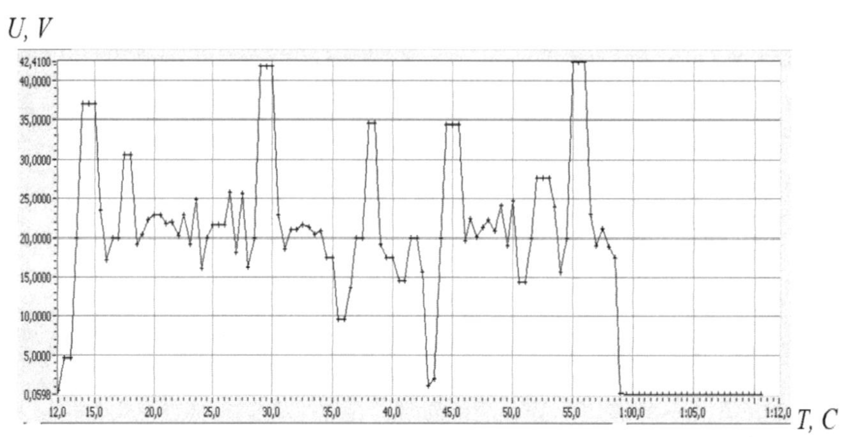

Figure 5. Capacity measures of WMPP at oscillation amplitude of 0.2 m and period of 1sec.

Sample of electro technical characteristics of generating unit at direct-current (DC) was shown in (Fig.5.). The empirical data, with wave simulation of different amplitudes and wave oscillation periods showed that the output power of one unit of WMPP was 15-60W.

Figure 6. Cluster scheme of WMPP of 4 micro units.

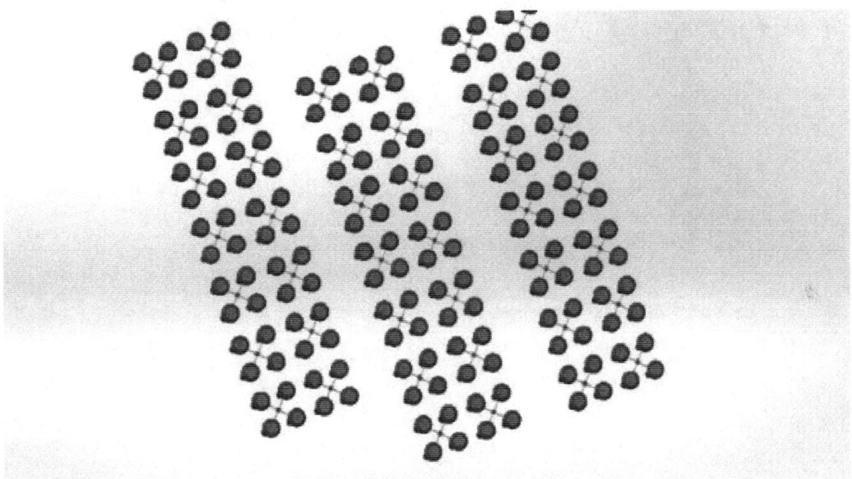

Figure 7. Scheme of energy farm on the basis of wave micro units.

Crank mechanism (CM) with a longitudinal mode of terminal segment – traction was used to simulate waves. CM transforms the engine drive rotation in back-and-forth movement of the traction. An induction motor with the capacity P = 1 kW and rotation frequency n_0 not less than 3000 r/min was selected as a drive. The gearbox was selected on the basis of reduction ratio Z = 25. The usage of wave simulation modes with the amplitude A = 20, A = 30, and the oscillation period T = 2, 3, 3.5sec., during the trials, let to receive some necessary technical values and characteristics to estimate the output power and determine optimal and effective operating modes of float type WMPP. The test stand operations were conducted in the laboratory of Eurasian centre renewable energy at the Ural Federal University. The test sample of WMPP was displayed in (Fig. 4).

The capacity of the projected unit can be upgraded to the level of several kW by using several micro units of WMPP combined in one single cluster (Fig. 6). The expansion of future capacity of WMPP to the level of several tens and hundreds kW can be achieved by connecting more micro units in cluster systems of wave micro units (Fig. 7).

4. Conclusion

The developed micro unit of WMPP showed the possibility of energy generation at wave amplitude from 20 cm and more for the wave oscillation period within 1-5 sec. The field research with the wave oscillation simulator confirmed the generation from one micro unit within the capacity limits of 20 to 60 W. Expansion of WMPP capacity to the level of several tens and hundreds kW is possible if the micro units are formed in clusters.

References

[1] Dyakov A F, Morozkina M V 1993 *Wave energy application problem* (Moskow: Energoatomizdat) 176 p

[2] Zharkov S V 2008 Sea wave energy and wave power plants *Energy: Economy, Technology, Ekology* **4** pp 11-18

[3] Sichkarev V I, Akulichev V A 1989 *Wave power plants in the oceans* (Leningrad: Nauka) 134 p

[4] Reznikova L N, Sichkarev V I 1984 Estimation of wave energy flow of the Pacific Ocean basin. Usage of tidal and wind waves energy in the ocean (Vladivostok: FESC AS USSR) 25-32

[5] Eliseev A V, Velkin V I, Shcheklein S E 2014 Development of research wave buoy for the world water area monitoring *Collected works of All-Russian general conference. Development of the Arctic and Subarctic regions* (Ekaterinburg) pp 214-216

[6] Reznikova L N, Sichkarev V I, Kukushkin I V 1983 Estimation of energy resources of the World ocean heaving" *Conference abstracts of IV All-Union conference. Research problems in the field of study and acquisition of the World Ocean: Ocean hydrophysical fields and methods of the exploration, part 2* (Vladivostok: Ministry of Higher and Specialized Secondary Education) FESC AS USSR 3-5

[7] Vissarionov V I, Volshanik V V, Zolotov L A 2002 *Usage of wave energy* (Moskow: Printing house MEI) 144 p

[8] Heath T V 2012 A review of oscillating water columns. Philosophical transection of the royal society, pp 235-245

[9] Alekseev A V 2006 Wave energy plant. Patent RF No. 023345, Patent of RF No. 2006121511

Analysis of Interdependence of Arctic Critical Infrastructures as Transportation Networks

S A Timashev[1,2,3,4], A V Bushinskaya[1,2,3]

[1]Sci.& Engng Center "Reliability and Safety of Large Systems and Machines", Ural Branch, Russian Academy of Sciences, Yekaterinburg 620016, Russia
[2]Ural Federal University, Yekaterinburg 620016, Russia
[3] N. Laverov Federal Center for Integrated Arctic Research, Arkhangelsk, Russia
[4]Old Dominion University, Norfolk, VA, United States

E-mail: timashevs@gmail.com

Abstract. The Arctic zone of the Russian Federation (AZRF) is characterized by the specificity of the development of its interconnected critical infrastructures' objects (ICI) - power generation and transmission systems, social infrastructure facilities, transportation (land, air, water) and communication systems, by the exceptional importance of their stable operation. The spatial isolation of the ICI objects predetermines, practically for each municipality, formation of its specific configuration, defining the range of needed capacities, corresponding set of ICI objects, etc. The main functional of Arctic ICI objects is to provide and sustain life support. Authors developed a model that permits high resolution description of ICI operation on the local level. The accuracy of the model is enough, especially for describing electrical and water supply grids and nets. Its substantial advantage is the generality of its mathematical description of the ICI of different physical nature and its capacity to account for the randomness of ICI parameters. The model permits simulation of the ICI behavior on the local level when exposed to ordinary, everyday conditions and in emergency situations (i.e., during an industrial disaster, Natural catastrophe or a terrorist attack). Several specific problems were solved using this approach for the AZRF region.

1. Introduction

In order to solve the problem outlined in the title of the paper, it is necessary to create a model of CI operation which would allow, for fixed external forces and actions, to assess the character and volume of damage, and produce numerical assessments of all possible consequences.

From the description of the problem it is obvious that the model should allow finding in a comparatively simple way and without big errors numerical values of all types of damage and how they change in time. The described model permits modeling ICI operation on the local level, which is very important when applying the "from-bottom-up" approach to assessing the resilience of the ICI [6], [7]. Most adequately fit models to solve these kinds of problems are the logic-structural models proposed by I.A. Ryabinin, different transportation network models, Petri networks and Bayesian nets [1], [3]. In this paper the advanced transportation network model [8] is used.

This synthetic practical model describes functioning of interdependent systems of critical infrastructures (ISCI), each of which is represented in the form of a transportation/services network comprised of nodes and directed links. Nodes typically represent physical infrastructure components (the so-called assets, such as electrical power grids and plants, oil, gas and water transmission pipeline systems, sewage systems, railway and highway systems, hospitals, industrial and office buildings, living quarters and the like), which are directly involved in supplying the population and different industries with different products or commodities, and services. The links of the network are modeling the flow (product transportation) between the nodes and may present electrical power lines, main and distribution pipelines of gas and oil; water supply and sewage systems, as well as railways and highways, etc.

Below, when describing the supply process, the generalized notion of "product" will be used for all types of consumption. Due to this factor it has a higher resolution as compared to ordinary methods, constructed for analyzing ICI on a national scale, as it is capable of accounting for local specifics of ICI, such as presence of product storages and emergency operators.

The model considers the conditions and specifics of supply, delivery, overhaul, and demand of resources, such as electricity, hot and cold water, and removal of the human activity waste or industrial waste [1].

It should be noted that the described approach ensures only the continuity of product flow in the nodes; at the same time, the physical laws, which define the patterns of product flow may not be complied with. But the accuracy of such simplified models is sufficient for most practical cases, especially for the electricity and water supply systems [4]. Considerable advantage of such models is the generality of the mathematical description of CI of different physical nature and their ability to account for the probabilistic character of the CI model parameters.

2. Generalized probabilistic model of ICI in the form of a network flow

The described below CI model is a generalization of existing models of transportation networks (flows). The infrastructure is presented in Fig. 1 by a digraph $G(V, E)$, i.e., by a set of nodes V, which are connected with directing edges (links) E. In the *standard* model nodes are the physical components of the infrastructure, which are responsible for the supply and, overhaul of the product, and the edges are responsible for the transportation (flow) of the product between the nodes [8].

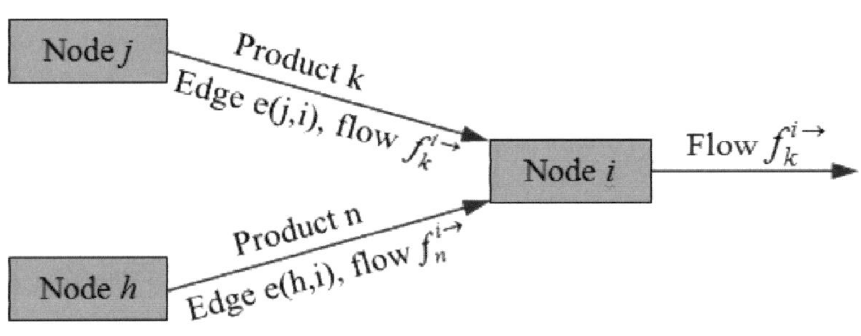

Figure 1. Basic parameters of the net flow model.

The model is developed for modeling behavior of the operating ICI on the local level in normal and emergency situations (such as industrial incidents, Natural catastrophes, or terrorist attacks). In order to reach this goal following modifications were introduced into the model [8]:

- if the CI is damaged, the modeling it network may become *unbalanced,* i.e., the overall demand of a specific product k may be larger than its on-hand supply. In order to be able to conduct the modeling some variables λ_k are introduced, which describe the cropped-up unsatisfied demand/deficit (the difference between the volume of actual demand of the product and the volume of its actual availability), as well as the corresponding to the cropped-up deficit rise of prices or fines c_k^{λ};
- resilience of the ICI systems may be increased by using local storages and/or production of needed products (like, spare pumps or generators).

Parameters that describe the quality of operation of CI physical components (*assets) of the infrastructure network* such as productivity, transportation capacity, demand) may be described at first approximation as random variables (RV). It is also possible to give to each network node (asset) scientifically justified/supported values of their failure rate. These probabilities are found using the statistical data on assets failures or through data obtained by corresponding probabilistic calculus/design.

To take these into account variables representing storage, s_k, and production, π_k (with the corresponding costs, c_{sk} and $c_{\pi k}$, respectively) have also been added to the model.

To take unmet demand/shortage into account variables s_k and π_k, representing storage and production, respectively, with the corresponding costs c_k^s and c_k^π are added to the model. Thus, the model distinguishes a commodity received from sources outside of the local infrastructure, from a commodity of local production and storage. However, to produce a commodity k another commodity (or commodities), n, is usually consumed.

To distinguish between *traditional supply* and internal production, the variable π_k is considered as the sum of variables:

- $\pi_{k,k}$, representing the traditional supply (i.e., supply from sources outside the boundaries of the local infrastructure);
- $\pi_{n,k}$, representing the internal production (i.e., consumption of commodity n to produce k is taken into account).

In a similar way, the variable κ_k, representing consumption of a commodity k, is the sum of variables:

- $\kappa_{k,k}$, which denotes traditional demand k;
- $\kappa_{n,k}$, denoting consumption of k for production of a commodity n.

The (links) edges of the considered model are analogous to the edges that are used in traditional transportation network model: each edge is a flow/supply channel of one product.

The volume of products stored in the node may change in time. Particularly, when the infrastructure is damaged due to an incident or catastrophe, and the available on-hand products are used to compensate for the cropped up undersupply. In order to account for this most important circumstance the time factor is explicitly introduced into the model.

The process of solving the problem starts from the moment of time t_0 when the initiating the CI failure event (infrastructure incident or catastrophe) occurs. The solutions are found for discrete moments of time with increment Δt (which could be variable) to the moment of full restoration of the damaged CI. On each time increment all the variables and constants do not change [5]. When forecasting the possible CI damage (i.e., assessment of its partial or full failure), independent from the character of the type of forces acting on the infrastructure, it is necessary to quantitatively assess the considerable uncertainty of parameters of the model equations. This assessment is performed either by establishing the probabilities of failure of the CI assets, as given by the nodes, or by considering the asset productivity and/or the volume of supply as random variable (RV). As the demand of products could be uncertain, it also is regarded as a RV.

3. A multifunctional nodes model

If the same node could simultaneously act both as a consumer and a producer, have storage facilities, emergency generators and, at the same time, belong to several different CIs, it is called a *multifunctional* node.

Let us describe the relationship between variables and constants, representing four functions of a multifunctional node of an infrastructure, as shown in Fig. 1.

The balance equations at a node i for commodity k are

$$f_k^{i\leftarrow} + \pi_k^i - \kappa_k^i - \tau_k^i - s_k^i = 0, \tag{1}$$

$$\tau_k^i - f_k^{i\rightarrow} = 0,$$

where $f_k^{i\leftarrow}, f_k^{i\rightarrow}$ are the commodities inflow and outflow rates; π_k^i is the production rate; τ_k^i is the transhipment rate; s_k^i is the rate of transferring the commodity to storage. If no cost is associated with the transhipment of the commodity and there is no *risk of damage of the transhipment function* of the node i, then these two balance equations can be replaced by a single one:

$$f_k^{i\leftarrow} + \pi_k^i - \kappa_k^i - f_k^{i\rightarrow} - s_k^i = 0 \tag{2}$$

It is worth noting that different commodities can be processed at a single multifunctional node. Hence, the balance equations should be satisfied at the node for each of these commodities.

The in-flow and out-flow rates, $f_k^{i\leftarrow}, f_k^{i\rightarrow}$ are the sums of the flow rates over all edges transferring commodity k into and out of the node i, respectively. The production rate of commodity k at a node i is expressed as

$$\pi_k^i = \pi_{k,k}^i + \sum_{n\neq k} \pi_{n,k}^i, \tag{3}$$

where $\pi_{k,k}^i$ is the supply rate, when the need in another commodity to produce the commodity k is not taken into account (usually, at nodes that represent sources of supply of the commodity k outside the boundaries of the local infrastructure network under consideration); $\pi_{n,k}^i$ is the rate of production of the commodity k that involves consumption of another commodity n. It is assumed that there is a linear relationship between the amounts of produced and consumed commodities so that

$$\pi_{n,k}^i = \alpha_{n,k}^i \kappa_{n,k}^i, \tag{4}$$

where $\alpha_{n,k}^i$ is the coefficient relating the production rate of the commodity k to the corresponding consumption rate $\kappa_{n,k}^i$, of the commodity n. Thus, Eq. (3) takes the form

$$\pi_k^i = \pi_{k,k}^i + \sum_{n\neq k} \alpha_{n,k}^i \kappa_{n,k}^i, \tag{5}$$

The consumption rate of commodity k at the node i is formulated as:

$$\kappa_k^i = \kappa_{k,k}^i + \sum_{n\neq k} \kappa_{n,k}^i, \tag{6}$$

where $\kappa_{k,k}^i$ is the rate of consumption of the commodity k to satisfy demand of consumers; $\kappa_{n,k}^i$ is the rate of consumption of the commodity k to produce another commodity n.

In order to be able to take into account the possibility that not all demands are met when infrastructure is damaged, the variable $\kappa_{k,k}^i$ is presented as:

$$\kappa_{k,k}^i = \kappa_k^{i,\max} - \lambda_k^i, \tag{7}$$

where $\kappa_k^{i,\max}$ is the actual demand rate, i.e., the maximum/required rate of consumption of the commodity k at the node i at a given time, a λ_k^i is the rate of unmet demand (or shortage).

Eq. (7) can then be written as:

$$\kappa_k^i = \kappa_k^{i,\max} - \lambda_k^i + \sum_{n\neq k} \kappa_{n,k}^i. \tag{8}$$

The introduction of storage function even in one of the nodes makes the whole model *time-dependent*. The amount of commodity k stored at the node i at time t is denoted as $\omega_k^i(t)$

$$\omega_k^i(t) = \omega_k^i(t-\Delta t) + s_k^i \Delta t \le \omega_k^{i,\max}, \tag{9}$$

where s_k^i is the rate of change in the amount of stored commodity; Δt is the time increment and $\omega_k^{i,\max}$ is the maximum storage capacity for the commodity k at the node i.

Taking into account Eqs. (5) and (8), Eq. (2) can be written as

$$f_k^{i\leftarrow} + \pi_{k,k}^i + \sum_{n\neq k} \alpha_{n,k}^i \kappa_{n,k}^i - \kappa_k^{i,\max} + \lambda_k^i - \sum_{n\neq k} \kappa_{n,k}^i - \tau_k^i - s_k^i = 0 \tag{10}$$

Eq. (2) remains unchanged, while Eq. (3) can be updated in a similar way as Eq. (1)

In this second equation of Eq. (1) remains unchanged, while Eq. (2) can be updated in a similar way as Eq. (1).

4. Formulation of the optimization problem

In the general case the problems related to the optimization of CI operation, can be solved by simple sorting of options (when the dimension of the problem is not large), and also by using different analytical methods (linear and nonlinear programming, *greedy* algorithms, etc.). The choice of the solving method is not, as yet, formalized and depends on specifics of the problem in consideration. The distinction of the developed in the paper method from the existing models is that it permits optimizing CI operation according to some risk criteria, and particularly, minimizing the overall cost of its operation in each specific case, taking into account possible human losses. In the latter case additional members that quantitatively describe the non-materiel losses (loss of limb, life), have to be introduced into the optimized cost functional.

The solution of the model is formulated as a linear optimization problem that minimizes the cost of maintaining the network infrastructure at the stage of its operation (or design) as follows:

$$\begin{aligned} &\min_{\mathbf{x}\in\Re^n} \mathbf{c}^T\mathbf{x} \\ &\text{subject to} \quad \begin{matrix} \mathbf{b}_1 \le \mathbf{A}\mathbf{x} \le \mathbf{b}_u & \text{(general constraints)} \\ \mathbf{x}_l \le \mathbf{x} \le \mathbf{x}_u & \text{(bounds on variables)} \end{matrix} \end{aligned} \tag{11}$$

where $\mathbf{x}$ is the vector of variables, $\mathbf{c}$ is the objective coefficient vector (i.e., costs); $\mathbf{A}$ is the coefficient matrix; $\mathbf{b}_1, \mathbf{b}_u, \mathbf{x}_l, \mathbf{x}_u$ are the vectors of lower and upper bounds on the constraints and the variables, respectively.

Costs C for operating the network infrastructure per unit time within the time interval $(t - \Delta t, t)$ can be expressed as:

$$C = \sum_{k=1}^{K} \begin{bmatrix} \sum_{e\in E} c_k^{f,e} f_k^e + \sum_{i\in V} c_k^{\pi,i} \pi_k^i + \sum_{i\in V} c_k^{\lambda,i} \lambda_k^i + \\ + \sum_{i\in V} c_k^{\omega,i} \frac{1}{2}\left[\omega_k^i(t) + \omega_k^i(t-\Delta t)\right] \end{bmatrix}, \tag{12}$$

where $c_k^{f,e}, c_k^{\pi,i}, c_k^{\lambda,i}, c_k^{\omega,i}$ are the costs associated with the flow, production, shortage and storage of commodity k per unit of time, respectively; K is the total number of commodities.

The cost of storage (the last sum in Eq. (12)) is based on the average amount of stored commodity within the time interval $(t - \Delta t, t)$. The total cost can be calculated by substituting Eqs. (5) and (9) into Eq. (12), which gives

$$C=\sum_{k=1}^{K}\left[\begin{array}{l}\sum_{e\in E}c_k^{f,e}f_k^e+\sum_{i\in V}c_k^{\pi,i}\left(\pi_{k,k}^i+\sum_{n\neq k}\alpha_{n,k}^i\kappa_{n,k}^i\right)+\sum_{i\in V}c_k^{\lambda,i}\lambda_k^i+\\+\frac{1}{2}\sum_{i\in V}c_k^{\omega,i}s_k^i\Delta t+\sum_{i\in V}c_k^{\omega,i}\omega_k^i\left(t-\Delta t\right)\end{array}\right],\tag{13}$$

where the last sum is a constant and should be excluded from the objective function; in order to obtain the total cost, including the full cost of storage, this term can be added after optimization.

The optimization problem can then be formulated as:

$$\min C=\min\sum_{k=1}^{K}\left[\begin{array}{l}\sum_{e\in E}c_k^{f,e}f_k^e+\sum_{i\in V}c_k^{\pi,i}\left(\pi_k^i=\pi_{k,k}^i+\sum_{n\neq k}\alpha_{n,k}^i\kappa_{n,k}^i\right)+\sum_{i\in V}c_k^{\lambda,i}\lambda_k^i+\\+\frac{1}{2}\sum_{i\in V}c_k^{\omega,i}s_k^i\Delta t\end{array}\right],\tag{14}$$

subject to the following general constraints:

$$\begin{array}{l}f_k^{i\leftarrow}+\pi_{k,k}^i+\sum_{n\neq k}\alpha_{n,k}^i\kappa_{n,k}^i-\kappa_k^{i,\max}+\lambda_k^i-\sum_{n\neq k}\kappa_{n,k}^i-\tau_k^i-s_k^i=\kappa_k^{i,\max},\ \forall i\in V\ u\ \forall k\in K\\ \tau_k^i-f_k^{i\rightarrow}=0,\ \forall i\in V\ u\ \forall k\in K\\ \pi_{k,k}^i+\sum_{n\neq k}\alpha_{n,k}^i\kappa_{n,k}^i\leq\pi_k^{i,\max},\ \forall i\in V\ u\ \forall k\in K\end{array}\tag{15}$$

and bounds:

$$\begin{array}{l}0\leq f_e\leq f_e^{\max}\forall e\in E;\\ 0\leq\pi_{k,k}^i,\ \forall i\in V\ u\ \forall k\in K;\\ 0\leq\kappa_{n,k}^i,\ \forall i\in V\ u\ \forall k\in K\ u\ \forall n\in K,\ k\neq n;\\ 0\leq\lambda_k^i,\ \forall i\in V\ u\ \forall k\in K;\\ 0\leq\tau_k^i\leq\tau_k^{i,\max},\ \forall i\in V\ u\ \forall k\in K;\\ -\frac{\omega_k^i}{\Delta t}\leq s_k^i\leq\frac{\omega_k^i-\omega_k^{i,\max}}{\Delta t},\ \forall i\in V\ u\ \forall k\in K,\end{array}$$

where $f_e^{\max}$ is the maximum flow capacity of the link e; $\pi_k^{i,\max}$ is the maximum production capacity of commodity k at the node i; $\tau_k^{i,\max}$ is the maximum transshipment capacity for commodity k at the node i.

Minimizing the total cost separately at each time step does not yield an optimal solution for the network performance over the whole time period under consideration. For example, when an infrastructure network is undamaged it is obvious that the optimal amount of a stored commodity is zero – if everything works properly there is no need for stored commodities because their storage incurs additional costs. However, when the network is damaged by a hazard, the availability of commodities from storage may reduce the total cost (since there are additional high costs penalizing for unmet demand). This problem can be resolved either by artificially setting negative costs for storage (that will prevent the use of stored commodities under normal conditions) or minimizing the total cost over all considered time increments, i.e., the objective function should become the sum of the costs associated with each time increment. The second approach seems preferable. However, in order to use this approach it is necessary to know the time required for restoring the infrastructure to its original undamaged condition (as new).

This time can serve as a good criterion for resilience and strategic readiness of the infrastructure, but it can be difficult to predict due to the considerable uncertainty of many parameters. Here, to obtain practical results, it may be necessary to use special probability models - fuzzy logic, interval probabilistic estimates, etc.

5. Accounting for the interdependency of critical infrastructures

This accounting is very important as from it depends the level of likelihood of modeling results, because quite often failures of elements belonging to the same CI or to two and more CI could be *interdependent*. Such dependency between elements of logic-structural CI models can be of two types:

- failure of one of the elements of a CI *changes the operational regime* of the whole CI (like, when failure of one controlling element changes the operational regime of other elements);
- when the whole CI in its entirety is exposed to some single random factor (wind, wave pressure, snow load, temperature, vibration etc.).

Specific calculations show that if a simple, without reserve elements and non-renewable system is considered, the dependence of the first type does not affect the reliability of the CI as a whole. If the CI has reserve elements and is renewable, the second-type interdependence must be taken into account. In general, failure to account for the strong dependence of failures in a multi-element CI can lead to large errors. Neglecting of dependent failures in series connection elements can lead to a significant *underestimation* of CI reliability. When the elements are connected in parallel, ignoring their interdependence leads, on the contrary, to an *overestimation* of the CI reliability.

If the operating mode of the CI is characterized by some continuous random variable V (say the wind speed) with a known distribution density function $f(v)$, then the reliability of the CI is determined by the formula [2]: $P(t)=\int P_{f|v}(v)f(v)dv$, where $P_{f|v}(v)$ is the conditional reliability of CI, provided that $V = v$. The integral extends over the entire range of possible values of V.

The larger is the number of reserve elements in the CI block, the higher is the overestimation of its reliability due to the neglect of the interdependence of the failures of its elements.

If the CI consists of elements connected in series and in parallel (only the most important components of the CI are duplicated), the neglect of the dependence of the failures can lead to both overestimation and underestimation of the reliability of the CI.

For more complex CIs that do not reduce to purely logical-structural schemes, it is necessary to conduct a series of numerical comparative calculations that allow estimating the amount of incident damage that depends on accounting for / not taking into account the interdependence of the elements of one, two or more CIs for each scenario of the development of an emergency situation.

6. Probabilistic analysis of a local interdependent infrastructure system. Real case study

Analyze, using the described above model, operation of a system which consists of two interdependent infrastructures (electricity and water supply) which are located in one of the Russian Far North cities (see. Fig. 2) during an extreme *winter snow storm with hurricane wind speeds*, which occur once in a 100 years (design event), with simultaneous rise of ambient temperature to levels that produce thick icing on the electrical power lines' wires.

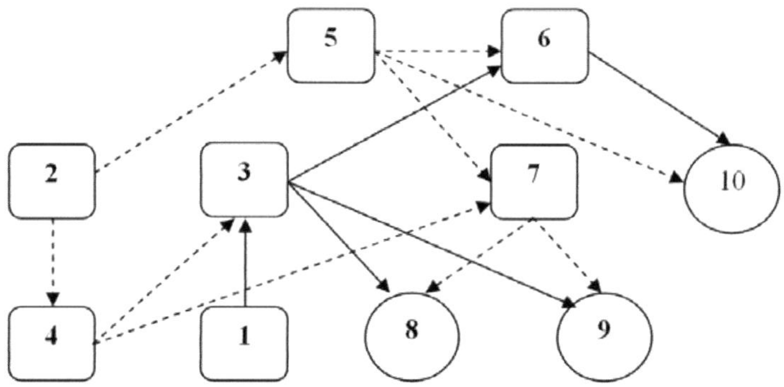

Figure 2. Local infrastructure network.
Links: continuous lines-water supply; dashed lines –power lines.

Modeling of the extreme winter snow storm per se is not considered in this paper, because it is a standalone problem which belongs to analysis of the consequences of natural forces on ICI structures class of problems. It is solved by considering the influence of the joint action of wind, and icing of electrical power grid (EPG) wires on the strength, stability and vibrations of the EPG foundation structures, supports and wiring of the EPG as a large mechanical system (Timashev S.A., 2016). As the result of solving this *stochastic mechanics* problem the so called *fragility curves* are produced, which *connect the physical properties of wind pressure* (average speed and variance), ambient air temperature, intensity and duration of snow fall, etc., with the *probability of any kind of damage* of *EPG supports* (warping, capsizing, buckling of individual strut elements, etc.) and wires (sagging, contacting of two adjacent wires, wire rupture).

The probability of failure (PoF) P_f of each EPG element can be assessed as

$$P_f(t) = P[Y(\tau) < 0;\ 0 \le \tau \le t]$$

where $Y(t)$ is the limit state function (LSF) of the EPG element. The LSF actually is the difference between the function, which describes the physical load vector on the element and the ultimate permissible value of it.

The *local* EPG (the first CI) is comprised of four electrical substations--one substation, which is an element of the Federal EPG (node 2), with maximal capacity of 20 MW hours/day and three distribution substations: nodes 4 and 7, with maximal distribution power of 5 MW hours/day, and node 5 with maximal distribution power of 10 MW hours/day. The second CI is comprised of: the water filter station (node 1) with maximal distribution power of 1000 m^3 drinking water per day; water tower (node 3), with maximal capacity 550 m3; and the pumping station (node 6) with maximal pumping power of 200 m^3/day. For pumping the water they use electricity. It is assumed that to pump 7,5 m^3 of water 1 KW hour of electricity is needed. Water consumers are: hospital (node 8), old age home and the surrounding it living quarters (node 9), and living quarters (node 10). Their daily consumption of electricity and water are given in Table 1. Nodes and connections between them are schematically shown in Fig. 2.

Table 1. Daily demand of electricity and water of CI elements in consideration.

Consumer	Daily demand	
	Electricity, MWhour/day	Water, m^3/day
Hospital (node 8)	1.3	35
Old age house & surrounding it living quarters (node 9)	3.1	75
Living quarters (node 10)	6.0	165

Two variants were analyzed: 1) CIs without the emergency generators; 2) CI with emergency generators in nodes 3, 6 и 8. Each generator of 12 KW power consumes one liter of diesel oil to produce 3 KW of electricity and is equipped with an oil tank with capacity of 100 liters.

In the considered case the expenditures for transportation, storage, production and the products deficit are not actual expenditures and are assigned in order to correctly distribute the products between the consumers as was intended. In order to prevent usage of the stored products (water and fuel) until the moment, when there would be no other source to satisfy the demand, the expenditures associated with their storage are set negative.

In the considered case following decreasing priority hierarchy was established for receiving the necessary resources: hospital, old age home, living quarters. The financial, materiel and social damages, related to the shortage/absence of water and electricity supply for these consumers are established according to the above priorities.

For the three substations in the case of high voltage wires rupture due to a combination of wind pressure and icing following probabilities of failure were obtained: node 4 – 0,4, node 5 – 0,6, and node 7 – 0,2. As the substations can be only in two states (operable or failure/switched off), hence, the whole network has only 23 states and the probabilities can be easily calculated. Infrastructure will operate being partially damaged until the EPG repair is complete, which can take up to several days. Results of the probabilistic analysis of the water and electricity supply process to consumers, as an illustration, are shown in figures 3 and 4, depending on the time of restoration of the EPG integrity (during one –six days, with one day increment). Particularly, the probabilities of delivery to consumer's necessary volumes (i.e., demand) of electricity and water without using the emergency generators are given.

It can be seen, that the interruption in delivering the electricity can start at the very first day. After that the percentage of the delivery and the corresponding probabilities do not change for the next six days. Interruptions in water delivery during the first day after the incident can take place only in the living quarters (node 10), due to the possible failure of electricity delivery for the pumping station (node 6). Keeping water in the water tower guarantees it will be delivered in quantities that are needed (full demand) to the hospital (node 8) and the old age home, including the surrounding it living quarters (node 9). Interruption of water delivery for these consumers may start at the third day and the situation gets worse on the sixth day.

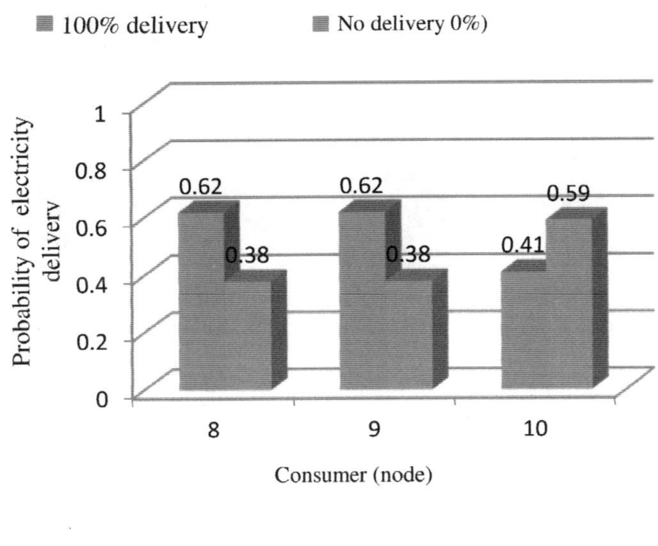

Figure 3. Probability of delivering electricity to consumers (without accounting for the emergency generators).

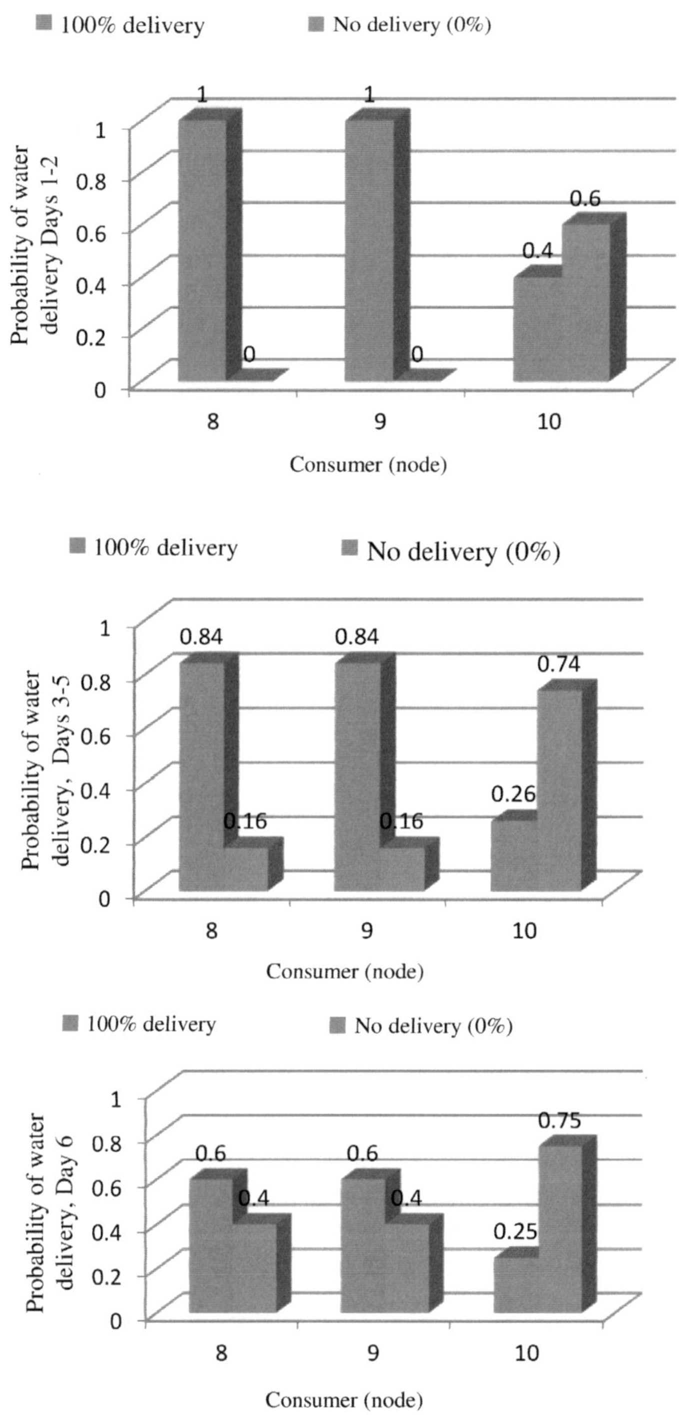

Figure 4. Probability of delivering water to consumers during the first six days after the incident occurrence (without accounting for the emergency generators)

7. Conclusion

The results of conducted research are urgent for decision makers, as they allow assessing the consequences of made decisions when managing interconnected CI. On a model but real-life example of two interdependent infrastructure systems of electricity and water supply during a winter snow storm it was shown how the uncertainties related to the damage/downtime of infrastructure elements can be accounted for, using variables which describe the originated unsatisfied demand (deficit) and the corresponding to it costs or fines.

The model can be used for studying productivity of interdependent CI (and via it, the *resilience*) of much more complicated interdependent network-type infrastructures, in much smaller time increments (every hour, minute or even second, if need be) and consider such parameters as demand and supply, production and/or products flow as non stochastic functions of time of random variables, functions or fields.

Acknowledgments

The study was carried out with the financial support of the Ural Branch of the Russian Academy of Sciences in the framework of the scientific project No. 18-9-17-37 "Modeling the environmental and economic scenarios of the spatial development of the Arctic regions of Russia".

References

[1] Ahuja R K and Magnanti T L and Orlin J B 1993 *Network Flows: Theory, Algorithms and Applications* (New Jersey: Prentice Hall)

[2] Bushinskaya A V and Timashev S A 2016 Generalized Probabilistic Network Flow Model of Interdependent Critical Infrastructures *J. of Problems of Safety and Emergency Situations* **3** pp 9-22

[3] Kachur S A 2011 Modeling of catastrophic situations of nuclear energy facilities using stochastic Petri networks *Collection of scientific research works* **4(40)** (Sevastopol National University of nuclear energy and industry) pp 22–29

[4] Manca A and Sechi G M and Zuddas P 2010 Water supply network optimisation using equal flow algorithms *J. Water Resource Management* **24 (13)** pp 3665–3678

[5] Svendsen N and Wolthusen S 2007 Connectivity models of interdependency in mixed-type critical infrastructure networks *Information Security Technical Report* **12(1)** pp 44–55

[6] Timashev S A 2013 Unified Quantitative Criteria for Management of Regional Risk *Proc. of the11-th International Conference on Structural Safety & Reliability*

[7] Timashev S A 2016 Infrastructures, Part 1. Reliability and Longevity (Yekaterinburg: Ural Branch, Russian Academy of Sciences) 530 p

[8] Val D V and Holden R and Nodwell S 2013 Probabilistic assessment of failures of interdependent infrastructures due to weather related hazards *Proc. of the 11-th International Conference on Structural Safety & Reliability*

IOP Conf. Series: Materials Science and Engineering **481** (2019) 012009 doi:10.1088/1757-899X/481/1/012009

Cyber Reliability, Resilience, and Safety of Physical Infrastructures

S A Timashev[1,2,3]

[1]Sci.& Engng Center "Reliability and Safety of Large Systems and Machines", Ural Branch, Russian Academy of Sciences, Yekaterinburg 620016, Russia
[2]Ural Federal University, Yekaterinburg 620016, Russia
[3]Old Dominion University, Norfolk, VA, United States

E-mail: timashevs@gmail.com

Abstract. The paper considers the problem of constructing a full group of failure scenarios for physical infrastructures when subjected to cyber attacks (CAs). Physical infrastructures actually are systems of systems, or network of networks [1]. The main idea of the research rests on the assumption, that in order to damage any *physical* infrastructure by a *cyber attack*, it has to be able to produce a powerful enough physical impact on the most vulnerable part(s) of the infrastructure. Only civil engineering and industrial structures and installations connected to Internet and World Wide Web are considered. Hence, all infrastructures discussed below have to be elements of the Enterprise IoT or IoT, namely: electrical grids, oil, gas and product pipeline systems, water supply and disposal (waste) systems, rail networks, air traffic control and telecommunications (finance, commerce, business) networks, etc. The paper discusses how to construct a full group of scenarios of physical impacts on an infrastructure and how to calculate reliability, resilience and safety of infrastructures exposed to CAs. This paper should calm down the legitimate concerns of lay people about disclosing vulnerabilities of critical infrastructures, because it raises the awareness and offers infinitely much more to the *armor/shield* than to the *canon/spear*.

1. introduction

The history of cyber attacks on infrastructures started in early 1980s, when the Internet wasn't existing, and is in the process of shaping itself. In one of the very first official reports on small-scale cyber attack attempts against various U.S. electrical utilities [2], results are described of surveys of 15 cyber attacks and incidents over a period of three decades (sic, since 1982). None of them have caused significant damage or disruption [2]. At the same time a group of hackers installed a Trojan into the SCADA system which controlled a Siberian pipeline, which resulted in a powerful blast. The identity of the hackers was identified only 22 years later, when DoD USA Secretary under R. Reagan, Thomas Read published his book "*At the Abyss: An Insider's History of the Cold War*" from which we learned that the attack was organized by the CIA. This and all cases without reference described below come from the Internet source [3].

The next incident is dated by 1992 when a fired Chevron worker penetrated computers of the company in New York and San Jose and reprogrammed them for allowing leaks of poisonous gases from Chevron installations. This man-made hazard exposed thousands of workers to potential danger for 10 hours, before the system was restored for normal safe operation. Other accidents involving cyber attacks were recorded in 1994 (the Salt River Project, where removal of files responsible for monitoring and logistics led to interruptions of serving electricity and water, and compromising personal and financial data), In 1997, the Worcester (Massachusetts, USA) airport was under attack that resulted in a six hour disruption of telephone communication at the dispatcher tower, fire fighters quarters and airport carriers' offices. In 1999 the Russian company Gasprom was attacked by hackers who, using help from an insider, used a Trojan to disrupt SCADA performance that controlled gas supply. The intrusion was curbed very early, without any serious consequences.

In 2000 a former employee of the Maroochy Water System (Australia) got two years in jail for hacking company's computers that controlled water supply that resulted in millions of liters of sewage entering an adjacent fresh water river and flooding of a hotel. In 2001 a US gas processing company was hacked by a supplier who did this to hide his logistics mistake. As a result gas supply was switched off in several European countries. In 2002 the PDVSA oil company in Venezuela was attacked during a strike. As result, its output went down from 3mln.b/day to 370 thousand b/day. In 2006 two engineers, experts in road traffic hacked out of protest the Los Angeles traffic lights, by making them stay red all the time that led to serious traffic jams. In 2008 a 14-year old student did the same with the tram (city cars) in Lodz, Poland. As a result, four trains went off track, 12 people were traumatized. In 2012 the offices of the world largest oil company, Saudi Aramco, was attacked grounding 30 thousand computers. Responsibility for this cyber attack was taken by a group that called itself "The Sword of Justice." In two weeks after that attack, the RamGas Company of Qatar was attacked by the same type of virus. As the result, the internal corporate web and its website were out of order for several days. Starting late in 2009, several U.S. natural gas pipeline operators came under a barrage of highly sophisticated cyber attacks related to industrial espionage, but they could be precursors to cyber attacks and/or physical attack [4]. In 2014 a German metallurgical facility was attacked by hackers who were able using social engineering to get access to an employee's computer through which they got access to the control system of a blast furnace. As the result the furnace could not be shut down, which resulted in great losses. In 2015 Ukraine's electrical grid was hacked and 600 thousand citizens were left without electricity.

The first in history massive cyber attack on a cyber system happened in 2007 in Estonia, when websites of its Parliament, ministries, banks, newspapers and other mass-media organizations, as well as the national system of processing telecommunication services and financial orders, went down. It was alleged that Russia is responsible for the incident.

The diagnosed malicious virus programs that were being used in the attacks were SQL Slammer that was tailored to attack data base servers (cases: US oil company, a major US automobile company). A hospital in Great Britain was attacked by virus Mytob. Virus Nimda was responsible for attacking a US food producing company. Air Canada, Mitsubishi Electric (sensitive inspection data about its two nuclear power plants were leaked as the result), Cook County, DOT, Illinois, USA were attacked by different viruses that were not made public.

The up to date first, largest and most elaborate cyber attack, that actually lead to significant physical damage of the infrastructure, was executed by a worm program called Stuxnet, on the Iranian centrifuge cascade for nuclear isotope separation in Natanz, Iran (Fig.1). The attack (which took place in 2008-2010) was preceded by extensive remote clandestine cyber diagnostics/monitoring of the target. Stuxnet gathered detailed information about the centrifuges and their control system, providing the basis for the development of a precisely-tailored worm attack tool. The general belief (not supported, as yet, by any fully trustable documents) is that Stuxnet was developed and inserted in the centrifuge cascade system by a concerted effort of the USA and Israel.

On the surface, Stuxnet was able to deceive the SCADA as to the true state of the centrifuge operating parameters at the same time that the centrifuges were forced to operate at speeds well over design values and, alleged, continuously changing the frequency of their spinning in time, thus causing the fatigue phenomenon in its mechanical parts and subsequent fast accumulation of high-cycle fatigue damage and rupture of its crucial details. Published damage estimates related to the centrifuges per se and the isotope separation process vary [5].

IOP Conf. Series: Materials Science and Engineering **481** (2019) 012009 doi:10.1088/1757-899X/481/1/012009

Figure 1. Cascade centrifuge facility in Natanz, Iran [3].

The motives in the above cases were industrial espionage, blackmail, reconnaissance for cyber or physical attack, and/or implanting code for later cyber attack [6]. Worldwide, there are no publicly known (as of 2018) military campaigns against the infrastructure of cyberspace, nor any military cyber attacks on physical infrastructures of any country. But a plethora of attacks were recorded by hackers of all hues and colors, and it is alleged that some of these attacks have been sponsored by (usually, not precisely identified) states.

Currently, already hundreds of cyber attacks are documented on critical infrastructures on all the inhabited continents (except, probably, Africa), but mostly, Europe, North America and Asia. Danger of a cyber attack on a critical infrastructure is real, and governments of all developed countries are well aware of this. After 03/11 2004 Madrid attack the EC developed *The European Programme for Critical Infrastructure Protection.* In May 2016 after the meeting of the G7, Energy ministers signed a joint Declaration in which they accented the need of implementing failure proof energy systems (including gas, electricity, and oil). Now G7 governments are creating centers for gathering data needed to improve safety of critical infrastructures. As a result, a complex strategy for solving this problem was developed, to be included into the national laws and ordnances of the G7 countries.

2. Some Definitions

For better understanding of the following, we start with a short description of some elements of infrastructure networks theory. From the point of topology, a network consists of nodes and links that connect the nodes in a specific fashion. The nodes represent the points of supply/origin and points of destination/ consumption. The links represent the routes of transmission/movement. The whole set of nodes and links comprises a network. A broad set of manmade, natural, and social systems can be represented and analyzed as this kind of transportation networks. Infrastructure networks are complex, irregular; and statistical in nature. Large networks can be classified as exponential networks (highly connected nodes are exponentially unlikely) and power-law networks (as they, typically, do not contain dominant nodes). In the first type of networks most of the nodes have approximately the same number of links. These kinds of networks are descriptively named as *uniform-random networks* URN (see Fig. 2).

In the second broad class of networks most nodes are connected to nodes that already have a considerable number of connections. This feature lead to describe them as scale-free networks (in the sense of number connections per node).The more descriptive term for this kind of networks is *hub-and-spoke random network* HASN (see Fig. 3).

Figure 2. A typical uniform-random network infrastructure.

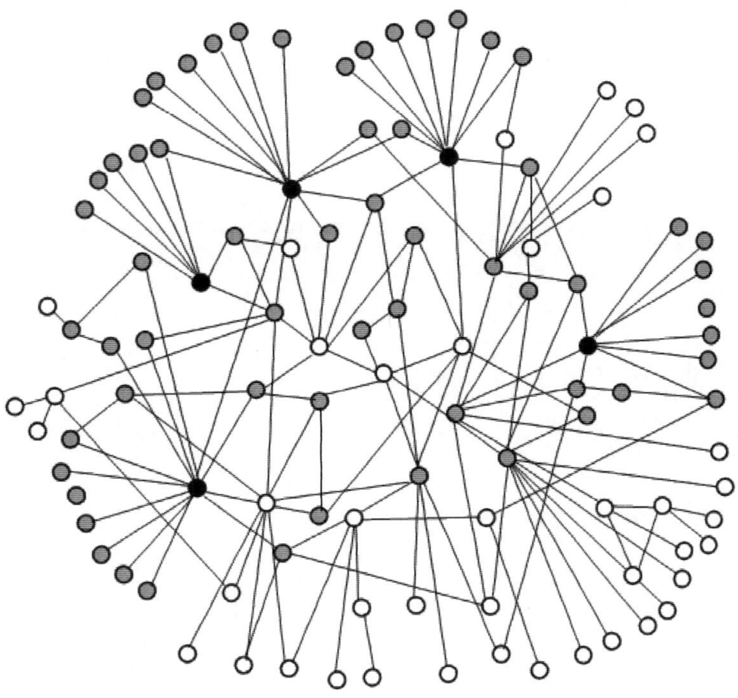

Figure 3. A scale-free infrastructure network.

A cyber attack that disables a randomly picked node from the URN (Fig.2) typically would disconnect only a few other nodes that are connected only to the disabled one. A random disablement in the hub-and- spoke network would likely do even less damage, since so few nodes have any other node that connects only through them. However, in the worst case scenario for the scale- free network (when several nodes are disabled, see Fig.3), the damage will be greater than in a similar scenario for the exponential network.

3. Cyber Networks

The Internet, like practically our entire society, is critically dependent on electric supply. Both the Internet and World Wide Web (WWW) are scale-free networks. WWW is the *very first truly planetary scale digital infrastructure* (Fig.4).

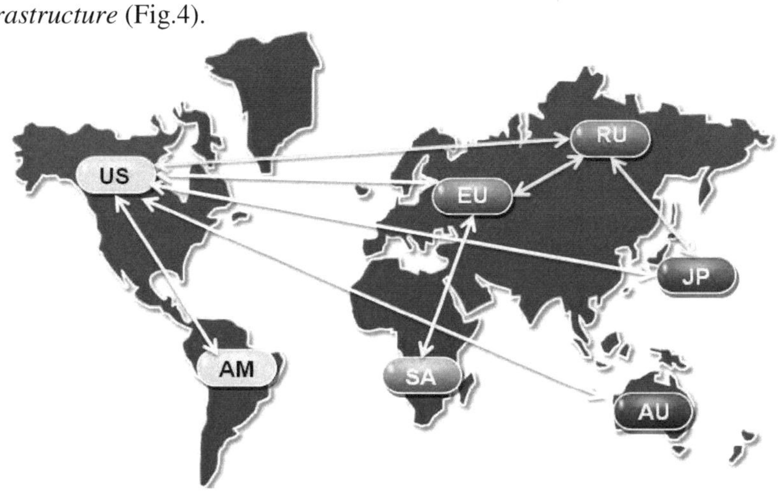

Figure 4. Russian computer network (RUNET) integrated into the global network [7].

Scale-free networks emerge usually through natural growth, as new nodes link preferentially to old nodes that are already highly linked. It is interesting that China didn't connect to WWW, it operates its own separate network, protected by a wall from WWW.

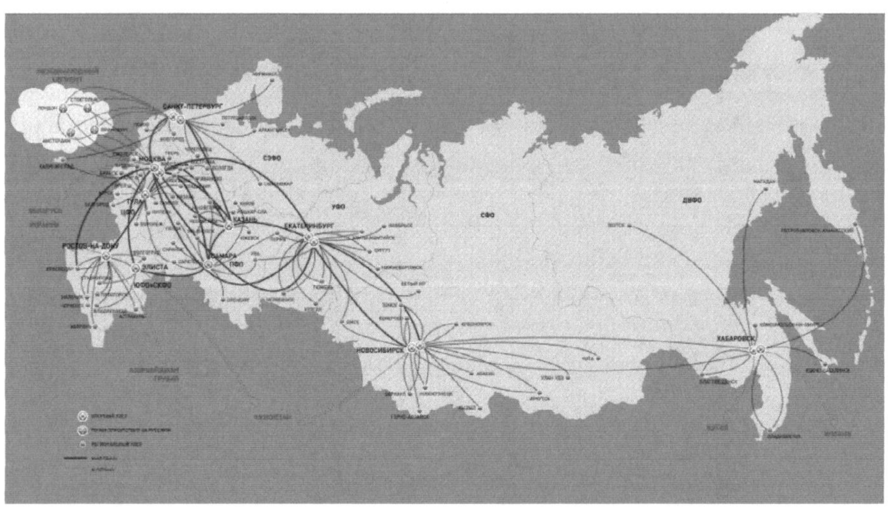

Figure 5. Backbone network of the Russian JSC Rostelecom [8].

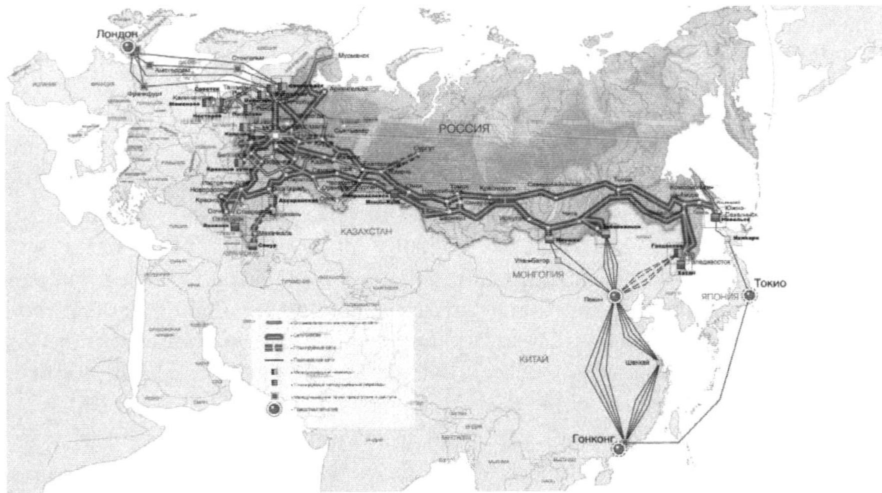

Figure 6. Backbone of a new Russian digital communication operator [9].

Currently, due to the observed weaponization of the Internet and WWW, they are subjected to particularly intense study of their capabilities, strengths and vulnerabilities [10]. Table 1 shows that the cyber content rests on a structure of physical elements that have physical properties and locations. It can be easily observed that cyberspace itself, like any other infrastructure, has its own geography and topology (the latter not identical with that of the network layers), and both affect its vulnerability and, hence, resilience.

Table 1. Schematic Description of Levels Involved in Cyberspace [1].

Level	Description	Examples
Cyber	Intellectual content	Data, commands, knowledge, ideas, mental models
Logical net	Services employing physical signals to carry logical messages	Telephones, broadcast radio and TV services, cable TV service, public Internet, private Internet protocol (IP)–based networks carried on common-carrier infrastructure, private-infrastructure IP- based networks, supervisory control and data acquisition networks
Hard net	Infrastructures formed from base elements that carry electrical or electromagnetic signals	Common-carrier telecommunications networks, tactical radio systems, dedicated wire line systems, community cable systems, cell phone systems
Base	Physical elements that underlie telecommunications services	Cable networks, optical fiber, coaxial cable, radio transmitters and receivers, radio transmission paths, communications satellites, Internet routers, modems

Internet nodes consist of computers (or devices that incorporate computers). Nodes are connected to an Internet service provider (ISP), which offers a connection to its hub or server bank (a cluster of high-speed computers) via some kilometers of telephone wire, coaxial cable, fiber optic cable, wireless cellular radio link, or satellite radio link. In their entirety, all these components comprise the current Internet architecture as a scale-free HAS network resembling that shown in Fig.2.

Scale-free networks are, for reasons described above, robust in the face of random or untargeted failures, which fall most heavily, according to laws of statistics, on the large numbers of nodes with only a few connections, and, hence, has scarcely any discernible effect on the overall network performance. Even more massive failures, due to widespread power outages, have been quite localized in their effects [11].

Successful attacks on many of the biggest hubs would have severe and pervasive effects. Hence, protection of major Internet hubs is a cornerstone of rational policy for cyberspace infrastructure defense, but keeping in mind that links that are logically and topologically separate may in fact be carried over the same physical communications infrastructure through multiplexing, via one fiber optic strand, or otherwise be vulnerable to the same damage agents. Thus, a single attack might take out thousands or tens of thousands of links, potentially cutting off multiple nodes from the network. The places where this can occur must be protected to assure cyberspace infrastructure integrity [1]. This is a particular concern for nodes located in geographically isolated sites (i.e., North Siberia, Far East of Russia), that are critical to national security. Loss of electricity does not ordinarily take down a major Internet hub – at least not at once, since most hubs have emergency backup power sources that can carry them for hours or even days.

Most modern infrastructure systems face profound transformation as a result of the fast moving technological and market innovations. Inevitably, such transformation involves greater reliance on cyber technology to improve the efficiency and effectiveness of infrastructure operation, and thus potentially further increases opportunities for cyber attack. This presents an ongoing challenge to regulators, developers, and operators.

Virtually all infrastructures of economic importance depend on information systems that are potentially the targets of cyber attacks. The latter can be classified by their intent as follows: 1) to provide information and garner specific knowledge needed for organizing a physical or cyber attack. The technique used here is similar to industrial espionage techniques; 2) to supplement a physical attack, in order to exacerbate the intended damage. These kind of attacks temporarily disable protective or corrective responses; 3) to damage or destroy critical physical nodes (for EG it would be electrical generators and large step-up and step-down transformers, for liquid pipelines--pumps, valves; for gas pipelines--compressors, valves, etc.).This kind of attack is the most difficult to organize, as it needs perfect timing, near-synchronous execution and extensive and precise knowledge of the intrinsic mechanical and material properties of the object being attacked.

With having in place operating personnel with high levels of diligence and compliance and security subsystems installed, the success of all types of cyber attacks on critical infrastructures can be severely limited and can be virtually nullified by rigorous reverse engineering. Reverse engineering is a way to design and operate existing infrastructures taking into full consideration system thinking of a malicious mind that is planning to execute a cyber attack on these objects.

To impose strategic-level and lasting physical damage to any major infrastructure system widely-dispersed over a territory (communication, electric, oil, gas, water, etc.), multiple quasi-simultaneous physical attacks on their most vulnerable nodes are required. Attacks on second-level nodes (transmission towers, separate single power generators, compressors, pumps, etc.), may be locally disruptive and costly to infrastructure providers and users, but yield only restricted damage and partial decrease of its productivity.

The design and manufacture of the damaged Iranian centrifuge cascade were inherently of marginal quality manufacture and inadequate materials for such highly stressed machines. It is doubtful whether Stuxnet or any other software worm would be successful against infrastructure systems with better design and more robust quality.

Government regulators and industry groups in general seek to provide balanced and integrated protection not only against cyber threats but the entire spectrum of natural as well as malicious threats. Infrastructure firms in general recognize their strong economic interest in protecting themselves and their investments, and for the most part are reasonably willing to comply with the state regulations [1].

4. Critical Infrastructure Vulnerabilities

Cyber Networks. The network levels that support and comprise cyberspace are run almost entirely remotely by computers without any direct human intervention. This opens widely the door to cyber exploitation and attack, to which they were more or less systematically subjected in the form of distributed denial of service (DDoS) attacks (the most common form of attack), as well as a variety of exploitation attempts. Motivations for the DDoS include *extortion, ransom, revenge, publicity for causes*, etc.

Reported intrusions [1, 3] have almost all been information-seeking, for criminal or obscure purposes. Remarkably, there have been no reports of coordinated cyber attacks attempting to exploit the damage and impede repair and recovery operations. But there is no guarantee that this will not change in the near future.

The Electrical Grid (EG) is the infrastructure of greatest concern in connection with cyber attack. The potential problems and solutions for other infrastructures broadly parallel those of the grid. The cyber security of the Russian electrical grid has been the subject of intense and broadly-based research for nearly two decades, allowing some fairly trustable conclusions. The topology of Russian EG resembles the HAS network of Fig.3, with some features of the UR network shown in Fig.2, with each generation source, transmission substation, or distribution substation as a node and each line connection of two nodes as a link. In this kind of transmission networks hubs connected directly to large numbers of nodes are rare and most nodes have more than one link, except for those in the fringes of the north-east parts of the Russian EG.

The fast and accelerating advent of the sixth tenor of technology will dramatically change the structure and layout of Russian electrical grids as it will become increasingly smart and hybrid, combining traditional energy generation (coal, oil, gas, atom) with renewable energy systems (hydro, sun, wind, ocean waves, bio-fuel, thermal, etc.), and some novel energy storage subsystems (high capacity batteries, flywheels, heat accumulators, etc.).

Currently the Russian state-owned unified EG is comprised of a relatively small number of large central station plants, usually located in the vicinity of their energy sources (Fig.6). Electricity is a bulk commodity that lacks specificity and is economically transmitted in the form of alternating current (AC) at high levels of energy and voltage, and most electrical use is AC at lower (220 v) voltages. Major production and transportation corridors are served by a few high-capacity HVAC lines along which distribution stations are located that feed local bulk users and local retail distribution networks, using the transformer as a passive device that allows economic tapping down high-voltage AC (HVAC) to lower voltage. The corridors follow the customers and, hence, are determined by economic geography (mostly, along the TransSiberian railroad).

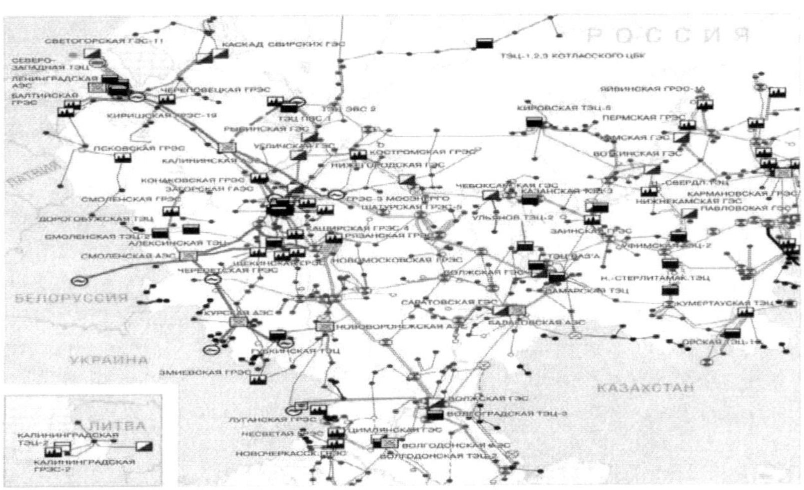

Figure 7. Unified national electric grid of the European part of Russia [12].

When the flow in an electricity network is near the limits of its capacity, the failure of one link could throw more load on remaining links than they can carry without overheating. This would lead to a cascade failure, due to overheating of links, one by one, or automatic /manual shutting them down to prevent damage.

On an AC network, the current must alternate precisely at the same frequency everywhere in synchronous operation. Any failure of this frequency synchronization would produce unbalanced forces that could literally tear equipment apart. If local overloading drags the frequency of a generator down, then it and the area it serves must immediately be disconnected from the grid [1]. Synchronization failures also can cascade, as generation or transmission equipment drops offline to avoid catastrophic failures. Within each of the regional grids the frequency must be the same at any given instant; misalignment of frequency between regions can be tolerated only if they are connected by DC interconnections, which serve as trip for a major blackout. This scenario, on a wide scale, can cut the grid up into isolated islands, many or all of which might fail under local load imbalances [1].

The loading on the grid varies from moment to moment. Users can randomly add loads by throwing a switch; generators and transmission equipment can go offline for a variety of reasons. Grid operators have a limited load-shedding capacity (temporarily cutting off customers who have bought "interruptible power" at reduced rates). In an emergency, a block of customers in a particular area may be blacked out to shed loads, but many systems are not set up to allow it to be done quickly, and utilities are reluctant to do this except as a last resort [1].

Could such a failure cascade engulf large regions of Russia and adjacent countries --former parts of the Soviet Union like Belorussia or Kazakhstan-- that are connected to the Russian EG? This scenario is highly unlikely, due to the fact that HVDC intertie lines isolate each region from frequency disturbances in other regions, and because disturbance from a major fault in the grid weakens as it disperses.

Because every part of the grid influences every other part, it has been difficult to construct a deregulation regime that would allow the truly independent operation necessary for fully effective competition. The same limitations that permit participants to impose costs on others without inherent limits (other than those interposed by the remaining regulators) equally allow serious technical problems to develop and spread without any individual participating firm or organization having a clear interest in taking corrective action.

The physics of electricity make it impossible for a fully disunited, every-entity-for-itself EG operating regime. If the system is to operate stably and safely, there must be some consistent set of operating rules that everyone is constrained to obey. This realization has been somewhat slow in emerging, perhaps in part because authorities were thinking in terms of analogies with networks that were not as tightly coupled as the electricity grid and thus less in need of highly disciplined operation.

SCADA and system management networks. All modern EGs and other infrastructure networks with *distributed equipment* generally have as their components *supervisory control and data acquisition* (SCADA) networks. The earliest SCADA nets (in 1950s) were immune to any types of cyber attack because of their fully isolated from the outer world dedicated transmission channels. EG equipment generally has *separate* control and limiting systems (inherited, by the way, from the eighteenth century steam engines). According to this strategy a modern generator has several trips: over-speed, overvoltage, overpressure, etc. The SCADA controllers enable operator command and provide feedback control to reach and maintain the commanded state. The limiters limit damage from inadvertent or malicious improper operator commands as well as system failures; hence, if need be, they can *override* the controller inputs [1].

Digitalization of SCADA (using programmable logic controllers PLCs that generally communicate digitally to a central computerized control system) while bringing important advantages of economy and efficiency also introduced potential vulnerabilities to cyber exploitation and attack. The energy management system (EMS) (or its equivalent in other types of infrastructure systems), which is at a level above SCADA, optimizes overall economics and feeds directions to the SCADA to adjust system operation accordingly. As both EMS and SCADA have been digitized the distinction between them has grown less clearly defined and they are often referred to as energy delivery control systems (EDCS). Standards for cyber security of EDCSs and their subsystems have been available since the mid-2000s. Nevertheless, examination and testing continue to reveal a series of common cyber vulnerabilities in many EDCSs [13].

Analysis of major outages and blackouts worldwide from the cyber security point of view revealed EG design defects and demonstrated how tightly coupled the grid is and what this implies for its operation and protection [14]. It also discovered a number of hardware and software failures, together with faulty operational procedures and operator errors on the local and regional levels.

The typical scenario of a blackout practically always includes a cascading effect. The trigger of such cascade of failures may vary. For instance, the immediate cause of the 2004 blackout in North America that involved 50 million people without electricity for a long time in winter was a series of instances in which high-voltage transmission lines contacted trees that had been allowed to grow too tall into the lines' rights of way. Autonomous safety systems sensed the resulting ground faults and automatically disconnected the lines to prevent more serious damage and fires. The prevention of the cascade was not achieved due to poor training of the operators and their excessive reliance on limited and fallible warning and diagnosis systems [1].

The future electric power grid, according to current research, will take the form of a cellular type smart grid with smart control, able to adapt to failures in real time with limited if any degradation [15], regulated by adaptive software rather than governmental agencies, and with provision to take advantage of distant power sources. This means that future EGs will depend even more on cyberspace. Hence, essential efforts are needed to improve the reliability and efficiency of power distribution without any increase of vulnerability to cyber exploitation and attack [16].

Cyber attacks against electrical grid targets can usually be expected only to exacerbate and/or impede response to physical defects or casualties, whether natural, accidental, or malicious in origin. For the most part, the potential of cyber attacks against *undamaged, robust* systems of modern design will remain limited to more or less temporary disruption of operation. Currently, the only known vulnerability (coined *Aurora vulnerability* AV) [17], [18], is the possibility to hack into the control system of an *electric generator or other rotating electrical equipment connected to the grid* and throw the equipment out of phase, causing severe physical damage to the equipment.

The practical difficulty of doing wide-scale damage to the whole EG is increased by the heterogeneity of the equipment and control software, produced by several different manufacturers at different times. To damage a substantial portion of such EG a set of highly coordinated attacks would be needed. It would be easier to attack only a small number of key highly- connected nodes [19] to produce a big damage that would take many days and even weeks to repair/replace, restart and reintegrate the EG, because the supply of replacement high-voltage step-up transformers (needed to raise electromotive force to the levels needed for long-distance transmission) is limited, as is the capacity for manufacturing additional ones quickly. No means to destroy transformers purely by cyber attack has yet been revealed, but it would be unwise to assume that it could not be done in the future.

Pipeline Networks. Two other Russian major energy-sector infrastructures, oil and natural gas – pipelines are also networked infrastructures, with about 71,000 km of oil pipelines and 179,000 km of natural gas transmission pipelines [21]. Water, waste and many non-hazardous and non-flammable fluids (CO_2, steam, coal, mineral and paper pulp, milk, etc.) are also transported by pipelines.

In a pipeline network, the driving force is provided by pressure which must be controlled, but not nearly so tightly as the electromotive force of EG network. The pumping/compressor stations of oil/gas pipelines and multiple valves that regulate input/ output as well as routing within the network are automatically and remotely controlled. The slow speed of the flow, rarely more than 5m/sec, permits using simpler than in EG, SCADA systems.

The registered low-level sabotage successful cyber attacks on pipeline SCADA systems (usually by individuals with economic or idiosyncratic motives), have the potential to inflict only temporary and local disruption, resulting in economic losses and nuisance to the pipeline operators and their customers [1].

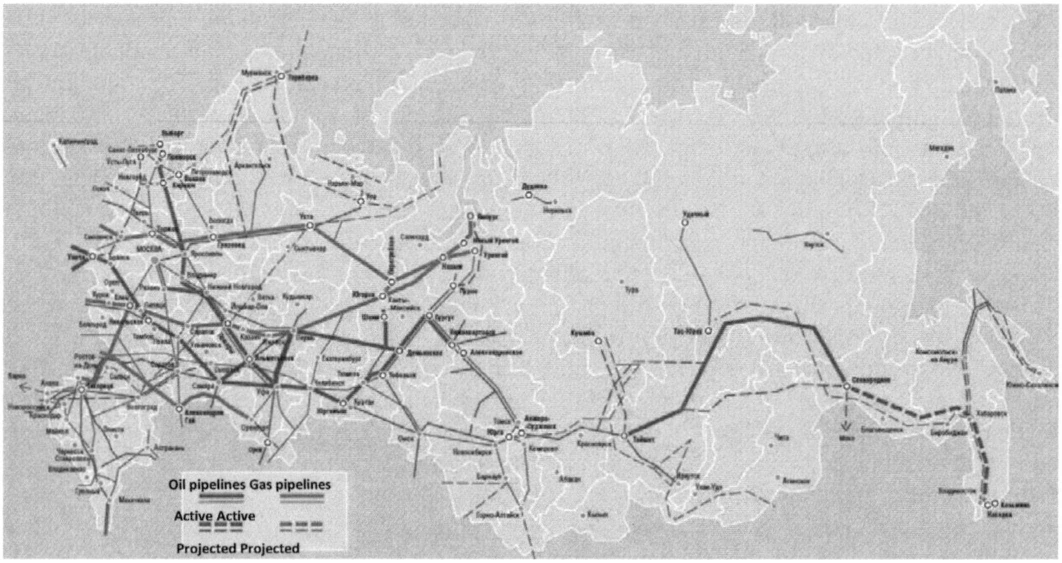

Figure 8. The main operating and planned oil and gas pipelines of Russia [20].

Central fresh water distribution and waste collection urban systems have been the target of sporadic attacks, almost always by individuals impelled by economic or idiosyncratic motivations. Like other pipeline systems, current fresh water systems use remote monitoring and control, and experienced some cyber attacks [22], which raised concerns about their cyber security [23]. For the most part, the low energies involved in local water systems makes them unlikely targets for severely destructive cyber attack. Moreover, they are engineered so as to make it all but impossible for waste water to enter the fresh water stream through *misalignment of valves* accomplished through cyber attack. In general, the damage that could be done by cyber attack seems to be limited to unpleasant and costly nuisances, by the exception of systems equipped by electrical pumping stations where the AV could be serious, or gas compressor stations, where surge of compressors could be organized. Specifics of many water/waste systems are that they are very old and deteriorated (Fig.9) which makes them absolutely not attractive to cyber attacks.

Research into the theory and experience of cyber incidents has resulted in a number of guides for sound engineering and operating practice. As we have seen, sound engineering and operations can go a very long way to protect infrastructure.

Human Factor is playing a leading role in critical infrastructure reliability, resilience, safety and security. In order to achieve reliability and resilience of CI rigorous application of existing guides and accumulated best practices is needed. In practice though, compliance has been considerably less than universal and thorough (due to ordinary inertia, reluctance to change, to master new knowledge and skills, or take on additional responsibilities) [1].

The private/public owned infrastructure systems that run on a for-profit basis usually pursue objectives that diverge sharply in financial aspects from the public interest regarding security against cyber attack. Corporate management tends to seek near-term profits, while the public interest is in spending on safety, not stakeholder's shares. On the other hand, infrastructure firms have strong incentives to safeguard themselves against threats that are a part of everyday business and may actually affect financial results. Threats that tend to be deprecated in management thinking can be categorized as follows: 1) Threats that occur at long, unpredictable intervals; 2) Threats whose financial impact is outweighed by the costs of protecting against them, viewed on a NPV basis; 3) Unimaginable (to the top management) threats even if rationally predicted to exist, because has no direct experience of similar threats; 4) Threats for which shareholders and the public (and their peers) are unlikely to hold managers personally accountable because they are held to be unforeseeable or unavoidable; 5) the black-swan type threats.

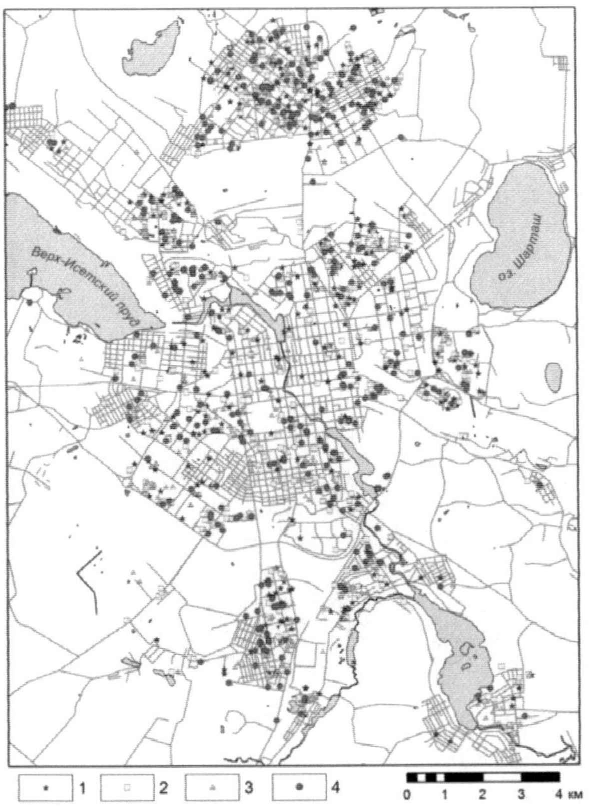

Figure 9. The City of Yekaterinburg: Scheme of emergency sections of urban water pipelines, (burst type of failure), for the period 2004-2007: 1 – 2004; 2 – 2005; 3 – 2006; 4 – 2007 [24].

In principle it is the responsibility of government to rectify imbalances between public and private needs by regulation backed by sanctions. In real life the advent of effective regulation almost always follows at least one catastrophe of a magnitude that draws widespread public interest. That's why progress on establishing effective regulations to avert cyber threats to infrastructure has been slow and halting. Moreover, the distinctions among threat sources—intrinsic, natural, criminal, or terrorist, often are not operationally meaningful: it can be difficult or impossible to discern the actual source of a threat in time to affect operations [1].

5. Problems of cyber reliability, resilience and safety of infrastructure systems

Generalizing all the above it is possible to formulate the hot spot problems of assessing and providing cyber reliability, resilience, and safety of modern infrastructure systems at all stages of their life time in the following form [25]-[27]. The cyber-problems are in a specific way inverse to ordinary reliability, resilience and safety problems, because the designer and operator of CI in order to meaningfully solve the above problems has to put himself in the shoes of the malicious adversary, whose goal is to damage or destroy the infrastructure being designed or in operation. Parsing the problem one would consider following inverse problems. Formulate new design schemes (mathematical operators of the CI under a CA) that take into account:

- existence of the SCADA subsystems (control, diagnostics, and maintenance) of the infrastructure proper;
- *full group of events-loadings*, that are generated by its own control subsystems, which were infested and/or influenced by viruses, and reflect the specifics of the cyber attack. For

instance, multiple simultaneous attacks on the most vulnerable components of CI, impossible at normal operation;
- new resilience and safety requirements to the personnel that operate critical infrastructures;
- imparting the CI with the «foolproof» quality against cyber attacks;
- formulate quantitative equations which allow assessing reliability, resilience and safety of different normal/usual operation, using classical and contemporary chapters of probability and uncertainty theories;
- include as obligatory quantitative assessments of reliability and safety of CI equipment that does not exist in ordinary design, but can take place during a cyber attack, i.e., non-stationary vibrations, resonance, gas turbine engines surge/pompage, water hammer, etc.;
- at the stage of design conduct assessment of vulnerability of the object as related to possible cyber attacks;
- explore vulnerability of CIs from the position of "wrong" control;
- conduct complex research of all possible new types of failures that are caused by cyber attacks.

6. Conclusion

1. The SCADA and EMS networks should be robust enough to continue providing accurate information and positive control even if subjected to coordinated cyber attack. Moreover, it is essential that operators be trained and adequately prepared to act resourcefully and decisively in response to casualties.

2. The foregoing examination of infrastructure protection issues has revealed that cyber attacks on physical infrastructures pose only a very limited strategic-level threat in and of themselves as long as right precautions are taken. This conclusion is broadly applicable to all well designed and manufactured major physical infrastructure systems.

3. Undependable software is one of the greatest vulnerabilities of infrastructure systems. The cost-driven trend to wide use of undependable and open-source software is exacerbating the risks. Software dependability will not achieve the necessary standards unless effective systems engineering is mandated for infrastructure systems.

4. While there is no clear limit to potential threats against infrastructures, there are limits to the resources that can be used for protection. Setting and keeping priorities for the allocation of financial and management resources are essential in order to provide effective protection.

5. The integrity of infrastructures affects everyone in our society, and the public will demand that its views be heeded. Hence, the public should be informed fully, clearly and accurately about all possible and happening cyber threats/incidents. This will ensure that the public will feel confidence in those who direct infrastructure protection efforts and will pay appropriate attention to their recommendations.

6. Policy direction for the various infrastructures should be tailored to their specific nature and needs.

7. Many important questions remain unsettled and more will arise as threats, technology, and economic conditions change. The Russian Federation policy and regulation institutions must have the authority, resources, and responsibility to sponsor and guide broadly conceived programs of research to serve their information needs. Knowledge can be expensive, but its absence can be much more so.

References

[1] Kramer F D and Starr S H and Wentz L K 2009 Cyber Power and National Security National Defense University (Center for Technology and National Security Policy) chapter 9
[2] Miller B and Rowe D 2012 A survey of SCADA and critical infrastructure incidents Proc. of the 1st Annual conference on Research in information technology pp 51-56
[3] Panda Security Rus November 30 2016 at 12:04 (https://habr.com/company/panda/blog/316500/)

[4] Lennon M 2013 Cyber Attacks Targeted Key Components of Natural Gas Pipeline Systems *Security Week*
[5] Kushner D 2013 The real story of Stuxnet *Spectrum IEEE* **50** 3 pp 48-53
[6] Gorman S 2009 Electricity Grid in US Penetrated By Spies *Wall Street Journal*
[7] Russian computer network (RUNET) integrated into the global network Available from: http://900igr.net/prezentacija/informatika/globalnaja-kompjuternaja-set-internet-99778/regionalnye-kompjuternye-seti-obedinennye-v-globalnuju-set-7.html [Accessed 28 Sept 2018]
[8] Backbone network of the Russian JSC Rostelecom Available from: https://www.expertsvyazi.ru/forum/printpage.php?forum=37&topic=4 [Accessed 28 Sept 2018]
[9] Backbone of a new Russian digital communication operator Available from: http://www.gudok.ru/newspaper/?ID=720435 [Accessed 28 Sept 2018]
[10] Krioukov D et al 2007 The Workshop on Internet Topology (WIT) Report *Computer Communication Review* **37** 1 pp 69–73
[11] 2003 Computer Science and Telecommunications Board *The Internet Under Crisis Conditions: Learning from September 11* (Washington, DC: National Academies Press)
[12] Unified national electric grid of the European part of Russia Available from: http://present5.com/vvedenie-v-elektroenergetiku-l-edinaya-energeticheskaya-sistema-rossii/ [Accessed 28 Sept 2018]
[13] 2011 Vulnerability Analysis of Energy Delivery Control Systems (Idaho Falls: Idaho National Laboratory)
[14] 2004 U.S. - Canada Power System Outage Task Force *Final Report on the August 14 2003 Blackout in the United States and Canada: Causes and Recommendations* (Washington DC and Ottawa: U.S. Department of Energy and Natural Resources Canada).
[15] Farhangi H 2010 The path of the smart grid *Power and Energy Magazine IEEE* **8** 1 pp 18-28
[16] Yan Y and Qian Y and Sharif H and Tipper D 2013 A Survey on Smart Grid Communication Infrastructures: Motivations, Requirements and Challenges Communications Surveys & Tutorials *IEEE* **15** 1 pp 5-20
[17] Markey E J and Henry A W Staff of Representatives *Idem* pp. 4-5.
[18] Staged cyber attack reveals vulnerability in power grid Available from: http://www.cnn.com/2007/US/09/26/power.at.risk/ [Accessed 20 Aug 2014]
[19] 2012 Structure of the North American Electric Power Infrastructure *IEEE Systems Journal* **6** 4 pp 616-626
[20] The main operating and planned oil and gas pipelines of Russia Available from: http://fedoroff.net/publ/geography/geografiya/gazoprovody_rossii/48-1-0-576 [Accessed 28 Sep 2018]
[21] Herberg M E et al. 2010 Pipeline Politics in Asia: The Intersection of Demand, Energy Markets, and Supply Routes (National Bureau of Asian Research)
[22] Slay J and Miller M 2008 Lessons Learned From The Maroochy Water *Critical Infrastructure Protection* **253** (New York: Springer) pp 73-82
[23] Dakin R and Newman R and Groves D 2009 The Case for Cyber Security in the Water Sector *Journal of the American Water Works Association* pp 30-32
[24] The City of Yekaterinburg: Scheme of emergency sections of urban water pipelines Available from: http://shkolaput.ru/kaknsa/%D0%98%D1%81%D1%81%D0%BB%D0%B5%D0%B4%D0%BE%D0%B2%D0%B0%D0%BD%D0%B8%D0%B5+%D0%B3%D0%B5%D0%BE%D0%B4%D0%B8%D0%BD%D0%B0%D0%BC%D0%B8%D1%87%D0%B5%D1%81%D0%BA%D0%BE%D0%B9+%D0%B0%D0%BA%D1%82%D0%B8%D0%B2%D0%BD%D0%BE%D1%81%D1%82%D0%B8+%D0%B3%D0%B5%D0%BE%D0%BB%D0%BE%D0%B3%D0%B8%D1%87%D0%B5%D1%81%D0%BA%D0%BE%D0%B9+%D1%81%D

1%80%D0%B5%D0%B4%D1%8B+%D0%B3+%D0%B5%D0%BA%D0%B0%D1%82%D0%B5%D1%80%D0%B8%D0%BD%D0%B1%D1%83%D1%80%D0%B3%D0%B0a/main.html [Accessed 28 Sept 2018]

[25] Timashev S 2015 Infrastructure Resilience: Definition, Calculation, Application *Proc. of the WEEF Conference*

[26] Timashev S 2018 Resilient Urban Infrastructures – Basics of Smart Sustainable Cities. *IOP Conf. Ser.: Mater. Sci.* Eng **262**

[27] Timashev S A, AlekhinV N, Poluyan L V, Fontanals I and Gheorghe A 2018 Transforming Yekaterinburg into a Safe, Resilient-Smart and Sustainable City *IOP Conf. Ser.: Earth Environ. Sci.* **177**

Assessing risk and resilience of critical infrastructures in the fuel and energy economy involving failure cascading effects

LV Poluyan[1,2]

[1] Science and Engineering Centre «Reliability and Safety of Large Systems and Machines», Ural Branch, Russian Academy of Sciences, Studencheskaya, 54-A, Yekaterinburg, Russia, 620049
[2]Ural Federal University, Mira, 19, Yekaterinburg, Russia, 620002.

E-mail: sec@wekt.ru

Abstract. A practical procedure of risk and resilience assessment, with cascading failures occurring in the fuel and energy economy, is proposed. It is based on the threshold values and multi-energy methods. The probability analysis of the equipment damage is carried out using vulnerability models for various equipment classes and in the process the probit-functional analysis is applied. Resilience indices-criteria are used for assessing the resilience of plants as complex mechanical systems. The procedure proposed enables calculating threshold values of cascading failure; safe distances for equipments; probability assessment of failure propagation. It also adds to identifying the facilities triggering cascading effects; probability of the equipment damage; to assessing the amount of damage and the damage loss rate. The strategy (procedure) allows ranking the equipment as related to the risk of domino effects' occurrence; analyzing consequences for each cascade direction and the possibility of its propagation on to the following level; evaluating the resilience state of the plant as a technical system. The damage amount and damage loss rate are evaluated comparing the load received and predetermined threshold values of cascading failure for various equipment classes. The strategy (procedure) developed will allow judging about safety of one or other type of a tank farm containing hydrocarbons; the results obtained in the process and the actual data converge. This strategy (procedure) is strongly recommended to be applied by experts in the field of industrial safety in developing industrial safety manifests, safety justification as well as in developing safety registration certificates of critical infrastructures.

1. Introduction

About 90% of substances involved in cascading failures are combustible while toxic substances make up only 4%. Currently, there are no universal quantitative techniques considering the domino effect, which would be certified, approved and accepted for risk analysis and control [1 - 3]. Analysis of scientific papers has revealed that the methods and techniques based on various vulnerability criteria (50%) and the method of threshold values (39%) are the most widespread methods for taking into account domino effects [4 - 10]. Their common feature deals with the fact that they all consider "domino" scenarios as a separate analysis, starting with the results obtained while making the quantitative risk assessment [2, 11]. These methods do not provide any indication as to how the "domino" effect has to be incorporated into the well-established risk analysis structure. That is why there is an urgent need in integrated methods which clearly define how the "domino" effect can be identified within the frame of the conventional structure of the quantitative risk assessment. To make sure the proposed strategy (procedure) is effective the following criteria have been deduced: it is to be applicable to the existing well-established methods of analysis; it must not involve complex algorithms of the events' chain analysis; it is to enable calculating risk indices for possible "domino" scenarios and safe distances to prevent property damage and "domino" effect occurrence, taking into consideration specific conditions for each plant. The strategy (procedure) reliability was evaluated comparing the results of the domino effect evaluation and results published in foreign resources. When comparing, symbols, designation, reference data, threshold values of cascading failures were taken from foreign resources.

2. Procedure stages

1. *Determination of threshold values of failure triggering and safe distances for the equipment.* Threshold values of failure triggering (TVFT) for various equipment classes and instructions for calculating safe distances to explosions and fires are taken from [2]. The multi-energy method is used to calculate safe distances (SD) [11]:

$$R_x = R/(E/P_0)^{1/3} \tag{1}$$

where R_x is the non-dimensional distance taken from [2]; R is the distance to the centre of explosion, m; P_0 is atmospheric pressure, $P_0 = 101325$ Pa, E is an effective mixture energy reserve, *J*, calculated by

$$E = M_т * E_{shc} \tag{2}$$

where $M_т$ is the dry-and-ash-free fuel (combustible substance) in the cloud, which contributes to the explosion, kg; E_{shc} is the specific heat of combustion, $E_{shc} = 44$ MJ/kg for typical hydrocarbons.

In order to determine a SD in case of a failure triggered by an explosion, it is necessary to determine the value of R in (1). If the size of destruction areas at TVFT is greater than the SD values, then the value at which the destruction effects' values are lower than those of TVFT will be taken as a facility- to-facility SD. If necessary, the SD value is calculated with a reserve in order to take into consideration all possible conditions at the plant.

2. *Assessment of the possibility for failure propagation* is made comparing destruction areas obtained as a result of the initial failure consequences' analysis, using TVFT for the equipment and other auxiliary facilities (plants).

3. *Identification of the equipment contributing to failure triggering.* In identifying danger, TVFT can be used as a basis for critical areas identification on the plant's territory with reference to the "domino" effect as it is with the help of TVFT that the facility to facility SD can be calculated. If the propagation is recognized to be possible, another step to be taken is to analyze which of the initial equipment unit can trigger the secondary failure.

4. *Equipment classification with respect to the risk of the "domino" effect occurrence.* It is necessary to rank all the equipment with respect to the "domino" effect occurrence. The «Domino index», DI_k where k is the equipment unit under consideration (k = 1...X), X is the number of equipment units, is used to estimate the triggering potential. Effective distance (d_e) is the distance at which TVFT is observed. It is compared with the actual distance (d_f) between two units of the equipment under consideration. Depending upon their relation, the «Distance factor» (DF_{ij}) is determined, where i is the initiating equipment under consideration, while j is the injured equipment being considered (i, j = 1,...,X). At $i = j$ $DF_{ij} = 0$.

$$DF_{ij} = \begin{cases} (d_e - d_f)/d_e * 100, d_f < d_e \\ 0, d_f > d_e \end{cases} \tag{3}$$

Based on the distance factors received for each pair of reservoirs, a matrix of distance factors is built and the domino index is determined using the formula:

$$DI_k = \sum_{i,j=1}^{X} (DF_{kj} + DF_{ik}) \tag{4}$$

The domino index enables identifying the most hazardous equipment with regard to its capability to cause the "domino" chain and become the secondary, tertiary and further objects of failure triggering. The domino index makes it possible to judge about the equipment in terms of the "domino" effect occurrence and propagation and take every effort to manage the risk. For a reservoir having the highest domino index and when it is not possible to ensure the facility-to facility SD, feasibility for lowering the failure probability is to be considered using safety barriers among other things.

To avoid a cyclic analysis of the equipment it is advisable to analyze the whole possible chain of events which can take place after the initial failure, starting from the unit equipment having the highest potential for triggering. The analysis should be continued up to the units exhibiting the lowest potential for triggering. If the analysis of the whole possible chain of events shows that the equipment having the highest domino index cannot be damaged by the primary failure, it is possible to leave out this equipment and continue analyzing that one the domino index of which is lower but which will be located in the destruction area of the initial event. To avoid a cyclic analysis of the equipment, a representative scenario for the analysis is chosen, where the probability for the equipment to be damaged by vectors of triggering is calculated using probit-functions. This choice is made basing on the triggering potential of each unit equipment. The equipment chain should also be inspected, starting with the facility which doesn't possess the highest index but which can turn out to be a failure precursor on the unit with a higher potential. Such chain of events will lead to the most hazardous consequences. The scenarios where failures occur on several reservoirs right after the primary failure will be characterized by the lowest frequency of occurrence and, consequently, are taken into account only for the reason of necessity, basing on the calculation aims. It is worth noting that the occurrence rate of sequences with the lowest amount of the equipment damaged will always be higher. It is recommended to stop the calculation when the multi-level failure sequence is realized at the rate value lower than 10^{-6} $year^{-1}$.

5. *Probability of the equipment damage* is analyzed applying vulnerability models for various equipment classes, with the probit-functional analysis being used [2].

6. *The damage amount and the loss rate* are evaluated comparing the load received and predetermined TVFT for various equipment classes. There are various rates of containment losses (LOC), which are related to different damage states (DS). Frequencies of damage states and loss intensity are given in [12].

7. *Identification of triggering vectors.* Using the data obtained, it is possible to approximate the secondary loss of containment and to estimate the triggering vector. Given the substance properties and storage conditions, the triggering vectors obtained will enable assessing the subsequent failures. Failure rates caused by the corresponding containment rates are presented in [12].

8. *Analysis of consequences made for each triggering vector.* Analyzing the consequences of triggering vectors identified earlier it is possible to determine an additional destruction area for each scenario of the secondary failure. In its general view, the occurrence rate of the subsequent failure is calculated as in [2]:

$$f = f_{source} * P_{esc}. \tag{5}$$

where f_{source} is the occurrence rate of the primary failure and the corresponding triggering vector; P_{esc} is the probability of triggering the subsequent failure due to the primary one, this probability is determined transforming the probit-function into probability percentage.

9. *Possibility for failure propagation on to the next level.* If additional damaged units are found in the destruction areas of the secondary failure triggering vectors, it is necessary to analyze the further level of triggering. The most quite clearly understandable and recommended approach to analyzing a chain of events is to analyze the tree of events for each representative scenario.

10. *Resilience (Robustness) assessment.* In our case partial resilience is assessed in the context of the domino effect occurrence for the equipment units under consideration. Index-criteria represent a widely-spread approach to assessing resilience of technical systems (TS) [13, 14]. In this research, the «Domino index» can be used to determine the triggering potential in accordance with (3).

In addition, the resilience index based on the ratio of direct and indirect damages can be chosen from the index-criteria list [4, 5]. A robust (survivable) system is considered to be the one where indirect risks do not contribute significantly to the total system risk. With this in mind, the following index of robustness I_{rob} is proposed, which measures the fraction of the total system risk resulting from direct consequences:

$$I_{\text{rob}} = R_{\text{dir}} \left(R_{\text{dir}} + R_{\text{ind}} \right). \qquad (6)$$

The index takes values from zero to one, depending upon the source of risk. If the system is completely robust and there is no risk caused by indirect consequences, then $I_{\text{rob}}= 1$. If all the risk is due to indirect consequences, then $I_{\text{rob}}= 0$.

Finally, this index can be easily extended to account for multiple exposures, or more complicated event trees.The robustness index will still be equal to the sum of direct risk divided by the sum of total risk:

Resilience is assessed in the context of the domino effect occurrence for the equipment units under consideration. Index-criteria represent a widely-spread approach to assessing resilience of technical systems (TS) [13, 14]. In this research, the «Domino index» can be used to determine the triggering potential in accordance with (3).

In addition, the resilience index based on the ratio of direct and indirect damages can be chosen from the index-criteria list [4, 5]. A robust (survivable) system is considered to be the one where indirect risks do not contribute significantly to the total system risk. With this in mind, the following index of robustness I_{rob} is proposed, which measures the fraction of the total system risk resulting from direct consequences:

$$I_{\text{rob}} = R_{\text{dir}} \left(R_{\text{dir}} + R_{\text{ind}} \right). \qquad (6)$$

The index takes values from zero to one, depending upon the source of risk. If the system is completely robust and there is no risk caused by indirect consequences, then $I_{\text{rob}}= 1$. If all the risk is due to indirect consequences, then $I_{\text{rob}}= 0$.

Finally, this index can be easily extended to account for multiple exposures, or more complicated event trees.The robustness index will still be equal to the sum of direct risk divided by the sum of total risk:

$$I_{\text{rob}} = \sum_i R_{\text{dir}_i} / \left(\sum_i R_{\text{dir}_i} + \sum_j R_{\text{ind}_j} \right). \qquad (7)$$

3. Example

Comparison analysis of the procedures offered and developed in [15] has been conducted. The initial data was taken from the procedure used to analyze cascading effect vulnerability of industrial enterprises. A failure in the tank farm was investigated; there were eight similar reservoirs of 200 m^3 (PBC-200) volume capacity at atmospheric pressure in the tank farm, containing benzol. The farm site plan is presented in Figure 1.

Constraints accepted. Spillage fire was dealt with. Heat radiation of $I = 15\ \text{kWt/m}^2$ was taken as TVFT. Outflow out of the hole of 15 c diameter at the constant speed was chosen as the most likely reason for initiating failure at every level of triggering. Areas of destructive factors are calculated from the wall of reservoir T_i to the wall of reservoir T_j. The calculations were made using software ALOHA 5.4.7 [16, 17].

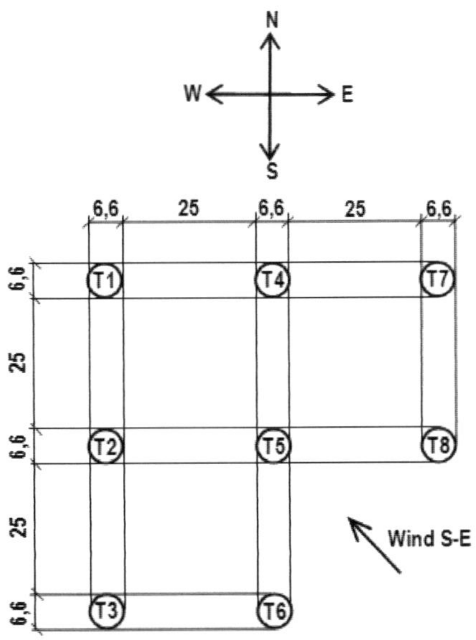

Figure 1. Farm site plan.

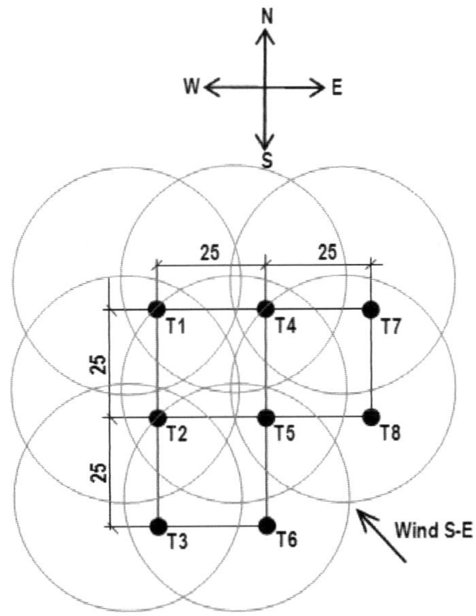

Figure 2. Destruction areas at $I = 15\ \text{kWt/m}^2$of heat radiation intensity.

1. Assessment of the possibility for failure propagation. Table 1 demonstrates intensity values for all the units of the farm equipment, where the values exceeding the triggering threshold are given in the boldface type. Making a pattern of areas destructed by the threshold values of triggering on the site plan will be the next step.

The results of patterning the destruction areas are shown in Figure 2, on the basis of which safe distances can be obtained with the aim of preventing the "domino" effect occurrence at the plant (farm). Thus, taking into consideration the minimum possible SDs between units of the equipment and using the destruction areas, we have $d = 32.4$ m.

The maximal value of $d = \max(32.4; 55) = 55$ m is taken as SD. In order to completely eliminate the possibility of failure triggering and allow for various directions and the speed of wind over the course of a year, it is recommended to use the value of d = 65 m instead of d = 55 m. The farm site plan obtained is demonstrated in Figure 3.

Table1. Heat radiation intensity (kWt/m^2) due to the influence of T_i on T_j

T_i	T_j							
	T_1	T_2	T_3	T_4	T_5	T_6	T_7	T_8
T_1	–	9.6	2.9	9.6	4.5	2.1	2.9	2.1
T_2	**24.8**	–	9.6	8.0	9.6	4.5	2.9	2.9
T_3	5.6	**24.8**	–	3.7	8.0	9.6	2.0	2.9
T_4	**24.8**	8.0	2.9	–	9.6	2.9	9.6	4.5
T_5	**16.0**	**24.8**	8.0	**24.8**	–	9.6	8.0	9.6
T_6	4.8	**16.0**	**24.8**	5.6	**24.8**	–	3.7	8.0
T_7	5.6	3.7	2.0	**24.8**	8.0	2.9	–	9.6
T_8	4.8	5.6	3.7	**16.0**	**24.8**	8.0	**24.8**	–

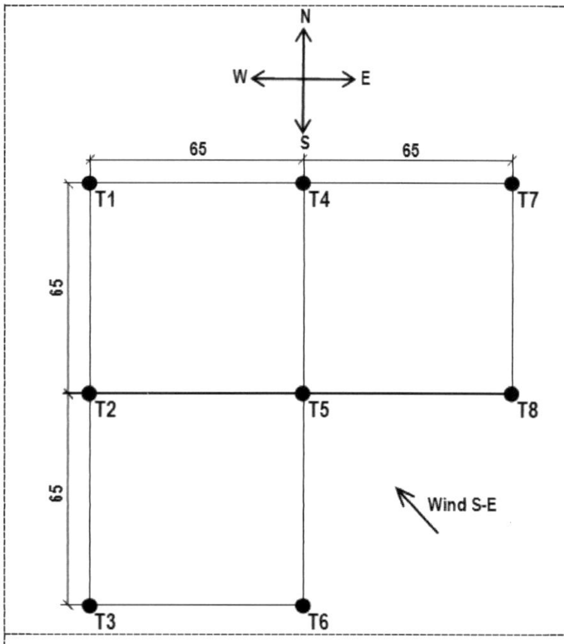

Figure 3. Site plan with regard to safe distances.

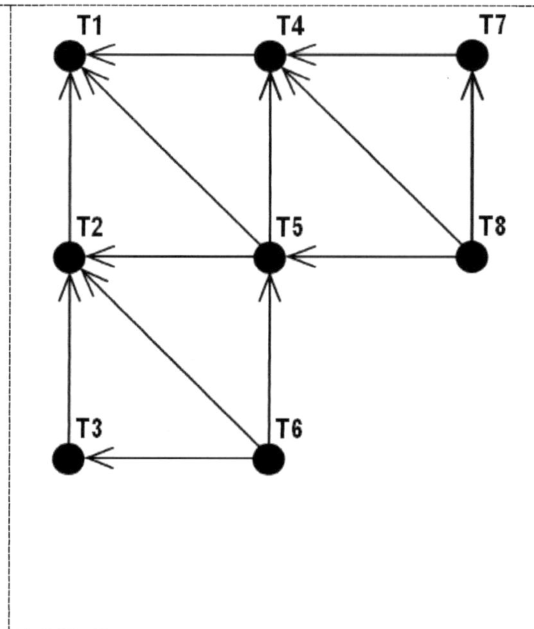

Figure 4. All possible triggering vector directions caused by spillage fire.

2. *Identification of the equipment contributing to failure triggering.* In order to find out which of the equipment units can contribute to the secondary failure occurrence, the data published in Table 1 and shown in Fig. 2 is used. The primary failure in the "domino" chain can be failures on reservoirs T_2-T_8. The failure on reservoir T_1 is not capable of becoming a trigger.

Comparing the data received it is possible to state that reservoirs T_5, T_6, T_8 will be the most hazardous units of the equipment as a failure on each of them can affect most of the similar units. All possible directions of the triggering vector caused by spillage fire on each equipment unit are designated by arrows in Figure 4.

3. *Equipment classification with respect to the risk of the "domino" effect occurrence.* Let's assess the equipment that creates destruction areas which affect other units of the equipment or is located in an area destructed by threshold values of failure triggering. Using 2.3 и 2.4, we rank the equipment with respect to the triggering potential.

Distance factors are then defined for reservoir T_5, using Figure 2: $DF_{51} = (36 - 35.4)/36 * 100 = 1.7$; $DF_{52} = (32.4 - 25)/32.4 * 100 = 22.8$; $DF_{54} = (32.4 - 25)/32.4 * 100 = 22.8$. Similarly, we define distance factors for other units of the equipment. The results are given in Table 2.

Table 2. Matrix of distance factors DF_{ij}.

T_i	T_j							
	1	**2**	**3**	**4**	**5**	**6**	**7**	**8**
1	0	0	0	0	0	0	0	0
2	22.8	0	0	0	0	0	0	0
3	0	22.8	0	0	0	0	0	0
4	22.8	0	0	0	0	0	0	0
5	1.7	22.8	0	22.8	0	0	0	0
6	0	1.7	22.8	0	22.8	0	0	0
7	0	0	0	22.8	0	0	0	0
8	0	0	0	1.7	22.8	0	22.8	0

Let's determine the domino index for every unit of the equipment: $DI_1 = DI_6 = DI_8 = 47.3$; $DI_2 = DI_4 = 70.1$; $DI_3 = DI_7 = 45.6$; $DI_5 = 92.9$. The event chain being analyzed is T5→T2/T4→T1. It's worth noting that the most hazardous scenario where the "domino" effect reaches the third level of triggering and propagates involving only one subsequent unit of the equipment at each level will be $T_6/T_8 \rightarrow T_5 \rightarrow T_2/T_4 \rightarrow T_1$ (Figure4). Scenarios dealing with the occurrence probability less than 10^{-6} year^{-1} are not considered.

4. Analysis of the equipment damage probability. The rate of the outflow from the hole at the constant speed is $f_i = 1 * 10^{-4}\ year^{-1}$. To assess the probability P_{esc} of T_2/T_4 reservoir damage caused by the spillage fire on reservoir T_5, the probit-function [67] $Y = 9.25 - 1.85 * \ln(ttf/60)$ is considered, where $\ln(\text{ttf}) = -1.13 * \ln(I) - 2.67 * 10^{-5} * V + 9.9$, ttf is the time in seconds till the failure, V is the reservoir volume (m^3), the heat flow is $I = 24.8\ kWt/m^2$. The calculations proved that $Y = 5.17, P_{esc} = 5.7 * 10^{-1}$.

5. Assessment of the damage amount and the loss rate. In accordance with the tabulated data on the relationship of threshold values and corresponding containment losses rates at the specified heat radiation, a reservoir will be damaged as a result of which its loss rate will be equal to LOC3 class, and the damage amount will be DS2 [18].

6. Identification of triggering vectors. The following triggering vectors [69] are possible for the reservoirs at atmospheric pressure and LOC3 containment loss rate: fire-flash of $P_{f.f.} = 0.8 * 10^{-1}$probability and spillage fire of $P_{s.f.} = 2 * 10^{-1}$ probability.

7. Analysis of consequences for each triggering vector. Due to the constraints accepted earlier, the spillage fire and heat radiation are considered to be the only triggering vector of the injured equipment. Thus, $T_5 \rightarrow T_2/T_4$ failure occurrence rate without further triggering vectors is calculated in the following way: $f_{5-2/4} = f_i * P_{п.п} * P_{esc} = 1 * 10^{-4}\ year^{-1} * 2 * 10^{-1} * 5.7 * 10^{-1} = 1.14 * 10^{-5}\ year^{-1}$.

8. Possibility for failure propagation on to the next level. When there is a failure on T_2/T_4 reservoir, T_1 reservoir is affected by $I = 24.8\ kWt/m^2$which is greater than the threshold value of 15 kWt/m^2. Calculations will be similar to the previous ones and the probability will be $P_{esc} = 5.7 * 10^{-1}$.

Trees of events are used to make the analysis of the representative chain of events strikingly visual. In accordance with the constraints accepted, the tree of events for the case under consideration is given in Figure 5.

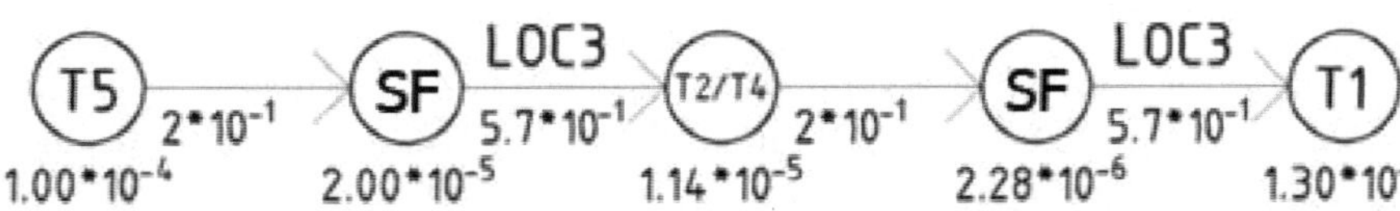

Figure 5. Tree of events for the "domino" chain under consideration.

In Figure 5 - T_5, T_2/T_4, T_1 nodes are reservoirs' designations; SF is spillage fire; LOC3 is the rate of the containment loss; the values given under the nodes represent rates of occurrence; the values located under the connecting ribs designate the probability of transition from the previous state to the following one. Thus, the estimated value of the representative $T_5 \rightarrow T_2/T_4 \rightarrow T_1$ scenario rate will be $f_{5-2/4-1} = f_i * P_{п.п.1} * P_{esc.1} * P_{п.п.2} * P_{esc.2} = 1 * 10^{-4}\ year^{-1} * 2 * 10^{-1} * 5.7 * 10^{-1} * 2 * 10^{-1} * 5.7 * 10^{-1} = 1.3 * 10^{-6}\ year^{-1}$. Comparison analysis of equipment ranking calculations based on the procedure developed and that presented in [2] correlate well with each other.

The ranking results agree very closely. The rates of the failure occurrence being considered are of one order $1.14 * 10^{-5}\ year^{-1}$ and $2.92 * 10^{-5}\ year^{-1}$ correspondingly, which is accounted for by different calculation procedures as well as by redistribution of probabilities in the Bayesian Networks as a result of treating even the least probable scenarios.

4. Future work

Of particular urgency is programmed implementation of the strategy which enables increasing the number of equipment units used for quick assessment of cascading failure. As mathematics used is simple for understanding, the strategy algorithm is not complicated as well.

The programming code in combination with GIS-technologies will allow considering all possible chains of events and, in doing so, to improve the calculation accuracy.

5. Conclusion

The strategy developed can be integrated in to the conventional structure of risk analysis and assessment; its mathematics is easy to utilize and it is independent of complicated algorithms of the events' chain analysis. Using it, it is possible to determine rates and consequences of the "domino" effect for equipment, safe distances to prevent property damage and triggering the "domino" effect.

This procedure can be used by experts for obtaining quick assessments in estimating the danger in tank farms having up to 10 units of equipment.

Unlike other more complicated techniques (weighted graphs, Bayesian Networks), the procedure developed does not require time consuming modeling for every particular set of conditions at the plant. It can be used by experts in the field of industrial safety in developing industrial safety manifests, safety justification as well as in developing safety registration certificates of critical infrastructures.

References

[1] Khakzad N, Reniers G, Abbassi R, Khan F 2016 Vulnerability analysis of process plants subject to domino effects *Reliability Engineering and System Safety* **154** pp 127-136

[2] Casal J 2018 *Evaluation of the Effects and Consequences of Major Accidents in Industrial Plants* (Elsevier) Second edition 570 p.

[3] Li J, Reniers G, Cozzani V, Khan F 2017 A bibliometric analysis of peer-reviewed publications on domino effects in the process industry *Journal of Loss Prevention in the Process Industries* **49** pp 103-110

[4] Clini F, Darbra R M, Casal J 2010 Historical Analysis of Accidents Involving Domino Effect *Chemical Engineering Transactions* **19** pp 335-340

[5] Darbra R M, Palacios A, Casal J 2010 Domino effect in chemical accidents: main features and accident sequences *Journal of Hazardous Materials* №**183** pp 565–573

[6] Zi-jian Ni, Yanzhang Wang, Zhigang Yin 2016 Relative risk model for assessing domino effect in chemical process industry *Safety Science* **87** pp 156-166

[7] Cozzani V, Gubinelli G, Antonioni G, Spadoni G, Zanelli S 2005 The assessment of risk caused by domino effect in quantitative area risk analysis *Journal of Hazardous Materials* **127** pp 14-30

[8] de Lira-Flores J, Vazquez-Roman R, Lopez-Molina A, Mannan M S 2014 A MINLP approach for layout designs based on the domino hazard index *Journal of Loss Prevention in the Process Industries* **30** pp 219-227

[9] Goodarzi M N, Farahani A F, Hosseini S E, Tavakoli M 2015 The Application of Domino Effect for Risk Analysis of Critical Assets in Oil Industry *International Journal of Biology, Pharmacy and Allied Sciences (IJBPAS)* №**4** pp 1850-1857

[10] Mukhim E D, Abbasi T, Tauseef S M, Abbasi S A 2017 Domino effect in chemical process industries triggered by overpressure – Formulation of equipment-specific probits *Process Safety and Environmental Protection* **106** pp 263-273

[11] The Russian State Standard Specification 12.3.047-2012. Occupational safety standards system. Fire safety of technological processes. General requirements. Methods of control

[12] Bernechea E J, Vilchez J A, Arnaldos J. 2012 A model for estimating the impact of the domino effect on accident frequencies in quantitative risk assessments of storage facilities *Process Safety and Environmental Protection* **91** pp 423-437
[13] Baker J W, Schubert M and Faber M H 2008 On the assessment of robustness *Structural Safety* **30** No.3 pp 253-267 doi:10.1016/j.strusafe.2006.11.004.
[14] Schubert M, Faber M 2008 On the modeling and analysis of robustness of systems. Inaugural *Int. Conference of the Engineering Mechanics Institute*
[15] Khakzad N, Reniers G 2015 Using graph theory to analyze the vulnerability of process plants in the context of cascading effects *Reliability Engineering and System Safety* **143** pp 63-73
[16] Jones R, Lehr W, Simecek-Beatty D, Reynolds R M 2013 ALOHA®(Areal Locations of Hazardous Atmospheres)5.4.4:Technical Documentation U. S. Dept. Of Commerce, NOAA Technical Memorandum NOS OR&R 43.Seattle, WA: Emergency Response Division, NOAA. 96 p
[17] The CAMEO Software Suite ALOHA® Example Scenarios National Oceanic and Atmospheric Administration Office of Response and Restoration Emergency Response Division Seattle, Washington. U.S. Environmental Protection Agency Office of Emergency Management Washington, D.C. 2016. 53 p.
[18] Bernechea E J, Vilchez J A, Arnaldos J 2012 A model for estimating the impact of the domino effect on accident frequencies in quantitative risk assessments of storage facilities *Process Safety and Environmental Protection* **91** pp 423-437

Influence of oil spills on tundra degradation

S I Chelombitko[1]

[1]Industrial University of Tyumen, Volodarskogo street 38, Tyumen, 625000, Russia

E-mail: sjlss@mail.ru

Abstract. Some aspects of the negative impact of oil spills on the tundra are considered. Oil pollution of the soil-plant layer leads to a significant increase in the absorption of radiant (solar) energy by the day surface, which leads to an increase in the power of the seasonally thawed layer. Swampy tundra conditions can lead to the development of thermokarst and subsequent degradation of the damaged area. Remediation of oil-contaminated areas in accordance with applicable regulations does not provide for the presence of permafrost. Planning and execution of the whole complex of works on remediation is not based on the simulation of heat and mass transfer processes for the period of complete restoration of the damaged area. The creation of vegetation cover of another species as a result of recultivation does not provide a similar absorption of solar energy by the day surface as in the natural (not polluted) territory, that can subsequently lead to the development of negative processes for the tundra. Taking into account the steady growth of oil-contaminated areas in the Northern regions and the severity of possible consequences, it is necessary to develop special regulatory documents for the remediation of oil-contaminated areas in the conditions of the cryolithozone.

Key words: oil spills, soil contamination, permafrost, reclamation, thermokarst, degradation.

1. Introduction

The main explored and developed hydrocarbon deposits of Russia are concentrated in its northern regions, which are characterized by harsh climatic conditions. The activities of oil and gas enterprises are complicated by low ambient temperatures, a long period of snow cover (more than six months), permafrost, as well as wetlands and a high moisture content in the soil. Despite the high reliability of the equipment used in the production process, the development of fields and the main transport accidental oil spills occur. Studies show that the areas of oil-polluted areas in the northern regions only increase every year and already exceed 1% of the total area of deposits [6]. As a result of long-term experience of development of oil and gas fields it is established that any technogenic influences are harmful for a northern country. So, for example, in [19] it is noted that "almost on any trace from the caterpillar there are ravines, sags and other violations of a relief. And ravines grow in the Subarctic region quickly 15-20 m/year... The Subarctic region nature as a peculiar resonator is capable to strengthen all actions: the pothole turns into the lake or a failure, a furrow - into a ravine". It has been established that self-recovery of damaged reindeer moss-the main feed of reindeer due to anthropogenic impact is possible only after a few decades [11]. Oil pollution of the soil-plant layer is a special type of technogenic impact, due to which soils acquire other physical characteristics that differ from the corresponding characteristics for natural (unpolluted) soil, sometimes very significantly[12,14]. It turns out that components of oil can move within all seasonal and thawed layer, being left during long time without essential changes, getting into underlying permafrost soil [23]. As a result of special laboratory researches, also, movement of oil in frozen samples of soil at a temperature of crystallization of steam moisture is established [25]. Various researchers note that pollution by oil of territories with permafrost breeds are pernicious, often causing irreparable injury to the nature because of specificity of course of processes of self-restoration of natural northern landscapes [3, 6,8]. The most essential damage to the environment is caused at accidents on the main oil pipelines because of which oil spreads on a day surface. In such cases easy fractions of hydrocarbons evaporate, and some part of the poured oil appears in a soil and vegetable layer. Components of oil are pernicious for any biological organisms. But the main negative impact on the nature of the tundra consists in development processes of degradation (a thermokarst, a thermoerosion, etc.) because of essential increase in absorption of solar energy the damaged day surface. It turns out that any technogenic effects on the natural vegetation cover (mechanical from the passage of technology, oil spills) affect the formation of a seasonally thawed layer in the cryolithozone due to increased absorption of solar energy by the day surface. The increase in thawing depth can cause the development of thermokarst and lead to irreversible degradation of the tundra.

2. The impact of remediation on the affected areas of tundra

In accordance with the decree of the Government of the Russian Federation organizations responsible for environmental pollution are obliged to carry out remediation in accordance with applicable regulations. As a rule, the complex of restoration work is carried out in two stages: technical and biological. At the 1st stage, the concentration of hydrocarbons in the surface layer is reduced to the set limit by plowing and harrowing, creating conditions for further decomposition (oxidation) of oil components in the soil. The works of the 2nd (biological) stage begin to be performed from the next vegetative season, while the application of various fertilizers is carried out with the subsequent sowing of herbs. Regional features of the contaminated area are taken into account when selecting the type of vegetation, which should subsequently form a continuous vegetation cover. Normative documents do not oblige to restore the natural vegetation cover in the damaged area, it is not required, also, the creation of soil and vegetation cover, which would have a similar "thermal function", as in the natural (unpolluted) territory. It is established that the damage (destruction) of vegetation or replacement with another type of vegetation for regions with frozen soils can lead to various negative phenomena (thermokarst, thermoerosion), and subsequently to the degradation site [9,14]. The creation of a continuous vegetation cover with other vegetation leads to changes in thermal and mass transfer processes in the soil. This will inevitably lead to a change in the thickness of the seasonally thawed layer. As you know, the tundra vegetation consisting of mosses and lichens, has specific heat and mass transfer properties, which means that any violations can lead to negative consequences. Therefore, for the actual reclamation of the territory in the tundra after oil pollution is not enough just to create a continuous vegetation cover. The choice of the type of vegetation and technology of work should be based on a long-term forecast of the processes of heat and moisture transfer in the damaged area. Various biological methods are used for remediation of damaged areas, allowing to start work on the creation of a new vegetation cover even in winter [2, 4]. All of them are designed to create a vegetation cover or using the sowing of herbs, or counting on its self-recovery in the future by stimulating the flow of microbiological processes in the surface layer. As the main indicator of the feasibility of using this or that method in the restoration of damaged areas are considered biological indicators of plants (their possible survival in these climatic conditions, the productivity of biomass and others), it does not take into account the possible development of subsequently negative processes in regions with permafrost. As a result of any work on remediation there is a change in the physical properties of the surface layer (density, albedo, heat capacity, thermal conductivity, etc.), which leads to a different course of heat and mass transfer processes in comparison with the natural (unpolluted) terrain. For areas that do not have permafrost, this is not a threat to the normal further functioning of the restored area. Observations at the experimental sites near Nadym (Tyumen region) showed, that the temperature of the day surface can differ by more than 10 degrees, and the depth of seasonal thawing - up to 120 cm depending on the type of vegetation [5]. In the process of remediation of the territory during the entire period of formation of sustainable vegetation cover (several years) will change the ability of the day surface to absorb solar energy. Therefore, the choice of technology for such work in areas with frozen soils should be based on a long-term forecast of the power dynamics of the seasonally thawed layer. Various biological methods of remediation are aimed at creating vegetation cover, rather than restoring the whole damaged area. As shown by observations, removal of vegetation or contamination of the day surface with oil in the Northern regions leads to an increase in the thickness of the seasonally thawed layer sometimes several times (depending on the physical properties and types of soils) [9]. Calculations of thawing depth for different types of soils for Ukhta region show [22] that the change only in the albedo of the day surface due to oil spills leads to an increase in thawing depth up to two times, and the combined effect of complete removal of vegetation cover and oil pollution leads to a corresponding increase of up to two meters and more for 5 years.

To clean the territories from oil pollution, other simplified methods of "restoration" are also used: burning the remaining oil together with damaged vegetation (prohibited), filling the site with a layer of imported soil (peat, sand), digging, harrowing, sowing herbs and others. These methods of restoration of territories can accelerate the flow of chemical and biological processes for the purification and restoration of soil. In the process of performing any such work is an external impact on the surface layer (burning the remaining oil together with contaminated vegetation, backfilling layer of imported soil, digging, harrowing, sowing herbs, etc.), leading to a change in the physical, mechanical and thermal properties of the treated layer. As a result of such measures, the functioning of the disturbed layer will differ from that which takes place in the intact nearby territory: a change in the physical and mechanical properties of the soil will lead to another mass transfer (moisture migration), and a change in its thermal and physical properties to another heat exchange. As a consequence, the depth of thawing during the summer period in the restored area will differ from the "due", namely in a big way due to the lack of grass cover for several years after recultivation (during the first two years, the day surface has virtually no vegetation). In General, currently used various methods of remediation of areas contaminated with oil spills, do not take into account the specific features of the terrain and lead to changes in the thermal properties of the surface layer. The absence of any recommendations to take into account the climatic and other features of the terrain (albedo, mechanical and physical properties of soil and soil) when performing such works in the Northern regions can lead to an increase in the capacity of the seasonally thawed layer in the years following the restoration. As you know, in such cases, under certain conditions, can begin destructive processes (increased humidity of the active layer, frost heaving, cracking, thermokarst), leading to the degradation of the tundra. The use of any of the known methods of remediation of oil-contaminated areas with permafrost does not guarantee a valid recovery, and can lead in the future to irreversible consequences for the damaged area.

3. Assessment of the consequences for not re-cultivated territories of the tundra

There are cases of the existence of contaminated oil territories without reclamation from various publications. The day surface polluted with oil has a much greater ability to absorb solar energy than the natural one, therefore, as a result of oil spill, the heat supply to the surface layer will increase significantly. It is known that in the cryolithozone the lower part of the seasonally thawed layer is characterized by high humidity. With the increase in solar energy absorption, the melting of frozen soil will be more intense due to the greater thermal conductivity of the wetter soil near the interface: thawed - frozen soil. This will lead to an increase in the power of the seasonally thawed layer, an even greater increase in the humidity of the thawed soil (due to the thawing of underground ice), and in winter-to the development of frost heaving, the formation of cracks in the moisture-saturated frozen soil and other destructive processes. Thus, non-restored oil-contaminated areas are subject to more intensive degradation in the conditions of permafrost soils, manifested in the subsidence of the surface, water filling of the resulting depressions, and subsequently to their further destruction. Examples of self-healing of oil-contaminated areas in the Northern regions of Western Siberia are not found in any publications.

4. About a possible prediction of degradation of the tundra because of its pollution

Mathematical modeling of heat and moisture transfer in the surface layer of the damaged area (as a result of oil spill, after restoration work) can be the basis for the forecast of possible degradation. The depth of the seasonally thawed layer is one of the most important factors affecting the development of thermokarst. The calculation of the thawing depth and the formation of the seasonally thawed layer for several years can be performed using the known numerical models of the Stefan problem (calculation of the freezing-thawing depth in the presence of permafrost) [18, 19, 22]. Estimation calculations to determine the thickness of the seasonally thawed layer for oil-contaminated areas can be performed, knowing the thickness and properties of the layer containing oil. These values can be determined experimentally in the laboratory by sampling and analysis of contaminated soil. Oil trunk pipeline ruptures lead to large-scale oil spills when sampling contaminated soil becomes a serious problem. In such cases, the depth of oil infiltration can be determined by calculation on the basis of the known Darcy filtration law of the following type [20]:

$$m\frac{dy}{dt} = k\frac{h + h_k + y}{y},$$

where: y – the depth of infiltration of oil; m – free soil porosity; t – time; k – filtration coefficient; h – the thickness of the layer of oil on the surface; h_k – the height of the capillary rise of oil.

It is known that most of the oils of Western Siberia contain paraffin, which at a temperature close to the ambient temperature, is in the oil in the form of solid fractions. It was established experimentally that in the process of oil filtration the pore channels [24], as a result of which the permeability of the porous medium decreases. This phenomenon can significantly affect the depth of oil infiltration depending on the content of solid paraffin fractions. To be able to account for changes in the permeability of the porous medium in the filtration of such oil, the corresponding dependence of [17] was used, as well as the experimental data from [24] were processed. The solution of the original differential equation (Darcy's law) taking into account the variable filtration coefficient is obtained in the following implicit form:

$$t = \frac{t_0}{1 - A\dfrac{us^2}{2D}t},$$

where:

$$t_0 = \frac{m}{k}\left[y - (h + h_k)\ln\left(1 + \frac{y}{h + h_k}\right)\right],$$

$$k = k_0 /(1 + A\frac{us^2}{2D}t),$$

k_0 – the permeability of contaminated soil; u – the initial rate of oil flow; s – the volume content of the dispersed phase in paraffin oil; D – the effective diameter of the pore channels of the soil; A – empirical coefficient (A = 0,003–0,018s).

5. Conclusion

When recultivating oil-contaminated areas in the tundra, it is necessary to restore the damaged area, and not the soil and vegetation cover or soil layer to prevent further possible degradation of the territory.

The increase in the area of oil-contaminated areas in the Northern regions of Russia indicates the need to tighten penalties for damage caused to nature.

To restore areas with permafrost after oil pollution, it is necessary to use specially created methods of reclamation, developed on the basis of mathematical modeling of heat and mass transfer processes. This would allow to take into account the specific features of the area when planning the necessary work, to prevent the subsequent degradation of the damaged area.

One of the methods of numerical simulation of the formation of the seasonally thawed layer can be used to predict the possible course of heat and mass transfer processes after recultivation of the territory. This would make it possible to justify the physical, technical and thermal parameters of the created layer, so that later the power of the seasonally melted layer in the restored territory would be close to the power of the seasonally melted layer in the non-polluted territory.

Areas contaminated with oil or petroleum products under cryolithozone conditions should be under special control of environmental services until they are truly fully restored.

References

[1] Kondratyev K Y 1981 *The albedo and angular characteristics of reflectance of the underlying surface and clouds* (Leningrad: Gidrometeoizdat) 231 p

[2] Balakirev I V and Nikishova A S 2012 Cleaning and restoration of disturbed soil and vegetation cover by biotechnological methods in the conditions of cryolithozone *Tenth international conference on permafrost study* **5** (Salekhard) pp 23-28

[3] Bobov V I, Gashev S N, Kazantseva M N and Panichev E A 1998 *Experience of ground survey and certification of oil-contaminated lands Forests and forestry of Western* Siberia (Tyumen: Publishing house of TSU) pp 172-178

[4] Dubrovin A V and Bidyn S A Patent RUS 2188529 A01 B79/02 Method of soil remediation contaminated with oil and oil-containing products in the North of the Russian Federation in the areas of oil fields and oil trunk pipelines No 2001119291/13 stated 12 07 01 published 10 09 02

[5] Kazantseva L A 2007 Influence of vegetation on temperature and depth of seasonal thawing of soil in the Northern taiga of Western Siberia *Proceedings of International conference "Cryogenic resources of the polar regions"* **2** (Salekhard) pp 87-89

[6] Ganeev I G and Kulagin A A 2009 Remediation and recultivation of technogenic degraded soils *Vestnik OGU* №6(100) pp 253-256

[7] Desyatkin R V, Desyatkin A R and Fedorov P P 2012 The soil temperature regime of the Central Yakutia *Cryosphere Of the Earth 16* № 2 pp 70-78

[8] Egelev Z S, Umarova A B, Goncharuk N Y, Zavgorodnyaya J A and Jeleva A S 2013 Properties remediated after oil pollution in tundra soils in the North of the European part of Russia *Nonparametric robust regression estimate Proc SPIE 13th Joint Int Symp on Atmospheric and Ocean Optics* **6522** (Tomsk) pp 130-139

[9] Ershov E D, NefedievaY A, Motenko R G and Parmuzin S Y 2007 Estimated forecast of the change in depth of seasonal thawing and freezing under the effect of oil pollution *Proceedings of International conference "Cryogenic resources of the polar regions"* **2** pp 217-220

[10] Kaverin D A, Pastukhov A V and Mazhitova G G 2012 On the current state of the soil-permafrost complex in the southern tundra (Syktyvkar) **2** (**10**) pp 51-56

[11] Kornienko S G and Yakubson K I 2011 Study of the transformation of vegetation in areas of the Tazovsky Peninsula based on satellite imagery *Arctic: ecology and economy* No 4 pp 46-57

[12] Nefedieva Y A, Motenko R G and Zykov Y D 2008 The Role of transformation of oil pollution in the formation of acoustic electric and thermal properties of frozen soils *Cryosphere of the Earth* No 4 pp 36-42

[13] Pavlov A V 1979 *Thermal physics of lands capes* (Novosibirsk: Science) p 285

[14] Trofimov S Y and Rozanova M S 2002 *Soil properties. Change under the influence of oil pollution* (Moscow: Publishing house of Moscow state University) pp 359-373

[15] Tartikov A P 1969 *Influence of vegetation cover on freezing and thawing soils* (Moscow: Publishing house of Moscow state University) p 192
[16] Shur Y P 1988 The Upper horizon of permafrost and thermokarst (Novosibirsk: Science) 213 p
[17] Shechtman Y M 1961 *Filtration of low concentration suspensions* (Moscow: USSR Academy of Sciences) 211 p
[18] *Methods for assessing the effects of climate change on physical and biological systems* 2012 (Moscow: Roshydromet) 511 p
[19] Grechishev S E, Costatino L V and Shur Y L 1980 *Cryogenic physical and geological processes and their forecast* (Moscow: Nedra) 383 p
[20] Polubarinova-Kochina P Y 1977 *Theory of ground water movement* (Moscow: Science) 664 p
[21] Vasiliev I S and Gerasimov E 2001 Variability of active layer thickness GAME study area Central Yakutia *Proceedings The Fifth International Study Conference jn GEWEX in ASIA and GAME* (Nagoya, Japan) **2** pp 440-444
[22] Mazhitova G, Malkova G ,Chestnyykh O et al 2004 Active layer spatial and temporal variability at European Russian Circumpolar Active-Layer-Monitoring (CALM) sites Permafrost and periglacial processes **15** pp123-139
[23] Biggar K W, Haider S, Nahir V and Jarrett P M 1998 Site investigations of fuel spill migration into permafrost *Journal of Cold Regions Engineering* **12** No 2 pp 88-104
[24] Kundin S, Allahverdiyev R, Borisov M, Izabakarov P and Rosenberg M 1973 *Experimental study of paraffin filtration at temperatures below the temperature of its saturation with paraffin Research in the field of oil field development and reservoir hydrodynamics* (Moscow: Nedra) vol 45 pp 40-52
[25] Stewart R S 1997 Distribution of multiple oil tolerant and oil degrading bacteria around a site of natural crude oil seepage *Tex J Sci* №4 pp 47-52

IOP Conf. Series: Materials Science and Engineering **481** (2019) 012012 doi:10.1088/1757-899X/481/1/012012

Statistical modelling of impacts on a pile-raft foundation

V O Efimov[1], A I Osokin[2], L N Kondrat'eva [3], N R Stepanova[4]

[1]Geotechnical department, Saint Petersburg State University of Architecture and Civil Engineering, 2-ya Krasnoarmeyskaya, 4, Saint Petersburg, 190005, Russia
[2]Geotechnical department, Saint Petersburg State University of Architecture and Civil Engineering, 2-ya Krasnoarmeyskaya, 4, Saint Petersburg, 190005, Russia
[3]Geotechnical department, Saint Petersburg State University of Architecture and Civil Engineering, 2-ya Krasnoarmeyskaya, 4, Saint Petersburg, 190005, Russia
[4]Department of Economy and management in Construction and Realty Market, Ural Federal University, Mira, 19 Ekaterinburg, 620002 Russia

E-mail: efimovvo@yandex.ru, a.osokin@geostroy.ru, kondratjevaln@yandex.ru, n.r.stepanova66@gmail.com

Abstract. This paper reviews performance of combined pile-raft foundations (CPRF). The goal of the study is obtaining the ability to design solid, reliable and cost-effective foundations allowing for minimum settlements. The task is to develop methods for calculating CPRF for optimum choice of the foundation raft size, dimensions and number of piles. To achieve these results, an analytical method was used, curves of the ratio of CPRF settlement to raft foundation settlement vs. total load *N* on the piles were plotted. The scientific significance of the study is in the method of engineering calculation of CPRF, which is better correlated to the experimental data and allows optimal distribution of loads between the raft and piles, ensuring reliability and operational integrity of buildings, reduction of settlements and the number of piles, which would lead to cost savings and reduced consumption of materials for foundations. A new coefficient showing the residual stress has been introduced, and a method for its calculation has been proposed. A new equation for determination of the share of total load *N* on the piles has been obtained. This equation is more general in comparison with the previously generated equations. It allows excluding the assumption that the stresses arising under the raft completely dissipate at the level of the pile end. Application of the developed methods will increase reliability and reduce settlements and the number of piles, which ultimately will lead to cost savings and reduced consumption of materials for foundations.

1. Introduction

Combined pile-raft foundation (CPRF) is a structure comprising a reinforced-concrete raft foundation and piles. The raft can be either constant or variable in thickness, and the piles can be either constant or variable in length, diameter and spacing [1, 2]. Methods of connection between the piles and the raft can also be different, e.g. rigid or hinged connection or one-sided contact support [3, 4, 5].

The structure is called combined, since the work of its elements is mutually complementary to each other through the foundation soil medium. This is manifested in partial transfer of the building weight at the upper level (under the raft bottom), allowing for increase in friction forces in the upper part of the piles [6], which, in its turn, leads to increase in the maximum load on a pile, while the raft allows the foundation remaining reliable even after the "failure" of a pile. This agrees with the emergence principle of the systems theory, stating that it is typical of a system to acquire properties of constituent components, as well as new properties.

The grounds stated in the article [7] are based on two basic provisions for comparison of pile foundations (PF) and raft foundations (RF):

- pile foundations are more expensive than raft foundations;
- pile foundations are more rigid.

Combined pile-raft foundation (CPRF) is assumed to be an intermediate option between PF and RF in terms of settlements, rigidity and costs.

Regulatory documents adopted in Russia, containing basic rules for design of bases and foundations [8, 9, 5, 10] do not consider the possibility of using piles in the breakdown zone, i.e. after their "failure". For this reason, the existing documents allow for transfer of no more than 15% of the total weight of the building via the raft to the base [8, 10], which rather corresponds to the classic pile foundation according to the classification [11], which properly accounts for the actual deformability of friction piles and the ability of the raft foundation to take up some load. The EU regulatory document on regulation of geotechnical design [12] classifies combined pile-raft foundations as foundations of Category 3. This means that standard provisions stated in this Eurocode are not applicable to them, and the design procedure should include alternative and specially developed rules and measures.

2. Methodology

The task is to develop methods for calculating CPRF for optimum choice of the foundation raft sizes, dimensions and number of piles, so that settlements would not exceed the ultimate values. The most cost effective and safe number of piles is the number which would provide settlements under the ultimate values. Load and settlement values of PF and RF have been determined for high-rise buildings under load *N*, with the assumption that the first settlement s_{pf} is below and the second settlement s_{rf} is above the ultimate value for the building. Thus, three foundation models are considered – pile (PF), raft (RF) and combined pile-raft (CPRF) – under the same conditions and the same static load *N*.

The following assumptions were introduced in the work [7]:

1. Bending and torque moments and transverse forces acting on foundation structures are considered negligible.
2. Settlement of piles under load is bilinear, i.e. the value of settlement is directly proportional to the corresponding load before achievement of the design load-bearing capacity; subsequent deformations develop without changes in internal forces of piles.
3. The raft foundation does not resists unless some ultimate load is achieved; all loads from the structure on the base are distributed among the piles. Developing settlement involves overlying soil layers (including the near-pile soil body and the structure as a whole).
4. When the ultimate load-bearing capacity of the piles is achieved, the piles slip along the shaft in relation to soil, and raft foundation resistance appears. It also takes up subsequent additional loads.
5. Deformability of the pile shaft is considered negligible.

Apart from clearly stated assumptions, there is a number of issues that are implicitly omitted: reliability issues (reserve factors), variability in characteristics of construction materials, soils and loads. This is due to the fact that the proposed method [7] does not imply final calculations for such a complex system, but it makes it possible to estimate the required number of piles in the first approximation with the help of a simple engineering calculation.

Analytical calculations performed for three models, taking into account the above assumptions, allowed obtaining an equation of the optimum share of loads from the structure, distributed by piles in CPRF, depending on the total load from the structure, in the work [7]:

$$\alpha_{cprf} = \frac{s_{rf} - s_u}{s_{rf} - s_{pf}} = \frac{\delta_{rf} - 1}{\delta_{rf} - \delta_{pf}}, \tag{1}$$

where α_{cprf} is the share of total load N on piles; s_u is ultimate settlement; s_{pf}, s_{rf} are total settlements of pile and raft foundations, respectively; δ_{pf}, δ_{rf} are relative settlements of pile and raft foundations.

Equation (1) was obtained after mathematical transformation, during which total settlement of pile and raft foundations s_{pf} and s_{rf} were divided by the ultimate settlement s_u, and relative settlements of pile and raft foundations were obtained. This conclusion implied an implicit assumption that the vertical stresses due to pressure transmitted along the raft bottom dissipated at the level of the pile bottom end. This assumption can be valid if the length of piles is much larger than the size of the foundation in plan, and the vertical stresses are significantly reduced.

It would be reasonable to assume that there is an opposite extreme assumption that the additional vertical stresses under the raft and at the level of the pile ends are equal. This assumption most closely describes the case when the length of piles is significant in comparison with the characteristic size of the raft foundation. As is the case in [7], taking into account the opposite assumption, we obtain the following:

$$k_{pf} = \frac{N}{s_{pf}}, \tag{2}$$

where k_{pf} is rigidity of pile foundation; N is the load on CPRF,

$$k_{rf} = \frac{N}{s_{rf}}, \tag{3}$$

where k_{rf} is rigidity of raft foundation.

The total load on CPRF can be determined when summing up the loads transmitted onto the piles and the raft:

$$N = N_{pf} + N_{rf},$$

where N_{pf} is the load applied to the piles; N_{pf} is the load applied to the raft.

The load N_{pf} is determined as follows:

$$N_{pf} = \alpha_{cprf} N \tag{4}$$

This being the case, N_{rf} is determined as follows:

$$N_{rf} = (1 - \alpha_{cprf})N \tag{5}$$

If the additional vertical pressure due to the load at the level of the pile bottom ends does not decrease due to pressure transmitted along the raft bottom, then the soil under the piles is loaded with the sum of forces N_{pf} and N_{rf}. Thus, the pile component of settlement $s_{cprf,1}$ (due to soil shrinkage below the pile ends) is equal to:

$$s_{cprf,1} = \frac{N_{pf} + N_{rf}}{k_{pf}} = s_{pf} \tag{6}$$

The second component of settlement (due to soil shrinkage under the raft, following the "failure" of piles) is equal to:

$$s_{cprf,2} = \frac{(1-\alpha_{cprf})N}{k_{rf}} = (1-\alpha_{cprf})s_{rf} \quad (7)$$

The total settlement of CPRF equals to the sum of the following components:

$$s_{cprf} = s_{pf} + (1-\alpha_{cprf})s_{rf} \quad (8)$$

If we assume that CPRF settlement equals to the ultimate settlement s_u, then α_{cprf} can be calculated from equation (8):

$$\alpha_{cprf} = \frac{s_{rf} + s_{pf} - s_u}{s_{rf}} = 1 - \frac{s_u - s_{pf}}{s_{rf}} = 1 - \frac{1-\delta_{pf}}{\delta_{rf}}. \quad (9)$$

Equation (9) allows determination of the load share taken up by the piles.

3. Results and Conclusion

In the work [2], materials on systematization of the monitoring results with regard to various objects constructed with CPRF are presented. Regularity of changes in the ratio of CPRF settlement to RF settlement $\frac{s_{cprf}}{s_{rf}} = \frac{\delta_{cprf}}{\delta_{rf}} = \frac{1}{\delta_{rf}}$ (assuming that design CPRF settlement is equal to ultimate settlement, i.e. $s_{cprf} = s_u$ and $\delta_{cprf} = \frac{s_{cprf}}{s_u} = 1$), depending on α_{cprf}, was determined in the course of the study. The values α_{cprf} obtained according to equation (1) in the work [7] were compared to the obtained results, accepting the value $0{,}18 \le \frac{s_{pf}}{s_{rf}} = \frac{\delta_{pf}}{\delta_{rf}} \le 0{,}47$ from experience (this ratio depends on physical and mechanical characteristics of soil layers composing the base, and their strata as well). This comparison is shown in Figure 1. The results obtained according to equation (9) are presented in Figure 2.

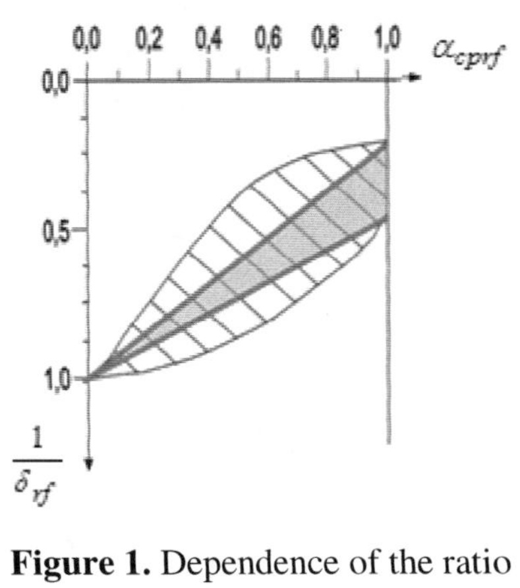

Figure 1. Dependence of the ratio of CPRF settlement to RF settlement vs. α_{cprf} calculated according to equation (1).

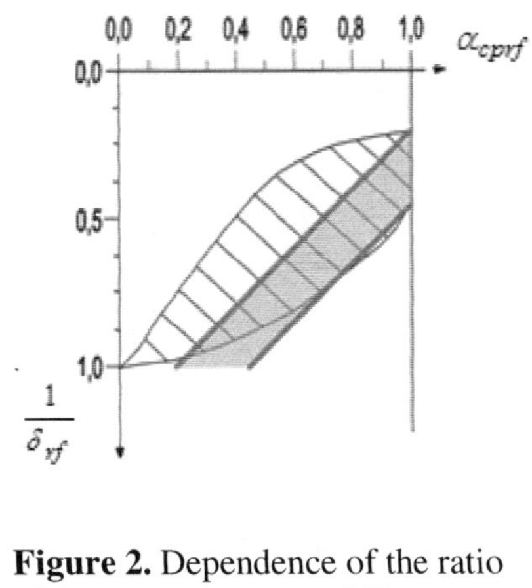

Figure 2. Dependence of the ratio of CPRF settlement to RF settlement vs. α_{cprf} calculated according to equation (9).

It can be seen in Figures 1 and 2 that two extreme hypotheses provide for distributions close to the limits of actual measurements, which is consistent with the assumption that actual foundation behavior lies between two abstract assumptions: the vertical stresses from the raft completely dissipate or do not dissipate at all between the levels of the raft bottom and pile bottom ends.

In order to account for a certain actual value of decrease in the vertical stress at the section from the raft bottom to the pile bottom ends, it is suggested to introduce value α representing the ratio of additional vertical stress in the soil body, occurred due to pressure below the raft at the level of the pile bottom ends, to vertical stress which is directly below the raft. In other words, α is the share of undissipated pressure, the value of which is known (for example, its tabulated values are given in Table 5.8 [9], depending on the ratio of the raft sides and the pile length as the depth parameter). In this case, equation (6) takes the following form:

$$s_{cprf,1} = \frac{\alpha_{cprf} N + \alpha(1-\alpha_{cprf})N}{k_{pf}} = \alpha_{cprf} s_{pf} + \alpha(1-\alpha_{cprf}) s_{pf} . \quad (10)$$

Considering the residual stress α $(0 < \alpha < 1)$, we will obtain α_{cprf} in the following form:

$$\alpha_{cprf} = \frac{\delta_{rf} + \alpha\delta_{pf} - 1}{\delta_{rf} - \delta_{pf}(1-\alpha)} . \quad (11)$$

It should be noted that under limit values $\alpha = 0$ and $\alpha = 1$, equation (11) turns into equations (1) and (9), respectively, confirming once again that the mathematical calculations are correct.

Figure 3 shows distribution at the arbitrary value $\alpha = 0.5$ (continuous line), and $\alpha = 0$ and $\alpha = 1$ represented for comparison (dashed lines). Analyzing the results obtained, we can conclude that equation (11) represents a more general expression, which coincides practically with the whole area of distribution drawn on the basis of observations and monitoring [1].

Figure 4 shows the area of possible location of all points if the value α constantly changes from 0 to 1, reflecting, at each α, the area of possible location of points in the graph in the same axes.

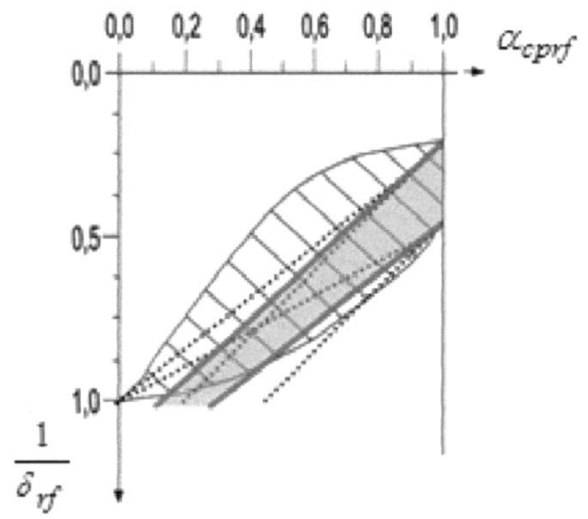

Figure 3. Dependence of the ratio of CPRF settlement to RF settlement vs. α_{cprf} calculated according to equation (11) at the arbitrary value $\alpha = 0.5$.

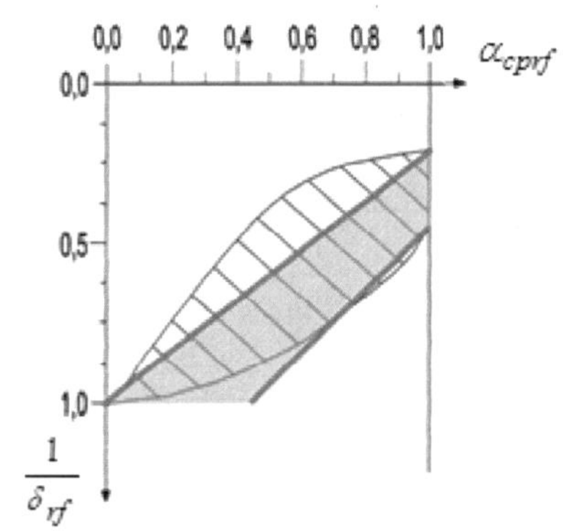

Figure 4. Dependence of the ratio of CPRF settlement to RF settlement vs. α_{cprf} calculated according to equation (11) under possible values of $0 < \alpha < 1$.

It should be noted that although only purely raft and pile foundations correspond to the limit values $\alpha_{cprf} = 0$ and $\alpha_{cprf} = 1$, and there is matching of distribution zones according to the proposed method and those obtained in the result of monitoring, significant deviation in the middle zone of the experimental distribution towards smaller actual settlements of CPRF is observed. This can be explained by the fact that internal interaction in the pile–raft–soil system is the most pronounced in the middle zone ($0.3 < \alpha_{cprf} < 0.7$), and behavior of this system has been studied to a less degree than behavior of separate components (raft or piles).

This fact implies that the proposed method for determining the optimum number of piles is conservative, that is, it can only understate the predicted settlement, which is consistent with the principles of reliability. Nevertheless, this also shows a lack of knowledge on properties of CPRF as a system with effective interaction of its elements.

The method of engineering calculation of CPRF has been proposed, which is better correlated to the experimental data.

The new equation for determination of α_{cprf} — the share of total load N on the piles — has been obtained. Proposed equation (11) is more general in comparison with equations (1) and (9), and excludes the assumption that the stresses arising under the raft completely dissipate at the level of the pile bottom ends.

The new coefficient α has been introduced, showing the residual stress, and the method for its calculation has been proposed.

References

[1] Badelow F and Poulos H 2016 Geotechnical foundation design for some of the world's tallest buildings (The 15th Asian Reg. Conf. on Soil Mechanics and Geotechnical Engineering) pp 96–108 https://www.researchgate.net/publication/292211606_Geotechnical_foundation_design_for_some_of_the_world's_tallest_buildings, accessed date August 12 2018

[2] Katzenbach R and Schmit A 2014 High-rise buildings in Germany soil-structure interaction of deep foundations (Int. Conf. on Case Histories in Geotechnical Engineering) pp 1–8 http://scholarsmine.mst.edu/icchge/5icchge/session01/23, accessed date: December 20 2017

[3] Penkovskii G F, Sakharov I I and Ershov A V 2008 Sposob vozvedeniya plitno-svainogo fundamenta [The method of construction of pile-raft foundation] Russian patent 2305154

[4] Samorodov A 2016 New structure of pile-raft foundation *Visnik Pridniprovs'koï derzhavnoï akademiï budivnictva ta arhitekturi* **1 (214)** pp 58–65

[5] SP 24.13330.2011 Svod Pravil. Svainye fundamenty. Aktualizirovannaya redakciya SNiP 2.02.03-85 [Russian Set of Rules SP 24.13330.2011. Pile foundations. Updated edition of Building Codes (SNiP) 2.02.03-85] 2011

[6] El-Mossalami Y M, Lutz B, Duerrwang R 2013 Special aspects related to the behaviour of piled raft foundations *International Journal Geotechnics* **2** pp 56-64

[7] Fadeev A B, Mangushev R A and Lukin V A 2007 Raschet plitno-svainogo fundamenta [Calculation of piled raft foundation] *Vestnik grazhdanskikh inzhenerov* **2 (11)** pp 64–67

[8] MGSN 2.07-01 Moskovskie gorodskie stroitelnye normy. Osnovaniya, fundamenty i podzemnye sooruzheniya [Moskow city building codes MGSN 2.07-01. Basements, foundations and underground structures] 2003

[9] SP 22.13330.2016. Svod Pravil. Osnovaniya zdanii i sooruzhenii. Aktualizirovannaya redakciya SNiP 2.02.01-83* [Russian Set of Rules SP 22.13330.2016. Foundations of buildings and structures. Updated edition of Building Codes (SNiP) 2.02.01-83*] 2016

[10] SP 50-102-2003 Svod Pravil. Proektirovanie i ustroistvo svainykh fundamentov [Russian Set of Rules SP 50-102-2003. Design and construction of pile foundations] 2003

[11] Poulos H 2001 Methods of analysis of piled raft foundations (Int. Society of Soil Mechanics and Geotechnical Engineering **51 (2)**, p 46 http://geo-soft.ru/userfiles/File/files/94_Pile_raft_foundation.pdf, accessed date July 5, 2016

[12] SNV SN EN 1997-1-2004 Eurocode 7 - Geotechnical design - Part 1: General rules

IOP Conf. Series: Materials Science and Engineering **481** (2019) 012013 doi:10.1088/1757-899X/481/1/012013

Structural safety of buildings in excess values of differential settlements

Y A Pronozin[1], L R Epifantseva[2], M D Kajgorodov[1]

[1]Department of Geotechnics, Tyumen industrial university, Tyumen, Russia
[2]Department of Building construction, Tyumen industrial university, Tyumen, Russia

E-mail: epifantsevalr@mail.ru

Abstract. The paper discusses the strength of load-bearing structures in a frameless building in case of excess differential settlements. A numerical simulation of the structural performance of load-bearing and enclosure structures has been made for the tilted block of flats erected in the city of Tyumen at the time of the maximum deviation of its skeleton from the vertical.

1. Introduction

In recent years, multi-storey housing is extensively developing in the Russian Federation and it is often necessary to erect foundations in complex engineering-geological conditions. For this reason, errors occur when choosing the type of foundation and excess absolute and differential settlements appear.

Taking into account geotechnical safety, it is most risky to erect multi-storey buildings on shallow foundations, slab foundations among them, with the beds composed of silt-clay soils. As practice shows, in buildings erected on such soils the actual settlements are often higher than the calculated values [2, 3 and 19]. The settlement itself is not as problematic as its variations which result in the deviation of the building skeleton from the vertical [1, 19 and 20].

The junctions of structural members are the most vulnerable zones in buildings subjected to the differential settlement: columns, party walls of lower floors from the side of extra loading with an additional moment of the tilt; joints of panels, walls, coverings, floor structures, junctions of walls and floor structures [7, 8]. Due to poor-quality erecting works and differential settlements of foundations, deviations from the vertical and misalignments of flat areas can cause redistribution of stresses and result in structural failures [19].

The tilted block of flats located in Belyaeva Street in the city of Tyumen is an example of a significant tilt which has provided a basis for studying the stress-strain state of its superstructure. The building constructed in 2014- 2015 consists of two blocks – 9-storey and 14-storey blocks respectively [12].

The slab foundations of both blocks are made of mass concrete B30, 1.0 m in thickness. The outer walls of the substructure are made of FBS panels, 600 mm in thickness, and the inner ones - 500 mm. The brick outer walls of the superstructure are insulated - 820 mm in thickness; the inner walls are 510 mm in thickness. The floor structures are prefabricated and made of reinforced concrete hollow core slabs.

The average pressure under the foundation slab footing of the 14-storey block with the area of S = 914.36 m^2 is p^s = 257 kPa from the specified characteristic load; p^d = 285 kPa - from the design load.

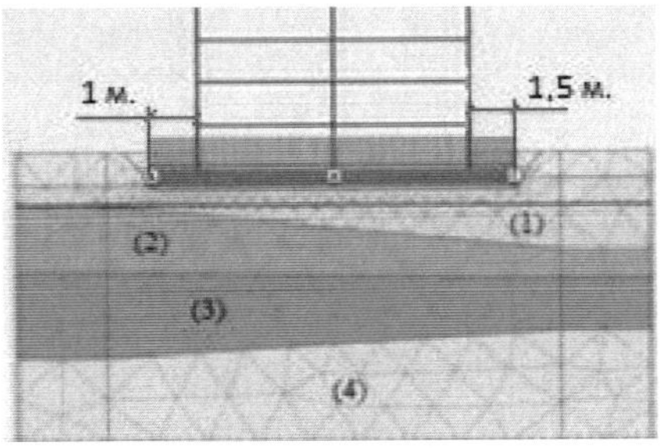

Figure 1. Soil formations.

The first geological surveys were carried out in 2012, and then control surveys were carried out during building site works in 2014. The comparison data of the results obtained after engineering and geological surveys are given in Table 1. Excavated soil and construction debris occur on the site up to 2.5 meters in thickness, and then soil formations are in accordance with Table 1. The depth of the constant groundwater level is 4.4-5.3 m from the ground surface.

Table 1. Comparison data of physical-mechanical properties of soils: *A* – engineering-geological surveys in 2012; *B* – control engineering-geological surveys in 2014.

№ of a layer	Layer thickness, m	Soils	*E*, MPa		γ, kN/m³		*c*, kPa		φ, degrees	
	A = *B*	*A* = *B*	*A*	*B*	*A*	*B*	*A*	*B*	*A*	*B*
1	2.1	High-plastic clay loams	13.7	8.4	19.1	19.5	22	18	21	12
2	4.7	Fluidal and fluidal-plastic clay loams	6.3	2.7	18.8	17.8	15	10	14	10
3	4.6	Alternation of fine sands with sand loams and clay loams	20.6	20.6	20.1	19.6	9	9	27	27
4	11.5	Fine sandy silt with layers of sand loams and clay loams	29	29	20.6	20.6	3	3	31	31

During building site works, both blocks suffered from excess differential settlements which developed rapidly. In here, cracks or failures of the foundation and walls were not observed. Hair-like vertical cracks were found only in the building basement, but deformations of the foundation bed were not caused by them.

Figure 2. Inclination of the building in completion.

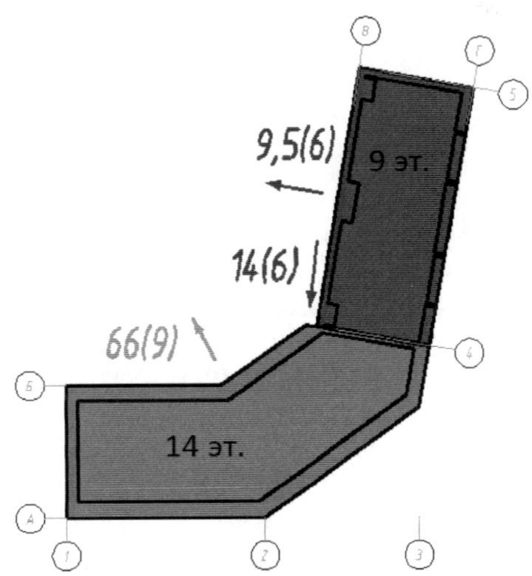

Figure 3. Horizontal deviation of the blocks from the vertical in (cm); the maximum permissible values are given in brackets in (cm).

IOP Conf. Series: Materials Science and Engineering **481** (2019) 012013 doi:10.1088/1757-899X/481/1/012013

The geodetic survey showed that the facade of the 14-storey block (Figure 2) deviated from the vertical in 660 mm at the maximum value - 9 cm after Building Regulations [16] and 11.5 cm after Building Regulations [17]. The relative difference in settlements reached $\Delta s/L = 0.019 > [\Delta s/L] = 0.0024$ – Building Regulations [17] (Table D1[4]). The 9-storeyed block suffered from a dramatic tilt in two directions (Figure 2).

The differential settlements were caused by the following factors: displacement of the center of gravity of the superstructure towards the inner angle of the building in plan (Figure 3), non-uniform soil formations with thinning of the layers (Figure 3, Table 1), and excavation works done partly in place of the former vegetable storehouse.

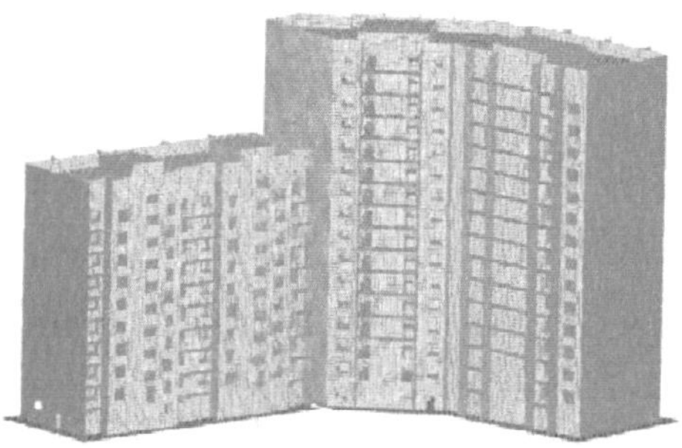

Figure 4. General view of the building design diagram in Lira SOFT 10.6.

Figure 5. Schematic illustration of the deformed model in Lira SOFT 10.6.

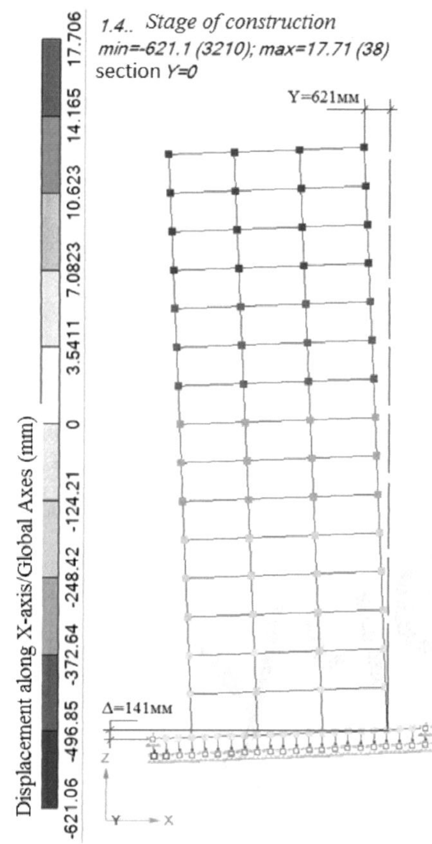

Figure 6. Horizontal displacements along *Y*- axis in Lira SOFT 10.6.

These factors were not taken into account at the design stage, and thus, it resulted in the differential settlement and the building tilt respectively.

In such cases, when the differential settlements are significant and exceed the recommended limit values dramatically, structural safety must dominate. To identify the reserve bearing capacity of the superstructure, the following issues were considered:

- extra loading of brick party walls of the ground floor with increase of the tilt;
- appearance of shearing forces in the junctions of floors structures and walls due to tensile forces in the horizontal disks of the building frame with increase of the tilt.

A numerical simulation of the building was created in order to reveal the regularities in the changed stress-strain state of the main load-bearing structures in the excess permissible tilt. The initial data were taken in accordance with the project, the design diagram was compiled in the Lira SOFT 10.6

software package (Figure 4), and the Mohr-Coulomb model of strength was used to simulate the foundation bed. Further, the results of calculation are given only for the 14-storey block. Based on the results of numerical simulation, the average settlement of the foundation was 47.3 cm $[S_u]$ = 10 cm, Building Regulations [17] (Figure 5). The relative difference in the settlements of the foundation extreme points was $\Delta S/L = 0.019 > [\Delta s/L] = 0.0024$, Building Regulations [17].

The maximum horizontal deviation of the building frame from the vertical along the B-axis was also obtained in the course of the numerical calculation of the 14-storey block. It was 621 mm (Figure 6).

Thus, when analyzing such parameters as maximum settlement, relative variations of settlements and horizontal displacement, the difference from the real deformations obtained from the geotechnical monitoring at the site is not above 6%.

2. Assessment of the stress-strain state of the most loaded brick party wall

In accordance with the Building Regulations [15], the ultimate load-bearing capacity of the 14-storey building brick walls in an upright position is $[N]$ = 3004 kN with regard to operational loads. It should be noted that when the building deviates from the vertical, eccentricity e is added to the real loads N. This results in the extra moment and shear force which appear in the base of the walls.

Based on the results of numerical simulation and design calculations, the reserve load-bearing capacity of the most loaded party wall of the 14-storey block is 27% in the vertical position, with $N^{i=0}$ = 2177 kN. Taking into account the variations of settlements $\Delta S/L$ = 0.019, the reserve load-bearing capacity of the most loaded party wall decreases from 27% to 5% ($N^{i=0.019}$ = 2848 kN).

To assess structural safety, the building model was given different values of the foundation bed performance to successively increase the difference in settlements. This technique has made it possible to analyze the influence of non-uniform deformations of the foundation bed on the overall stability of the system and assess the structural safety of the building.

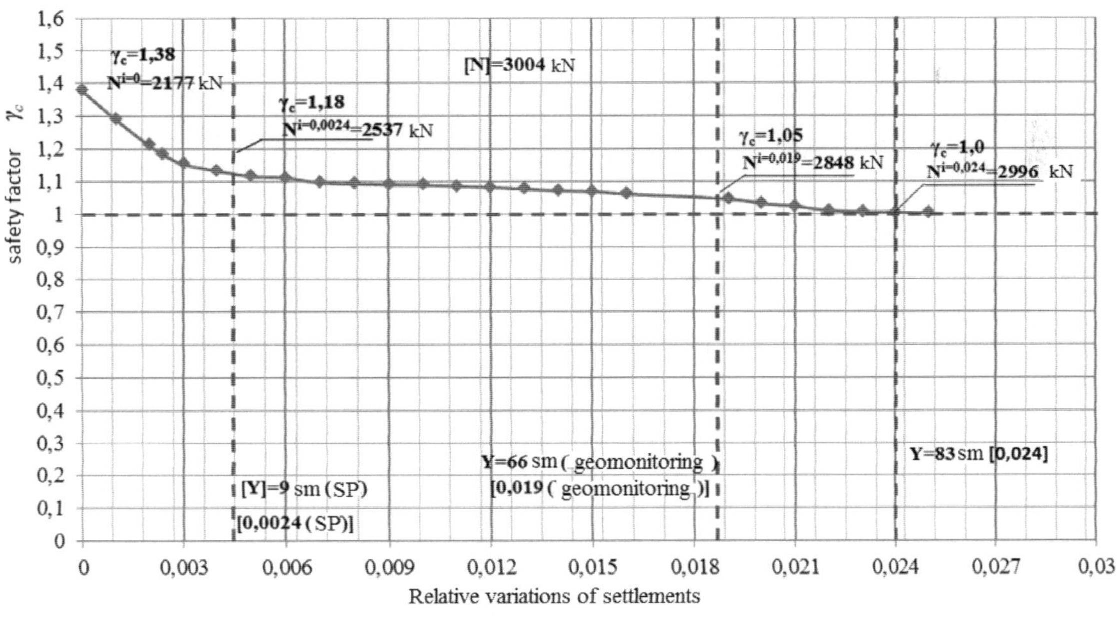

Figure 7. Dependency graph of the reliability factor on stability of the system versus the values of the differential settlements of the foundation bed.

It is generally accepted that the reliability factor on stability of the system for structures calculated using certified software packages should not be less than γ_c =1.3 [18]. The system is considered to be unstable when the coefficient approaches γ_c = 1.0 [18].

Figure 7 shows the dependency graph of the reliability factor on stability of the building system versus the values of the differential settlements of the foundation bed. In accordance with the results of geotechnical monitoring and numerical simulation $\Delta S/L$ = 0.019 in relative variations of settlements, the coefficient γ_c = 1.05 is at least equal to the ultimate value $[\gamma_c]$ = 1. After the results of geomonitoring, cracks in the party walls due to deformations were not found given the excess values of the differential settlements. This is confirmed by the results of numerical simulation (Figure 7), and stability of the system is γ_c =1.05.

In accordance with the results corresponding to the ultimate values of Building Regulations [17], namely: the value of the relative deformation $[\Delta S/L]$=0.0024 and horizontal displacement of the building frame [Y] = 9 cm [16], the reserve of the load-bearing capacity of the brick walls should be 15%, and stability of the system $[\gamma_c]$ = 1.18.

The results of numerical simulation show that the stress in the brick party walls will reach the ultimate load-bearing capacity if the values of the differential settlements are $\Delta S/L$ = 0.024 (Figure 7). Deviation of the building frame from the vertical is Y_{max} = 83 cm.

3. Assessment of shear forces in junctions of floor structures and walls

In order to assess the stress-strain state of the floor structure to wall joint from the horizontal tensile stresses arising from the non-uniform deformations of the foundation bed, it is necessary to determine the stress N_{max} when the floor structure breaks away from the wall [13].

In order to solve this problem, a 3D model of the floor structure to brick wall joint was created in a numerical software package.

The construction was approximated by the 3D FE-36 - universal spatial eight-joint isoparametric finite elements. The horizontal tensile stresses between the floor structure slabs were set with regard to the stages of variations in settlements (Figure 8).

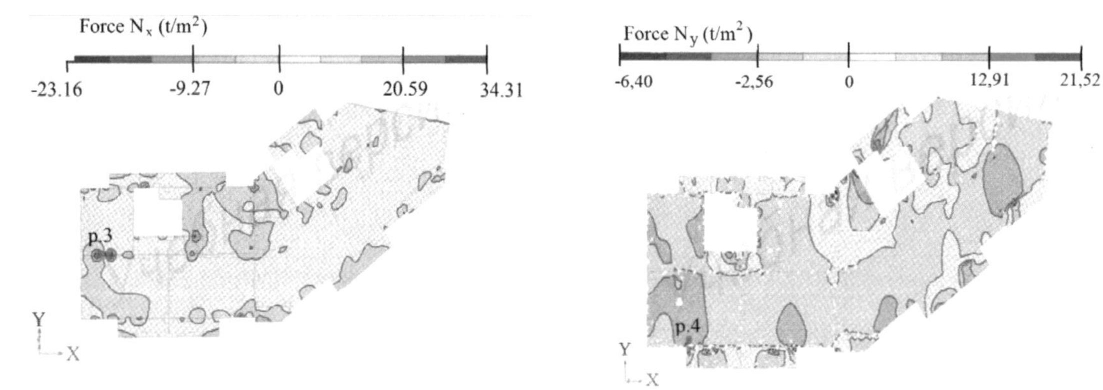

Figure 8. Distribution of stresses in the floor structure of the 14th floor along *X*-axis (p.3), along *Y*-axis (p.4) in Lira Soft 10.6

The tensile stresses in the floor structure slabs of the 14th floor were assessed; the relative variations in settlements - i_1 = 0.0024, i_2 = 0.007, i_3 = 0.019 [11].

Since the graph in Figure 9 is a linear relationship, it is necessary to find the horizontal stress N_{max}, and then determine the relative variations in settlements which refer to this stress.

$$N_{\max} = N_c + N_o$$

where N_c – longitudinal compressive force at the bearing stress zone;
N_o- crimping force;

$$N_o = \tau \cdot A_0$$

where A_0 – bearing surface of the brick work onto the slab;

$$\tau - \sigma \cdot tg\varphi + c$$

where σ – normal stresses from the brick work proper weight

$$\varphi = \mathrm{asin}(\frac{R-R_b}{R+R_b});\ C = \frac{R_b*(1-sin\varphi)}{2*cos\varphi};\ [10]$$

$$N_c = K_{\text{CT}} \cdot \psi \cdot d \cdot R_c \cdot A_c\ [15]$$

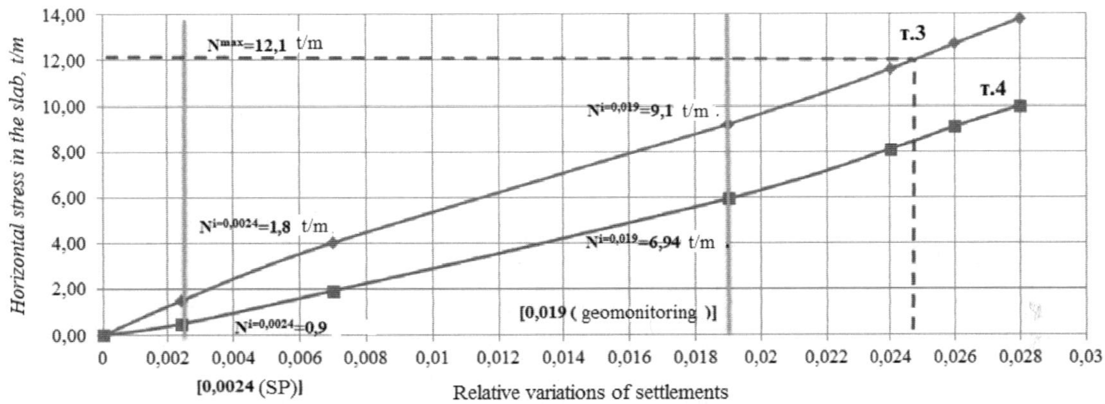

Figure 9. Dependency graph of the horizontal stress in the floor structure versus relative variations of settlements

According to the results of calculation N_c= 8.5 ton-meters. N_o = 3.6 ton-meters.

Thus, N_{max} = 8.5 + 3.6 = **12.1** ton-meters. $\Delta S/L$ = 0.025 in such force value from the graph (Figure 9).

4. Conclusion

These two controlling cases have made it possible to conclude that the loss of the load-bearing capacity of the party wall is most likely to occur with the relative variations of settlements - i = 0.024 and deviation of the building frame from the vertical by Y = 830 mm, and then the floor structure will possibly break away from the wall due to the horizontal force if i = 0.025 and Y = 870 mm.

5. References

[1] Brylla H, Niemiec T and Zotov V 2004 Berichtuber die Horizontierung lines Hochauses in Kotowice4 (Polen) DMW Mazksceidewesen 111 No 1 pp 10-15

[2] Burland J B, Jamiolkowski M B and Viggiani C 2009 Leaning Tower of Pisa: Behaviour after Stabilization Operations *International Journal of Geoengineering Case Histories* vol **1** Issue 3 pp 156-169

[3] Burland J B, Jamiolkowski M B and Viggiani C 2003 The stabilisation of the Leaning Tower of Pisa *Soils and Foundations* vol 43 pp 63-80

[4] Jamiolkowski M B 2001 The Leaning Tower of Pisa: End of an Odyssey *15th Int. Conf. on SMGE, Third Millenium Terzaghi Oration* (Istanbul Turkey) vol 4 pp 2979-2996

[5] Ravaska O 2002 A sheet pile wall design according to Eurocode 7 and Plaxis *In Numerical methods in geotechnical engineering* ed P Mestat (Paris: Presses de l'ENPC/LCPC) pp 649–654

[7] Schweiger H F 1997 Berechnungsbeispiele des AK 1·6 der DGGT-Vergleich der Ergebnisse fur Beispiel 1 (Tunnel) und 2 (Baugrube) *Tagungsband Workshop 'Numerik in der Geotechnik'* DGGT/AK 1 6 pp 1–29

[8] Tamez E, Ovando E and Santoyo E 1997 Underexcavation of Mexico City's Metropolitan Cathedral and Sagrario Church *14th Int. Conf. on SMFE (*Hamburg Germany) vol. **4** pp 2105-2126

[9] Zotov M V and Zotov A M 2009 The levelling of multistorey buildings by means of hydraulic flat jack systems *ISA Int. Housing Conf. 1-4 September 2009* (Glasgow: University of Glasgow) p 260

[10] Klovanovich S V 2009 Metod konechnyh ehlementov v nelinejnyh rabotah inzhenernoj mekhaniki [Klovanovich S V Finite element analysis in nonlinear works of engineering mechanics] (Zaporozh'e) (rus)

[11] Korolev K V 2015 Ploskaya zadacha teorii predel'nogo ravnovesiya gruntov [Plane problem of Coulomb's wedge theory] (Novosibirsk: Izd-vo SGUPS) p 251 (rus)

[12] Pronozin Ya A and Epifanceva L R 2015 Kreny zdanij, problemy i puti reshenij [Building tilts, problems and solutions] *Sbornik materialov 15 nauchno prakticheskoj konferencii molodyh uchenyh, aspirantov, soiskatelej i magistrantov v dvuh tomah* (Tyumen) pp 116-120 (rus)

[13] Rezvan I V, Mailyan D R and Rezvan A V 2012 Postroenie diagrammy «napryazheniya-deformacii» betona v usloviyah passivnogo bokovogo obzhatiya [Stress-strain state diagram of concrete under conditions of passive lateral compression] *Inzhenernyj vestnik* (Dona) No 4 chapter 1 (rus)

[14] Samohvalov M A, Zazulya YU V and Kajgorodov M D 2017 Rezul'taty issledovaniya napryazhenno-deformirovannogo sostoyaniya gruntovogo massiva pri formirovanii kontroliruemogo ushireniya na konce buroin"ekcionnoj svai [Results of study on the stress-strain state of the soil massif in formation of controlled widening at the end of the drill-injection pile] *Nauchnyj zhurnal stroitel'stva i arhitektury* No 2 (46) pp 11-17 (rus)

[15] SP 15.13330.2012 Kamennye i armokamennye konstrukcii. Aktualizirovannaya redakciya SNiP II-22-81* [Building Regulations 15.13330.2012 Stone and reinforced stone structures. Updated version of SNiP II-22-81*] (Moscow Minregion Rossii) (rus)

[16] SP 20.13330.2016 Nagruzki i vozdejstviya. Aktualizirovannaya redakciya SNiP 2.01.07-85* [Building Regulations 20.13330.2016 Loads and effects. Updated version of SNiP 2.01.07-85*] (Moscow Minregion Rossii) p 85 (rus)

[17] SP 22.13330.2016 Osnovaniya zdanij i sooruzhenij. Aktualizirovannaya redakciya SNiP 2.02.01-83* [Building Regulations 22.13330.2016 Foundations of buildings and structures. Updated version of SNiP 2.02.01-83*] (Moscow Minregion Rossii OAO «CPP») p 161 (rus)

[18] SP 16.13330.2017 Stal'nye konstrukcii. Aktualizirovannaya redakciya SNiP II-23-81* [Building Regulations 16.13330.2017 Steel members. Updated version of SNiP II-23-81*] (Moscow Minregion Rossii) p 142 (rus)

[19] Ulickij V Mm, Shashkin A G and Shashkin G K 2010 Gid po geotekhnike (putevoditel' po osnovaniyam, fundamentam i podzemnym sooruzheniyam) [Guide on geotechnical engineering (guide on bases, foundations and underground structures)] *PI«Georekonstrukciya»* (St.Peterburg) p 208 (rus)

[20] Shulyat'ev S O 2013 Vliyanie nesushchego karkasa zdaniya na napryazhenno-deformirovannoe sostoyanie fundamentnoj plity [Effect of the bearing building frame on the stress-strain state of the foundation slab: PhD thesis] *dissertaciya ... kandidata tekhnicheskih nauk* 05.23.02 (Moscow) p 142 (rus)

IOP Conf. Series: Materials Science and Engineering **481** (2019) 012014 doi:10.1088/1757-899X/481/1/012014

Constitutive relations for isotropic materials allowing quasi-linear approximation of the deformation law

A A Treschev[1], A A Bobrishev[2], L N Shafigullin[2]

[1]Tula State University, Lenin Avenu, 300012, Tula, Russia
[2]Kazan (Volga region) Federal University, 423812, Süümbike Avenue, Naberezhnye Chelny, Russia

E-mail: taa58@yandex.ru

Abstract. Variants of deformation potentials for isotropic materials that do not obey the "single curve" hypothesis are considered. The advantage of the potentials formulated in two normalized stress spaces in comparison with other variants of quasi-linear models of the constitutive relations for isotropic materials is demonstrated. It is shown that the claims of some authors on the originality and universality of the deformation potentials proposed by them are not justified, and their variants of relations are reduced to two forms of writing in normalized spaces.

1. Introduction

Two centuries of comprehensive use of the generalized Hooke's law approved it as an inviolable physical law. However, a wide range of deformation experiments on materials such as cast iron [1 – 5], graphite and graphite plastics [6 – 18], technical and construction ceramics [19], concrete [20 – 24] most composite materials [25 – 30], confirm the fact that the linear representation of the equations of the relationship between stresses and strains with the establishment of the elastic modulus E and the Poisson's ratio ν even in a very narrow range of deformation can not be called correct. Much more accurately represented the analytical interpretation of the experimental results to test the uniaxial tension and uniaxial compression in the form of linear functions with the module definition, E^+ (secant modulus of elasticity) under uniaxial tension and the modulus E^- under uniaxial compression. At the same time, Poisson's ratios ν^+ and ν^- (transverse strain coefficients) are calculated. In this key characterizes the properties of an isotropic elastic material, the mechanical characteristics of which depend on the type of stress state. Such a representation in the 60s of the twentieth century was called the multi-modulus theory of elasticity, based on quasi-linear deformation equations.

The indisputable primacy in the transformation of the fundamental law of Hooke in the case of materials that are sensitive to the type of stress state, was assigned to Russian scientists and their colleagues from some republics of the former Soviet Union. Research in this area of materials science and deformation theory, which occurred before the 1960-s, were random and did not attract much interest. In subsequent years, the development of regular research was stimulated by the widespread use of new materials, which obviously caused the emergence of major fundamental results in the formulation of equations of state for materials whose mechanical characteristics depend on the type of stress state. The increased interest in such a physical and mechanical phenomenon, which has arisen in recent years, is caused by the extensive use of polymer and composite materials, the most characteristic feature of the mechanical resistance of which, as many authors note, is a significant deviation of the deformation characteristics from the seemingly unshakable "hypothesis of a single curve" [1–30].

2. Analysis of quasi-linear potentials

A review of most works devoted to the construction of equations of state, taking into account the dependence of the deformation and strength characteristics of isotropic materials to the type of stress state, is available in [31, 32]. In the same works, within the framework of two normalized stress spaces, rather general potential relations between strains and stresses in a quasi-linear form are formulated. And from these relations, as a special case, follows almost all known variants of the physical equations proposed by other authors.

The deformation potential is derived from the first space in the form of:

$$\begin{aligned} W = 0.5\cdot[(A+B\cdot\alpha_1)\sigma_1^2+(A+B\cdot\alpha_2)\sigma_2^2+(A+B\cdot\alpha_3)\cdot\sigma_3^2] \\ +[C+E\cdot\alpha_3+D\cdot(\alpha_1+\alpha_2)]\cdot\sigma_1\cdot\sigma_2 \\ +[C+E\cdot\alpha_1+D\cdot(\alpha_2+\alpha_3)]\cdot\sigma_2\cdot\sigma_3 \\ +[C+E\times\alpha_2+D\times(\alpha_1+\alpha_3)]\times\sigma_1\times\sigma_3, \end{aligned} \quad (1)$$

and from the second – its analog in the following form:

$$W=(\tilde{b}_1+\tilde{b}_3\cdot\xi)\cdot\sigma^2+(\tilde{b}_2+\tilde{b}_4\xi+\tilde{b}_5\cdot\eta\cdot\cos 3\phi)\cdot\tau^2, \quad (2)$$

where $\alpha_k=\sigma_k/S$ – main normalized stresses; $S=\sqrt{\sigma_k\cdot\sigma_k}$ – norm of the first vector space; $\cos\psi=\xi=\sigma/S_0$, $\sin\psi=\eta=\tau/S_0$ – normalized stresses of the second space; $\cos 3j=\sqrt{2}\times\det(S_{ij})/\tau^3$ – phase invariant; $S_0=\sqrt{\sigma^2+\tau^2}$ – norm of the second space; $\sigma=\delta_{ij}\cdot\sigma_{ij}/3$ – normal octahedral stress; $\tau=\sqrt{S_{ij}S_{ij}/3}$ – tangent octahedral stress; $S_{ij}=\sigma_{ij}-\delta_{ij}\cdot\sigma$ – deviator of the stress tensor, δ_{ij} – Kronecker symbols; $\tilde{b}_1=1.5\cdot(A+2\cdot C)$, $\tilde{b}_2=1.5\cdot(A-C)$, $\tilde{b}_3=1.5\cdot(B+4\cdot D+2\cdot E)/\sqrt{3}$, $\tilde{b}_4=4.5\cdot(B-E)/\sqrt{3}$, $\tilde{b}_5=0.75\cdot(B-2\cdot D+2\cdot E)\cdot\sqrt{2}/\sqrt{3}$ – constants, which are determined by the mechanical characteristics of the material.

Two normalized spaces are interconnected through their invariants:

$$S_0=S/\sqrt{3};\mathrm{I}_\alpha=\sqrt{3}\cdot\xi;\mathrm{III}_\alpha=(3\cdot\xi^3+9\cdot\xi\cdot\eta^2+1.5\cdot\sqrt{2}\cdot\eta^3\cdot\cos 3\phi)/\sqrt{3}, \quad (3)$$

where $\mathrm{I}_\alpha=\alpha_k=\alpha_1+\alpha_2+\alpha_3$, $\mathrm{II}_\alpha=\alpha_k\cdot\alpha_k=1$, $\mathrm{III}_\alpha=\alpha_k\cdot\alpha_k\cdot\alpha_k$ – invariants of the first space.

Of the proposed potentials derive generalized laws of deformation:

a) volume changes

$$e=\sigma/(3\cdot K_0)+\tau/(3\cdot D_0); \quad (4)$$

b) changes of form

$$Э=\sqrt{1+tg^2\omega}\cdot[\tau\cdot/(2\cdot G_0)+\sigma/(3\cdot D_0)]; \quad (5)$$

c) when the phase characteristics

$$\mathrm{tg}\omega=3\cdot\tilde{b}_5\cdot\eta\cdot\sin 3\phi/[3\cdot\eta/(2\cdot G_0)+\xi/D_0], \quad (6)$$

where K_0 – functional «module» of volume deformations; G_0 - functional «module» shift; D_0 – functional «module» dilatancy; $\omega=\phi-\beta$ – the phase difference between the stress and deformation; $K_0=1/[2\tilde{b}_1+\tilde{b}_3\xi(2+\eta^2)-\tilde{b}_5\eta^3\cos 3\phi]$; $D_0=1/(\tilde{b}_4\eta^3)$; $2G_0=3/[2\tilde{b}_2+\xi^3(2\tilde{b}_4-\tilde{b}_3)+\tilde{b}_5\eta^3(2+\xi^2)\cos 3\phi]$; $3\cdot e=\theta=\delta_{ij}\cdot e_{ij}$ – volumetric strain; $Э=\gamma/2$; $\gamma=\sqrt{4\cdot\varepsilon_{ij}\cdot\varepsilon_{ij}/3}$ – octahedral shear; $\varepsilon_{ij}=e_{ij}-\delta_{ij}\cdot\theta/3$ – tensor deviator strain.

In [31, 32] it is demonstrated that the obtained quasi-linear equations of state take into account the dilatation properties of materials. In addition, as follows from the equations (5), the medium stresses affect the shape change, but so that if the shear stresses $\tau = 0$, the shape change is absent: $Э = 0$, that is purely hydrostatic stress does not lead to a shift. The existence of these properties in materials with different resistance has been repeatedly confirmed in all theoretical and experimental studies [1 – 30].

Two forms of strain potential (1) and (2) have five constants to be determined. It is obvious that experiments on uniaxial tension and uniaxial compression, of which four constants are established, are not enough. The following parameters are determined from these experiments:

$$A = 0.5\cdot(1/E^+ + 1/E^-);\ B = 0.5\cdot(1/E^+ - 1/E^-);$$
$$C = -0.5\cdot(\nu^+/E^+ + \nu^-/E^-);\ D = -0.5\cdot(\nu^+/E^+ - \nu^-/E^-). \qquad (7)$$

An additional option was to consider the possibility of finding the fifth constant from the experiments on the net shift [31, 32]. From the processing of the shear test results, the value of the material constant G_q (shear modulus at pure shear) follows, and from the representation of the potential (2) the following dependence

$$\tilde{b}_2 = 1.5\cdot(A - C) = 0.75/G_q. \qquad (8)$$

On the other hand, from the experiments on uniaxial tension and uniaxial compression potential (2) requires equality:

$$\tilde{b}_2 = 0.75\cdot[(1+\nu^+)/E^+ + (1+\nu^-)/E^-]. \qquad (8^*)$$

Obviously, this confirms that the considered quasilinear relations (1) and (2) describe the states of materials with different resistance, in which the shear modulus G_q is not an independent mechanical characteristic, but is defined as an medium parameter:

$$1/G_q = 1/G = (1+\nu^+)/E^+ + (1+\nu^-)/E^-. \qquad (9)$$

The problem of calculating the fifth constant is solved [31, 32] by processing the phase characteristic, as recommended in the nonlinear theory of elasticity [33]. However, in publications concerning experimental studies on deformation of materials with different resistance, there are only fragmentary, insignificant information about the values $tg\omega$. Therefore, [31, 32] a priori variants of determining the fifth constant were proposed, which do not contradict the results of experiments and the fundamental laws of mechanics.

In General, three quite acceptable variants of a priori calculation of the fifth constant are recommended.

Option No. 1. It is assumed that the states of the isotropic material are independent of the stress phase, and this leads to equality $\tilde{b}_5 = 0$ (2) and from the experiments on uniaxial tension and compression the relations follow

$$\tilde{b}_1 = 0.75\cdot[(1-2\cdot\nu^+)/E^+ + (1-2\cdot\nu^-)/E^-];$$
$$\tilde{b}_2 = 0.75\cdot[(1+\nu^+)/E^+ + (1+\nu^-)/E^-];\ \tilde{b}_3 = -1.5\cdot\sqrt{3}\cdot(\nu^+/E^+ - \nu^-/E^-);$$
$$\tilde{b}_4 = 0.375\cdot\sqrt{3}\cdot[(3+2\cdot\nu^+)/E^+ - (3+2\cdot\nu^-)/E^-]. \qquad (10)$$

In this case *E*, the constant in the form (1) is determined by the formula:

$$E = -0.25\cdot[(1+2\cdot\nu^+)/E^+ - (1+2\cdot\nu^-)/E^-]. \qquad (11)$$

Then, in the form of potential (1), five calculated constants are stored, and the form (2) becomes four constant:

$$W = (\tilde{b}_1 + \tilde{b}_3\cdot\xi)\cdot\sigma^2 + (\tilde{b}_2 + \tilde{b}_4\cdot\xi)\cdot\tau^2. \qquad (2^*)$$

Option No. 2. Mathematically consistent is the assumption of the smallness of the component in the polynomial decomposition in the construction of the potential (1), in the expression of which the cofactors are simultaneously three main normalized stresses – quantities of small order in comparison with the norm of vector space (normalized space № 1). At the same time, we note that the simultaneous consideration of three normalized stresses among the elements of the second order of smallness aggravates the smallness of this element. Hence, in form (1) can be taken $E = 0$. Then the potential (1) turns into a four-constant:

$$W = 0.5\cdot[(A+B\cdot\alpha_1)\cdot\sigma_1^2 + (A+B\cdot\alpha_2)\cdot\sigma_2^2 + (A+B\cdot\alpha_3)\cdot\sigma_3^2] \\ +[C+D\cdot(\alpha_1+\alpha_2)]\cdot\sigma_1\cdot\sigma_2 + [C+D\cdot(\alpha_2+\alpha_3)]\cdot\sigma_2\cdot\sigma_3 \\ +[C+D\cdot(\alpha_1+\alpha_3)]\cdot\sigma_1\cdot\sigma_3 \qquad (1^*)$$

and experiments on uniaxial tension and uniaxial compression are sufficient to calculate constants under the conditions (7). At the same time, the potential record (2) stores five constants and they are calculated by formulas:

$$\tilde{b}_1 = 0.75\cdot[(1-2\cdot\nu^+)/E^+ + (1-2\cdot\nu^-)/E^-];\ \tilde{b}_2 = 0.75\cdot[(1+\nu^+)/E^+ + (1+\nu^-)/E^-];$$
$$\tilde{b}_3 = 0.25\cdot\sqrt{3}\cdot[(1-4\cdot\nu^+)/E^+ - (1-4\cdot\nu^-)/E^-];\ \tilde{b}_4 = 0.75\cdot\sqrt{3}\cdot(1/E^+ - 1/E^-); \qquad (12)$$
$$\tilde{b}_5 = 0.125\cdot\sqrt{6}\cdot[(1+2\cdot\nu^+)/E^+ - (1+2\cdot\nu^-)/E^-].$$

Option No. 3. The fifth constant can be hypothetically obtained if we assume that the forms of the law of volume change of quasilinear material with different resistance under hydrostatic compression and tension coincide with the forms arising from the classical generalized Hooke's law:

$$\theta^-/3 = \sigma^-\cdot(1-2\cdot\nu^-)/E^-;\ \theta^+/3 = \sigma^+\cdot(1-2\cdot\nu^+)/E^+. \qquad (13)$$

A comparison of the laws of volume change (4) and (13) in hydrostatic compression ($\xi = -1$, $\eta = 0$) and tension ($\xi = 1$, $\eta = 0$) gives the following results:

$$E = 0.25\cdot\sqrt{3}\cdot[(1-2\cdot\nu^+)/E^+ - (1-2\cdot\nu^-)/E^-] - 0.25\cdot[(1-4\cdot\nu^+)/E^+ - (1-4\cdot\nu^-)/E^-];$$
$$\tilde{b}_1 = 0.75\cdot[(1-2\cdot\nu^+)/E^+ + (1-2\cdot\nu^-)/E^-];\ \tilde{b}_2 = 0.75\cdot[(1+\nu^+)/E^+ + (1+\nu^-)/E^-];$$
$$\tilde{b}_3 = 0.75\cdot[(1-2\cdot\nu^+)/E^+ - (1-2\cdot\nu^-)/E^-]; \qquad (14)$$
$$\tilde{b}_4 = 0.375\cdot\sqrt{3}\{[(3-4\cdot\nu^+)/E^+ - (3-4\cdot\nu^-)/E^-] - \sqrt{3}\cdot[(1-2\cdot\nu^+)/E^+ - (1-2\cdot\nu^-)/E^-]\}.$$

The third option leads to the preservation in both forms of the deformation potential (1) and (2) of five constants with their unambiguous values arising only from two uniaxial experiments.

In article [34] N.M.Matchenko and I.N.Matchenko tried to formulate more general approaches and forms of construction of quasi-linear four-constant deformation potentials than proposed in [31, 32]. This attempt to build new forms of quasi-linear four constants tion deformation potentials [34] proved only by replacing the invariant signs of the potentials (1) and (2). What is the matter here, explain. The existing quasi-linear models of deformation of isotropic materials are based on the introduction of certain quantitative and qualitative parameters of their stress state. In [34] N.M.Matchenko and I.N.Matchenko write down three groups of such parameters:

a) quantitative parameters of the first group –

$$H_I = \sigma_1 + \sigma_2 + \sigma_3;\ H_{II} = \sigma_1^2 + \sigma_2^2 + \sigma_3^2; \qquad (15)$$

b) quantitative parameters of the second group –

$$I_I = \sigma_1 + \sigma_2 + \sigma_3;\ I_{II} = (\sigma_1 - \sigma_2)^2 + (\sigma_2 - \sigma_3)^2 + (\sigma_3 - \sigma_1)^2; \qquad (16)$$

c) quantitative parameters of the third group –

$$J_I = \sigma_1 + \sigma_2 + \sigma_3;\ J_{II} = \sigma_1\sigma_2 + \sigma_2\sigma_3 + \sigma_3\sigma_1. \qquad (17)$$

The quantitative parameters of the stress state (15) – (17) introduced in this way are nothing but another interpretation of the values used in [31, 32], and namely, for the first group the introduced parameters [34] are equivalent to the following [31, 32]:

$$H_I = 3\sigma = \delta_{ij}\sigma_{ij}\,;\; H_{II} = S^2 = \sigma_k \cdot \sigma_k = 3 \cdot S_0^2\,; \tag{15*}$$

for the second group these parameters [34] coincide with invariants [31, 32]:

$$I_I = 3 \cdot \sigma = \delta_{ij} \cdot \sigma_{ij}\,;\; I_{II} = 9 \cdot \tau^2\,; \tag{16*}$$

for the third group the recorded parameters [34] are replaced by expressions [31, 32]:

$$J_I = \delta_{ij} \cdot \sigma_{ij};\; J_{II} = 3 \cdot (S_0^2 - 1.5 \cdot \tau^2\,). \tag{17*}$$

The qualitative characteristic of the stress state for all these groups in [34] is proposed to use one parameter presented in the form of:

$$\chi = (\sigma_1 + \sigma_2 + \sigma_3\,) / \sqrt{\sigma_1^2 + \sigma_2^2 + \sigma_3^2} = 3 \cdot \sigma / S = 3 \cdot \sigma / (\sqrt{3} \cdot S_0\,), \tag{18}$$

and this is nothing but a combination of invariants considered in [31, 32]:

$$\chi = 3 \cdot \sigma / S = 3 \cdot \sigma / (\sqrt{3} \cdot S_0\,). \tag{18*}$$

Thus, when using the considered H.M.Matchenko and I.N.Matchenko's parameters (15) – (18) allowed them to [34] record three forms of quasi-linear potential:

$$W = (A_h + a_h \cdot \chi) \cdot H_I^2 + (B_h + b_h \cdot \chi) \cdot H_{II}\,; \tag{19}$$

$$W = (A_i + a_i \cdot \chi) \cdot I_I^2 + (B_i + b_i \cdot \chi) \cdot I_{II}\,; \tag{20}$$

$$W = (A_j + a_j \cdot \chi) \cdot J_I^2 + (B_j + b_j \cdot \chi) \cdot J_{II}\,. \tag{21}$$

It is easy to notice that all these forms of potential (19) – (21) taking into account the equations (15*) – (18*) are unambiguously transformed and completely coincide with the representation (2*). Therefore, the expressions (19) - (21) have no independent meaning, but represent only a different form of writing. At the same time, in [31, 32] it was found that the five-constant potential has a higher accuracy when compared with experimental deformation diagrams of most materials with different resistance under complex stress conditions.

In another work [35], A.V.Berezin argues that he obtained a more General quasilinear potential than formulated in [31, 32] and this potential is as follows:

$$\Phi = \Phi_1 \cdot [\sigma_i \cdot \phi_1(u,\zeta)] + \Phi_2 \cdot [\sigma_0 \cdot \phi_2(u,\zeta)] + \Phi_3 \cdot [S_3^{1/3} \cdot \phi_3(u,\zeta)]\,, \tag{22}$$

where σ_i - the stress intensity; σ_0 - hydrostatic stress; $u = \sigma_0 / \sigma_i$ - function, taking into account the type of stress state (disadvantages of this qualitative parameter discussed in [32]); $\phi_1(u,\zeta)$, $\phi_2(u,\zeta)$, $\phi_3(u,\zeta)$ – functions to be experimentally determined; S_3 – the third invariant of the stress deviator tensor.

At the same time, the author himself [35] without specifying the functions $\phi_1(u,\zeta)$, $\phi_2(u,\zeta)$, $\phi_3(u,\zeta)$, argues that « the greatest difficulty is the definition of functions $\phi_i(u,\zeta)$ $(i = 1,2,3)$. However, this was done in [31] under some assumptions». These assumptions are presented here above (see expressions (7) – (14)).

3. Summary

From the given analysis of the presented potential dependences, it can be concluded that at present the most common quasi-linear constitutive relations for isotropic materials suitable for problem-free use in solving applied scientific and engineering tasks are two forms of deformation potential (1) and (2).

Acknowledgments
This work was supported by the research grant of Kazan Federal University.

References

[1] Leonov M Ya Panaev V A and Rusinko K N 1967 Dependence between the deformations and stresses for semi-fragile bodys *Engineering journal of solid Mechanics* 6 pp 26 – 32
[2] Pisarenko G S Lebedev A A and Romaszewski V P 1968 Experimental study of the regularities of deformation of carbon steel in conditions of complex stress state at low temperatures *Probl. strengths* **5** pp 42-47
[3] Pisarenko G S and Lebedev A A 1976 *Deformation and strength of materials in a complex stress state* (Kyiv: Naukova Dumka) 416 p
[4] Stetsenko V A 1972 *Researches on the mechanics of deformable environment* (Tula: Tula Polytechnical Institut) pp 103-109
[5] Jamroz L and Mechanizne I 1971 Wytzymalosciowe Wlasnosci Zeliwa Sferoidalnegon *Prace Instytutu Odlewnictwa* R 21/3 pp 283 – 302
[6] Berezin A V 1990 *Influence of damages on deformation and strength characteristics of solids* (Moscow: Science) 135 p
[7] Berezin A V Strokov V I and Barabanov V N 1976 *Structural materials based on carbon* vol **11** (Moscow: Metallurgy) pp 102-110
[8] Friedman A M Anufriev J P and Barabanov V N 1972 Some features of methods of research of strength properties of graphite under a plane stress state *Factory laboratory* 9 pp 1137-40
[9] Berezin A V 1979 Resistance to deformation and destruction of isotropic graphite materials under complex stress conditions *Strength Problems* **2** pp 60-65
[10] Strokov V I and Barabanov V N 1974 Methods of research of strength and deformation properties of graphite in conditions of complex stress state *Factory laboratory* **9** pp 1141-44
[11] Friedman A M Anufriev J P and Barabanov V N 1973 Investigation of the destruction of carbon-graphite materials under complex stress state *Problems of strength* **1** pp 52-55
[12] Jones R M 1980 Modeling Nonlinear Deformation of Carbon-Carbon Composite Materials *AIAA Journal* **18/8** pp 995–1001
[13] Jones R M 1977 Stress-Strain Relations for Materials with Different Moduli in Tension and compression *AIAA Journal* **15/1** pp 16–25
[14] Jones R M 1971 Buckling of Stiffened Multilayered Circular Shells wiht Different Ortotropic Moduli in Tensione and Compression *AIAA Journal* **9/5** pp 917–923
[15] Jones R M 1977 A Nonsystemmetric Compliance Matrix Approach to Notlinear Multimodulus Ortotropic Materials *AIAA Journal* **15/10** pp 1436–43
[16] Jones R M and Nelson D A R 1975 Further Characteristics of a Nonlinear Material Model for ATJ-S Graphite *Jounal Composit Materials* **9/7** pp 251–265
[17] Jones R M and Nelson D A R 1976 Theoretical-experimental correlation of material models for non-linear deformation of graphite *AIAA Journal* **14/10** pp 1427–35
[18] Jones R M and Nelson D A R 1976 Material for non-linear Deformation *AIAA Journal* **14/6** pp 709–716
[19] Romanov V V 1984 *Phisics-cemistry researches according to the technology of glass and sitals* (Moscow: Science) pp 78-81.
[20] Matchenko N M Tolokonnikov L A and Treschev A A 1999 The governing relations of isotropic differen resistant environments. Part 2 A Nonlinear correlation *Izvestia of Rusian Academy of Sciense Mechanic of solid body* **4** pp 87-95
[21] Bazant Z P and Bhat P D 1976 Endochronic Theory of Inelasticity and Failure of Concrete *Journal of the Engineering Mechanics Division ASCE* **102** EM4 pp 701–722
[22] Kupfer H B 1973 Das nicht-linear Verhalten des Betons bei Zweiachsinger Beanspruchung *Beton und Stahlbetonbau* **11** pp 269–274
[23] Kupfer H B Hilsdorf H K and Rusch H 1969 Behavior of Cancrete Under Biaxial Stresses *ACI Journal* **66/8** pp 656–666

[24] Tasuji M E Slate F O and Nilson A H 1979 Stress-Strain Response and Fracture of Concrete in Biaxial Loading *ACI Journal* **7** pp 806–812
[25] Goldman A Ya Savelyev N F and Smirnova V I 1968 Investigation of mechanical properties of fabric fiberglass under tension and compression is normal to the plane of reinforcement *Mechanics of polymers* **5** pp 803-809
[26] Derevyanko N So 1968 Properties of reinforced polystyrene in short-term tension, compression and bending *Mechanics of polymers* **6** pp 1059-64
[27] Elsufiev S A 1968 Investigation of the deformation of fluoroplastic-4 in the linear and plane stress conditions *Mechanics of polymers* **4** pp 742-746
[28] Elsufiev S A Chebanov V M 1971 Ressoches by elasticity and plasticity Issue 8 (Leningrad: Publishing house of Leningrad state university) pp 209-213
[29] Kalinka Y A and Borovikova S M 1971 Study of physico-mechanical properties of randomly filled with fiberglass *Mechanics of polymers* **3** pp 411-415
[30] Schwartz R T and Schwartz H S 1967 Characteristics of Boron Fibers and Boron-Fiber-Reinforced Plastic Composites *AIAA Journal* **5/2** pp 119–126
[31] Matchenko N M Tolokonnikov L A and Treschev A A 1995 The governing relations of isotropic differen resistant environments Part 1. A quasilinear correlation *Izvestia of Rusian Academy of Sciense Mechanic of solid body* **1** pp 73-78
[32] Treschev A A 2016 *Theory of deformation and strength of materials with initial or induced sensitivity to the type of stress state. The governing relations* (Tula: Tula state university) 328 p
[33] Tolokonnikov L A 1979 *Mechanics of deformable solid body* (Moscow: Higher scoole) 318 p
[34] Matchenko N M and Matchenko I N 2016 To build chetyrehkantnyj equations are quasilinear isotropic elastic medium *Vestnik of Chuvash state pedagogical University Ser. Mechanics of the limit state* **4** pp 73-80
[35] Berezin A V 2007 On the laws of deformation of multimodular dilating environment *Problems of mechanical engineering and automation International journal* **2** pp 70-72

An indirect method for determining the dynamic properties of soils at modeling the vibrating effects of technological equipment on the building structures

I Sh Alirzaev[1], A Sh Rakhimov[1] and N S Sova[1]

[1]Voronezh State Technical University, 20-letiya Oktyabrya St, 84, Voronezh, 394006, Russia

E-mail: sova1950@yandex.ru

Abstract. The results of experimental studies of the elastic characteristics of soil bases are presented in the simulation of dynamic loads from machinery and equipment to building structures. In the course of work on the inspection of the technical condition of building structures, the authors developed an experimental and theoretical method for determining the elastic characteristics of soil bases. The proposed method involves carrying out full-scale dynamic tests in the mode of natural and forced oscillations.
Keywords. Foundations of machines, dynamic effects, structural vibration, dynamic properties of soils, full-scale dynamic tests.

1. Introduction

While inspecting the technical condition of buildings and structures, it is necessary to study the propagation of oscillations from the foundations of machines with dynamic loads that cause vibration of building structures and foundations [1–4]. The period and logarithmic pitch decrement of the building's natural oscillations are important calculation parameters for conducting surveys and monitoring the technical condition of buildings and structures [5–7].

Under certain conditions, resonance oscillations of bearing and enclosing building structures cause a decrease in their serviceability. One of the reasons for the increased vibration of foundations with dynamic loads and adjacent building structures is an inaccurate determination of the initial dynamic characteristics of the base soil.

Currently, the main normative document regulating the determination of dynamic properties of soils is [8]. This standard is relatively new and establishes three methods of laboratory testing of soils to determine their dynamic properties: dynamic triaxial compression, low-amplitude dynamic tests by resonance column and torsional shear. All three methods require specialized expensive equipment and certain skills to work on it [9–11].

In addition, during the inspection of existing buildings and structures, laboratory methods for determining the elastic characteristics of the soil base have a number of drawbacks, which include not taking into account the actual technical condition of the foundation slabs under equipment and soil bases. In this regard, the development of field methods for assessing the dynamic properties of soil bases of existing buildings and structures by the full-scale tests is a vital task [12–13].

The complexity of this problem is one of the reasons for the use of empirical methods in other normative documents [14–17]. According to [14], the main elastic characteristic of the soil bases is the coefficient of elastic uniform compression C_z, which is determined by the results of tests or on the basis of data on the velocities of longitudinal and transverse waves. In the absence of experimental data, the value for foundations with a sole area of not more than 200 m^2 can be determined by the empirical formula:

$$C_z = b_0 E \left[1 + \left(\frac{A_{10}}{A} \right)^{1/2} \right], \tag{1}$$

where b_0 – coefficient for sandy soils taken equal to 1.0; E – the deformation modulus of the soil under the foundation bed; A_{10} = 10 m^2; A – area of the foundation bed.

According to [14], the coefficients of elastic non-uniform compression, elastic uniform shear, and elastic non-uniform shift are derived characteristics and are determined by the coefficient of elastic uniform compression. Thus, one of the main factors that can affect the results of determining the elastic characteristics of soil base is the value of the modulus of deformation of the soil. According to [18], the soil deformation modulus is a variable value depending both on the stress level in the soil bases and on the method of its determination [19–20].

2. The Essence of the Method

The essence of the method is to establish the dynamic properties of soils in full-scale tests under the conditions of the operating load. Tests are carried out by changing the inertial properties of the foundation in the modes of natural and forced oscillations. Varying inertial properties is accomplished by changing the mass of the foundation. On the basis of the obtained natural frequency values, the elastic characteristics of the soil bases are calculated, and the reaction of building structures is determined. With this approach, empirical methods are used in combination with the results of measurements obtained by field tests.

3. Object of Research

The object of the study is a one-story industrial building with a metal frame. Two technological lines for drying and classification of sand are mounted in the building (Fig. 1). In the process of visual inspection, it was found that all the building structures of the building are affected by the harmonic dynamic load from the technological equipment. According to the results of instrumental surveys, it was found that the amplitude of oscillations of wall sommers and profiled flat for wall fencing exceeds the normative values.

Figure 1. Fragment of building structures and technological equipment.

The framing of the building is made according to the wall-frame scheme and is the main supporting structure, which perceives and transmits to the foundations all vertical and horizontal loads. Elements of the frame are steel through two-branch columns, rafter trusses, vertical and horizontal links and spacers. The conjunction of trusses with steel columns is made in a hinged variant. The cross frames of the framing provide the rigidity of the building in the transverse direction, and the vertical links and spacers between the columns – In the longitudinal direction. The structures of working platforms and cover are leaned to the columns on different labels. Wall fences and roof coat are made of profiled flat bed, which is attached to the sommers. Wall sommers and binding rafters are designed for installation of profiled flat bed, which provide transmission to the framework of snow and wind loads, increase the rigidity and stability of the entire structure. The profiled flat bed is attached to the sommers of the coating and walls by self-tapping bolts with joint washers.

The construction site is composed of the following soils: filled-up ground with depth 0.8 m, medium-sized sand, average density with depth 2.3 m and medium-sized sand, dense with depth 7 m.

Separately standing isolated footing for steel columns are made of concrete class B25. The height of the first stage and the bases of columns of all foundations is 600 mm, and the height of the middle stage, depending on the location of the foundations ranges from 100 to 3700 mm. The depth of the foundations, respectively, ranges from 2.2 m to 4.9 m. The foundations are prepared from concrete class B10 with a thickness of 100 mm.

Each production line consists of two types of equipment: a furnace-drying unit and a vibrating pneumatic classifier. The main characteristics of the process equipment are presented in Tables 1–2.

Table 1. Technical characteristics of the furnace-drying unit.

Position	Parameter	Value
	Vibro-duct	
1	Oscillation amplitude, mm	≤ 4
2	Oscillation frequency, Hz	12.5–16.7
	Drive (Electromechanical vibrator MVE 13000/1)	
3	Amount	2
4	Capacity, kW	10
5	Oscillation frequency, Hz	16.7
	Overall dimensions	
6	Length, mm	11228
7	Width, mm	2308
8	Height, mm	13950
9	Weight, kg	13950

Table 2. Technical characteristics of the vibrating pneumatic classifier.

Position	Parameter	Value
	Vibro-duct	
1	Oscillation amplitude, mm	≤ 4
2	Oscillation frequency, Hz	12.5–16.7
	Drive (Electromechanical vibrator MVE 13000/1)	
3	Amount	2
4	Capacity, kW	10
5	The driving force of one vibrator, kg	13009
6	Oscillation frequency, Hz	16.7
	Overall dimensions	
7	Length, mm	7123
8	Width, mm	1838
9	Height, mm	3365
10	Weight, kg	7503

The equipment of technological lines is installed on the Foundation slab of reinforced concrete. The monolithic slab with a thickness of 400 mm is made of concrete of class B25 and reinforced with two flat grids of reinforcing bars with a diameter of 12 mm (A400). The step of bars in both directions is 200 mm. Under the monolithic plate a concrete blinding coat is made with class B10 and 100 mm thick.

4. Testing

Before the measurement of vibrations, structures with increased amplitudes were identified. Measurements were carried out in the modes of natural and forced oscillations. The frequencies of the natural oscillations of the system "Foundation-Machine" were determined with stopped technological machines. In order to fixate large dynamic displacements of building structures, forced oscillations were carried out in resonance regimes. Measurement points were selected in the immediate vicinity of the reference nodes, at the interface of structural elements and in the middle of the spans of beam structures.

To determine the extent of influence of driving forces that arise during operation of the furnace-drying unit and pneumatic classifiers on the amplitude of the oscillation, measurements have been carried out in the operating and idle modes. In order to establish a connection between the fluctuations of the building structure and the operation of specific equipment, measurements were carried out by alternately disconnecting and turning on two production lines.

During the test, in addition to fixing the dynamic movements of building structures, measurements of vibrations of the frame and slab parts of the foundations for the equipment were carried out. In order to identify the separation of the frame part of the foundation and the loosening of the anchor bolts, measurements are made for two points of the foundation plate located in opposite directions.

The escalation in the weight of the foundation was achieved by loading the frame part of the foundation with sandbags (Fig. 2–3). During the test, five stages of loading were provided. All stages of the test were carried out under the same conditions and in the absence of snow loads to cover the building.

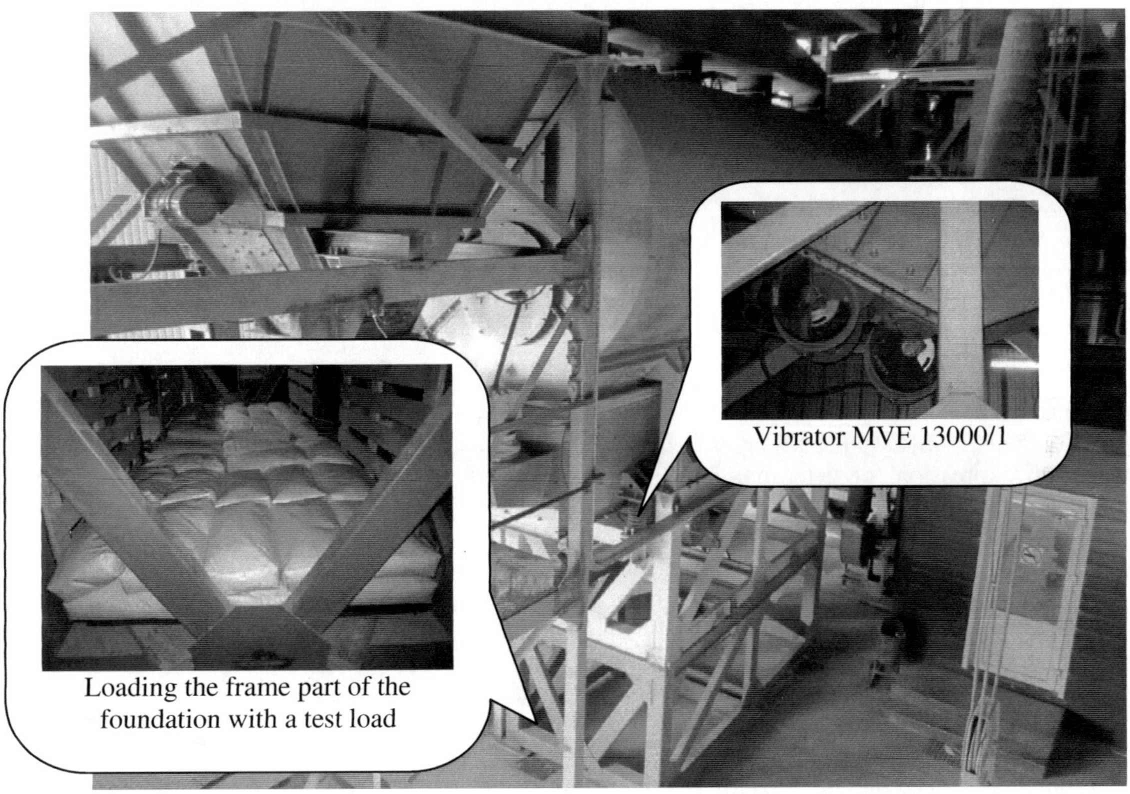

Figure 2. General view of the unit and change of own weight of the frame part of the foundation by the test load.

5. Results and Discussion

According to the results of the research, the values of the natural frequencies of the system "Soil Base-Foundation-Machine" and the amplitude of oscillations of building structures were obtained. Determination of elastic characteristics was carried out by solving the inverse problem by the method described below.

IOP Conf. Series: Materials Science and Engineering **481** (2019) 012015 doi:10.1088/1757-899X/481/1/012015

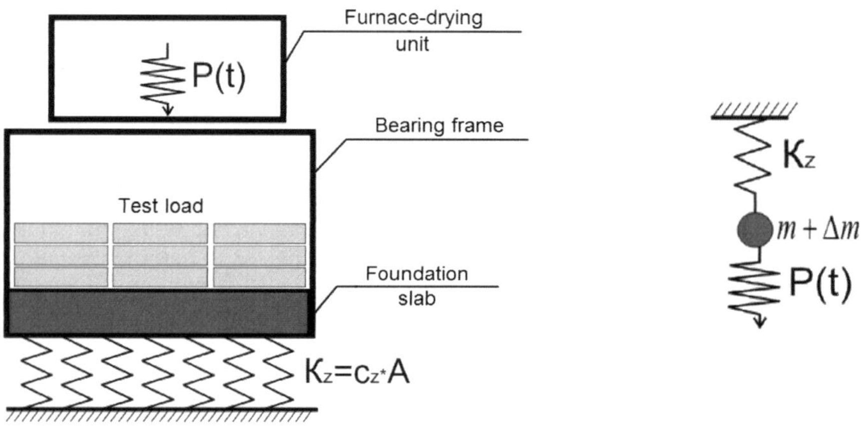

Figure 3. Design model of the system "Soil Base-Foundation-Machine".

It is known that for a one-mass system (figure 3) the eigen frequency is determined by the formula:

$$\omega = \left(\frac{K_z}{m + \Delta m} \right)^{1/2}, \quad (2)$$

where K_z – the stiffness coefficient of the soil base under uniform compression; m – the weight of the system "Foundation-Machine"; Δm – testing weight.

The stiffness coefficient for the earth foundation is determined in accordance with clause 6.1.4 [9] as for the model of Winkler:

$$K_z = C_z \cdot A, \quad (3)$$

where C_z – the coefficient of elastic uniform compression; A – area of the foundation bed.

With the help of formulas (2) and (3), taking into account the change in the mass of the system, we obtain an expression for determining the eigen frequencies of the system "Soil Base-Foundation-Machine":

$$\omega = \left(\frac{K_z}{m + \Delta m} \right)^{1/2} = \left(\frac{C_z \cdot A}{m + \Delta m} \right)^{1/2}. \quad (4)$$

From the obtained formula we express the coefficient of elastic uniform compression:

$$C_z = \frac{(m + \Delta m) \cdot \omega^2}{A}. \quad (5)$$

From the obtained ratio taking into account (1) determine the modulus of deformation of the soil:

$$E = \frac{\omega^2 \cdot (m + \Delta m)}{b_0 \cdot A \cdot \left[1 + \left(\frac{A_{10}}{A} \right)^{1/2} \right]}. \quad (6)$$

Some results of tests and determination of characteristics according to the above procedure are given in the summary Table 3.

Table 3. Results of tests in the mode of natural oscillations at work of the furnace-drying unit of the first line.

Stage	Test load Δm, t	Frequency of the system "Soil Base-Foundation-Equipment" ω, sec^{-1}	Coefficient of elastic uniform compression C_z, kN·m^{-3}	Deformation modulus of the soil E, MPa
I	0 (idle mode)	128.3	$32.66 \cdot 10^3$	18.29
II	1.5	128.5	$34.35 \cdot 10^3$	19.24
III	3.0	132.1	$37.98 \cdot 10^3$	21.27
IV	5.0	138.7	$44.33 \cdot 10^3$	24.83
V	8.0	143.4	$51.34 \cdot 10^3$	28.75

According to the test results, the increase in the values of the deformation modulus as the test load increases at high frequencies was established. The difference between the strain modulus values is due to different stress values at different stages. At low frequencies, no significant changes in elastic characteristics were observed.

According to the results of testing the foundation of the first line furnace-drying unit, it was found that the increase in the mass of the foundation at low frequencies leads to an increase in the amplitude of vibration fluctuations of building structures.

According to the results of the research it was decided to change the stiffness of the wall sommers in order to withdraw them from the resonance zone.

6. Conclusion

- The proposed method makes it relatively quick and easy to obtain a clear relationship between the vibration effects on the building and the dynamic characteristics of the soil bases.
- All vibrations to determine the elastic characteristics of soil bases are caused by the action of existing equipment, so in the process of testing additional dynamic effects on the soil bases are not made.
- The dynamic properties of soils, in particular, the coefficient of elastic uniform compression are variable, which depends on both the level of stresses in the bases and the parameters of external dynamic effects.
- An increase in the weight of the foundation at low frequencies leads to a raising in the amplitude of vibratory oscillations of building structures.

References

[1] SP 13-102-2003 Rules of inspection of bearing structures of buildings and structures. 2003 (Moscow: Gosstroy Rossii) 27.

[2] GOST 31937-2011 Buildings and constructions. Rules of inspection and monitoring of the technical condition. 2014 (Moscow: ISC) 55.

[3] GOST R ISO 10137-2016 Bases for design of structures. Serviceability of buildings against vibrations. (ISO 10137:2007, Bases for design of structures – Serviceability of buildings and walkways against vibrations, IDT). 2017 (Moscow: ISC) 36.

[4] RD 34.21.306-96 Methodical instructions on inspection of dynamic condition of building structures and the bases of the equipment of power enterprises. 1998 (Moscow: SPO ORGRES) 48.

[5] GOST R 54859-2011 Buildings and constructions. Determination of the parameters of the basic tone of free oscillations of buildings. 2012 (Moscow: ISC) 18.

[6] Eisenberg J M and Mamaeva G V 2009 Determination of dynamic characteristics of frame buildings for practical calculations on seismic influences. *Structural Mechanics and Analysis of Constructions* **6** 44–53.

[7] Lapin S G 2005 Transfer of vibration from machine foundations to adjacent foundations supporting buildings and structures. *Soil Mechanics and Foundation Engineering* **5** 29–31.
[8] GOST R 56353-2015 Soils. Laboratory methods for determination of soil dynamic properties. 2015 (Moscow: ISC) 35.
[9] Voznesensky E A 2013 Dynamic testing of soils. The status of this question and standardization. *Engineering Survey* **5** 20–26.
[10] Kholmyanskii M L 2006 Method for determination of dynamic parameters of the "foundation-soil" system from amplitude-frequency characteristics. *Soil Mechanics and Foundation Engineering* **3** 14–19.
[11] Bondar I S 2012 Shear tests of cohesive soils at different loading trajectories. *Magazine of Civil Engineering* **7** 50–77.
[12] Boldyrev G G, Melnikov A V et al 2014 Comparison of laboratory and field soil testing methods. *Engineering Survey* **5** 98–108.
[13] Boldyrev G G, Melnikov A V et al 2014 Part I. Interpretation of field test results to determine soil strength characteristics. *Engineering Survey* **5–6** 68–77.
[14] SP 26.13330.2012 Updated edition of SNiP 2.02.05-87. Foundations for machines with dynamic loads. 2011 (Moscow: Minregion Rossii) 66.
[15] SP 22.13330.2016 Updated edition of SNiP 2.02.01-83*. Soil bases of buildings and structures. 2016 (Moscow: Minregion Rossii) 225.
[16] Guidelines for the design of foundations of machines with dynamic loads 1982 (Moscow: Stroyizdat) 207.
[17] Instructions for determining the dynamic loads from machines installed on the coverings of industrial buildings. 1970 (Moscow: Stroyizdat) 288.
[18] Goldstein M N 1982 Mechanical properties of soils and improvement of their research methods. *Soil Mechanics and Foundation Engineering* **3** 21–23.
[19] Bellendir E N, Berzov V F and Kutergin V N 2001 Consideration of dynamic impact on soil strength parameters. Vedeneev VNIIG vol. 239 pp. 143–153.
[20] Kholmyanskii M L 2005 Stress state of soil subjected to a periodic system of strip loads. *Soil Mechanics and Foundation Engineering* **2** 2–6.

IOP Conf. Series: Materials Science and Engineering **481** (2019) 012016 doi:10.1088/1757-899X/481/1/012016

Mathematical modeling technology of a foundation slab concreting on a frozen base

V VMolodin[1], M S Gorlatov[1], S M Zaika[1]

Novosibirsk State University of Architecture and Civil Engineering, 630099 Novosibirsk, Russia

E-mail: molodin@sibstrin.ru, max 20-97@yandex.ru, stanislav1198@inbox.ru

Abstract. The current regulations prohibit to concretebuildings sights on a base with low temperature. The base warming is a long and energy-intensive process. The substantiation of concreting technology was carried out to concrete the foundation slab at an air temperature of 20°C on a frozen base. The first stage was the use of mathematical method modeling for the feasibility of a base heating. After making sure that the heating had been inefficient, the possible to provide favorable temperature holding conditions due to the heat release of the cement, compensating for heat loss through the side faces of the slab or by the concrete heating on contact with the base and on the exposed surface was estimated. Mathematical modeling, confirmed by the results of real concreting, allowed to determine the technology of heat treatment, which provided the necessary quality of concrete, hardened on a frozen base.

Key wads: modeling technology, foundation slab, frozen base, surface, heat release of the cement

1. Introduction

Winter conditions in the concrete works production in Siberia, in the edition of SP 70. 13330.2012. [1. p.5.11.1], last from 5 to 8 months in a year. This leads to a significant rise in the construction cost. Norms [1, paragraph 5.3.3] directly indicate the need to ensure the base temperature in the process of a concrete mixture laying at least 5°C. This means that the frozen base needs to be heated before the concrete laying, or to offer a method of laying that will prevent the possibility of concrete mixture freezing in the contact zone with the base [1, item 5.11.4].

In the winter construction 2017-18 seasons, in the process of the foundation slab concreting in the closed parking lot at Fabrichnaya Street in Novosibirsk, the construction company NTBild LLC was faced with the need to lay concrete on the frozen foundation. Up to the beginning of the slab concreting, in autumn 2017, according to the project, at the excavation bottom, a layer-by-layer cushion compacted sand of medium size with 300 mm thickness and a concrete preparation on a compacted base of B 7.5 class with a thickness of 100 mm were made. With the beginning of winter, site's construction work was stopped. The pit was filled with snow, which worked as a heater in the future. In February 2018, it was decided to continue work. The depth of ground freezing in the base, determined after snow removal by pilot drilling, was 500 mm. The work was scheduled for 01.03.2018. At the expected minimum daily temperature in Novosibirsk -20°C, the foundation plate with dimensions of 47100 × 19200 mm and a thickness of 500 mm was to be concreted, using the concrete grade of B20 F75 W6 plate with cement consumption of PC 400 - 400 kg/m^3. Heat treatment of concrete was assumed to be a heating wire fixed to the upper and lower reinforcing mesh.

The substantiation of a slab concreting technology was carried out in two stages.

1. Estimation of energy inputs and base heating time.

2. Evaluation of temperature conditions and strength of concrete foundation plate subjected to a wire heat treatment in various combinations of its location.

2. Heated base

According to environmental requirements, the ground heating in the urban area by coal massif burning on the surface is strictly forbidden. In addition, an open fire on the surface of the frozen concrete can lead to its destruction. The only possible way to warm the base remains heating with a heating wire or thermo-mats [2-4]. The slab area is about 900 m^2. It is rather problematic to collect such a number of mats even within the concreting into 2 captures. Therefore, preliminary there was evaluation of the ground heating with the use of the heating wire or cable under a dry sand layer.

Physical formulation of the problem. In accordance with the design scheme (Fig. 1), the heating wire is laid on the surface of concrete preparation with step b and covered with a sand layer padding of δ_n thickness. In order to ensure the uniformity of the temperature field at the base, the maximum pitch of the heating wire is b_{max}=250 mm. To simplify the calculations, fine concrete preparation and sandy pouring, considering the proximity of their thermophysical characteristics and small dimensions, can be included in the calculation scheme with a single layer of frozen soil. At the time the heating cable is turned on, the temperature of the sandy substrate is assumed to be equal to the temperature Θ.

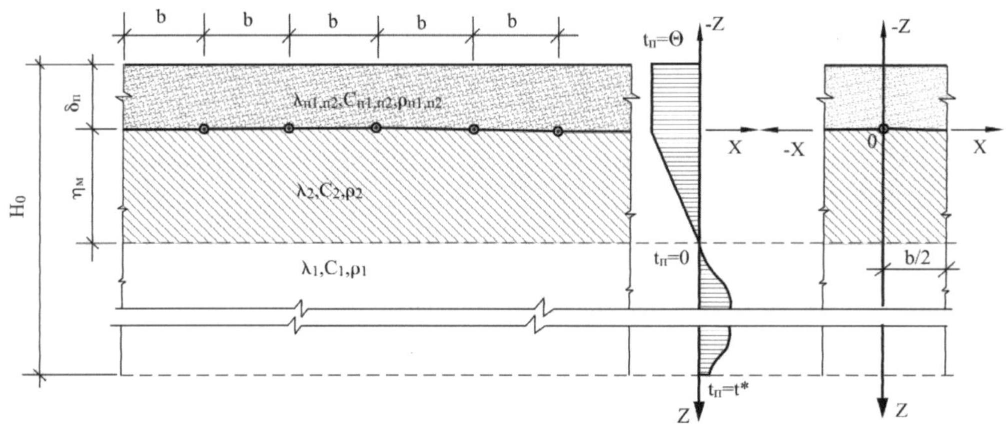

Figure 1. Calculation scheme for thawing and warming up the soil base of the pit under the slab foundation.

The freezing boundary of the unbound base ($\eta_м$) is clearly fixed by the plane of phase transitions [5-8]. Physicomechanical and thermophysical characteristics: sandy priming: λ_{n2}, C_{n2}, ρ_{n2}– in the frozen state and λ_{n1}, C_{n1}, ρ_{n1} - in the melted state; the characteristics of the base: the foundation soil, the priming and the concrete: respectively λ_2, C_2, ρ_2 и λ_1, C_1, ρ_1 with the moisture components (in fractions of unityω_{nsv} and ω_{sv}, due to the amount of unfrozen, respectively, unbound and bound moisture. The heating wire of the PNSV brand with a 1.2 mm steel core diameter allows providing a heat output of 8.9 w/m.

The mathematical formalization of the physical model has the form:

$$1.\ \frac{\partial t_{n1,n2}}{\partial \tau} = \frac{\lambda_{n1,n2}}{C_{n1,n2}} \left[\frac{\partial^2 t_{n1,n2}}{\partial x^2} + \frac{\partial^2 t_{n1,n2}}{\partial z^2} \right], x \in \left(\frac{b}{2} \right) \wedge z \in \left(-\delta_n \right)$$

$$2.\ \frac{\partial t_{1,2}}{\partial \tau}=\frac{\lambda_{1,2}}{C_{1,2}}\left[\frac{\partial^2 t_{1,2}}{\partial x^2}+\frac{\partial^2 t_{1,2}}{\partial z^2}\right]\pm\frac{W_{\phi}(\tau)}{C_{1,2}},x\in\left(\frac{b}{2}\right)\wedge z\in H_0$$

$$3.\ -\lambda_{n1,n2}\frac{\partial t_{n1,n2}(z=-\delta)}{\partial z}=\alpha\left[t_{n1,n2}(z=-\delta)-\Theta\right]$$

$$4.\ -\lambda_{n1,n2,1,2}\frac{\partial t_{n1,n2,1,2}(z=\eta(\tau))}{\partial z}+\lambda_{n2,n1,2,1}\frac{\partial t_{n2,n1,2,1}(z=\eta(\tau))}{\partial z}= \quad (1)$$

$$=\varepsilon\omega_{нсв}\cdot\rho_{n1(n2),1(2)}\frac{\partial\eta}{\partial z}$$

$$5.\ -\lambda_{n1,n2}\frac{\partial t_{n1,n2}(z=-\delta_n)}{\partial z}+\lambda_{1,2}\frac{\partial t_{1,2}(z=-\delta_n)}{\partial z}=q_{\omega_{ГЕК}}$$

$$6.\ t_1\left(x,z=H_0,\tau\right)=\text{const}=t^*$$

$$7.\ \frac{\partial t_{n1,n2,1,2}(x=0\vee x=b/2)}{\partial x}=0$$

$$8.\ t_{n2,1,2}(x,\tau=0)=\psi(z)$$

where 1.1 is the differential equation of the thermal conductivity of the sandy priming; 1.2 is differential equation of the thermal conductivity of cohesive soil, sandy priming and the base with a volumetric-distributed source of phase type heat; 1.3 is PG III kind on the surface of sandy priming; 1.4 is GU Stefan on the boundary of freezing (thawing) of unbound moisture; 1.5 is the thermal connection condition of the heating electrical cable to the sandy priming (from above) and to the ground of the base; 1.6 is PG of the first kind at the zero amplitude level of annual temperature variations; 1.7 is the thermal symmetry condition at the location of the cable core (x = 0) and at a distance b/2; 1.8 is the initial condition.

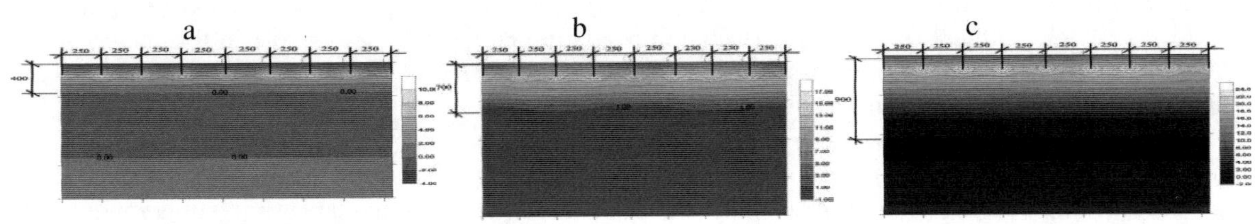

Figure 2. Temperature field of the base after a – 100, в– 200, c – 300 operating hours heaters. The horizontal chain sizes - step of laying the heating wire, vertical – depth base thawing.

Numerical implementation of the mathematical model. The dynamics of the change in the temperature field in the frozen ground mass after 100, 200 and 300 hours of the heaters operation is shown in Fig. 2.

It can be seen from the calculation results that at least 300 hours or 12.5 days will be needed to thaw the concrete, sandy priming and the frozen base. The decision is unacceptable in time and energy costs.

2.1. Temperature conditions and strength of concrete

Physical problem formulation. In accordance with the design scheme (Figure 3), the bottom of the pit was frozen to the depth h by the time of concreting. Estimated temperature Θ. Material of concrete and

Sandy priming managed to take the temperature of the ambient air. Peripheral heating of the plate can be carried out by electric heating wires with running heat q_l.

The intensity of the heat flux from the heating wires is determined by the formula

$$q_{\text{пр}} = \frac{q_l \cdot L}{1\,\text{m}^2}, \frac{\text{Wt}}{\text{m}^2} \tag{2}$$

where L is the length of the heating wire per 1 m^2 of the area.

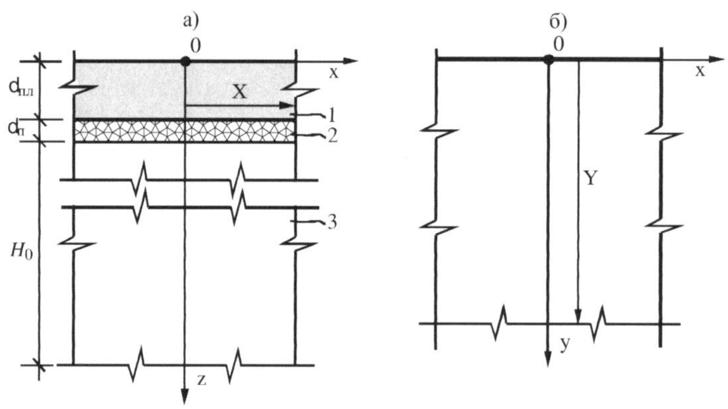

Figure 3. The calculated fragment of a concrete foundation slab: a) in a section; b) top view: 1 – foundation plate; 2 – concrete + sandy priming; 3 – foundation soil

With numerical realization of the differential equation of thermal conductivity of concrete foundation

$$Q_{\text{пр}} = \frac{q_{\text{пр}}}{C_{\text{б}} \cdot l}, \tag{3}$$

where l is the step of numerical integration along the axis normal to the vertical face in question.

Despite the fact that the foundation slab proposed for concreting refers to massive structures (M_n=0.47≤3 m^{-1}), where the volume-distributed heat source $\omega_{экз}$ plays a big role in the formation of the concrete temperature, it is necessary to verify the sufficiency of this source. If it is not enough, then you can implement warm-up methods:

- only the warming up process with electric heating wires of the side slab faces, which is concreted on the unheated soil base;
- heating the concrete of the slabs with electric heating wires at the level of the upper and lower reinforcing mesh and side faces.

In the process of a concrete mixture laying on a non-heated ground base, it is compulsory to make a computational confirmation that the concrete temperature at the contact with the concrete and sandy priming does not fall below 0 °C (for conventional concrete, without chemical additives) or below the freezing point of the mixing solution (concretes with antifreeze additives) .

The boundaries of the thermal influence zone of peripheral warming along the Y axis should be taken from the preliminary calculations from the $Y \geq 5\delta_{пл}$, condition of the boundary along the X axis should be taken from the condition $Y \geq 5\delta_{pl}$,, where b is the heating cable laying step. The lower boundary of the base calculation soil is at the point H_0, where the temperature t^* does not change during the year.

The mathematical formalization of the physical model has the form:

$$1.\ \frac{\partial t_{пл}(x,y,z,\tau)}{\partial \tau}=\frac{\lambda_{пл}}{C_{пл}}\left[\frac{\partial^2 t_{пл}(x,y,z,\tau)}{\partial x^2}+\frac{\partial^2 t_{пл}(x,y,z,\tau)}{\partial y^2}+\right.$$

$$\left.+\frac{\partial^2 t_{пл}(x,y,z,\tau)}{\partial z^2}\right]+W_{экз}+Q, x\in X\wedge y\in Y\wedge z\in\delta_{пл}\wedge\tau\in T$$

$$2.\ \frac{\partial t_{n1,n2}(x,y,z,\tau)}{\partial \tau}=\frac{\lambda_{n1,n2}}{C_{n1,n2}}\left[\frac{\partial^2 t_{n1,n2}(x,y,z,\tau)}{\partial x^2}+\frac{\partial^2 t_{n1,n2}(x,y,z,\tau)}{\partial y^2}+\right.$$

$$\left.+\frac{\partial^2 t_{n1,n2}(x,y,z,\tau)}{\partial z^2}\right], x\in X\wedge y\in Y\wedge z\in\left(\delta_{пл},\delta_{пл}+\delta_n\right)\wedge\tau\in T$$

$$3.\ \frac{\partial t_{1,2}(x,y,z,\tau)}{\partial \tau}=\frac{\lambda_{1,2}}{C_{1,2}}\left[\frac{\partial^2 t_{1,2}(x,y,z,\tau)}{\partial x^2}+\frac{\partial^2 t_{1,2}(x,y,z,\tau)}{\partial y^2}+\right.$$

$$\left.+\frac{\partial^2 t_{1,2}(x,y,z,\tau)}{\partial z^2}\right]\pm W_{ф}, x\in X\wedge y\in Y\wedge z\in\left(\delta_{пл}+\delta_n,\delta_{пл}+\delta_n+H_0\right)$$

$$4.\ -\lambda_{пл}\frac{\partial t_{пл}(x,y,z=0,\tau)}{\partial z}=K_{пр1}\left[t_{пл}(x,y,z=0,\tau)-\Theta\right]$$

$$5.\ -\lambda_{пл}\frac{\partial t_{пл}(x,y=0,z,\tau)}{\partial y}=K_{пр2}\left[t_{пл}(x,y=0,z,\tau)-\Theta\right] \quad (4)$$

$$6.\ -\lambda_{пл}\frac{\partial t_{пл}\left(x,y,z=\delta_{пл},\tau\right)}{\partial z}=\lambda_{n1,n2}\frac{\partial t_{n1,n2}\left(x,y,z=\delta_{пл},\tau\right)}{\partial z}$$

$$7.\ -\lambda_{n1,n2}\frac{\partial t_{n1,n2}\left(x,y,z=\delta_{пл}+\delta_n,\tau\right)}{\partial z}=\lambda_{1,2}\frac{\partial t_{1,2}\left(x,y,z=\delta_{пл}+\delta_n,\tau\right)}{\partial z}$$

$$8.\ -\lambda_{n1,n2,1,2}\frac{\partial t_{n1,n2,1,2}(z=\eta(\tau))}{\partial z}-\lambda_{n2,n1,2,1}\frac{\partial t_{n2,n1,2,1}(z=\eta(\tau))}{\partial z}=$$

$$=\varepsilon\omega_{нсв}\cdot\rho_{n1,n2,1,2}\frac{\partial\eta(\tau)}{\partial z}$$

$$9.\ t_1\left(x,y,z=\delta_{пл}+\delta_n+H_0,\tau\right)=\mathrm{const}=t^*$$

$$10.\ t_{пл}(x,y,z,\tau=0)=t_{бс}$$

$$11.\, t_{n2}\left(x, y, z \in \left(\delta_{nл}, \delta_{n} + \delta_{nл}\right), \tau = 0\right) = \Theta \wedge$$

$$\wedge t_{1,2}\left(x, y, z \in \left(\delta_{n} + \delta_{nл}, \delta_{n} + \delta_{nл} + H_{0}\right), \tau = 0\right) = \psi(z)$$

where 4.1 is the three-dimensional differential equation of the concrete thermal conductivity of a foundation slab with a volumetric-distributed heat source: exothermic (W_{ex}) type, acting throughout the concrete entire thickness ($x \in X \wedge y \in Y \wedge z \in \delta_{pl}$) and an artificial heat source Q_{pr} with fixation of its area actions for numerical approximation of the equation; 4.2 is three-dimensional differential equation of concrete thermal conductivity and sandy priming; 4.3 is three-dimensional differential equation of the base soil thermal conductivity with a volumetric-distributed heat source of phase type: W_f at $t_2 \in (0 \div 5$ °C) $\wedge W_f = 0$ at $t_{2(1)} > 0$ °C$\wedge t_2 < -5$ °C; the action of the source (sink) $W_f \neq 0$ is due to the freezing (thawing) of the bound moisture in the temperature range $t_2 \in (0 \div 5$ °C); 4.4 is PG III kind on the surface of the plate (allows for any insulation of open surfaces to be taken into account); 4.5 is PG III kind on the vertical edge of the plate; 4.6 is PG IV kind (the condition of thermal slab abutment to the concrete and sand priming); 4.7 is PG IV kind (the condition of sandy priming thermal contiguity to the foundation bottom); 4.8 is classical GU Stefan on the boundary of freezing (thawing) and unbound moisture; 3.13.9 is PG of the first kind at the zero amplitude depth of the soil temperature annual variations; 4.10 is initial condition for a concrete slab; 4.11 is initial condition for the basis. In order not to clutter up the article with mathematical transformations for the convenience of the reader, the numerical realization of the mathematical model is not given in the article.

Numerical implementation of the mathematical model. As it was already noted, in the case of expensive heating, it is more appropriate to use heating wires [9-15] Considering the massiveness and powerful heat release of cement, we calculate the most economical option: only peripheral heating of the concrete laid vertical faces on the unheated soil base, to compensate for heat losses through the side formwork and insulation and vapor barrier of exposed surfaces. On top of the heater in the form of a sawdust layer, K_{ym} =6 0.9 W / (m^2·K), on the vertical faces of uninsulated plywood formwork, K_{on} = 6 W / (m^2·K). The concrete mixture is laid with a temperature of t_{bs} = 7°C. Only the side faces of the slab are heat treated. The calculated air temperature is Θ = -20°C. The temperature fields in the slab concrete at the moment of the end of its aging, the average temperature change of the concrete temperature in time, as well as the temperature of the concrete at the contact with the concrete and sandy pouring, is shown in Fig. 4. Concrete did not score critical strength in the contact zone.

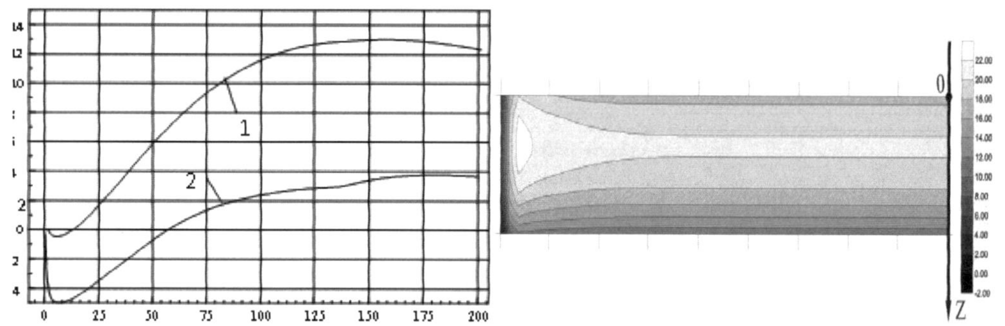

Figure 4. Results of calculations for concrete peripheral warming on the unheated soilbasis. 1 – volume-weighted average temperature of concrete according to calculations; 2 – concrete temperature in the place where it joins the sandy priming.

As can be seen from the forecast, the concrete loses its temperature at the boundary with the frozen surface after laying sharply, but at the expense of exothermal it quickly restores the average volume, but on the contact with the base it freezes at an early age and remains in the zone of negative temperatures for a long time, which can not but affect its quality. After 200 hours the average volume strength will reach 65 ÷ 70 % of the branded strength. However, the border zone is in danger.

In order to ensure the normal hardening of concrete and the acquisition of guaranteed strength, we calculate the more energy-intensive option. It is heating the concrete slab with electric heating wires at the level of the upper and lower reinforcement grids and peripheral heating of the vertical faces of the concrete laid on the unheated soil base (Fig. 5). The received intensity of internal heating by heating wires has been accepted equal: on the top reinforcement grid of 65 W/m^2 and on the bottom reinforcing grid 200 W/m^2. The time of critical strength concrete set was 75 hours, and the maximum temperature of concrete was 45°C.

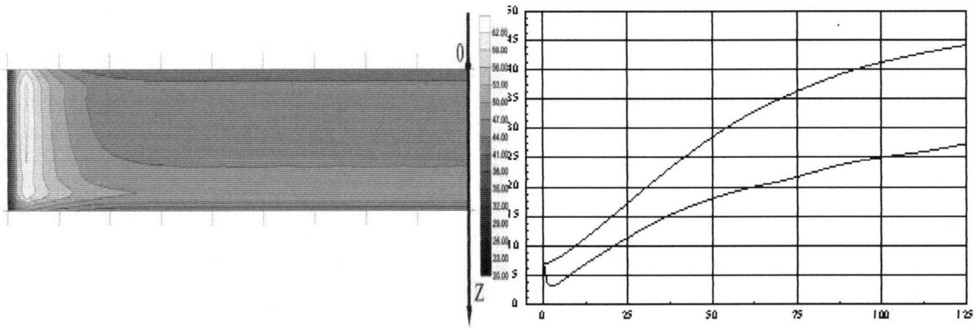

Figure 5. Concrete peripheral warming in combination with its heating at the level of the upper and lower reinforcement grids on the unheated soil base: 1 – weight-averaged volume of concrete temperature; 2 – temperature of concrete in the place where it joins the sandy priming.

3. Results

As a result of these studies it is argued:

- for thawing the base with a heating wire on a sand layer it will take 12.5 days, which is completely unacceptable;
- foundation plate thickness of 0.5 m in the process of concrete laying on a cold base, including 0.1 m of concrete preparation, 0.3 m of sandy soil and 0.5 m of frozen ground, due to intensive accumulation of heat, the base rapidly loses temperature and exothermic heat is clearly not enough to compensate for heat losses and conditions ensuring the purchase of concrete on contact with the base of the required strength;
- Calculations show the need for an additional heat source of 65 W/m^2 on the upper and 200 W/m^2 on the lower reinforcing mesh, which, together with heating of slab sides, ensures the purchase of concrete with specified strength.

References

[1] JV 70. 13330.2012. Set of rules. Bearing and enclosing structures. Updated version of SNiP 3.03.01-87СП 70. 13330.2012

[2] Zhu Bo-Fang, Long-Sheng Wu, Yue Li, and Guo-Xin Zhang 2007 The thermal stress and superficial thermal insulation of concrete dam under construction in winter in cold regions. *Shuili Shuidian Jishi* (*Water Resources and Hydropower Engineering*) **38** (**8**) pp 34-7

[3] 2005 *Guide to heating concrete in monolithic structures* Ed. Krylov B A, Ambartsuman S A, Zvezdov A I (Moskow: NIIZhB Gosstroy RF) p 275

[4] Popov Yu A 2005 Mathematical modeling of the dynamics of the temperature and strength fields under differentiated concrete heating in building structures of a prismatic shape *Izv. universities* **4** pp 73-78

[5] Pekhovich A I 1976 *Calculations of the thermal regime of solids* (Leningrad: Energy) p 52

[6] Kowatari T H, Ishikawa T, Tanno and Madokoro K 1988 Construction of mass concrete and winter concreting on foundation of bridge. *Concrete Journal* **26** (**12**) pp 22-9

[7] Palkin Yu S 1970 To the calculation of a two-dimensional nonstationary heat transfer process in a freezing ground *Izv. universities* **3** pp 51-57

[8] 1989 *Recommendations for electric heating of monolithic concrete and reinforced concrete by heating wires* (Moskow: Publishing house TSNIIOMTP) p **67**

[9] Sugiyama M, Takamori H, Fukai A and Ishimori F 2003 Curing method of the winter construction of the housing base concrete *Concrete Journal* **41** (**7**) 23-8

[10] Liu, Rong-Su 2011 Inverted siphon project of south zhulong winter temperature control of concrete culvert *Water Conservancy Science and Technology and Economy* **17** (**7**) 93-4

[11] Chen, You-Zhi, Tao Sun, Fu-Kui Ma, Chang-Chun Yuan, Qi-Long Li, and Zhi-Long Qu 2007 Research on the temperature control measure in the winter construction of C20 large volume concrete of zhijing river bridge. *Wuhan Ligong Daxue Xuebao/Journal of Wuhan University of Technology* **29** (**3**) 110-3

[12] Krylov B A 1978 Methods of production of concrete works with the application of heating and heating of structures *Second International. simp. for winter concreting: gener. doc.* (Moskow: Stroiizdat) pp 101-122

[13] Kilpi, Eero, and Heikki Kukko 1982 Properties of hot concrete and its use in winter concreting. Nordic Concrete Research

[14] Lysov V P 1982 *The effectiveness of concrete works in construction* (Minsk: Publishing house of Belarus) p 90

IOP Conf. Series: Materials Science and Engineering **481** (2019) 012017 doi:10.1088/1757-899X/481/1/012017

Geotechnical monitoring of a residential building being erected on strip-shell foundations in "Aquarel" housing complex in Tyumen

Y A Pronozin[1], **L R Epifantseva**[2], **M A Stepanov**[1]

[1]Department of Geotechnics, Tyumen industrial university, Tyumen, Russia
[2]Department of Building construction, Tyumen industrial university, Tyumen, Russia

E-mail: epifantsevalr@mail.ru

Abstract. The actual issues of expanding the field of application of foundations for high-rise buildings are considered. Traditional foundations are not always effective on construction sites composed of normally and strongly compressible soils; thus, in certain conditions thin-walled strip-shell foundations (TSSF) can be used for cost reduction of the zero cycle. The paper describes the geotechnical monitoring program for GP1.1 built on TSSF - an alternative type of foundation as compared to traditional shallow foundations. The paper presents theoretical modelling of interaction of soil with the building; it also describes the technology of its manufacture using standard materials of STO SRO 001-2015 - "Design requirements for strip-shell foundations". The modelling results have shown the efficiency of foundations with a curvilinear convex upwards contact surface. The increased stiffness of the soil bed is caused by additional side squeezing of soil due to the specific features of the contact surface shape. The paper discusses the efficiency of TSSFs as to material consumption and cost as compared to pile-slab foundations. The paper gives the comparative data of geotechnical monitoring and approximate analysis.

1. Introduction

At present, the tendency to use the systems with controlled properties is appearing in geotechnical engineering [1-6]. Thin-walled strip-shell foundations (TSSF) are an example [7, 8]. TSSFs make it possible to form the given stress in the soil bed under strips and shells during the design phase and improve soil behavior under loading [9-11]. The curvilinear shape of the loaded surface of TSSF affects the characteristics of soil deformability by additional horizontal squeezing of soil in the upper zone of the bed resulting in stiffness increase [12, 13].

TSSF consists of the elements differing in stiffness and peculiar features of behavior (Fig. 1). The first one - *strip foundation* – is a supporting structure for bearing walls or columns possessing a certain stiffness in the longitudinal direction, taking random eccentricities of load transfer [14-16] and forming the desired cantilever widening on the outer contour of the building. The second one is a *flexible element* in the span joining the nearest supporting strips as a slab foundation. It functions as a transmitter of the certain (given) load from the building on the soil bed in the span. The convex upwards shell which is good in tension is considered to be the most appropriate flexible connection between strip foundations. The shell may actively influence the soil "response" due to such parameters as: width ratio of the strip and span, shell height, tensile stiffness, flexural stiffness, fixing in the strip foundation (flap hinge, embedding), function curvature of the shell.

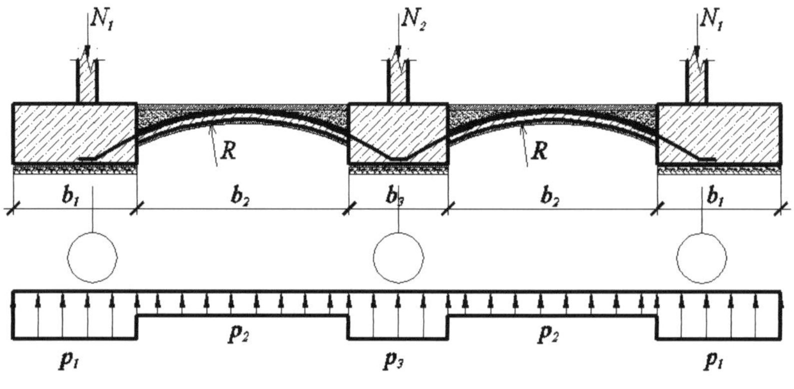

Figure 1. Strip foundations with flexible connection.

Structurally, the convex upwards shell is good in central tension as energetically the most favorable state of stress [17, 18] and possibility to form almost equal-stress state in length; technologically, it is good because of small thickness, single-layer reinforcement and low complexity of work.

The economic advantages, as compared to slab foundations, are as follows: significantly smaller amount of concrete, slightly smaller amount of reinforcement and hence, lower cost.

The building for geotechnical monitoring - GP 1.1 is located in "Aquarel" housing complex in Tyumen. The 22-storeyed monolithic frame building with dimensions of the main axes to be 25.75 m x 26.0 m in plan is a space-planning solution for GP 1.1. The structural diagram of the building is a frame. The monolithic reinforced concrete frame is made of concrete of class B25 in strength. Spatial rigidity and stability of the building is provided by the joint performance of the monolithic stiffening diaphragms (the walls of the staircase-elevator unit) with horizontal stiffness discs, monolithic slabs and columns of the building strictly interconnected with foundations and slabs.

Engineering-geotechnical elements XV-XV (Figure 3) are shown in Table 1. The laboratory data and engineering-geological section indicate the complexity of the geological structure; that is necessary to be considered in designing.

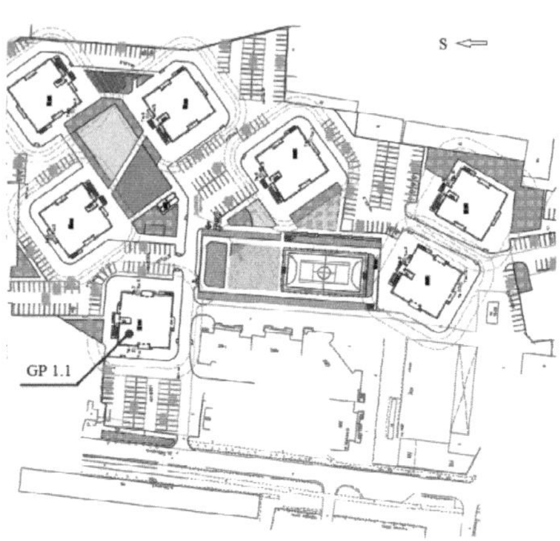

Figure 2. General schematic view.

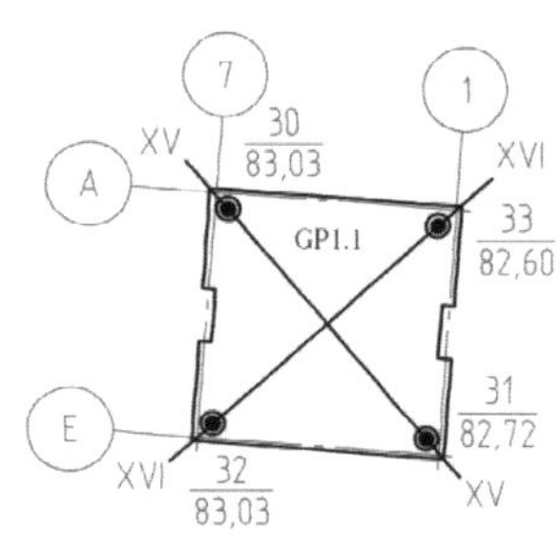

Figure 3. Schematic view of drill holes location.

Table 1. Engineering-geological conditions of the construction site.

Description of soil	Depth H, m	γ_I, kN/m^3	φ_{II}, degrees	c_{II}, kPa	E, MPa
1. Semi-solid-solid loam with clay and sand layers	0-1.7	19.3	21.6	24.3	22.58
2. Stiff loam with sand layers	1.7-4.9	17.31	16.1	26.4	11.32
3. Semi-solid-solid clay with sand layers	4.9-7.7	17.31	19.4	63.1	22.29
4. Sandy, medium-density, water-saturated silt with loam layers	7.7-14.8	19.32	29.3	4.4	12.56
5. Semi-solid-solid clay with sand layers	14.8-23.5	17.31	19.4	63.1	16.34

Groundwater is considered to be glacial-lacustrine deposits of the IV terrace above the flood-plain. The depth of groundwater occurrence is 3.9-5.2m. The depth of the standing-groundwater level is 3.2-4.8 m.

The standard depth of the seasonal freezing of soils calculated after [19] is 1.8 m.

2. Methods

The analysis was carried out in SCAD software programme using the contact model based on Winkler's hypothesis; the 3D analysis took into account all the major elements of substructure and superstructure [20]. The analysis was performed according to the algorithm presented in STO SRO 001-2015 "Design requirements for strip-shell foundations" [21].

The average settlement of TSSF was analyzed according to [21] and Building Regulations 22.13330 [22] using the total calculated deformation modulus.

The total calculated deformation modulus Ej * needed to find k_d (see. p. 6 of the Algorithm [21]) was determined in PLAXIS 2D AE geotechnical software programme; the 2D analysis was done on the basis of numerical implementation of the problem of soil-shell interaction, the soil being simulated with elastic half-space [23-25].

In order to determine k_l (see. p. 6 of the Algorithm [21]), the layering summation method dividing shear and bulk soil deformations [23] was taken for analysis. Poisson's ratios for soils within the depth of the compressible strata were taken in accordance with the Building Regulations [19] due to lack of the source data. Limits of the depth of the compressible strata were also taken according to the Building Regulations [19]. The total specified characteristic load was 230 kPa.

Implementation of the analysis algorithm [18] for interaction of the foundations under study with the soil bed is given further.

1. The foundation depth is taken according to the architectural solutions, taking into account the requirements of p. 5.5 [22], equal to 2.1 m from the level of the design elevation.

2. Preliminary determination of TSSF dimensions (including consoles) is done. The outline of the shell is compatible to a square parabola and criterion of flatness (Figure 4). As a result, the following TSSF active design parameters are obtained: f = 0.75 m, EA = $4.6 \cdot 10^6$ kN/m, EI = const = 17050.3 kN$\cdot$ m.

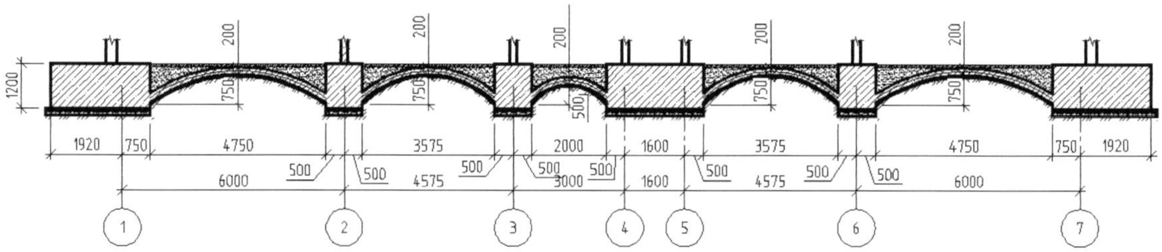

Figure 4. Overall dimensions of TSSF characteristic section and reactive pressures under the individual elements of foundation.

3. Let us determine the depth of the compressible strata H_{comp}, according p.5.6.41 [22]. Given the total specified characteristic load - $H^{230}{}_{\text{comp.}} = 18.2$ m.

Let us evaluate the initial coefficient of soil bed reaction k_i, by the formula:

$$k_i = \frac{p_{aver}}{s_{aver}} \qquad (1)$$

where p_{aver} – average pressure under the TSSF footing, kPa; S_{aver} – average settlement of TSSF bed, m, determined according to p.5.6.31 [22].

Thus, $k_i^{230} = 1.084\,\text{MN/m}^3$

4. Let us simulate the foundation and the building superstructure frame in 3D and perform the static analysis of the *bed*-foundation-*superstructure* system in the total specified characteristic load with the specified number of floors for adequate consideration of the superstructure stiffness. Let us determine the settlements s_i, m, and contact pressures p_i, kPa under the individual elements of TSSF in normal loading. The average settlement values are $s^{230}{}_{aver}$ =137.0 mm. The reactive pressure diagram in normal loading is shown in Figure 5.

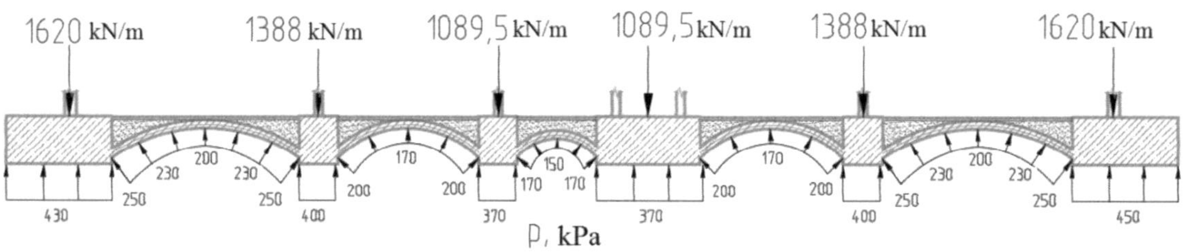

Figure 5. Reactive pressure diagram in normal loading.

5. Let us check the contact pressure values under the TSSF strips in accordance with the condition (2) in the final stage of loading:

$$p_{sf} = (0.5 \div 0.9)R \qquad (2)$$

where R – calculated resistance of soil under the strip foundation, kPa, determined in accordance with p.5.6.7 [22], taking into account the counterweight from pressure of the neighboring shells for average strip foundations and the counterweight from the shell and backfill soil for edge strip foundations.

In non-compliance with the condition (2), the geometric dimensions of the strips and shells in plan and the design parameters are adjusted: shell depth f, shell tensile stiffness EA, its bending stiffness EI and the system is reanalyzed.

Thus, in the last stage of loading $P_{sf}^{aver} = 0.3R_{sf}^{aver}$, $P_{sf}^{edge} = 0.61R_{sf}^{edge}$.

Changes in the geometry of the system are not preformed.

6. Let us integrate the static analysis and evaluate the corrected values of the coefficients of soil bed reaction MN / m3, under the separate TSSF elements by the formula:

$$k_{i+1} = p_i / S_i \tag{3}$$

where p_i – pressure under the separate zones of loading – TSSF elements in the first approximation, kPa; S_i – average settlements under the separate zones of loading – TSSF elements in the first approximation, m.

If the difference between k_i and k_{i+1} exceeds 10%, the system is reanalyzed with the corrected values of the coefficients of soil bed reaction – the second approximation, etc. n times up to conformance with pressures p_n arising from the defined coefficients of soil bed reaction k_n and settlements s_n under the separate zones of TSSF loading.

7. Let us perform the static analysis taking into account the deformable characteristics of soil bed loaded along the curvilinear convex surface in normal loading.

The calculated values of the coefficient of soil bed reaction, k_n^*, MN/m^3, are evaluated by the formula:

$$k_n^* = k_n \cdot k_f \tag{4}$$

where k_n – coefficients of soil bed reaction under the individual TSSF elements, MN/m^3, determined by the iterative approach in the last approximation; k_f – coefficient considering soil bed stiffness increase under the shells of TSSF due to the curvilinear form of loading is evaluated by the formula:

$$k_f = 1 + \frac{k_d A_{shell}}{A \cdot k_1} \tag{5}$$

where k_1 – the settlement ratio of soil bed layers from additional vertical stresses σ_z as a result of soil bed loading by the entire area of TSSF and a separate shell within the limits of the j-layer of soil; k_d – coefficient taking into account compressibility decrease of the soil bed loaded along the curvilinear surface with respect to loading along the flat surface:

$$k_d = 1 - \frac{S_{shell}}{S_{flat}} \tag{6}$$

where S_{shell} – soil bed settlement along the shell axis, loaded along the curvilinear surface, with the distribution diagram of the contact pressures specified from the approximate analysis considering E_j^*; S_{flat} – soil bed settlement along the load axis on the flat surface, with a uniform distribution diagram of the contact pressures considering E_j.

Thus, the coefficient of the form for normal loading is $k_f^{230} = 1.192$

8. Let us check the conditions:

$$S_n < [S] \tag{7}$$

$$\frac{\Delta S_n}{L} < \left[\frac{\Delta S}{L}\right] \tag{8}$$

where s_n- average settlements under separate zones of loading – TSSF elements in the last approximation, m; $[S]$– ultimate limit settlement, m, after [22]; ΔS_n – difference in settlements of the neighbouring strip foundations, m; L – distance between the axis of the principal strip foundations, m;

According to the analysis done after the geomechanics model using E_j^*, the settlement is 135 mm, after the contact model using k_n^* the settlement is 124 mm, which is less than the ultimate limit value – [s] = 150mm.

Unevenness of settlement is almost 1.6 times less as compared to the standard settlement – 0.0018 < $[\Delta s/L]$ = 0.003, L – distance between the axis of the principal strip foundations, m.

It should also be noted that the standard analysis for settlements by the method of summation by layers [22], using the total calculated deformation modulus E expects the settlement to be within 21.2 cm; this exceeds the ultimate limit values and makes it difficult to use shallow foundations in these conditions.

9. The checking analysis is required for the soil bed under strip foundations in order to prove the validity of using the theory of linear deformability of soils in normal loading in accordance with (9), taking into account the last iteration considering soil bed stiffness increase under the shells.

$$p_{st} \leq R \tag{9}$$

where p_{st} – average pressure under strip foundations of TSSF, kPa, considering reanalysis; R – the same as in formula (2).

Due to some decrease in pressure under the strip foundations, the analysis can be ignored and taken as the final results presented in p. 5.

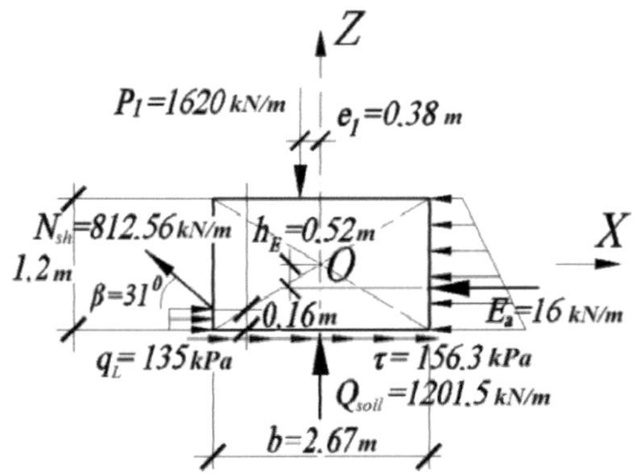

Figure 6. Design diagram for the edge strip foundation of 22-storeyed building.

10. Let us perform the analysis of the edge uncompensated strip foundations for possible torsion. The analysis is made for displacement of the resulting load vector from the section gravitational center in order to eliminate torque, taking into account the whole of forces acting on strip foundations [14-16]. The side reaction of the natural bed at the contact zone with the side surface of the strip foundation (Figure 6):

$$q_L = q_{vert} \cdot \xi = 250 \cdot 0.54 = 135 \text{ kPa},$$

$\xi = \dfrac{\nu}{1-\nu} = \dfrac{0.35}{1-0.35} = 0.54$ – coefficient of side soil pressure; $q_{vert} = 250$ kPa – vertical pressure under the shell at the contact point with strip foundation.

Force due to friction of soil along the strip foundation footing:

$$\tau = \frac{Q_{soil}}{b} \cdot \text{tg}\phi + c = \frac{1201.5}{2.67} \cdot \text{tg}16.1° + 26.4 = 156.3 \text{ kPa}$$

where running reactive soil pressure under the strip foundation:

$$Q_{soil} = p_1 \cdot b = 450 \cdot 2.67 = 1201.5 \text{ kN/m}$$

Longitudinal stress in the shell:

$$N_{shell} = \frac{p_1 - Q_{soil}}{\sin\beta} = \frac{1620 - 1201.5}{\sin 31°} = 812.56 \text{ kN/m}$$

The resultant of the active pressure: $E_a = 16$ kN/m.

If the value of eccentricity is $e_1 = 0.38$ m, $h_e = 0.52$ m, the torsional load is $\Sigma M_0 \approx 0$. The maximum deflection of the edge strip foundation in maximum distance of the transverse edges – $l = 5.8$ m is 0.15 cm, which is considerably less than the limit value - [1 / 150l] = 3.86 cm.

Strip foundations are constructed in the zones of load-bearing walls and columns and take the load from the building. The shells are formed in the span between the loaded zones and represent gently sloping cylindrical convex upwards reinforced shells (Figure 7).

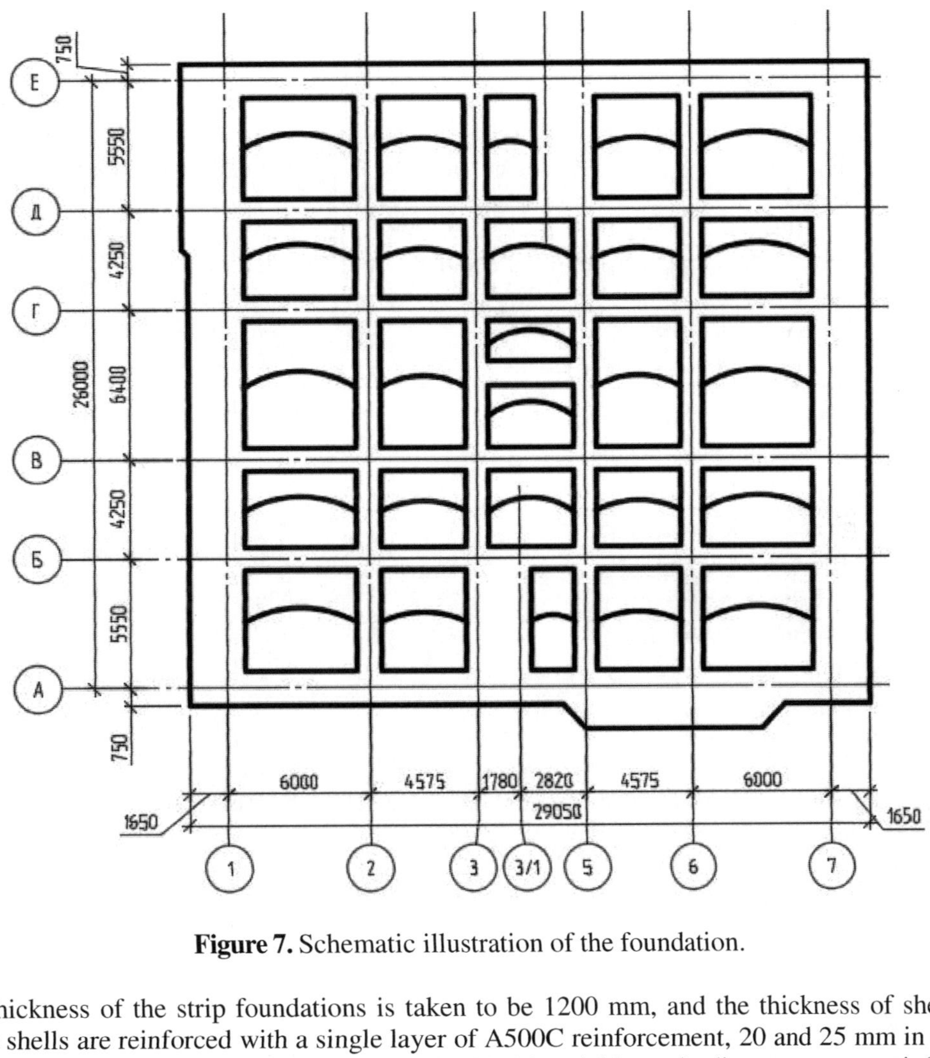

Figure 7. Schematic illustration of the foundation.

The thickness of the strip foundations is taken to be 1200 mm, and the thickness of shells - 200 mm. The shells are reinforced with a single layer of A500C reinforcement, 20 and 25 mm in diameter. The strip foundations are reinforced with A500C - 20, 22 and 25 mm in diameter, cross reinforcement - A240, 8 mm in diameter. Heavy concrete - B30, F200 and W8 was taken for concrete works.

The foundation for GP-1.1 was carried out through line system. Construction of each section was conditionally divided into 5 phases: ground excavation (Figure 8), reinforcing and timbering (Figure 9), concrete works (Figure 10).

Figure 8. Ground excavation. Digging trenches and forming curvilinear surfaces.

Figure 9. Reinforcing and timbering.

Figure 10. Concrete works.

Table 2 shows materials consumption and the estimated cost for raw version of GP-1.1 erected on pile-slab foundation with driven piles - 300x300 mm, 12 m in length, incorporated with a slab - 1m in thickness and TSSF.

Table 2. Materials consumption and the cost of foundations for GP-1.1.

Kinds of works	GP-1.1	
	Pile-slab	TSSF
Ground excavation, m^3	3200	2740
Fine excavation, m^3	97	191.1
Driven piles, 1 m^3 of piles	348.8	-
“Cutting” of pile heads, 1 pile	320	-
Crushed stone bedding, m^3	137	64.3
Concrete bedding, m^3	86.7	43.3
Heavy concrete, m^3	841	581.1
Reinforcement, kg	57720	74501
Estimated total labour costs, man/hours	5550	3448
Estimated cost in prices of the 3d Q, 2013, thous.rub.	15154	7674

3. Results

The settlement points (Figure 11) were installed on perimeter in the base course of the building [27]. The stationary part of the settlement points is rigidly fixed in the base course, and the inventory part with a truncated spherical head makes it possible to install the leveling board at the middle of its toe when re-leveling strictly on the same point. Monitoring was carried out with a high-precision level - H-0.3 using a barcode leveling board, thus the accuracy of measurements (0.5 mm) was ensured at the given facility [28].

Monitoring was being carried out from commencement of construction. The foundation was concreted in October, 2014. In December 2015, construction of GP 1.1 was approximately 90% complete based on 100% of the specified characteristic load - p_{aver} = 230 kPa (Figure 12). According to the developed algorithm (p.2) the calculated value of the final settlement is within 12.5-14.0 cm for the considered facility, in ultimate limit values - 15cm [22].

In the middle of August 2016, the average settlement was 83.5 mm, the maximum one – 97.9 mm, the minimum one - 64.3 mm. The maximum settlement unevenness was $\Delta s/L$ = 0.00093. The diagram of settlement points located on perimeter of the building and diagram of settlements are shown in Figure 13.

Figure 11. Installation of a leveling board on the settlement point.

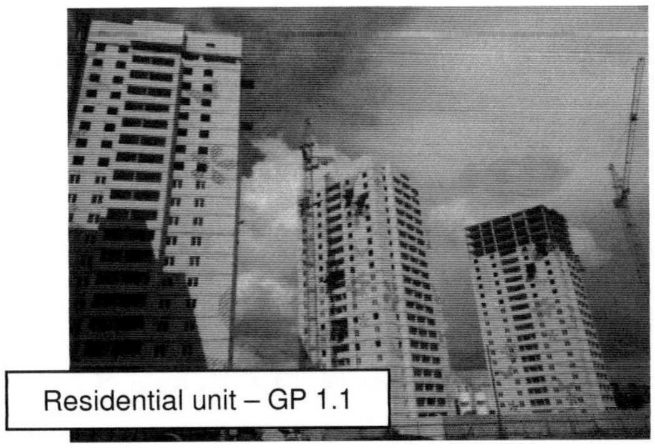

Figure. 12. General view of the residential unit – GP 1.1 under construction.

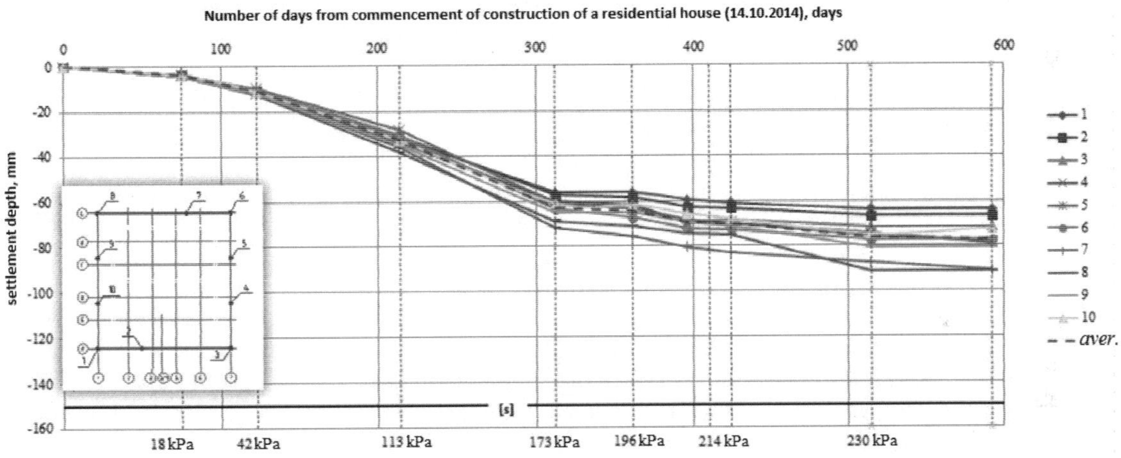

Figure13. Diagram of settlements developing in time for the residential unit – GP 1.1.

Taking into account the results of monitoring, soil formations in the bed of the given construction site, it is needed to predict the final settlement value within 11-13cm, i.e. up to 10-15% in accordance with the calculated data [29-31].

4. Conclusion

The modeling results have shown the efficiency of foundations with a curvilinear convex upwards contact surface. The increased stiffness of the soil bed is caused by additional side squeezing of soil due to the specific features of the contact surface shape.

Reasonable use of effective types of shallow foundations will reduce the basic normalized the design parameter - settlement, up to several tens of percent, increasing the carrying capacity of the ground base, while reducing the consumption of materials, namely steel and concrete by 20-50%, with respect to the traditional, such as slab or pile-slab foundations.

References

[1] Goncharov B V, Galimnurova O V, Gareeva N B and Bashlykov A V 2011 Estimation of bearing capacity and settlement of shell foundations on tamped beds *Soil Mechanics and Foundation Engineering* Vol **48** No 2 pp 62-66 (rus)

[2] Ter-Martirosyan Z G, Pronozin Ya A and Stepanov M A 2012 Feasibility of pile-shell foundations with prestressed soil beds. *Soil Mechanics and Foundation Engineering* Pp 1-5 (rus)

[3] Samokhvalov M, Zazulya J, Melnikov R and Mironov V 2016 Design Calculation of Drill-Injection Piles with Controlled Broadening and Silty-Clayed Soil Foundation Basic Interaction Parameters *MATEC Web of Conferences* 73 2016 01009 (rus)

[4] Kiselev N, Pronozin Y, Stepanov M, Bartolomey L and Keck D 2016 Theoretical and Experimental Substantiation for Applicability of a Damping Layer in a Foundation Slab Placed on Soil Bed *MATEC Web of Conferences* 73 2016 01017 (rus)

[5] Ikonin S V and Suhoterin AV 2015 Konstruksia fundamentnoy plity c reguliruemymi svoistvami [The design of the foundation plate with adjustable effort] *Magazine of Civil Engineering* No **3** (**55**) pp 10-20 (rus)

[6] Goncharov Yu M 1990 Experience with construction and operation of residential buildings on shell-foundations in Igarka *Soil Mechanics and Foundation Engineering* **27** No **3** pp 91-96 (rus)

[7] Goncharov B V and Rybakov A V 2001 Shell foundations on tamped soil beds *Soil Mechanics and Foundation Engineering* **3** pp 167-171 (rus)

[8] Luo Y F and Teng J G 1998 Stability analysis of shells of revolution on nonlinear elastic foundations *Computers & Structures* **69** No **4** pp 499-511

[9] Hong T, Teng J G and Luo Y.F 1999 Axisymmetric shells and plates on tensionless elastic foundations *International Journal of Solids and Structures* **36** No **34** pp 5277-5300

[10] Paliwal D N, Pandey R K and Nath T 1996 Free vibrations of circular cylindrical shell on Winkler and Pasternak foundations *International Journal of Pressure Vessels and Piping* **69** No **1** pp 79-89

[11] Kolesnikov A G 2015 Issledovanie napryazhenno-deformirovannogo sostoyania pologih obolochek na uprugom osnovanii [Investigation of stress-strain state of shallow shells on elastic foundation] *Science and Peace* vol **1** pp 83-85 (rus)

[12] Abdel-Rahman M 1996 Geotechnical behaviour of shell foundations *Ph. D Thesis, Department of Civil Engineering, Concordia University* (Montreal, Canada)

[13] Bogomolov A N and Ushakov A N 2013 Stress-strain state of an elastic half plane under a system of inclined piecewise-linear loads *Soil Mechanics and Foundation Engineering* **50** No **2** pp 43-49 (rus)

[14] Bogomolov A N, Ushakov A N and Bogomolova O A 2014 Stress Distribution in the Bed of an Inclined Absolutely Rigid Platewith Consideration of Friction Along the "Plate-Soil" Contac *Soil Mechanics and Foundation Engineering* **51** No **4** pp 165-172 (rus)

[15] Ter-Martirosyan Z G, Pronozin Y A and Kiselev N Y 2014 Shallow Strip Foundations Joined by Gently Inclining Envelopes on Highly Compressible Soils *Soil Mechanics and Foundation Engineering* **51** No **4** pp 157-164 (rus)

[16] SNiP 2.02.01–83* Osnovaniya zdaniy i sooruzheniy [Soil bases of buildings and structures] (Moscow: Stroyizdat) 1995 (rus)

[17] Utkin V S 2016 Raschet nadezhnosty gruntovyh osnovaniy fundamentov zdaniy I sooruzheniy po kriteriu deformasiy pri ogranichennoy informasii o nagruzkah I gruntah [Calculation of the reliability of soil basement foundations of building and structures by deformation criterion with limited information about the load and soil] *Magazine of Civil Engineering* pp 4-13 (rus)

[18] *Standart organizasii STO SROP 001-2015* Trebovaniya k proektirovaniyu I ustroistvu lentochno-obolochechnyh fundamentah [Requirements for the design and construction of strip-shell foundations] (Tyumen) p 43 (rus)

[19] SP 22.13330.2011 Osnovaniya zdaniy i sooruzheniy. Aktualizirovannaya redaksia SNiP 2.02.01-83* [Building Regulations 22.13330.2011 Foundations of buildings and structures. Updated version of SNiP 2.02.01-83*] (Moscow: OAO "SPP) 2011 p 161 (rus)

[20] Goncharov B V, Gareeva N B, Galimnurova O V and Bashlykov A V 2010 O raschete fundamentov-obolochek na vytrambovannom gruntovom osnovanii [About calculation of shell foundations on compressed building base] *Proceedings of the Kazan State Architectural University* No **2** (**14**) pp 143-148 (rus)

[21] Bashlykov A V, Bogomolov A N and Goncharov B V 2007 Issledovanie raboty fundamentov-obolochek na vytrambovannom osnovanii I metod osenki nesushey sposobnosti [Research of the shell-foundations work on compressed building base and method for estimating the bearing capacity] *Herald of Volgograd State University of Architecture and Civil Engineering. Series: Construction and architecture* No **8** pp 18-21 (rus)

[22] Emelyanov I G and Kuznetsov V Y 2002 Oprdelenie uprugoplasticheskogo napryazhennogo sostoyania kontaktiruyushei s osnovaniem silindricheskoy obolochki [Determination of elastic-plastic stress state in contact with the base of the cylindrical shell] *Herald of Russian Peoples Friendship University. Series: Engineering studies* No **1** pp 42-46 (rus)

[23] Ukhov S B, Semenov V V, Znamenskiy V V and Ter-Martirosyan Z G etc. 2005 Mechanika gruntov, osnovaniya I fundamenty [Soil mechanics, foundation and Basement: Textbooks for high schools building specialties] (Moscow: ASV Publishing) p 528 (rus)

[24] Bogomolov A N and Ushakov A N 2013 Stress-strain state of an elastic half plane under a system of inclined piecewise-linear loads *Soil Mechanics and Foundation Engineering* **50** No **2** pp 43-49 (rus)

[25] Bogomolov A N, Ushakov A N and Bogomolova O A 2014 Stress distribution in the bed of an inclined absolutely rigid plate withconsideration of friction along the plate-soil contact *Soil Mechanics and Foundation Engineering* **51** No **4** pp 165-172 (rus)

[26] Ter-Martirosyan Z G, Pronozin Y A and Kiselev N Y 2014 Shallow strip foundations joined by gently inclining envelopes on highly compressible soils *Soil Mechanics and Foundation Engineering* **51** No **4** pp 157-164 (rus)

[27] Ter-Martirosyan Z G, Telichenko V I and Korolev M V 2006 Problemy mechaniki gruntov, osnovaniy I fundamentov pri stroitelstve mnogofunksionalnyh vysotnyh zdaniy I kompleksov [Problems of soil mechanics, bases and foundations in the construction of multi-functional high-rise buildings and complexes] *Herald of MGSU* No **1** pp 18-27 (rus)

[28] Yun H-B and Reddi L N 2011 Nonparametric Monitoring for Geotechnical Structures Subject to Long-Term Environmental Change Advances in Civil Engineering 275270

[29] Pronozin Y A, Naumkina U V and Epifantseva L R 2015 Lentochnye fundamenty, obedinennye pologimi silindricheskimi obolochkami, dlua zdaniy poviyshennoy etazhnosty [Strip foundation, combined shallow cylindrical shells for high-rise buildings] *Industrial and civil construction* No **12** pp 58-62 (rus)

[30] Huat B B K and Mohammed T A 2006 Finite Element Study Using FE Code (PLAXIS) on the Geotechnical Behaviour of Shell Footings *Journal of Computer Science* vol **2** No **1** pp 104-108

[31] Rinaldi R 2012 Inverted Shell Foundation Performance In Soil *A Thesis in the Department of Building Civil & Environmental Engineering* (Montreal: Concordia University)

IOP Conf. Series: Materials Science and Engineering **481** (2019) 012018 doi:10.1088/1757-899X/481/1/012018

Empirical Period-Height Relationship for reinforced Concrete Moment Resisting Buildings in India

S P Kewate[1], M M Murudi[2]

[1] Department of Civil Engineering, Vishwaniketan's Institute of Management Enterpreneurship & Engineeringn, Kumbhivali, Khalapur, Maharashtra 410203, India
[2] Department of Civil Engineering, Sardar Patel College of Engineering, Andheri (West), Mumbai, Maharashtra 400058, India.

E-mail: spkewate@vishwaniketan.edu.in, m_muudi@spce.ac.in

Abstract. The fundamental period of vibration of buildings is the unique parameter that has a major influence on the earthquake induced lateral forces. Many design codes provide simple empirical period -height relationships for the determination of the fundamental period of vibration but using these equations time period and the resulted base shear are conservatively estimated. In most design cases the results obtained using conservative estimates formulae are largely deviating from the required engineering accuracy. Therefore a relationship is sought that will give accurate prediction of period of vibration for structural assessment of Indian reinforced concrete moment resisting frame buildings. On the basis of fundamental period of vibration of an Indian reinforced concrete moment resisting frame (RC MRF) building, obtained by finite element analysis, new equation has been proposed. In this study 21 existing reinforced concrete moment resisting frame buildings in India ,with varying length ,width and height are considered. The fundamental time periods of all the selected buildings were calculated using finite element method and IS 1893(Part 1) : 2002 based empirical formula. The comparison of proposed equation with Indian code shows that finite element analysis yielded 100 % higher periods as compared to the code approach.

1. Introduction

The period of vibration is a key parameter in the seismic design of structures. The spectral acceleration to be resisted by structure is calculated using the fundamental period of vibration. Moreover, the estimation of fundamental period of buildings is useful to recognize possible resonance phenomena between structure and soil vibration. Therefore the reliable evaluation of the fundamental period is an essential step in calculating the seismic response mutually in seismic design and assessment.

The fundamental time period depends upon mass, strength and stiffness of structure [1]. It is affected by many other factors such as structural irregularity, number of storeys and bays, and section properties including dimension. Generally, empirical formula is given as function of only height since even if other factors such as horizontal dimension has been used in regression analysis, height plays important role in fundamental period prediction [2]. For seismic design of a reinforced concrete frame, the period of vibration will not be identified a prior and hence simplified equations are employed in the seismic design codes to express the fundamental period to the height of the structure. Equations employed in the seismic design codes have been acquired by carrying out regression analysis on the periods of vibration measured during earthquakes [6].

For determination of the fundamental period of vibration many design codes provide simple empirical period-height relationships, which leads into conservative estimates of period and hence resulting base shear force will be conservatively evaluated. The results obtained using conservative estimates formulae are largely deviating from the required engineering accuracy. Therefore a relationship is sought that will give accurate prediction of period of vibration for structural assessment of Indian RC buildings.

The objective of this study is to find an empirical relationship between fundamental period-height with analytical methods instead of measured periods of vibration during earthquakes. This pragmatic choice stems from two major causes; (i) deficient of an adequate number of Indian buildings equipped with frequency measuring devices as well as exposure to earthquakes of adequate intensity and (ii) the significantly increased flexibility and control of structural response accessible when using analytical methods .The present work deals with determination of an improved empirical period height relationship to estimate the fundamental period of vibration of Indian reinforced concrete moment resisting frame multi storey buildings for structural assessment in India. First an evaluation of the empirical equation from current building codes IS 1893(Part I):2002 [15] for the calculation of fundamental periods of buildings was carried out. Then fundamental period of vibration was calculated using dynamic analysis. On the basis of these investigations new period- height relationship is proposed.

2. Empirical formulae

The empirical formula relating height of structure and period of building was first time presented in building code ATC3-06 (1978) and had the form of following Eq.(1):

$$T = C_t H^{0.75} \tag{1}$$

where C_t was a regression coefficient ,was taken to be 0.025 for reinforced concrete moment resisting frame and H represented the height of the building in feet. It was theoretically derived using the Rayleigh's method and with the assumption of linear distribution of equivalent static lateral forces over the height of the building and generation of uniform drift by the distribution of stiffness with height along with the linearly distributed lateral forces. The base shear force was supposed to be inversely proportional to $T^{2/3}$. The numerical value of the constant Ct was acquired from the measured periods of buildings in the 1971 San Fernando earthquake. In SEAOC88(1996) it was found that a value of C_t = 0.03 (where H is measured in feet) was more appropriate for reinforced concrete buildings and has been implemented in various design codes since 1978, for instance in UBC-97 (1997), SEAOC-96 (1996),NEHRP-94 (1994). The NEHRP-94 provisions also suggested an another formula as mentioned in Eq. (2) for reinforced concrete moment resisting frame buildings.

$$T = 0.1\, N \tag{2}$$

where N = number of stories. This formula is constrained to buildings not greater than 12 stories in height and having a least story height of 10ft. This expression is given in many of the codes, e.g. IS 1893:1984 (1984), NEHRP-94 (1994), Korea 2005 (2005), ASCE 7-05 (2006), ASCE 7-10 (2010).Indian Code IS 1893-84 (1984) recommended using this Eq. (3) only for regular outline building having even allotment of mass and stiffness in horizontal and vertical direction.

$$T = 0.075\, H^{0.75} \tag{3}$$

In EC8(2004), this C_t coefficient has simply been altered considering that the height is measured in meters ($0.03 \times 3.281^{0.75} = 0.073$ and has also been rounded up for simplicity), leading to $C_t = 0.075$.

Taiwan and Iran have used more conservative value by rounding the value on the lower side and hence as $C_t = 0.07$ in their seismic codes [19, 14]. Goel and Chopra [5] collected data measured from eight Californian earthquakes, with 1971 (San Fernando earthquake) ending with 1994 (Northridge earthquake) and different formulas as given in Eq.(4)-(5) were proposed follow-on from semi empirical analysis, by the best-fit plus 1 standard deviation recommended for displacement-based evaluation, while the best-fit minus 1 standard deviation recommended for conservative force-based design.

$$T_U = 0.067\, H^{0.9} \tag{4}$$

$$T_L = 0.0466\, H^{0.9} \tag{5}$$

where H is the height of the structure in meter.The latter period-height formula has been included in ASCE 7-05 (2006).

3. Period of vibration of reinforced concrete moment resisting frame buildings

Goel and Chopra carried out the regression analysis of measured data to develop formulas to estimate fundamental periods of the buildings [5]. Database contained a total of 106 Californian buildings including 21 of them with peak ground acceleration (PGA) higher than $0.15g$. It was noticed that the calculated code periods were shorter than the measured periods from the recorded motions. For buildings up to 36 meters high, the code formulas produced lower-bound values compared to measured period data, on the other hand, the same formulas resulted in 20% to 30% shorter periods compared to the measured ones for buildings taller than 36 meters. For many buildings, measured period values were greater than $1.4T$, where T was described as the fundamental period obtained from the empirical equation.

Hong and Hwang experimentally determined the fundamental time period of 21 reinforced concrete moment resisting frame buildings, located in Taiwan through vibration measuring instruments [2]. Based on the experimental results, an empirical relationship between building period and height was derived. However, the obtained relation was different from that of U.S. building code formula. On comparing the numerical values of fundamental time period it was observed that the formula based on data pertinent to Taiwanse buildings under predicted the period value as compared to U.S. Code proposed formulae. This implies that Taiwanese buildings are stiffer than the Califonian buildings.

Helen Crowley et al. studied existing buildings from five different European countries exposed to earthquake forces [9]. The majority of the existing buildings were designed and built between 1930 and 1980. The study led to a simplified period versus height equation in the assessment of existing reinforced concrete framed buildings taking due account of the presence of infill panels.

Pinho and Crowely took a critical look at the way in which seismic design codes around the world have endorsed the designer to estimate the period of vibration for use in both linear static and dynamic analysis [6]. Based on this review, some preliminary proposals are made for updating the clauses related to the estimation of the periods of vibration in EC8(2004). They suggested that when presenting the period-height equations for reinforced concrete moment resisting frames there should be two formulae for the infill panels are to be isolated from the moment resisting frame and the infill panels are to be rigidly connected to the moment resisting frame for the estimation of the periods of vibration in EC8(2004).

Paolo et al. carried out modal analyses on 3D numerical RC MRF building models, varying structure morphology (height, surface area, ratio between plan dimensions) and infill characteristics. Simplified formulas based on regression analysis of obtained numerical data were presented and discussed [23]. These relationships were also compared with similar literature numerical expressions and empirical data from experimental measurements on existing buildings.

Chiauzzi et al. focused on twelve reinforced concrete framed buildings, located in the cities of Victoria and Vancouver (British Columbia, Canada) and carried out a field campaign to estimate their fundamental periods using in-situ ambient vibration measurements [10]. They concluded that building periods estimated based on simple equations provided by earthquake design codes in Europe (EC8) and North America (UBC-97 and NBCC-2005) are significantly greater than the periods computed using ambient vibration records on the monitored buildings.

Ditommaso et al. investigated 68 reinforced concrete moment resisting frame buildings in Italy with different characteristics, such as age, height and damage level, by performing ambient vibration measurements that provided their fundamental translational periods [4]. Four different damage levels

(DL0,DL1, DL2, DL3) were considered according to the definitions in EMS 98 (European Macro seismic Scale), trying to regroup the estimated fundamental periods versus building heights using damage level as a key parameter. The two suggested experimental period height relationships were obtained.

Elgohary carried out a parametric study using finite element analysis to study the effect of the major parameters influencing in the fundamental period [8]. He concluded that the code's formulae, in most cases, underestimate the fundamental period with a large deviation from finite element results. This large deviation is a result of considering only the effect of frame height and neglecting of remaining major parameters in the codes formulae.

Al-Nimry et al. investigated the fundamental period of vibration of infilled reinforced concrete frame buildings using measurements of ambient vibrations and numerical analyses [22]. Ambient vibrations were measured at the roof level of 29 selected buildings with heights of one to six stories. Using Nakamura technique, the horizontal-to-vertical spectral ratio curves were obtained in the two orthogonal building directions. The estimated period values ranged between 44% and 91% of elastic periods suggested by the local code. Preliminary period–height relations were proposed using regression analysis of the measured periods. Analytical period values showed large differences with both the measured and code values.

Asteris et al. investigated the fundamental period of vibration of reinforced concrete buildings by means of finite element macro-modelling and modal eigenvalue analysis [24]. 14-storey RC buildings were considered "according to code designed" and "according to code non-designed". They studied various parameters including the number of spans; the span length in the direction of motion; the stiffness of the infills; the percentage openings of the infills and; the location of the soft storeys. From the analysis of the results it has been found that the span length, the stiffness of the infill wall panels and the location of the soft storeys are crucial parameters influencing the fundamental period of reinforced concrete buildings.

Magdy proposed improved formulas for estimating the fundamental period of vibration of reinforced concrete moment-resisting frame buildings by taking the effect of both building height (*H*) and number of stories together (*N*) [3]. The improved formula is based on regression analysis of the existing data for the fundamental vibration period of reinforced concrete moment resisting frame buildings measured from their motions recorded during eight California earthquakes.

Prakash and Dubey predicted the fundamental period of vibration of reinforced concrete buildings with moment resisting frames by considering the effect of building base width in both the directions and stiffness of the structure [7]. Values of the time period of vibration differ substantially, and the variation comes out to be on higher side than those derived from code base formula, which results in reduction in base shear and makes the structure economical.

4. Existing period- height relationship in India

The fundamental natural period of vibration for a moment resisting frame building without brick infill panels may be estimated by the empirical expression as mentioned in Eq. (3) as per of IS 1893 (Part 1): 2002.

The fundamental natural period of vibration of all other buildings, as well as moment resisting frame buildings with brick infill panels, estimated by the empirical expressions per IS 1893 (Part 1):2002 is shown in Eq. (6):

$$T_a = \frac{0.9\hbar}{\sqrt{d}} \tag{6}$$

where h = height of the building in meter, d = Base dimensions of the building at the plinth level in meter, along the considered direction of lateral force.

5. Structural modeling

Modeling a building involves the modeling and assemblage of its various load-carrying elements. The model must ideally represent the mass distribution, strength, stiffness and deformability. The Finite element analysis of the selected building is carried out using ETABS 2015 (Integrated Building Design Software).

In this study 21 reinforced concrete moment resisting frame buildings in India with varying length, width and height (ranging from 6.6m to33m) are considered. Compressive strength for M20 and M25 grade of concrete are 20MPa and 25 MPa respectively. Modulus of Elasticity for M20 and M25 grade of concrete are 19.3364 x103 MPa and 22.364x103 MPa respectively. Modulus of Elasticity for Fe 415 grade of steel is 2 X 105MPa. Density of Reinforced Concrete is 25 KN/m^3 and density of steel is 76.82 KN/m3. Live load intensity is considered as 3kN/m2 and floor finishes intensity as 1kN/m^2. Live loads were considered as 25 % of imposed load, as a part of the seismic weight.

Beams and columns were modeled as 3D frame elements. The members were represented through the assignment of properties like cross sectional area, reinforcement details and the type of material used. It has 6 Degrees of Freedom (*Ux*, *Uy*, *Uz*, *Rx*, *Ry*, and *Rz*) for each node. Slabs have been modeled as rigid diaphragms. Shell type elements were used to model slabs. Auto meshing was carried out. Auto Line Constraints were imposed on these slabs. Supports were considered as fixed at the base of the structure. Buildings under consideration do not include infill walls.

6. Estimation of fundamental period for Reinforced concrete moment resisting frame buildings

In this study 21 existing reinforced concrete moment resisting frame buildings in India with varying length, width and height as shown in Table 1 are considered. The fundamental time periods of all the 21 selected buildings were calculated using finite element method and code based empirical formula. The analytical results of fundamental period of these structures using analytical methods are compare with those using empirical formula IS 1893(Part 1) : 2002 . The results are presented in table 1.

Table 1. Period data for RC moment resisting frames.

Model number	Dimensions (m)			Analytical Time period (s)			Fundamental Natural period obtained from IS 1893 (Part 1):2002 (s)	% Conservative
	Height	Length	Width	Longitudinal Direction (T_x)	Transverse Direction (T_y)	Fundamental Time Period (T)	T	
1	6.6	12.78	9.735	0.622	0.501	0.622	0.309	101
2	8	10.14	9.085	0.645	0.591	0.645	0.357	81
3	8.6	6.045	28	0.638	0.764	0.764	0.377	103
4	10.5	15.05	9.98	0.902	0.785	0.902	0.437	106
5	10.5	7.48	11.26	0.732	0.838	0.838	0.437	92
6	10.8	14.87	16.15	0.793	0.661	0.793	0.447	77
7	11	5.698	13.43	0.886	1.115	1.115	0.453	146
8	11.5	7.06	13.44	1.004	0.864	1.004	0.468	115
9	13	7.21	18.1	1.152	0.856	1.152	0.513	125
10	13	16.27	8.77	1.043	0.897	1.043	0.513	103
11	15.05	7.11	11.41	1.062	1.052	1.062	0.573	85
12	14	10.5	7.25	0.906	0.807	0.906	0.543	67
13	15.1	12.6	12	0.785	1.134	1.134	0.575	97
14	15.65	17.39	8.37	0.995	0.831	0.995	0.59	69

15	18	51.201	15.118	1.315	1.204	1.315	0.655	101
16	19.97	9.2	18.9	1.415	1.324	1.415	0.709	100
17	15	51.201	15.118	1.087	0.989	1.087	0.572	90
18	21	51.201	15.118	1.542	1.422	1.542	0.736	110
19	24	51.201	15.118	1.772	1.643	1.772	0.813	118
20	30	51.201	15.118	2.079	1.936	2.079	0.961	116
21	33	51.201	15.118	2.204	2.093	2.204	1.033	113
							Average	100

The results of fundamental period calculation of reinforced concrete moment resisting frame buildings using IS 1893(Part I):2002 are shown in Fig. 1. It can be noticed that the fundamental period of vibration is proportional to the height of reinforced concrete moment resisting frame buildings as given by Eq.(3):

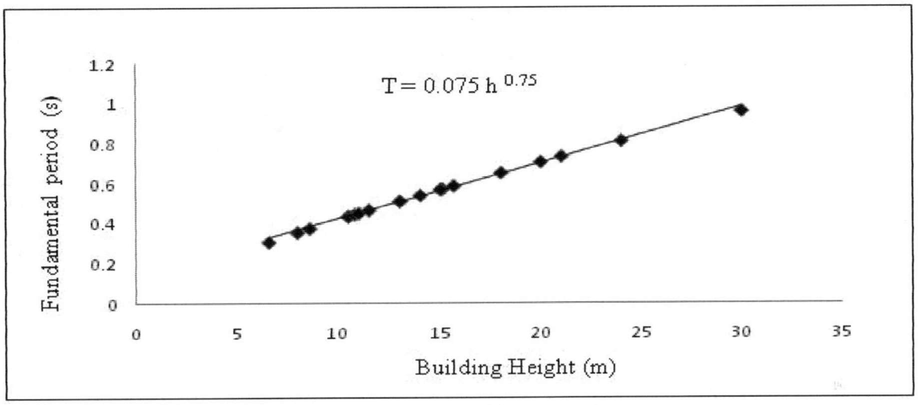

Figure 1. Relationship between fundamental period and height of reinforced concrete moment resisting frame buildings using IS 1893(Part I :2002).

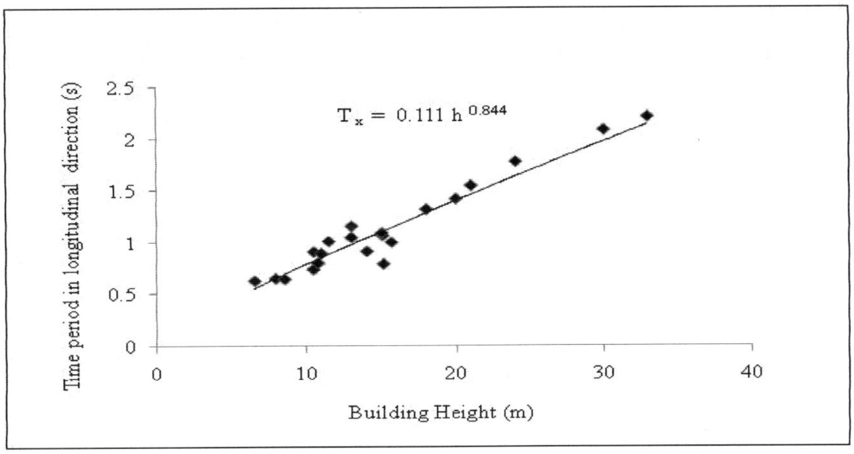

Figure 2. Relationship between time period in longitudinal direction and height of reinforced concrete moment resisting frame buildings using finite element analysis.

The results of time period calculation in longitudinal (x) direction of reinforced concrete moment resisting frame buildings using Finite element analysis are shown in Fig. 2 with best fit curve. Dots in figure indicates the fundamental time period in longitudinal direction by analytical method. The straight line is drawn using best fit curve, which gives the relationship between fundamental time period and height of building as given by following Eq. (7) :

$$T_X = 0.11\, h^{0.844} \tag{7}$$

The results of time period calculation in transverse (y) direction of reinforced concrete moment resisting frame buildings using Finite element analysis are shown in Fig. 3 with best fit curve. Dots in figure indicates the fundamental period in longitudinal direction by analytical method. The straight line is drawn using best fit curve which gives the relationship between fundamental time period and height of building as given by following Eq.(8):

$$T_y = 0.107 h^{0.837} \tag{8}$$

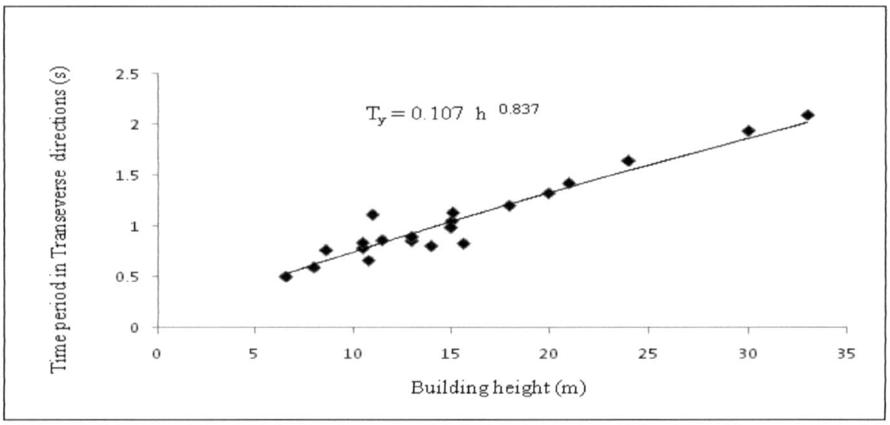

Figure 3. Relationship between time period in transverse direction and height of reinforced concrete moment resisting frame buildings using finite element analysis.

The fundamental time period is the highest time period among the two directions. The fundamental time period is plotted vs height of building as shown in Fig. 4. The dots in figure represents these values. Using best fit curve the straight line is drawn as shown in Fig. 4 which indicates the period - height relationship.

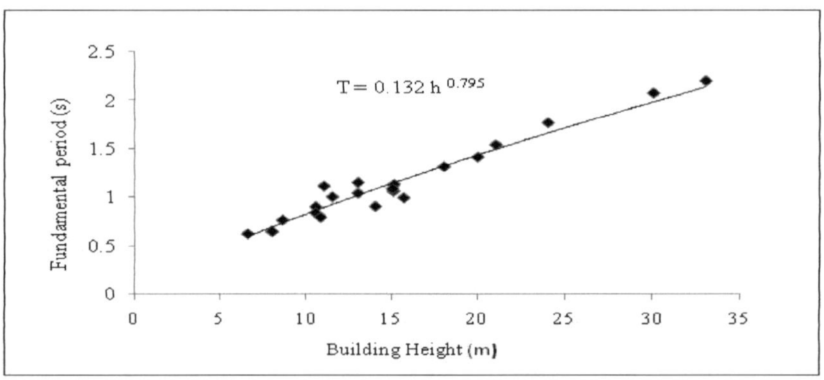

Figure 4. Relationship between fundamental period and height of reinforced concrete moment resisting frame buildings using finite element analysis.

Comparison of fundamental period between empirical relationship using IS 1893(Part I):2002 and finite element analysis method is shown in Fig. 5. It can be noticed that proposed equations yielded higher periods as compared to the code approach. The two curves are seen to be very similar so it provides confidence towards use of proposed equations. Therefore, the proposed period -height relationship could be utilized for the estimation of the fundamental vibration period of reinforced concrete moment resisting frame buildings in India.

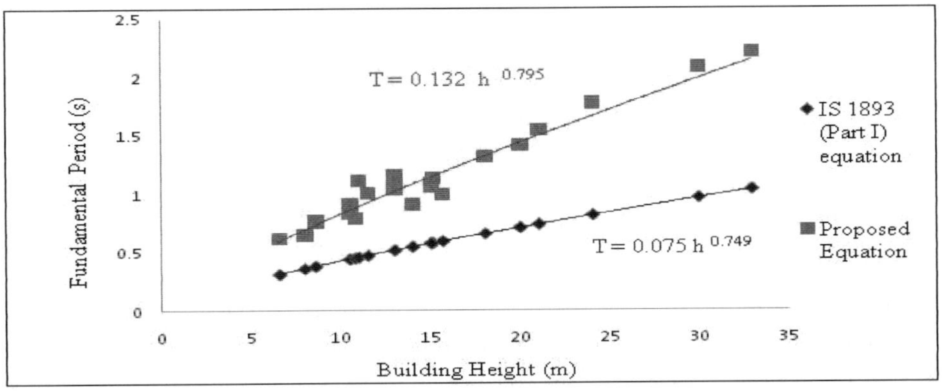

Figure 5. Comparison of proposed equation results and IS 1893(Part I):2002 equation results for reinforced concrete moment resisting frame buildings in India.

Comparison of Empirical and Analytically calculated time periods for reinforced concrete moment resisting frame buildings in India is shown in Table 1 .It can be noticed that IS 1893(Part I):2002 yielded average 100 % conservative periods as compared to the analytically calculated results.

7. Proposed equation

In the present study of the fundamental period of various existing reinforced concrete moment resisting frame buildings in India with varying number of stories and number of bays is carried out for RC moment resisting frames. Analysis is done using finite element method using software ETABS 2015 V15.2. On the basis of these evaluation and investigation following new Eq. (9) is proposed for fundamental time period for reinforced concrete moment resisting frame buildings in India. Comparing this equation with IS 1893(Part1):2002 formula is highly conservative.

$$T = 0.132\, h^{0.795} \tag{9}$$

where h = height of reinforced concrete moment resisting frames in meter, T = fundamental period of reinforced moment resisting frames in seconds

Proposed analytical - based relation provides an insight into the period-height relation but needs further investigation before being adopted in the local seismic code.

8. Conclusions

In this study 21 existing reinforced concrete moment resisting frame buildings in India with varying length, width and height are considered. The fundamental time periods of all the 21 selected buildings were calculated using finite element method and code based empirical formula. The analytical results of fundamental period of these structures using analytical methods are compare with those using empirical formula IS 1893(Part 1) : 2002. The code proposed equations are based on experimentally determined periods of certain set of buildings measured during earthquakes.

- It is observed from this study that the fundamental time period increases with increase in height of structures.

- New period -height relationship is proposed based on the study. Compare to this relationship IS1893(Part 1) : 2002 relationship is found to be conservative.

References

[1] Anastasia K E and Karabinis A I 2013 Correlation of structural seismic damage with fundamental period of RC buildings *Open Journal of Civil Engineering* **3** 45-67

[2] Li L H and Hwang W L 2000 Empirical formula for fundamental periods of vibration for Reinforced concrete buildings in Taiwan *Earthquake Engineering and Structural Dynamics,* **29** 327-337

[3] Magdy I S 2015 Estimation of period of vibration for concrete moment-resisting frame buildings *HBRC Journal* **11** 16–21

[4] Ditommaso R, Vona M, Gallipoli M R and Mucciarelli M 2013 Evaluation and considerations about fundamental periods of damaged reinforced concrete buildings *National Hazards Earth System Sci*ence **13** 1903-1912

[5] Goel R K and Chopra A K 1997 Period formulas for moment—resisting frame buildings *Journal of Structural Eng*ineering, *ASCE* **123**(11) 1454-1461

[6] Pinho R and Crowley H 2010 Revisiting Eurocode 8 formulae for periods of vibration and their employment in linear seismic analysis *Earthquake Engineering and Structural Dynamics* **39**(2) 223-235

[7] Prakash S and Dubey S K 2015 Effect of base width and stiffness of the structure on period of vibration of RC framed buildings in seismic analysis *Open Journal of Earthquake Research* **4** 65-73

[8] Elgohary H 2013Empirical formula for the fundamental period of vibration of multi-storey RC framed buildings Proceeding of *Vienna Congress on Recent Advances in Earthquake Engineering and Structural Dynamics* Vienna, Austria August.

[9] Crowley H and Pinho R 2006 Simplified equations e the period of vibration of existing building *Proceeding of First European Conference on Earthquake Engineering and Seismology,* Geneva, Switzerland ,September

[10] Chiauzzi L Masi A and Mucciarelli M 2012 Estimate of fundamental period of reinforced concrete buildings: code provisions vs. experimental measures in Victoria and Vancouver(BC,Canada) *Proc. of the 15th World Conference on Earthquake Engineering*, Lisbon, September

[11] Applied Technology Council 1978 *Tentative provisions for the development of seismic regulations for buildings*, Report No. ATC3-06 Applied Technology Council, Palo Alto, California.

[12] ASCE/SEI 7-05 2006 *Minimum design loads for buildings and other structures*, American Society of Civil Engineers, Reston, Virginia

[13] Eurocode 8 2005 Design of structures for earthquake resistance - Part 1 : General rules, seismic actions and rules for buildings, European Committee for Standardization (CEN) Brussels.

[14] Iran 2007 *Iranian code of practice for seismic resistant design of buildings (standard no. 2800)*, Building and Housing Research Centre, Ministry of Housing and Urban Development

[15] IS 1893 Part I 2002 Criteria for earthquake resistant design of structures. Part I :General provisions and buildings (Fifth Revision), Bureau of Indian Standards; New Delhi, India.

[16] KBC-05 2005 *Building code for structural regulations*, Ministry of construction and transportation; Korea

[17] NEHRP-94 199 Recommended Provisions for the Development of Seismic Regulations for New Buildings, Building Seismic Safety Council; Washington, D. C.

[18] SEAOC-96 1996 *Recommended lateral force requirements and commentary*, Seismological Committee Structural Engineers Association of California; San Francisco, California.

[19] Tiwan 2005 *Seismic design code for buildings in Taiwan*, Construction and Planning Agency, Ministry of the Interior, R.O.C.

[20] UBC-97 1997*Uniform building code*, International Conference of Building Officials. Whittier, California.

[21] NBCC-95. 1995 *National Building Code of Canada*. Associate Committee on the National Building Code, National Research Council of Canada.

[22] Al-Nimry H, Resheidat M and Jamal M 2014 Ambient vibration testing of low and medium buildings in Jordan *Soil Dyn. Earthq. Eng.* **59** 21-29

[23] Paolo R Verderame G M and Manfredi G 2011 Analytical investigation of elastic period of infilled RC MRF buildings", *Engineering Structures* **32**(2) 308-319

[24] Asteris P G, Repapis C C Cavaleri L and Athanasopoulou A 2015 On the fundamental period of infilled RC frame buildings *Structural Engineering and Mechanics* **54**(6) 1175-1200

Natural frequencies of a vertical pipeline element

A A Sekacheva[1], L G Pastukhova[1], V N Alekhin[2] and A S Noskov[1]

[1]Hydraulics Department, Ural Federal University, 19 Mira Street, Yekaterinburg 620002, Russia

[2]Department of CAD Systems in Civil Engineering, Ural Federal University, 19 Mira Street, Yekaterinburg 620002, Russia

E-mail: tonechka_marakulina@mail.ru

Abstract. This work is devoted to the problem of noise and vibration from the pipeline systems in multi-storey buildings. A method for determining the probability of the occurrence of increased vibrations using modal analysis in the ANSYS Workbench software package is proposed. Numerical analysis of straight sections of the pipeline, each differing in diameter and length, is performed. The results of the numerical analysis of the influence of the length, diameter, and wall thickness of a pipe system element on the change in the frequencies of its natural oscillations are presented with a view to predicting the risk of possible resonance regimes. Pairwise and multiple regression analyses are performed, and regression equations are obtained describing the functional relationship between the first frequency of the natural oscillations and the length, diameter, and wall thickness of the pipeline section.

1. Introduction

The 21st century has become the century for the intensive development of building technological systems and the introduction of effective innovative technologies in the construction industry with the goal of creating buildings that have the highest degree of comfort – multi-comfort.

A building technological system should be understood as a set of interrelated elements in an engineering system, united by a single constructive and technological solution aimed at improving the quality, reliability, durability, and efficiency of the construction [1].

The newest technological ideas differ from older ones, first of all, by the fact that they are oriented towards creating the most comfortable working and living conditions. The embodiment of such ideas in building (engineering) systems is, as a rule, of a qualitative nature.

One of the tendencies in the modern development of large cities is the constant growth in the number of storeys, which has arisen in connection with the need to increase the spaciousness of cities, a consequence of the growth of the urbanization index.

At present, the majority of unsolved engineering tasks requiring considerable expenditure on empirical research are an important component of scientific and technological progress. Therefore, computer modeling in combination with the use of numerical methods is an efficient and affordable way for analyzing similar problems.

This article is directly related to the solution of one such problem in the design and operation of the engineering systems of multi-storey apartment buildings - the emergence of noise and vibration in pipeline systems.

The Rospotrebnadzor Office often receives complaints from apartment owners located above individual heat points (ITPs) in the cellars of multi-storey buildings and residents of apartments adjacent to communication and ventilation mines. As a rule, prolonged exposure of the human body to noise and vibration causes functional changes in internal systems and organs. It also causes significant psychological stress, which ultimately affects the general condition of the body. It is also worth noting that, due to vibration, communications at home may suffer.

In building design, the modal analysis method for studying the vibrations of building structures has been widely used. Nevertheless, only the experimental and analytical modal analysis methods, described in the works of G. Mikota, A.F. D'Souza, G.M. Makaryants, D.J. Ewins, J.C. Wachel, J.D. Tison, X. Li, J.J. Nieter et al., were used to solve the vibration problems of hydraulic systems [2-16].

G. Mikota's results provide a theoretical basis for the experimental modal analysis of hydraulic pipelines. They demonstrate how standard experimental modal analysis procedures can be adopted for hydraulic pipelines if mobility input is replaced with the frequency response functions between flow rate excitation and pressure response. Following these considerations, the experimental modal analysis of hydraulic pipeline systems shall be carried out in practice [2-5].

For example, in [8] it is shown that the study of the pipeline system's vibroacoustic response is prone to fail the reliable causation of elevated dynamic loads when based only on Fourier analysis of the working fluid pulsation and pipeline vibration. It is suggested that we complement vibroacoustic response analysis with the calculation of modal parameters. A mathematical computation of the modal frequencies and shapes was implemented instead of experimental observation in order to exclude the undesired signal introduced with attached mechanical equipment. On top of this, the mathematical computation of the pipeline modal parameters allowed us to perform a pipeline modal frequency shift from the water-hammer effect frequency range through cautious relocation of additional supports.

Paper [17] presents estimates of the acceptability of vibrations, together with methods for troubleshooting piston compressors, pumps, and / or pipelines to determine whether problems are caused by pulsation or mechanical resonances. The main principles of pulsation generation and control are presented. The authors believe that the key to designing and operating safe piping systems is to control the pulsation levels and separate the mechanical natural frequencies from the pulsation excitation frequencies.

In this paper, a method is proposed for determining the probability of the occurrence of increased vibrations in pipeline systems using numerical modal analysis in the ANSYS Workbench 12.0 software package. This program complex is used for conducting numerical experiments in various fields of industry (mechanical engineering, agrarian industry) [18-21]. However, it has not been used for carrying out experiments of this kind for pipeline systems.

Through conducting ANSYS-based modal analysis for a ditch device, the authors of [18] obtained the natural vibration frequency and response effect of the ditch device (during field work, the vibration frequency of the ditch device was 8-9HZ). The natural frequency of the ditch device completely avoided the machine vibration frequency in the seeding period, and thus avoided the poor working conditions caused by resonance.

The method of numerical modal analysis for analysing the frequencies of fluctuations in rod basic insulators is described in [19]. Using the method of free fluctuations as a form of diagnosis, it was found that the main difficulty is the choice of the informative frequency ranges of the spectrum characterizing the existence of defects. The ANSYS software package was used for final element modeling.

The fundamentals of the ANSYS Workbench software in engineering analysis are described in [22, 23]. A modal analysis on the numerical model of the pipeline system element using the ANSYS PC for the purpose of determining the frequencies and modes of the element's natural vibrations was considered in [24].

In the most modern software systems, modal analysis is the solution to the problem of free oscillations in a discrete system by the finite element method [25-27].

The purpose of this work was to study the influence of the geometric parameters of an element of the pipeline system on its resistance to vibrations.

To achieve this goal, the following tasks were set:

1) Conduct a numerical modal analysis of the dynamic parameters of the elements of a pipeline system of various geometric parameters.
2) Determine the dependence of the first critical frequency of natural oscillations on the geometric parameters of an element of the pipeline system.
3) Perform regression analysis and obtain regression equations that describe the functional relationship between the first natural frequency and the length, diameter, and wall thickness of the pipeline section.

2. Method

The elements of a pipeline system, like most structures, make mechanical fluctuations. The modal analysis method is used to calculate the dynamic characteristics of pipeline elements: the frequency and shape of natural oscillations. The oscillations take place with an initial deviation from the equilibrium state of one of the system parameters and the subsequent absence of an external variable effect on the system. The "dynamic individuality" of the system is determined by its behavior with free vibrations, which characterizes the behavior of the system in other conditions.

A determination of these kinds of parameters makes it possible to assess the degree of danger of possible resonance regimes occurring when dangerous harmonics enter the operating range of the acting external loads (harmonic analysis). As a rule, the source of the forced oscillations in this case is that the pumping units have large capacities.

Modal analysis of the processes for determining N's frequencies and shape modes is performed on the basis of the oscillation equation (1):

$$[\mathbf{M}]\{\ddot{x}\} + [\mathbf{C}]\{\dot{x}\} + [\mathbf{K}]\{x\} = 0 \tag{1}$$

where [M], [C] and [K] are the mass, damping and stiffness of the structure in the matrix form [20].

If we neglect the damping effect, we obtain an equation of the form (2):

$$[\mathbf{M}]\{\ddot{x}\} + [\mathbf{K}]\{x\} = 0 \tag{2}$$

3. Results

In this paper, numerical modal analysis of the dynamic parameters of the pipeline system elements of various geometric parameters was carried out in ANSYS Workbench PC. The first ten frequencies and forms of natural oscillations for different sections of the pipeline were determined as shown in Figure 1.

The values of the outside diameter of the pipes (mm) are taken in accordance with GOST 10704-91 "Pipes steel electrically welded straight. Assortment": 15 × 1.0; 15 × 1.2; 15 × 1.4; 15 × 1.5; 15 × 1.6; 20 × 2.0; 25 × 2.0; 32 × 2.8; 40 × 3.0; 48 × 3.0; 60 × 3.8; 70 × 4.0; 102 × 5.0; 114 × 5.0; 140 × 5.0; 168 × 5.0; 180 × 5.0; 219 × 6.0. The lengths of the calculated straight sections of the pipeline system lie in the range 1.0 ÷ 6.0 m.

To determine the equations (mathematical models) that most closely describe the experimental data, two types of regression analysis (pair and multiple) were performed using MS Excel.

For the regression analysis, only the first (critical) frequency was used since it has the smallest value; therefore, it is the most dangerous in the case of resonance regimes.

The main purpose of regression analysis is to determine the analytical form of communication, in which the influence of one (simple regression analysis) or several (multiple regression analysis) factors cause a change in the outcome.

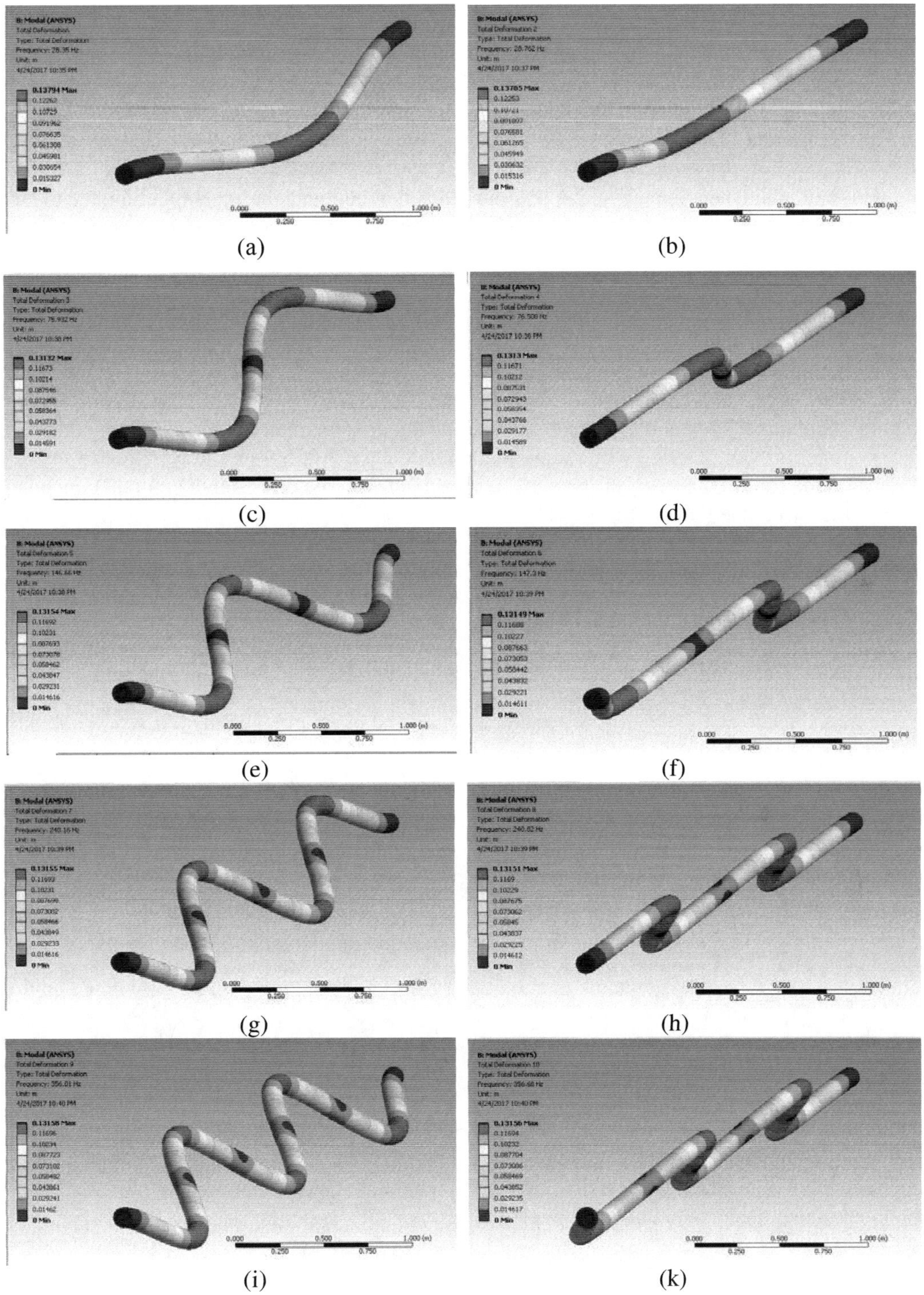

Figure 1. Forms of the natural oscillations of a 6-meter straight section of the pipeline system with an external diameter of 114 mm and a wall thickness of 5.0 mm filled with water: (a) the 1st form; (b) the 2nd form; (c) the 3rd form; (d) the 4th form; (e) the 5th form; (f) the 6th form; (g) the 7th form; (h) the 8th form; (i) the 9th form; (k) the 10th form.

In this paper, a simple (pairwise) regression analysis was performed for two cases: determining the dependence of the value of the first frequency of natural oscillations on the diameter of the pipeline section ($F_1 = f(d)$) and on the length of the pipeline section ($F_1 = f(l)$).

Preliminary analysis showed that the dependence in both cases is non-linear.

The coefficient of determination is the main criterion characterizing the degree of quality of the regression model describing the relationship between variables. Its value lies in the interval: $0 \le R^2 \le 1$. Thus, when $R^2 = 1$ - the coupling is functional and the regression model explains 100% dependence and change of the investigated parameter Y from independent factors X. It is considered permissible to accept the regression model for $R^2 \ge 0.8$.

Analyzing the results of simple (paired) regression analysis, we can draw the following conclusions:

1. There is a dependence of the first natural oscillation frequency on the diameter and length of the pipeline section.
2. The regression equation describing the dependence of the first frequency of natural oscillations on the diameter of a pipeline section has the form of a sixth-order polynomial function. For pipeline sections with a length of 1.5; 3.0; 6.0 m, the equation of the dependence of the first frequencies of the natural oscillations on the diameters (Figure 2), respectively, has the form of equations (3), (4), (5):

$$F_1^{1.5} = 7 \cdot 10^{-10} d^6 - 5 \cdot 10^{-7} d^5 + 0.0001 d^4 - 0.02 d^3 + 1.4416 d^2 - 47.107 d + 684.56 \quad (3)$$

$$F_1^{3.0} = 3 \cdot 10^{-10} d^6 - 2 \cdot 10^{-7} d^5 + 6 \cdot 10^{-5} d^4 - 0.0088 d^3 + 0.6254 d^2 - 20.924 d + 299.41 \quad (4)$$

$$F_1^{6.0} = 1 \cdot 10^{-10} d^6 - 1 \cdot 10^{-7} d^5 + 3 \cdot 10^{-5} d^4 - 0.0038 d^3 + 0.267 d^2 - 9.0641 d + 129.4 \quad (5)$$

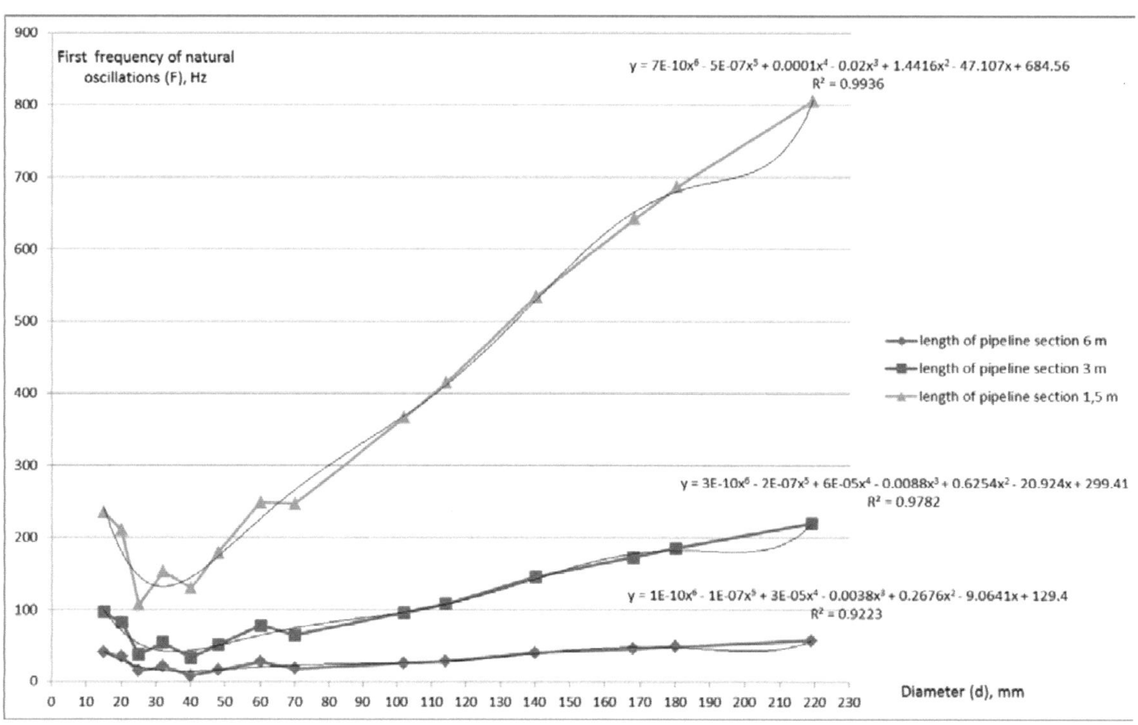

Figure 2. Graphs of the dependencies of the first frequencies of natural oscillations on diameters for sections of pipelines 1.5, 3.0, 6.0 m in length and their regression models.

IOP Conf. Series: Materials Science and Engineering **481** (2019) 012019 doi:10.1088/1757-899X/481/1/012019

3. The regression equation describing the dependence of the first frequency of natural oscillations on the length of the pipeline section has the form of a power function. The equation of the dependence of the first frequencies of natural oscillations on the lengths (Figure 3) has the following form of equations (6), (7), (8), (9), (10), (11) for the diameters 32, 48, 70, 114, 168, 219 mm, respectively:

$$F_1 = 283.74l^{-1.479} \quad (6)$$

$$F_1 = 370.12l^{-1.773} \quad (7)$$

$$F_1 = 533.05l^{\ 1.905} \quad (8)$$

$$F_1 = 901.94l^{-1.931} \quad (9)$$

$$F_1 = 1365.2l^{-1.889} \quad (10)$$

$$F_1 = 1693.5l^{-1.876} \quad (11)$$

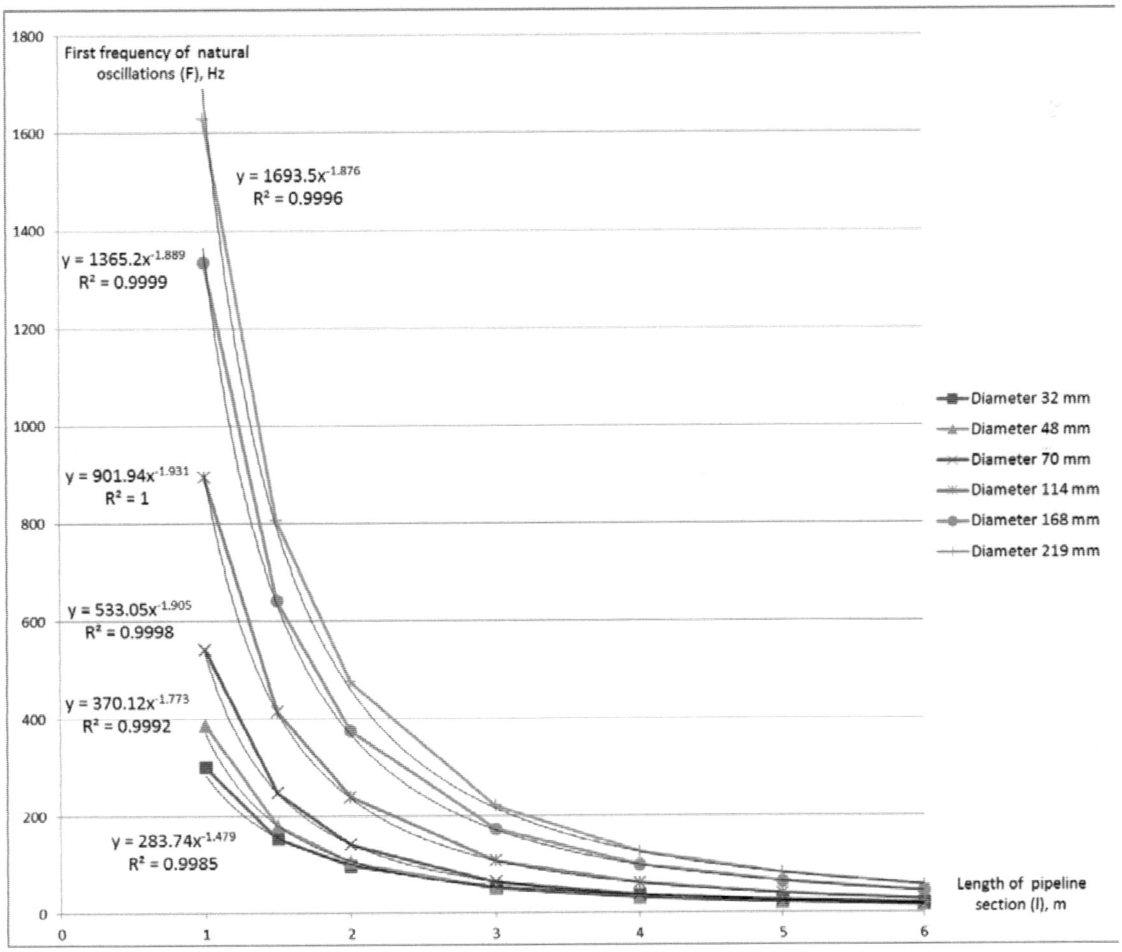

Figure 3. Graphs of the dependencies of the first natural frequencies on lengths for pipeline sections with diameters of 32, 48, 70, 114, 168, 219 mm and their regression models

4. When the length of a pipeline section is shortened, the coefficient of determination of the regression equation describing the dependence of the first frequency of natural oscillations on the diameter tends towards unity. This means that the shorter the pipeline section, the better the regression model describes the experimental values.
5. The coefficient of determination of the regression equation describing the dependence of the first natural frequency on the length, when the pipeline section is reduced, behaves randomly. Nevertheless, its significance for all the cases studied is close to unity.

A linear multiple regression model was also considered in which the variables (length (l), diameter (d), and wall thickness (δ) of the pipeline section) had a first degree. It does not qualitatively describe the experimental data obtained. It was concluded that the functional dependence under investigation is described by a nonlinear function.

Therefore, we believe that the regression function is a power-law model. This model refers to models that are non-linear in their parameters, where the linear least-squares method cannot be applied directly.

Since MS Excel tools allow only linear multiple regression to be performed, it was necessary to select an appropriate model conversion to bring it into a linear form. For a power-law model, this transformation is the logarithm of both parts of the model.

After analyzing the results of multiple regression analysis, we can draw the following conclusions:

1. The functional relationship between the value of the first natural frequency and the length, diameter, and wall thickness of the pipeline section is nonlinear (power law).
2. The regression model of the dependence of the first frequency of natural oscillations on two parameters (the length and diameter of the pipeline section) has the form:

$$F_1 = 80.592 \cdot l^{-1.688} \cdot d^{0.468} \phi_k(\vec{r}) = (2\pi)^{2/3} \exp\left(i\vec{k} \cdot \vec{r}\right) \quad (12)$$

   The coefficient of determination in this case is $R^2 = 0.879$.
3. The regression model of the dependence of the first frequency of natural oscillations on three parameters (the length, diameter, and wall thickness of the pipeline section) has the form:

$$F_1 = 27.569 \cdot l^{-1.658} \cdot d^{1.093} \cdot \delta^{-1.25} \quad (13)$$

   The coefficient of determination in this case is $R^2 = 0.893$.
4. The value of the coefficient of determination (the quality of the correspondence of the regression model to the experimental data) is the higher when the number of factor attributes is greater.

4. Conclusions

A method that allows excluding resonant phenomena in the operation of hydraulic systems of residential buildings using modal analysis in the ANSYS Workbench software package is proposed. It consists in determining the first critical frequency of natural oscillations of the pipeline for its further extraction beyond the region of forced frequencies by other methods.

Modal analysis of straight sections of the pipeline system filled with water, length 1.0; 1.5; 2.0; 3.0; 4.0; 5.0; 6.0 and diameters of 15 × 1.0; 15 × 1.2; 15 × 1.4; 15 × 1.5; 15 × 1.6; 20 × 2.0; 25 × 2.0; 32 × 2.8; 40 × 3.0; 48 × 3.0; 60 × 3.8; 70 × 4.0; 102 × 5.0; 114 × 5.0; 140 × 5.0; 168 × 5.0; 180 × 5.0; 219 × 6.0 mm in the software package ANSYS Workbench was conducted.

A paired regression analysis was performed for two cases: determining the dependence of the value of the first natural oscillation frequency on the diameter of the pipeline section and on the length of the pipeline section.

It is revealed that the regression equation describing the dependence of the first frequency of natural oscillations on the diameter of a pipeline section has the form of a sixth-order polynomial function. The regression model of the dependence of the first frequencies of natural oscillations on diameters for pipeline sections with a length of 1.5, 3.0; 6.0 m is determined.

It was also revealed that the regression equation describing the dependence of the first frequency of natural oscillations on the length of the pipeline section has the form of a power function. The regression model of the dependence of the first frequencies of natural oscillations on the lengths for pipeline sections with diameters of 32, 48, 70, 114, 168, 219 mm is determined.

It is established that when the length of the pipeline section is shortened, the coefficient of determination of the regression equation describing the dependence of the first natural frequency on the diameter tends to unity.

Multiple regression analysis was performed and regression equations describing the functional relationship between the first frequency of natural oscillations and the length, diameter, wall thickness of the pipeline section were obtained. It is established that the functional relationship between the value of the first natural frequency and the geometric parameters (length, diameter, wall thickness) of the pipeline section is nonlinear (power law). The regression model of the dependence of the first frequency of natural oscillations on two parameters (length and diameter of the pipeline section is determined.

A regression model of the dependence of the first frequency of natural oscillations on three parameters (length, diameter and wall thickness of the pipeline section) is determined.

It is revealed that the value of the coefficient of determination (the quality of the regression model's correspondence to the experimental data) is higher the more the number of factor attributes.

Preliminary modal analysis becomes very important in the modeling of vibrations and transient processes in hydraulic system designs, since its main purpose is to predict and eliminate the occurrence of possible resonance regimes in them.

The frequency of the natural oscillations of the piping system designs obtained as a result of the modal analysis should be considered throughout the calculated frequency spectrum (and not only within the first critical frequency), since noise and vibration normalization is also performed in the frequency spectrum.

If the calculated natural frequencies are located in the operating range of the existing external loads, then there are dangerous resonant oscillations and the design will not satisfy the requirements for vibration strength. Then it is changed in such a way as to output the frequencies of natural oscillations beyond the range of forced frequencies of equipment of hydraulic systems.

The evaluation of the stress-strain state of the structure under the forced oscillation variables and the verification of the resonance regimes are performed with harmonic analysis, which is the next stage of the complete vibration study, and hence the safety of the engineering systems.

References

[1] Badin G M and Sychev S A 2013 *Sovremennyye tekhnologii stroitelstva i rekonstruktsii zdaniy* (Saint Petersburg: BKhV-Peterburg)

[2] Mikota G, Manhartsgruber B, Kogler H and Hammerle F 2017 Modal testing of hydraulic pipeline systems *J. Sound Vib.* **409** 256–73

[3] Mikota G 2013 Modal analysis of hydraulic pipelines *J. Sound Vib.* **332** 3794–805

[4] Mikota G 2004 Experimentelle modal analyse hydraulischer netzwerke (Experimental modal analysis of hydraulic networks) *Proc. of the VDI-Schwingungstagung* (Wiesloch) vol 1 (Düsseldorf: VDI Verlag GmbH) pp 15–26

[5] Mikota G 2004 Duality applied to experimental modal analysis *Proc. of the Bath Workshop on Power Transmission and Motion Control* (Bath) vol 1, ed C R Burrows, K A Edge and D N Johnston (London and Bury St. Edmunds: Professional Engineering Publishing Limited) pp 111–122

[6] D'Souza A F and Oldenburger R 1964 Dynamic response of fluid lines *J. Basic Eng.* **86** 589-98

[7] Li X, Wang S and Liang R 2011 Modal analysis of two typical fluid-filled pipes in aircraft *Proc. of the Int. Conf. on Fluid Power and Mechatronics* (Beijing) vol 1 (New York: Curran Associates, Inc.) pp 468–472

[8] Makaryants G M, Prokofiev A B and Shakhmatov E V 2015 Vibroacoustics analysis of punching machine hydraulic piping *Procedia Eng.* **106** 17–26

[9] Yigang C, Qi Z, Jingchao S and Shichao Y 1987 Experimental modal analysis of metallic pipeline conveying fluid *Proc. of the Fifth national congress on pressure vessels and piping* (San Diego) vol 1, ed M K Au-Yang and S S Chen (New York: American Society of Mechanical Engineers) pp 87–90

[10] Nurkkala P, Haapaniemi H, Luukkanen P, Saarenheimo A and Vepsä A 2004 Operational modal analysis of a piping system *Proc. of the IMAC XXII Conference and Exposition on Structural Dynamics* (Dearborn) vol 4, ed P Sas and M De Munck (New York: Curran Associates, Inc.) pp 2168–2174

[11] Augusztinovicz F and Sas P 1996 Acoustic modal analysis at low frequencies: similarities and differences in formulations *Proc. of the Int. Conf. on Noise and Vibration Engineering* (Leuven) vol 3, ed P Sas (New York: Curran Associates, Inc.) pp 1685–1699

[12] Nieter J J and Singh R 1982 Acoustic modal analysis experiment *J. Acoust. Soc. Am.* **72 (2)** 319–26

[13] Kung C-H and Singh R 1985 Experimental modal analysis technique for three-dimensional acoustic cavities *J. Acoust. Soc. Am.* **77 (2)** 731–38

[14] Rossetto G D, Arruda J R F and Huallpa B N 2000 Experimental modal analysis of a cavity using a calibrated acoustic actuator *Proc. of the Int. Conf. on Noise and Vibration Engineering* (Leuven) vol 1 (New York: Curran Associates, Inc.) pp 1415–1422

[15] Yoshimura T, Saito M, Maruyama S and Iba S 2012 Modal analysis of automotive cabin by multiple acoustic excitation *Proc. of the Int. Conf. on Noise and Vibration Engineering* (Leuven) vol 1, ed P Sas, S Jonckheere and D Moens (New York: Curran Associates, Inc.) pp 4061–4069

[16] Ewins D J 2000 *Modal Testing: Theory, Practice and Application* (Baldock: Research Studies Press Ltd.)

[17] Wachel J C and Tison J D 1994 Vibrations in reciprocating machinery and piping systems *Proc. of 23rd Turbomachinery Symposium* (College Station) vol 1, ed J C Bailey and D W Childs (College Station: Turbomachinery Laboratory) pp 242–272

[18] Yu Y, Zhang S, Li H, Wang X and Tang Y 2017 Modal and Harmonic Response Analysis of Key Components of Ditch device Based on ANSYS *Procedia Eng.* **174** 956–64

[19] Zagretdinov A R, Vankov Y V and Ivshin I V 2016 The Analysis of Frequencies of Own Insulators Fluctuations for the Solution of a Problem of their Control *Procedia Eng.* **150** 2334–9

[20] Vakhidov U Sh, Sogin A V, Shapkin V A and Shapkina Yu V 2014 Chislennyye issledovaniya kolebaniy uzlov avtomobiley *Transactions of Nizhni Novgorod State Technical University n. a. R. Y. Alexeev* **3 (105)** 145–53

[21] Bezborodov S A and Ulanov A M 2017 Calculation of vibration of pipeline bundle with damping support made of MR material *Procedia Eng.* **176** 169–74

[22] Madenci E and Guven I 2006 *The finite element method and applications in engineering using ANSYS* (New York: Springer Science + Business Media LLC)

[23] Basov K A 2002 *ANSYS v primerakh i zadachakh* (Moscow: Kompyuter Press)

[24] Pastukhova L G and Sekacheva A A 2016 Chislennyy analiz dinamicheskikh kharakteristik elementa truboprovodnoy sistemy mnogoetazhnogo zhilogo doma *Int. J. for Computational Civil and Structural Engineering* **12 (4)** 137–46

[25] Kattan P I and Voyiadjis G Z 2012 *Damage Mechanics with Finite Elements: Practical Applications with Computer Tools* (New York: Springer)

[26] Zienkiewicz O C, Taylor R L and Fox D D 2005 *The Finite Element Method for Solid and Structural Mechanics* (Amsterdam: Elsevier)
[27] Mironova T B, Prokofiev A B and Sverbilov V Y 2017 The finite element technique for modelling of pipe system vibroacoustical characteristics *Procedia Eng.* **176** 681–8

IOP Conf. Series: Materials Science and Engineering **481** (2019) 012020 doi:10.1088/1757-899X/481/1/012020

Concept of Organic Architecture in the Second Half of the XXth Century in the Context of Sustainable Development

T Y Bystrova[1]

[1]Ural NII project RAASN, 50a Lenina st., Ekaterinburg, 620075, Russia

E-mail: taby27@yandex.ru

Abstract. The article defines the most essential elements of the organic architecture concept. Features of organic architecture appearing in the works of architects of the second half of the twentieth century are specified in the context of sustainable development. We consider T. Alberts to be the representative of this period and the study of his works enables us to define these elements more precisely. The authors refer to the following features of organic architecture of the second half of the twentieth century: current development of regional types of organic architecture, which are opposed to globalistic and postmodern trends; organization of international community of organic architecture followers; dominance of complex rather than single structures in the projects of organic architecture representatives including landscape, water, greenery and other elements; intensification of the attention to a human as a consumer of architecture; utilization of organic architecture as the element for building up the image of large European companies; intensification of the architects' attention to technological and economic indices in projects in contrast to the emphasis on the aesthetics of forms in the early twentieth century. These regularities would help understand the principles of organic architecture in connection with sustainability.

1. Introduction

Urban environment of the twentieth and twenty-first centuries is becoming more chaotic and disharmonic, and particular objects of architecture may contrast sharply with the environment, as cities grow and construction is carried on a large scale [7; 20]. Classical ensemble resulting in the creation of harmonious paces is rarely achieved today. In some cases, innovative architectural forms can be created exclusively based on technologies, that is by means of engineering rather than architecture; their absolutization makes people estrange themselves from material environment. Disintegration of architecture in cities, in particular, actualizes the creation of architectural objects that fit human spirit and mentality and restore the sense of harmony. New paradigm of urban development, for example, the concepts of creative city, intelligent city, open city or neo industrial city, allows to turn to socially responsible position of architects. This position, in turn, helps realize organic architecture.

The concept "organic architectural form" relies on the idea of necessary integrity and inner structuredness of an object of any scale designed by an architect. L. H. Sullivan, F. L. Wright, A. Gaudi and R. Steiner are considered to be the founders organic morphogenesis in contemporary architecture. Both ideological and professional [9] and technological factors highly influence the concept development.

IOP Conf. Series: Materials Science and Engineering **481** (2019) 012020 doi:10.1088/1757-899X/481/1/012020

2. Methodology

Object of the article: to define fundamental concepts of organic morphogenesis and emerging features of organic architecture of the second half of the twentieth century analyzing the works of Ton Alberts (1927–1999), one of the most typical representatives of this approach.

Research methods: *System-genetic analysis* allows to define the object of study and to follow its evolution in terms of specific manifestations of organic approach in architecture. *Interdisciplinary* nature of study is defined by the necessity to attract theoretical material of humanities and philosophy of organicism starting from Aristotle. *Morphological* analysis allows to evaluate the realization sequence of organicism as designed methodological guidelines for architects.

3. Argumentation

Organic architecture held itself out as certain trend on the cusp of the nineteenth and twentieth centuries, although the approach itself is mature [5; 14]. Philosophic theoretical argument for the necessity and possibility to create organic forms in architecture took place, first, in cultural environment of Western Europe. The concept of organic architecture was realized in Antiquity but was not constantly developed in the subsequent epochs. One may point to a number of discontinuous historical periods when the interpretation of architectural form as an integral, dynamic and inner potential realizing principle – Gothic, Art Nouveau, Bionics of the 1960s and 1970s [10; 12; 18; 24] – predominated.

3.1. The profile of organic architecture

Organic architecture is the architecture of integral and structured forms, similar (isomorphic) in its structure to forms of nature. However, the accents in the interpretation of organic architecture are changing under the influence of scientific data as well. For Antiquity, it is the possibility to attain *rationality* of the form, consequentiality of unified idea realization, no matter, whether it is the idea of Demiurg or an architect. For the Medieval and Renaissance, it is human moving on the way of divine creation and possibility to escape chaos owing to the *primacy of the whole* rather than parts [4; 8]. For organic architecture of the twentieth century, it is the idea of the necessity of relationships with *natural environment* attained by means of application of new technologies and materials as well. For Art Nouveau architects, from F. Schechtel to A. Gaudi, first, it is *aesthetic* idea of harmony of the form being integral and unified in its elements. For R. Steiner and his followers, it is the idea of human body *isomorphism* and the "body" of architecture [1; 3; 13; 21] filled with ecological contents in the second half of the twentieth century, mainly thanks to the activity of F. Hundertwasser [4, *130–140*].

Note that the issues of organic form appeared in the works of authors – theologians, naturalists and others - indirectly related to architecture but reflecting upon world structure. Theorization was not always realized in practice because of the lack of construction process development. The development of construction technologies in the second half of the nineteenth century resulted in the formation of the architectural trend – "organic architecture".

3.2. Identification of basic elements and development of theoretical model of organic architecture

The analysis of the idea genesis shows:

- there is no unique interpretation of organic architecture, nowadays it has diverse interpretations and manifestations. Nevertheless, extended interpretation of organic architecture as activity *imitating* natural forms and even their structure is not correct.
- at the end of the nineteenth and the beginning of the twentieth centuries the architects from the USA and Western Europe suggest similar views on organic architecture as the creation of forms on the principles closely related to those of the development of wildlife objects. In the works of F. L. Wright, for example, pragmatic view on nature prevailed initially, being typical of American culture of this period.
- L. H. Sullivan and F. L. Wright typically understood organic architecture through the category of *interrelation* of the exterior and interior of a building and landscape.

- although high degree of reasoning, morphogenesis by R. Steiner may seem to be irrational resulting in semantic "unreadability" of a function of a building. R. Steiner did not aim himself for the totality of organic architecture and suggested it was used for completely specific conditions and tasks to create educational, religious and scientific centers and accompanying objects.
- in the twentieth century organic architectural form was created by various processes and utilization of various materials from earth and clay to metal ware (R. Piano, S. Calatrava [23] et al.) and novel materials (foil ETEF, for example, by N. Grimshaw [4, *232*]), which provide architects with various opportunities. Their utilization effects the structure of an architectural object.

One can specify the features of an organic architectural form in the course of studying the texts on architecture. Prominent urbanistic researcher K. A. Lynch fairly pointed out, "A beautiful and delightful city environment is an oddity, some would say an impossibility. Not only American city larger than a village is of consistently fine quality, although a few towns have some pleasant fragments. It is hardly surprising, then, that most Americans have little idea of what it can mean to live in such an environment. They are clear enough about the ugliness of the world they live in, and they are quite vocal about the dirt, the smoke… and yet monotony of it. But they are hardly aware of the potential value of harmonious surroundings…" [16, *2*].

K. A. Lynch used exactly a string of correlated words, which helped him interpret basic quantities of aesthetic urban environment. These are *continuity* (regularity, *непрерывность), integrity, harmony, giving "delight" (enjoyment наслаждение) every day*. According to K. A. Lynch, opposed to them are randomness, stuffiness, crowd, and monotony – more physical and special features than psychological ones. From the scientific point of view, a man in a city is lost in discrete, disordered and uneven environment, insufficiently influencing man's feelings and emotions.

The analysis of organic architecture genesis [6; 10; 13] and representations of professional thinking (of architects [4; 16] generally affords a basis for an author's model of organic architecture (Fig.1). The qualities emphasized in each historical period remain significant and can become actual in the creativity of the following generations of architects and at a new problem and technological spiral. All qualities at the same time can be either unrecognized or unused by an architect. Their presence in objects enables not so mush classification, as conception understanding.

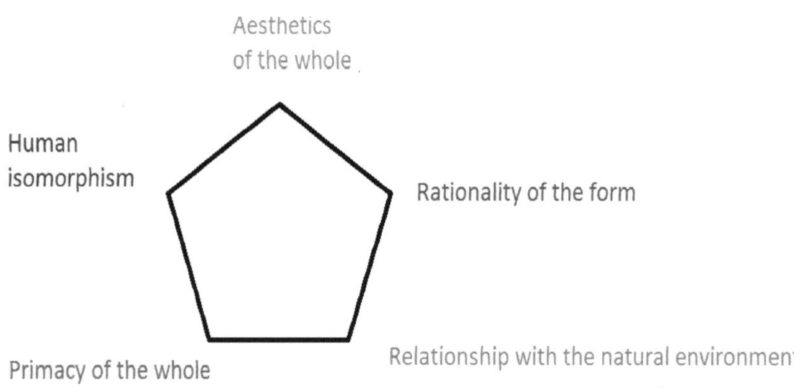

Figure 1. Model of organic architecture, author T. Bystrova.

3.3. Specific character of organic architecture of the second half of the XXth century: T. Alberts

The concept of sustainable development emerged in the 1980s emerged as a result of combining the three main dimensions: economic, social and environmental. Sustainable development is understood by specialists as a development that meets modern needs and does not threaten the opportunities of future generations to meet their needs. The economic aspect of the Concept of Sustainable Development implies the optimal use of limited resources and the use of environmentally friendly, energy and

IOP Conf. Series: Materials Science and Engineering **481** (2019) 012020 doi:10.1088/1757-899X/481/1/012020

material-saving technologies, including the extraction and processing of raw materials, the creation of environmentally acceptable products, minimization, processing and disposal of waste. It's not about refusing to increase production, but about stopping the growth in the use of environmental resources. The existence of this aspect actualizes the ideas of organic architecture. In this part we illustrate the relationship with examples from the legacy of Ton Alberts, one of the pioneers of the organic approach in Belgium architecture.

Antony (Ton) Alberts (1927–1999) is recognized to be the representative of organic architecture of the second half of the twentieth century [15] in spite of few works done by this unusual architect. It was merely ten years before his sudden death when T. Alberts became internationally recognized for having designed the headquarters building of the bank ING in the south of Amsterdam. Besides, during previous years, the architect was working under the banner of organicism.

Ton Alberts suggested abandoning functional interpretation of geometric patterns peculiar to organic architecture of F. L. Wright and his followers. Such version of organic architecture seems to be appropriate for Holland. Routine standards of this region tend to restraint and rationality inherited from protestant value system. Being involved in constant, rather dramatic interaction with nature, the Dutch have to comprehend the dilemma of natural and civilized, to solve the problem of following the nature or confronting it.

T. Alberts was born in Belgium and studied at Higher Technical School and and Ecole des Beaux-Arts in Paris [2]. Then he went to Amsterdam to teach in the Academy of Architecture for more than twenty years. In 1963, T. Alberts opened his own architect bureau. In 1993, he was invited to lecture at Harvard but continued practicing in architecture. In the course of teaching Allberts was likely to arrive at the interpretation of architecture that was similar to that of an Austrian architect F. Hundertwasser. They almost at the same time formulated the metaphor of architecture as «*the third skin*» of a man [4, *133*] having the structure similar to that of natural skin and clothes.

The objects of T. Alberts may resembled in appearance the works of R. Steiner. Nevertheless, analysys [11] emphasize the very originality of the path of a Dutch architect. If the creation of structures fitted in with world structure was of primary importance for R. Steiner focused on anthroposophy, T. Alberts who prioritized a man, wrote, "It seems we have forgotten that we have to live in buildings that are over rationalised. Besides this rationality we also need fantasy, intuition and emotion" [11]. That defined the steady interest of the author as a consumer of both architecture and its technical facilities, that, in its turn, allows to enhance the ecological potential of objects. Thus, Dutch version of organic architecture approaches the issue of sustainable *development being* relevant for the beginning of the 2000s.

The headquarters' building of the bank Internationale Nederlanden Group (ING) was built in 1983 году in the southeastern part of Amsterdam; recently the headquarters moved to a new building. We would analyze the building in terms of the presence of features typical to organic trend in its architecture. splay

3.3.1. The relation with natural environment

Ten-tower complex has sloping and expanding downwards patterns. They provide the feeling of stability, reliability and the action in life (Figure 2). Brick specific for Dutch structures is used with brown concrete curtains of piles and shaded by the blue colour of roof and aluminum window frames. The traditional image of a respectable bank remained untouched and was even accentuated. In addition, the slope of the walls intensifies the possibilities of natural lighting and acoustic efficiency as well as decreases the air mass drag. Air spaces accumulate heat and increase energy efficiency. The connection with the environment is not only the connection with landscape but with socio-cultural context and climatic conditions as well influencing the technology solutions.

Figure 2. Group of headquarters' buildings of the bank ING. Amsterdam, Holland. 1983. Architects: T. Alberts, M. van Huut. From site R. Gorter. URL: http://robert-gorter.info/ton-alberts.

3.3.2. Primacy and aesthetic qualities of the whole

The complex was planned as the whole rather than as mechanical structure from ten separate elements. The buildings are located S-shaped with the street between members. One can see a great number of patterns and perspectives when walking along the street. Though, ten "towers" are of similar structure, there is no feeling of monotony. The complicated configuration makes similar patterns look alike. At the same time, the architects breaking "vernacular architecture" of Modernism [17], use fivefold symmetric turning of the principle element and simultaneously tie form to landscape. In other words, it is not referred to the imitation of natural forms but to the creation of a structure possessing the degree of flexibility and peculiarity similar to that of life forms.

3.3.3. Form rationality

Seven out of ten towers are south-oriented thus providing a maximum of light and solar heat. Employees work 80% of their working hours at natural lighting. It was possible to save 96 million US dollars and make the construction cheaper because of recoverable standard version of the principle element of the complex.

The complex is equipped with computer-controlled solar batteries and shutters. Energy saving was converted from abstract slogan to economic and human equivalent. For example, costs on battery planting are reported to have been covered during the first year of operation. Exhaust air is kept in pentagonal glass-covered structures at the top of each tower; 70% of such air is used to heat fresh air. These structures serve as monitors. Cool air is caught at night and used then in the daytime.

"Towers"-clasters are of various height with five offices on each story. Similar "cut" geometric patterns are used in the interior. Such patterns are close to the architecture of R. Steiner but look sometimes somewhat deliberately. The finishing from natural materials – stone and wood – intensifies the feeling of dynamism and naturalness.

3.3.4. Social aspect of organic architecture of T. Alberts

What is more important, organic conception is realized to the full at all project stages including a social one firstly mentioned by the theorist of architecture B. Dzevi [15]. The conception is opposed to post modernistic trends of that time with their aspiration for uncommon exterior forms. Social aspect of organic architecture increases in the second half of the twentieth century.

In the complex ING, offices are equipped with devices allowing to control light and air coming. The number of plants used as humidifiers is calculated. They are rain watered; the rainwater is collected from outside water basins and oxygen enriched when coming through special installations. Furniture is from natural wood, stair railings are made of birch, handles are of metal – additional attention was paid to tactile contact with architecture. Building are not very high, nevertheless, employees preferred to use stairs instead of elevators. People felt better due to utilization of water, greenery, and fresh air. Consequently, the decreased number of seek leaves days enabled the company to reduces expenditures by 1.4 million US dollars a year. In that period, in the middle of the 1980s, these indices were novel and extremely positive as they strengthened the position of organic architects.

Conclusions

The project analysis of T. Alberts chosen as a typical organic architecture representative of the second half of the twentieth century allowed formulating specific features of that period concretizing the understanding of possibilities and task of organic architecture and shows its connections with the concept of sustainable development.

We can underline:

- parallelism in the development both concepts and practices caused with ecological and socio economic problems:
- the development of regional versions of organic architecture in the second half of the twentieth century in contrast to internationalization tendencies prevailing in that period;
- direct and indirect connection of regional projects in the area of organic architecture with projects in other countries and regions assigned by the conception integrity in the whole;
- continuous attraction of architects to the complex development rather than single structures with the element of landscape, greenery, water and other elements that allow to create an original architectural microcosm based on the laws of nature;
- growing attention to a consumer of architecture resulted from philosophy and socio cultural situation in the second half of the twentieth century; abandoning the design for an abstract "consumer" and the emergence of conceptual human system models including the system of values, life style, emotional, cultural and other parameters;
- possibility to combine image-oriented interests of a company with organic, human-oriented solutions approved by the projects of T. Alberts and other representatives of the trend;
- increased attention to economic indices of a project, opportunities to achieve economic effect thanks to application of new technologies;
- conversion of energy-saving, resource-saving, health and safety characteristics into integral characteristics of organic architecture as the public interest in the problems of sustainable development, ecology, and healthy life style grows in the second half of the twentieth century and the beginning of the twenty-first century.

Acknowledgments

The article is written with financial support of the institute Ural NII project RAASN Board of Construction and HUI RF on the theme 1.6.5. "Mastering the New Concepts of Formation in the Architecture of Russia in the 21st Century".

References

[1] Auer W M Was ist Kunst? Zur Ästhetik Rudolf Steiners *Erziehungskunst. Waldorfpaedagogik heute.* Retrieved from http://www.erziehungskunst.de/artikel/was-ist-kunst-zur-aesthetik-rudolf-steiners. Accessed 04 April 2018.

[2] Bachman L R 2003 *Integrated Buildings: The Systems Basis of Architecture* (John Wiley&Sons, Inc)

[3] Baur S «Fremder Planet – Architekturfuehrer Anthroposophenhuegel» *Literatur&Kunst / A magazine of literature and art.* Retrieved from http://www.literaturundkunst.net/fremder-planet-architekturfuhrer-anthroposophenhugel/ Accessed 28 March 2018.

[4] Bystrova T J 2013 *From modernism to neoclassicism: architectural concepts XX-XXI centuries* (Ekaterinburg: Webste) 288 p.

[5] Bystrova T J 2014 Integrity of civilized criteria *Izvestija UrGU. Ser.: Education, science, culture* **3** pp 172–182.

[6] Definition Organische Architektur *Volkania. Organische Architektur. Eine Sammlung.* Retrieved from http://volcania.wordpress.com/definition-organische-architektur. Accessed 03 March 2013

[7] Dobricyna I A 2007 *From postmodernism to nonlinear architecture. Architecture in the context of modern philosophy and science* [Abstract of a PhD thesis] Retrieved from

http://niitiag.ru/publications/auto_referat/37-dobricyna-ia-ot-postmodernizma-k-nelineynoy-arhitekture-arhitektura-v-kontekste-sovremennoy-filosofii-i-nauki.html. Accessed 15.11.2015.

[8] Eco U 2002 *Art and Beauty in the Middle Ages. Yale Nota Bene* 131 p. Retrieved from https://sites.google.com/site/kunstfilosofiesite/Home/texts/eco-art-and-beauty-in-the-middle-ages Accessed 23 March 2015

[9] Fumara L and Gorbik E Introduction in the principles of organic architecture. Retrieved from https://goetheanum.org/fileadmin/_migrated/content_uploads/Prinzipy_Org_Arx.pdf Accessed 03 September 2018

[10] Hess A 2006 *Organic Architecture: The Other Modernism* (Salt Lake City: Gibbs Smith)

[11] Holdsworth B 1999 Ton Alberts, Dutch organic architect, dies at 72 *Architects Journal* 2.09.1999. Retrieved from http://www.architectsjournal.co.uk/home/ton-alberts-dutch-organic-architect-dies-at-72/773534.article Accessed 10 September 2018.

[12] Ikonnikov A V 2001 *Architecture of the XX century. Utopia and Reality* (Moskow)

[13] Koehler M 2011 Anthroposophenkolonie Dornach – Rudolf Steiner und die organische Architektur *Baunetzwoche`243* Retrieved from http://www.baunetz.de/dl/1249021/baunetzwoche_243_2011.pdf Accessed 06 April 2018

[14] Kruft H-V 2004 Geschichte der Architekturtheorie: Von der Antike bis zur Gegenwart. Studienausgabe. Funfte Auflage (Verlag C. H. Beck oHG, Munchen 1985)

[15] Kurbatov J The age of organic architecture *Form. Architecture and design for those who understand.* Retrieved from http://www.forma.spb.ru/magazine/articles/d_0010/main.shtml Accessed 02 October 2014.

[16] Lynch K 1960 *The Image of the City* (MIT-Press) 194 p

[17] Motro R 2006 *An Anthology of Structural Morphology* (World Scientific Publishing Co.Pte. Ltd.) 212 p

[18] Pearson D 2001 New organic architecture: the breaking wave (University of California Press) 223 p.

[19] Rappoport A G and Somov G J 1990 *Form in architecture: Problems of theory and methodology* (Moskow: Strojizdat) 344 p

[20] Rjabushin A V 2005 *Architecture on the cusp of millenium* (Moskow: Art. XXI century)288 p

[21] Rudolf Steiner als Architekt *Anthromedia. Internetportal Anthroposophie.* Retrieved from http://www.anthromedia.net/de/themen/anthroposophie/goetheanum/der-goetheanum-bau/architektur-des-ersten-und-zweiten-goetheanum/ Accessed 10 March 2013.

[22] Schräge Bank *Volcania. Organische Architektur. Eine Sammlung.* Retrieved from http://volcania.wordpress.com/2011/10/27/schrage-bank/ Accessed 06 September 2018.

[23] Tzonis A and Lefaivre L 2001 *Santiago Calatrava´s creative process. Part 1: fundamentals* (Basel: Birkhäuser Verlag)

[24] Van der Ree P *Mensch und Natur als Inspirationsqelle. Ursprung, Entwicklung und Aktualitaet der organischen Architektur* Retrieved from http://www.organische-architektur.org/pags/docs/77.1336550571.pdf. Accessed 16 March 2013

Efficiency of Installation of Acoustic Structures in Urban Environment

A V Kolmakov[1], V O Ignatyeva[1]

[1]Ural State University of Architecture and Art

E-mail: Kolmakov_AV@mail.ru

Abstract. The article reviews publications within the framework of all-Russian conferences devoted to the problems of increasing the efficiency of acoustic screens in the urban environment. The proposed engineering solutions are identified and analyzed. The principles of the most rational placement of acoustic structures in the urban environment are formulated.

Keywords: acoustic panels, anti-diffraction devices, acoustic hoods, noise.

1. Introduction

Noise exposure leads to increased fatigability either of the whole body, or one particular organ (including brain). While working in noisy environment, a person gets more tired faster, and while resting in such an environment, a person's body cannot fully restore itself. The lack of proper rest causes the body to tense, which, in turn, could lead to organ or body system failure. This is the way, in which noise could influence a vulnerable spot in a person's body (complaints of discomfort vary among people exposed to noise).

In Russia, the population of major cities is exposed to excessive noise. Noise could be regarded as a byproduct of human activities. Most people may suffer from noise exposure, though lack social activism concerning solving this problem: people do not know what could and (or) should be done to negate the harmful effect of noise.

2. Importance

Today, there is a fair amount of various sound-absorbing installations and other measures, which help decrease noise exposure of residential areas and their inhabitants. However, implementation of designed antinoise solutions may not always ensure the actual noise reduction, because the taken measures are ineffective or difficult to accomplish due to various reasons (high cost, design flaws, less than responsible attitude to construction and so on) [1-5]. Different technical design projects aimed at noise protection (primarily, installation of acoustic panels (AP)) are being implemented in major Russian cities, although their efficiency and esthetics may be initially inadequate or unable to withstand a "challenge" even within a short period.

3. Research objective

A review of publications on AP efficiency (proceedings of conferences on Protection Against Excessive Noise and Vibrations, held in D.F. Ustinov Baltic State Technical University VOENMEH,

Saint Petersburg, 2006-2017) made it clear that there is no shared vision among experts regarding the efficiency of the proposed design concepts aimed at protection against noise (e.g. AP).

It is widely acknowledged that the efficiency effect of an acoustic panel depends on its size, which, in turn, depends on the size of the noise source (NS), and their relative position in respect to the shielded object. When designing an AP, it should be noted that its efficiency would majorly depend on height of the future structure. The tests show that "the effect of height on AP efficiency is nonlinear in nature: the higher the AP, the fewer is the efficiency improvement with further height increase. For example, if the height is increased from 2 to 3 m, the efficiency improves by 4 dBA, whereas, if the AP height is increased from 5 to 6 m, the efficiency only improves by 1 dBA. A significant improvement of shielding efficiency (up to 3-5 dBA) is observed when the AP height is doubled" [1]. However, even the tallest APs (6 m) have no effect on protection of upper floors inhabitants against noise, which means that either the noise sources have to be shielded, or windows of the building fronts, which are exposed to excessive noise, have to be soundproof [6, p. 433; 7-12].

To improve cost-effectiveness of installation of tall APs, and to enhance the efficiency of smaller size APs, a number of researchers claim it possible to arrange various types of additional structures (anti-diffraction devices, AD), which help compensate for part of the AP height (see figure 1) [1, 13-20]. However, their publications contain no record of efficiency of such additional structures: dimensional ratio (width of construction and AP height) and physical dimensions are not provided. There are no layouts of relative positioning of an AP and a noise source, or information on a reference point for calculations or tests. In the work of T.E. Klyupa, the author attempted to determine AD efficiency in practice [19]. However, structures of different height (AP h=4 m and AP+AD h=4.5 m) were compared in the research. There is no data on the efficiency of a 4.5 m AP without an AD, and no references to theoretical description of the stated approaches, on which AD performance is based (impedance increase on the free edge and diffraction angle increase). Consequently, the issue of AD utilization efficiency remains open.

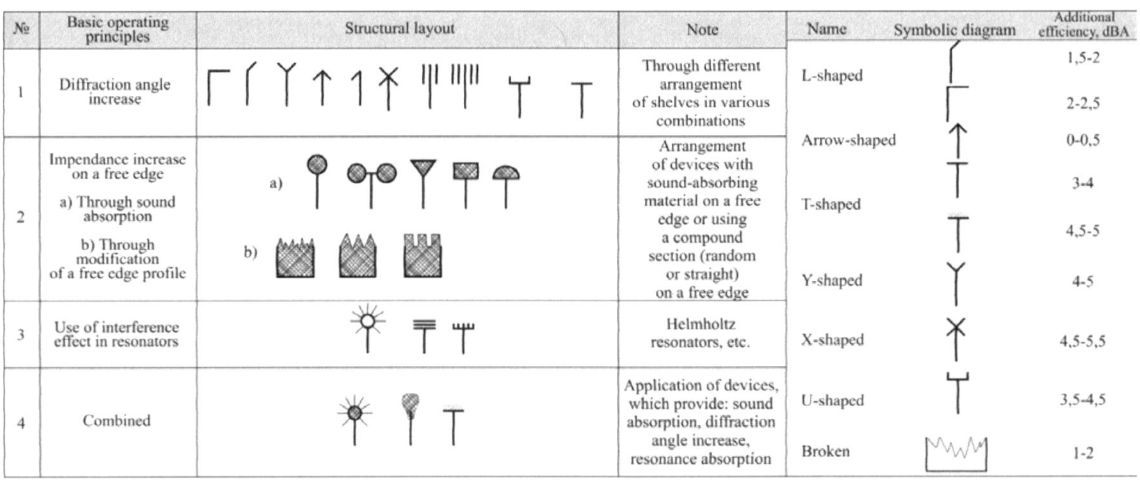

№	Basic operating principles	Structural layout	Note
1	Diffraction angle increase		Through different arrangement of shelves in various combinations
2	Impendance increase on a free edge a) Through sound absorption b) Through modification of a free edge profile	a) b)	Arrangement of devices with sound-absorbing material on a free edge or using a compound section (random or straight) on a free edge
3	Use of interference effect in resonators		Helmholtz resonators, etc.
4	Combined		Application of devices, which provide: sound absorption, diffraction angle increase, resonance absorption

Name	Symbolic diagram	Additional efficiency, dBA
L-shaped		1,5-2
		2-2,5
Arrow-shaped		0-0,5
T-shaped		3-4
		4,5-5
Y-shaped		4-5
X-shaped		4,5-5,5
U-shaped		3,5-4,5
Broken		1-2

Figure 1. Additional Acoustic Efficiency of APs of Different Shapes Compared with Straight AP without Additional Structures on the Free Edge [15].

4. Theoretical Part

A "panel" could mean any obstruction on the sound propagation path. In urban environment, a role of such an obstruction could be played by walls made of various materials (concrete, metal, wood, stone, plastic, glass and so on) that are erected intentionally, including walls of buildings and constructions, or by natural or man-made terrain (earthworks, banks, hills, cut slopes, ravines, etc.), or a combination of both [6, p. 425; 21-22]. Main function of a panel is to prevent noise propagation into the protected area. Considering diffraction and sizes of audible sound waves (17 m < λ < 17 mm), the AP efficiency depends on its physical dimensions, soundproofness (acoustic transitivity), and spatial relationship between a noise source, AP and reference point (protected area).

The principle of AP performance is widely known. It is based on the reflection effect (rereflection) of the acoustically hard surface, the absorption effect of the acoustically soft surface, and the scattering effect of the contoured surface [1, 10-11, 15, 23-30]. When giving grounds for efficiency of AD application, the authors of researches should first provide a functional description of the anti-diffraction device, mainly, how it succeeds to manage one of the fundamental properties of a wave – diffraction.

Efficiency of AD structure geometry is yet to be studied further. Calculation methods that are being developed should be described theoretically and justified by proper experimental measurement data. Today, the authors of this paper believe, numerous AD structures proposed for application only have an ornamental function. Any AD structure operates as a wide screen. This means that a narrow AP (compared to a wavelength, λ), which has a broadening atop (AD, shape is irrelevant), imitates a gain in its geometric width, forcing a sound wave to undergo double diffraction from the sides of a panel resulting in additional noise reduction. A similar approach is used for estimation of shielding effect of buildings, depth shapes, earth banks, slopes, hollows, etc. (see figure 2). Sound-absorbing linings of the AD intended to reduce diffracted noise are useless, unless their design is based on Helmholtz resonators. It is understood that sound absorption properties of a material are most effective at sound incidence onto the front surface of the sound-absorbing material. When a sound wave interacts with the AD, especially cylindrical in form, planes of the sound beams and sound-absorbing surface may be considered parallel. Therefore, sound-absorbing effect should be considered null [3, p. 320].

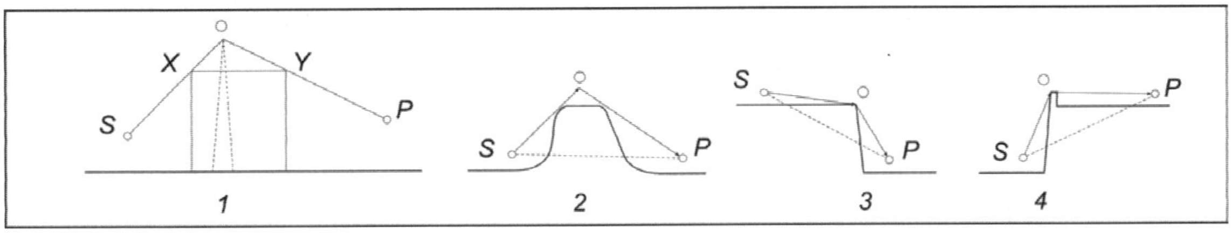

Figure 2. Noise Reduction at Double (1-2) and Single (3-4) Sound Wave Diffraction. 1 – building; 2 – earth bank; 3 – terrace; 4 – hollow road [6, p. 435].

5. Practical relevance

If no special measures are taken for acoustic suppression, residential development should be far removed from a traffic way (at least 150 m from expressways and truck routes) [6, p. 205]. Acoustic comfort of people living in close proximity to residential construction sites and main transport routes could be ensured through erection of antinoise buildings, which could be used as powerful acoustic panels. Reduction of noise in the acoustic shadow of buildings reaches up to 25 dBA. Taking into account that multi-storied buildings are frequently erected under modern residential development projects, their shielding efficiency may reach an even higher level [31]. Though it should be mentioned that, just like sound-reducing walls, their efficiency depends on length, soundproofness, and height. Drive-throughs into yards and spacings of buildings reduce sound insulation letting noise reach residential areas. Sometimes, due to rereflection of sound waves, noise level may increase beyond acceptable in the rooms of neighboring buildings located deep in a residential area. To prevent soundproofing of antinoise buildings from decreasing, these buildings should be as long as possible. In case the buildings are erected on constricted spaces within existing development, either they should abut neighboring buildings, or APs of required height should be designed (see figure 3). To lower diffraction on end walls, it would be rational to design a U-shaped noise-protective structure. To limit noise propagation through drive-throughs, A.V. Kolmakov recommends designing them as complex labyrinths with sound-absorbing facing, or plan a driveway into a yard through an underground car park [32-33].

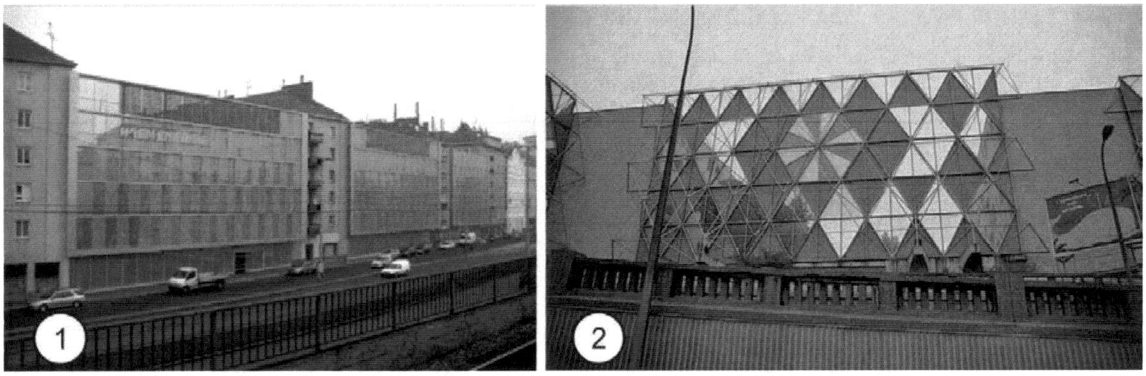

Figure 3. An example of tall AP structures. 1. Vienna, highway A23; 2. Parisian suburb [34].

Installation of canopies (AD) with larger overhangs (reaching up to the center of a roadway) could help improve the efficiency of acoustic panels (AP) (up to 18 dBA). In this case a noise source could be compared with a sound source placed into an acoustical shell that opens away from the protected object [8, 35]. As an alternative for such kind of AP structures, N.K. Kiryushkina proposes to cover up the roads completely. This would minimize noise emission and eliminate spatial incoherence created by a traffic route [8]. Open road sections are proposed to be fenced off by multi-storied car parks. In this case, APs should be regarded as acoustic hoods (AH).

Today, cycling or walking along highways with heavy traffic is decidedly problematic. Especially bad acoustic environment is found in enclosed pedestrian overpasses due to increased noise coming through entrances into the enclosed area of the overpass [36].

It is conceivable that implementation of an idea proposed by N.K. Kiryushkina could lead to construction of a network of bicycle routes and electronic transportation devices, which would ensure slower yet soundless transportation, in place of covered-up highways. Apart from noise reduction, this could help lessen primary and secondary dust formation (production and volatilization). Then it may be possible that a part of the population would change their means of transportation (by choosing bicycle and individual e-transport (self-balancing scooters, regular scooters, etc.)) and spend more time outside.

6. Conclusion

Installation of APs, especially of tall panels, should conform to requirements for efficiency, ease of maintenance and long-term operation, due to their cost. When defining and comparing cost-performance ratio of projects designed to shield noise sources, all possible noise sources (NS) found in the urban environment should be considered, including those, which may appear over the course of time. Taking into account that major noise sources in the city are motor- and railways, the most beneficial and cost-effective solutions may be those proposed by N.K. Kiryushkina and A.V. Kolmakov. Covering main traffic routes with acoustic hoods and construction of sound-absorbing drive-throughs (gates) could help reduce the size of urban areas exposed to noise pollution.

References

[1] Tyurina N V 2015 Relevant Problems of Acoustic Transport Shields Design *Protection against Elevated Noise and Vibration: Book of Conf. Repp* (St.-Petersburg) pp 685–93

[2] Pospelov P I 2006 Experience in the use of Acoustic Screens on the Moscow Ring Road and Prospects for their use on Roads *Protection of the Population from Increased Noise Exposure: Book of Conf. Repp* (St.-Petersburg) pp 77–82

[3] Kalashnikova N K, Rudneva E A, Klimenkova O I, Goncharenko I A and Ponofedin I A 2009 Mistakes of Acoustic Design of Transport Facilities with Noise Sources *Protection of Population against Elevated Noise: Book of Conf. Repp* (St.-Petersburg) pp 424–9

[4] Pospelov P I and Shchit B A 2011 Improving of Design of Noise Protection Measures on the Streets and Roads *Protection of Population against Elevated Noise: Book of Conf. Repp* (St.-Petersburg) pp 626–36

[5] Ivanov N I, Tyurina N V, Kudayev A V, Kuklin D A, Matveyev P V and Oleynikov A Yu 2013 Experience in Using Acoustic Screens to Reduce Noise on the October Railway *Protection of Population against Elevated Noise: Book of Conf. Repp* (St.-Petersburg) pp 438–42

[6] Osipov G L, Yudin Ye Ya and Khyubner G 1987 *Noise Control in Buildings and Residential Areas* ed G L Osipova and Ye Ya Yudin (Moscow:Stroiizdat) p 558

[7] Shubin I L and Nikolskiy A I 2006 Experience of Noise Reduction in Moscow *Protection of the Population from Increased Noise Exposure: Book of Conf. Repp* (St.-Petersburg) pp 58–65

[8] Kiryushina N K, Puzakova A I and Stepanov V N 2009 Application of Means and Methods of Noise Protection during Reconstruction and Construction of Main Transport Routes in Moscow *Protection of Population against Elevated Noise: Book of Conf. Repp* (St.-Petersburg) pp 372–81

[9] Ivanov N I and Kuklin D A 2011 The Problem of Railway Noise and Ways to Solve it *Protection of Population against Elevated Noise: Book of Conf. Repp* (St.-Petersburg) pp 108–23

[10] Kalinichenko M V, Kirillov I N and Bulkin V V 2015 Analysis of Possible Application of Sound-Reflecting Screens based on Helmholtz resonators *Protection against Elevated Noise and Vibration: Book of Conf. Repp* (St.-Petersburg) pp 278–80

[11] Kalinichenko M V and Bulkin V V 2017 Noise-Absorbing Panels based on Helmholtz resonators: Possible Application for Noise Reduction *Protection against Elevated Noise and Vibration: Book of Conf. Repp* (St.-Petersburg) pp 412–9

[12] Butorina M V, Minina N N, Tyurina N V and Anisimov E P 2017 Development of Noise Protection Activities for Highways using Information Modeling *Protection against Elevated Noise and Vibration: Book of Conf. Repp* (St.-Petersburg) pp 69–78

[13] Petrova S S 2006 Selection of Acoustic Structure for Protection Against Traffic Noise *Protection of the Population from Increased Noise Exposure: Book of Conf. Repp* (St.-Petersburg) pp 345–52

[14] Gribov S A 2011 Design and Production of Noise Protection Panels *Protection of Population against Elevated Noise: Book of Conf. Repp* (St.-Petersburg) pp 326–31

[15] Ivanov N I, Semenov N G and Tyurina N V 2013 Issues of Acoustic Panels Design and Their Application for Noise Reduction from Railway and Motor Transport *Protection of Population against Elevated Noise: Book of Conf. Repp* (St.-Petersburg) pp 52–88

[16] Shashurin A E, Tyurina N V and Kornilov V A 2015 Noise-proof Screens with Superstructure on the Free Edge *Protection against Elevated Noise and Vibration: Book of Conf. Rep* (St.-Petersburg) pp 580–3

[17] Ivanov N I 2015 *Acoustical engineering. Theory and Skills for Noise Pollution Control* ed N I Ivanov (Moscow:Logos) p 432

[18] Ivanov N I 2017 *Protection against Noise and Vibrations* ed N I Ivanov (publishing house NIC ART) p 268

[19] Klyupa T E 2017 Research of Acoustic Efficiency of Additional Structures of Cylindrical Form Atop a Free Edge of Noise Protection Panels *Protection against Elevated Noise and Vibration: Book of Conf. Repp* (St.-Petersburg) pp 576–86

[20] Roshchina N V 2017 The Peculiarities of Installing Noise Barriers in Human Settlements *Protection against Elevated Noise and Vibration: Book of Conf. Repp* (St.-Petersburg) pp 635–40

[21] Methodological Recommendations for Estimation of Required Noise Reduction in Residential Areas and Determination of the Required Efficiency of Acoustic Panels with Regard to Sound Absorption 2003 (Rosavtodor) p 130

[22] Shashurin A E, Tyurina N V, Semenov N G and Kornilov V A 2011 Application of Acoustic Panels with Sound-Absorbing Screens Made of Imposted Wood *Protection of Population against Elevated Noise: Book of Conf. Repp* (St.-Petersburg) pp 332–6

[23] Tyurina N V 2006 The use of Acoustic Screens to Reduce Noise in Residential Buildings *Protection of the Population from Increased Noise Exposure: Book of Conf. Repp* (St.-Petersburg) pp 97–112

[24] Minina N N 2006 Noise Reduction in Construction *Protection of the Population from Increased Noise Exposure: Book of Conf. Repp* (St.-Petersburg) pp 225–34

[25] Ivanov N I, Kuklin D A and Tyurina N V 2009 Experience of Application and Prospects of use of Acoustic Screens for Reducing Noise of Automobile Streams and Trains *Protection of Population against Elevated Noise: Book of Conf. Repp* (St.-Petersburg) pp 96–111

[26] Shubin I L 2011 Calculation of Noise Penetrating the Screen *Protection of Population against Elevated Noise: Book of Conf. Repp* (St.-Petersburg) pp 580–8

[27] Kolmakov A V 2013 Use of Fold Structures to Increase Acoustic Shield Functionality and Aesthetics *Arkhitekton: Izvestiya Vuzov* **42** (Yekaterinburg: Arkhitekton)

[28] Kolmakov A V 2013 Eliminating Monotony Created by Acoustic Panels *Protection of Population against Elevated Noise: Book of Conf. Repp* (St.-Petersburg) pp 420–5

[29] Shashurin A E 2015 Small Noise Shield *Protection against Elevated Noise and Vibration: Book of Conf. Rep* (St.-Petersburg) pp 458–60

[30] Kudayev A V, Kuklin D A and Matveyev P V 2017 Small Acoustic Screens as a Means of Reducing Railway Noise *Protection against Elevated Noise and Vibration: Book of Conf. Repp* (St.-Petersburg) pp 448–92

[31] Kochnev A P 2015 City Planning Residential Complexes with Noise Control Functions. Urban Areas Rehabilitation in terms of Noise Pollution. Roadmap *Protection against Elevated Noise and Vibration*: *Book of Conf. Repp* (St.-Petersburg) pp 160–79.

[32] Kolmakov A V 2017 Sound Insulation Structures of Yard Passages *Protection against Elevated Noise and Vibration: Book of Conf. Repp* (St.-Petersburg) pp 389–96

[33] Kolmakov A V 2017 City Noise Control with Architectural Means *Academic Bulletin UNIIP RAACS* **1** (Yekaterinburg) pp 51–6

[34] Kotzen B 2009 *Environmental noise barriers. A guide to their acoustic and visual design* ed B Kotzen and C English (London, New York:Spon Press) p 256

[35] Pimenov E V and Musin I G 2015 Application of Acoustical Shell in an Event Hall Protection *against Elevated Noise and Vibration*: *Book of Conf. Repp* (St.-Petersburg) pp 256–60

[36] Knyazev D A and Elkin Yu I 2013 Acoustic Effect in Enclosed Pedestrian Overpasses *Protection of Population against Elevated Noise*: *Book of Conf. Repp* (St.-Petersburg) pp 698–703

Dependency of noise level indicators on the heating power of rooftop boiler room

A Plotnikov[1], T Zhilina[1]

[1]Institute of Civil Engineering, Tyumen Industrial University, Volodarskogo Street, 38, Tyumen, 625000, Russia

E-mail: plotnikovas@tyuiu.ru

Abstract. The paper presents results of practical measurements and results of data processing of the noise level from rooftop boiler rooms used in monolithic framed buildings. Three types of rooftop boiler rooms with power of N=0.8 MW; N=1.0 MW; and N=1.2 MW were used for data recording and analysis with further obtaining of dependencies regarding various levels of their heating efficiency. A distinctive feature of monolithic framed house building is a stiffening core formed of staircase and elevator section. The basic load-bearing capacity of the frame is provided by the joint work of metal reinforcement (metal) and elements made from concrete. During observation of operating boiler rooms noise measurements were carried out in order to collect and analyze information characterizing operation mode of rooftop gas boiler house under various climatic conditions per heating period. Based on the obtained dependencies it is possible to predict future noise levels from boiler rooms while designing further territory development. Authors supplement the existing method to evaluate noise pollution of the city districts. Result of experiment and application of the obtained dependencies is predicting of harmful physical factors occurrence at different levels of heating efficiency with further reduction of negative impacts on building structures; predicting noise levels while noise mapping of the territories under development; increasing comfort of living. The given conclusions can be applicable for further modeling of noise transmission into the environment.

Keywords: noise, noise maps, noise prediction, independent heat supply sources, boiler room facilities, analytical dependencies

1. Introduction

Urban development in Russia determines high level of urban growth, increasing housing density and rising average noise level.

Acoustic comfort inside the buildings and on the surrounding territories relies upon the indicators combining both noises from external and internal sources. In order to evaluate the indicators of noise radiated from rooftop boiler rooms under various environmental conditions and transmission of acoustic energy into the environment during heating period one needs predictive methods for selection of optimal solutions in choosing building structures and noise mapping of the territories under development. The issue of capital investment into construction and installation works has always been relevant. Climatic aspects of regions for construction in our country [1,2] determine application of rational methods in construction industry. The advantages of using rooftop boiler rooms include decreased heat losses during heat carrier transportation, the minimum losses in "quality" of temperature chart and lowering of expenditures for heating. The necessity of noise mapping of the city districts (both existing and the ones planned for development) can be referred to its drawbacks [18]. Evaluation of each of the separate noise source takes long time, thus requiring significant time expenditures during estimation of residential areas of various scopes. Detection of noise level dependency on the productive power of boiler room is an opportunity to predict the noise level. Authors consider that performance analysis of operating independent heat supply sources is the key to predict development of areas being favorable for construction.

The paper aims to evaluate the noise indicators from engineering equipment on the territory of the city as a condition of steady noise in different periods.

2. Materials and Methods

Theoretical foundation for the study includes acoustic laws, and laws of the nature of sound and vibration, the known research works on acoustics and noise distribution in residential and industrial buildings, noise evaluation methods from street and road network in urban development, statistical data processing techniques and methods to record the noise level by measuring equipment [6, 9, 10, 11].

Evaluation of ambient noise level in urban development is a multivariable process including accumulation of various kinds of information. The intensity of traffic in street and road network, presence of detached linear and local noise sources are being evaluated upon multiple variables, which complicates participatory rapid noise assessment of existing and designed territories. Noise evaluation of boiler room facilities requires performing noise measurements, conducting calculations on noise generation from equipment and sound spectral overlap. The noise sources from equipment are rotating mechanisms, flow pulsations in the elements of air channels and liquids in the pipes. During urban development planning of future territories the level of reliable estimate of future noise pollution from rooftop boiler rooms should be based on the dependencies obtained while observation and recording of indicators in operating boiler rooms.

3. Experiment

Indicators of equivalent continuous sound level were recorded on three independent heat sources – rooftop boiler rooms with power of N = 0.8-1.2 MW. Sound level meter ASSISTANT TOTAL was used for data recording. The basic noise sources in boiler rooms are [3]: circulating-water and booster pumps, burner units, moving of heat carrier along the pipelines, operating peculiarities of boiler plants. These boiler rooms are equipped with burner units CIB UNIGAS [7]. These burner units are marked by correspondence of the noise level during operation with the requirements of national standards [4, 5, 6]. The noise level was recorded in four various climatic seasons, as independent sources work to cover the needs in heat carrier heating for hot water and heat supply. During observation of boiler rooms [8] the level of heating efficiency was recorded in percentage ratio regarding the maximum possible efficiency taken as 100%.

For replication of the obtained data on actual measurements the methods of physical similarity theory were applied. Regression equation is presented by a second degree polynomial and is given as (1)

$$y = 0.1548x + 70.333 \tag{1}$$

Determination coefficients of regression equations of the obtained dependencies are R_2 = 0.98. Reproducibility of experimental data was proved by Cochran's Test. Regression equation obtained during experiment to define the design noise level while determining dependencies of operation of various kinds of rooftop boiler rooms with power of N = 0.8–1.2 MW. The obtained data can be found in Tables 1-3.

IOP Conf. Series: Materials Science and Engineering **481** (2019) 012022 doi:10.1088/1757-899X/481/1/012022

Table 1. Boiler room No.1, power $N = 0.8$ MW

Recorded acoustic load, L [dBA]	Acoustic load obtained from regression equation, L [dBA]	Discrepancy between the actual and predicted acoustic loads, %	Heating efficiency of boiler room, %
72	71.88	-0.17	10
74	73.43	-0.78	20
75	74.98	-0.03	30
77	76.53	-0.62	40
77	78.07	1.37	50
79	79.62	0.78	60
81.5	81.17	-0.41	70
82	82.72	0.87	80
84	84.27	0.31	90
87	85.81	-1.38	100

Table 2. Boiler room No.2, power $N = 1.0$ MW.

Recorded acoustic load, L [dBA]	Acoustic load obtained from regression equation, L [dBA]	Discrepancy between the actual and predicted acoustic loads, %	Heating efficiency of boiler room, %
71	71.88	1.23	10
74	73.43	-0.78	20
75	74.98	-0.03	30
77	76.53	-0.62	40
77	78.07	1.37	50
80	79.62	-0.48	60
82	81.17	-1.02	70
82	82.72	0.87	80
84	84.27	0.31	90
88	85.81	-2.55	100

Table 3. Boiler room No.3, power $N = 1.2$ MW

Recorded acoustic load, L [dBA]	Acoustic load obtained from regression equation, L [dBA]	Discrepancy between the actual and predicted acoustic loads, %	Heating efficiency of boiler room, %
69	71.88	4.01	10
72	73.43	1.95	20
73	74.98	2.64	30
75	76.53	1.99	40
75	78.07	3.94	50
78	79.62	2.04	60
80	81.17	1.44	70
81	82.72	2.08	80
83	84.27	1.50	90
86	85.81	-0.22	100

Resulting from the obtained regression equation a diagram was built (Figure 1), describing the relation of actual measurements of the noise level [12, 13, 14] and the results received empirically. These values lie within the 5% discrepancy interval. Reproducibility of results has the accuracy of 0.98.

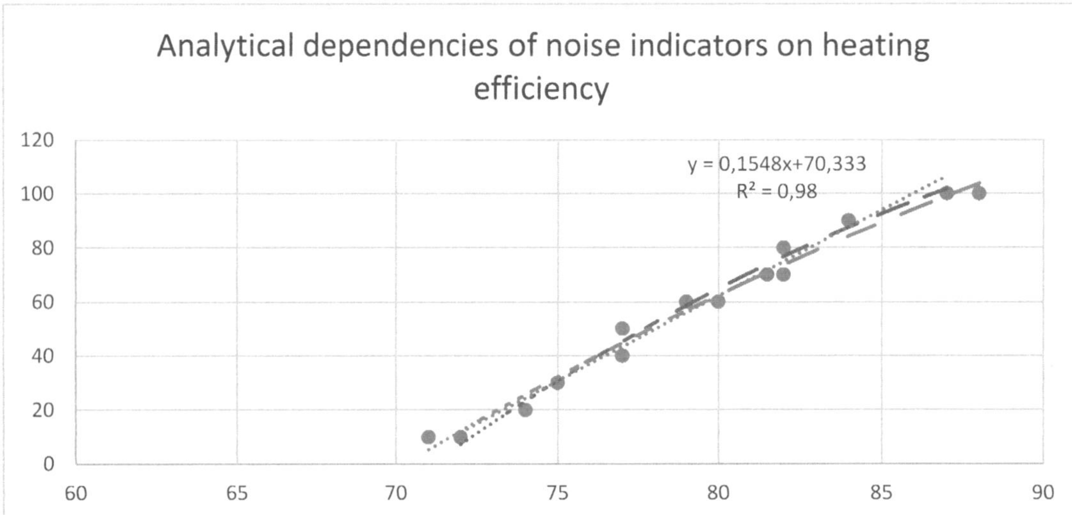

Figure 1. Analytical dependencies of noise indicators on heating efficiency

4. Conclusions

The findings of this study can be understood as:

- analysis of noise parameters of existing independent heat supply sources enables obtaining valuable analytical dependencies with accuracy of 0.98 and maximum deviation of no more than 5;
- the obtained regression equation is applicable for predicting noise level from rooftop boiler rooms during noise mapping of the territories under development;
- accuracy of obtained results is applicable to technically similar rooftop boiler rooms.

The given conclusions can be applicable for further modeling of noise transmission into the environment [15,16,17].

References

[1] Code of Practice 2012 Building climatology № 131-13330-2012 Revised edition of SNIP 23-01-99*
[2] Code of Practice 2016 Boiler installations №89-13330-2016 Revised edition of SNIP II-35-76
[3] Bottega W J 2015 *Engineering Vibrations* (Taylor & Francis Group) chapter 2
[4] State standard 2013 Gas burners, gas/oil burner and pulverized solid fuel/gas burners of steam and hot-water boilers №55851-2013
[5] State standard 1998 Industrial gas burners №21204-97
[6] State standard 2015 Noise Methods of noise measurement in residential areas and in the rooms of residential public and community buildings №23337-2014
[7] DIN EN 2017 Blower burner for gaseous fuels №676
[8] Zimmermann R K, Skjelmose O, Jensen K G, Jensen K K and Birgisdottir H 2018 *Categorizing Building Certification Systems According to the Definition of Sustainable Building* (WMCAUS)
[9] VDI 2012 Sound insulation in buildings Apartments Assessment and proposals for increased sound insulation №4100

[10] EN ISO 2013 Acoustics Rating of sound insulation in buildings and of buildings elements Part 1 Airborne sound insulation Part 2 Impact sound insulation №717
[11] EN ISO 2016 Acoustics – Measurement of sound insulation in buildings and of building elements – Part 3 Field measurements of impact sound insulation of building elements№16283
[12] Preis A, Honorata H, Malina S and Jedrzej K 2016 Noise Control *Noise Control Engineering Journal* **1** (Reston INCE-USA) p 64
[13] Fu Y, Yaob J, Wanc Z and Zhaod G 2018 Vibration analysis of laminated composite rectangular plates with general boundary conditions *Proceedings of the scientific conference INTER-NOISE Impact of Noise Control Engineering* (Chicago)
[14] Dresig H and Holzweißig F 2010 *Dynamics of Machinery Theory and Application* (Springer-Verlag Berlin) pp 177-222
[15] Ziaran S 2013 *Technical diagnostics. Scientific monograph* (Bratislava: STU)
[16] Schönfelder T 2017 *Dissertation Influence of Thermal Operational Burdens on the Safety of Type III Compressed Gas Containers* (Bergische Universität Wuppertal)
[17] Hambric S A, Sung S H and Nefske D J 2016 *Engineering Vibroacoustic Analysis- Method and Applications* (John Wiley & Sons New York)
[18] Minina N N 2012 *The problem of decreasing acoustic who-action on the residential building at design, building and operation of transport facilities* [Doctoral Dissertation] (BSTU «VOENMEH» named D.F. Ustinov St. Petersburg) pp 60-102

IOP Conf. Series: Materials Science and Engineering **481** (2019) 012023 doi:10.1088/1757-899X/481/1/012023

Infranetics: The New MAICS-convergent Technology Science

S A Timashev[1,2,3]

[1]Sci.& Engng Center "Reliability and Safety of Large Systems and Machines", Ural Branch, Russian Academy of Sciences, Yekaterinburg 620016, Russia
[2]Ural Federal University, Yekaterinburg 620016, Russia
[3]Old Dominion University, Norfolk, VA, United States

E-mail: timashevs@gmail.com

Abstract. The paper describes the basics of a new convergent science- Infranetics- based on the MAICS-convergent technology (Digital Stochastic **M**echanics, **A**rtificial Intelligence, **I**nformation Theory, **C**ognitive and **S**ocial Sciences). The name Infranetics comes from I*nfra*structures + Cyber*netics*. Infranetics is being constructed for solving the central problem of safe innovative development of a region/territory/municipality by creating a methodology of harmonized regulation of regional risk. This methodology is based on optimal management of systems of interdependent critical infrastructures (ICI) that play crucial role in the sustainable development of a modern society of risk. To this end following four aggregate optimization criteria are proposed: 1) regional average life expectance at birth (RALE); 2) regional (municipal) life quality index (RLQI); 3) resilience of regional systems of interdependent smart infrastructures (RICI) and 4) spatial entropy of creation and destruction of a regional system of ICI. Infranetics is a useful tool of managing and governance for regional decision makers (DMs), as they can track how their decisions influence the RALE, quality of life and the wellbeing of their citizens/constituents. The paper demonstrates how Infranetics is used to solve real life problems related to regional resilience and governance.

1. Introduction

Current basic research develops simultaneously in two directions: 1) according to the internal logic of fundamental science, when solution of one problem usually automatically creates another problem for the researcher to tackle. In this case the form and time (if it ever comes) of applying obtained results is unknown; 2) following the demands of the society and the logic of technology development, when obtaining any fundamental or applied result is put into a specifically designed research plan (e.g., deciphering of the human genome, exploration of the surface of Mars, discovery of the Higgs boson). This kind of research is financed considering the assigned time table for receiving the needed results. Though both directions are equally important, the academic science suffers ever increasing pressure of the modern society to satisfy its requirements not only in the long-term, but also in the mid-range and even short-term perspective.

The infrastructures, facilities and machines to be designed or already in operation must have such crucial characteristics as reliability, *resilience*, safety, strategic preparedness and moral durability. It is not yet quite clear how to correctly formulate and solve variable problems of mechanics which account for all the enumerated above aspects of a future critical infrastructure, but efforts in this direction should be undertaken without delay.

IOP Conf. Series: Materials Science and Engineering **481** (2019) 012023 doi:10.1088/1757-899X/481/1/012023

In this aspect the approach based on a pre-fixed objective, that dictates the volume and composition of the needed basic research seems to be promising. To obtain in time an adequate and fast solution of the reliability and safety problems of complex systems it is necessary to be able to synthesize the already existing and proven theories. More and more often the problem definition must simultaneously consider the technological and social environment in which the presumed scientific result is planned for application, all other considerations being subjugated to this principle. Here we are talking about the methodology of solving *interdisciplinary* problems. Currently, there are two equal approaches to this problem. According to the first approach a scientific or scientific-technological problem of fundamental science (for example, Moon expedition or green development of the Arctic) is divided into components, each of them to be studied by a separate group of scientists. These groups are either independent from each other, or weakly coordinated. This kind of activity, though leads to new knowledge about basic patterns of the phenomenon under study, not necessarily leads to solution of the problem in consideration. This could be due to the lack of an umbrella-unified approach to the problem and, hence, inconsistencies between directly related results both in essence and in form. Here we purposefully do not consider usefulness of independent explorations exercised in smaller-scale or intrinsically narrow researches.

The second approach is logistically more complicated both in the scientific and organizational aspects and has not yet been widely used. Its essence is in that a unified concept is designed of how the considered problem can be solved. In the frame of this concept, the problem is divided in a way that a chain of interrelated tasks is arranged (Fig. 1). Solution (output) of the first task is the input data for the second task, and so on, until solution of the last task delivers the necessary new knowledge/product. This approach is likely to be the only one that allows obtaining new *basic* knowledge that is of immediate practical or commercial value.

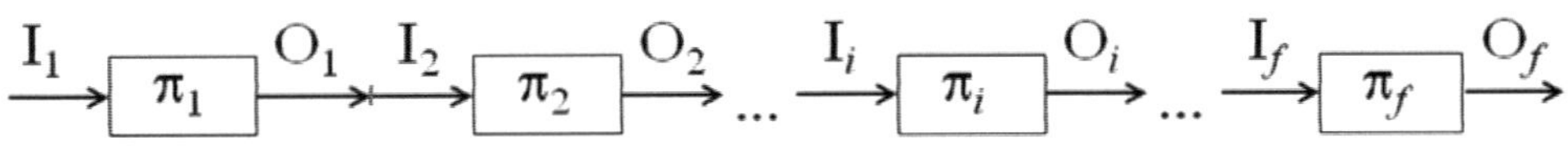

Figure 1. Flowchart of a novel solution of interdisciplinary problems.

I_i – *i*-th input; O_i – *i*-th output; π_i – *i*-th problem of the solution chain; O_f – final output that delivers the sought new knowledge/solution of a technical innovative problem. Each problem could be from a different science or branch of engineering

Recently, a new concept of solving interdisciplinary problems that shores up the second approach emerged. It started, unnoticed, much earlier within the community of engineering sciences, out of necessity, by default, when different types of specialists were brought together to form a team for solving practical scientific problems (say, quantification of the notion of resilience), or create a technological innovation (e.g., an smart phone or an electric car). But officially it was consciously brought to life in the natural sciences, that picked up this idea in 2002, when M. Rocko and V. Bambridge introduced the notion of convergent technologies (sciences) [1].Convergence of technologies is the integration of various disciplines in the process of conducting scientific and/or research activities. The historically first basic convergent technology was dubbed NBIC-technology, as it refers to the process of interpenetration of four groups of technologies: nano-, bio-, info- and cognitive. Russian science added to NBIC one more component – social and humanitarian technologies, and named it the NBICS technology [2]. These are rapidly developing sciences and technologies with high economic potential and practical applications. The defining principle for them is the creation of a new, Nature-like technology-based economy, crucial to social development and national security.

2. Contemporary convergent technologies/sciences

In modern natural sciences several convergent technologies are used. Currently, NBIC is only one of the most common basic concepts in convergent technologies. For example, American NSF supports three such concepts:

- NBIC (Nanotechnology, Biotechnology, Information technology, + Cognitive Science),
- GRAIN (Genetics, Robotics, Artificial Intelligence + Nanotechnology) and
- BANG (Bits, Atom, Neurons, Genes).

The European Union is funding the concept called *Convergent Technology for the European Knowledge Society*. It wouldn't be an exaggeration to say that main features of modern development of the natural sciences are: transition to nano-scale; paradigm shift from analysis to synthesis; convergence of nonorganic and organic chemistry; *interdisciplinary approach* vs. narrow specialization. Some representatives of natural sciences believe that the future of science belongs to the development of multidisciplinary and interdisciplinary NBIC technologies [2, 3].

Modern technology needed for creating competitive products and providing support for high quality of life for the modern "risk society" is implemented using *infrastructure systems* which consist of: 1) *smart hardware* (instruments, appliances, machines, computers, robots, structures, buildings and facilities); 2) *software* (Internet of Things (IoT), software systems that implement the logistics of production, services and livelihoods, as well as systems using institutional knowledge, social and cultural practices, for creating industrial and social regulations), and 3) A *cohort of professionals* who manage and maintain those systems. Together, this *triad* forms the space of global *infrastructure*, distributed across the whole planet. In the apt words of the famous Russian writer, M. Gorky, it is the *second, manmade, Nature*, and in contrast to the first Nature, is a continuously and fast-growing global phenomenon. This, second universe, needs its own set of convergent technologies that could create synergy among the global infrastructure triad.

This trinity of infrastructures as an object of study is essential, as it naturally connects the materiel production, organization of its efficient operation, and the social and the cultural human factors. This fact directly links the infrastructures operation indicators with the quality of life and the gross domestic product (GDP) of a municipality, region, country and humanity as a whole.

For a full picture it is necessary to give a very short revue of modern infrastructures. As a scientific object critical infrastructure may be defined as a multi-component distributed bio-geo-technical system "society - man – critical infrastructure – environment" (SMCIE), consisting of a set of interdependent and interacting assets and groups of people considered in a particular area within a certain period of time. Such distributed system of ICI must: 1) provide stable operation of any potentially dangerous object (further PDO) or a whole industry sector; 2) sustain normal social processes and regional growth within a municipality.

Usually CIs are materialized as networks, distributed over a certain area. In general, there are hardware, software and social infrastructures. As separate infrastructure is a system, it is possible to speak of *system of systems*. The most important infrastructures created to provide safety and life security, stable growth of economy and society are termed critical infrastructures (CI). The most essential of them ensuring energy and economic security of a country as well as defending its core interests are termed strategic infrastructures (SI). Modern systems of infrastructures are specified as *interdependent* (SICI). SICI are a *conductor and mediator* between natural environment and the society (population of a certain region) and at the same time, the main means to maintain life support, and the main source of accidents that damage the environment, and disasters that negatively influence the population residing in the area.

Examples of critical infrastructures: mineral extraction/production, transportation and distribution of gas, oil and oil products, generation and transmission of electricity; chemical productions; water supply and heating; railway network, motor roads, airports, inland and coastal shipping; dams and levees; telecommunications, internet and world wide web including (enterprise) internet of things; agriculture, food production and distribution; scientific-educational institutions and public health; security services (Ministry of Internal Affairs, Ministry of Emergency Situations); banking and finance; postal services; commercial services, etc.

Examples of strategic infrastructures: nuclear reactors, materials and waste storage; defense industrial base; space infrastructures; critical manufacturing; information technology systems; central state institutions; some national monuments. Modern infrastructures are the backbone of the economy, the means of sustainable GDP growth, extension of the average life expectancy of a healthy population and an indicator of the country's power.

The convergent technologies are designed to respond to all the challenges of modern society, especially in the developing and poor countries. It is necessary to find effective solutions to refurbish such mega-objects as the energy and transportation infrastructures, the *construction sector*, as well as the basic production assets, networks, logistics, the sector of municipal services, and the like. In addition, there is a need to develop new industries of preservation and reproduction of human health, creating a highly organized living environment. This group of challenges *requires its own special set of convergent technologies.* Among them there is a separate group of unique challenges – the *need for a complex infrastructure upgrade* of developed areas, especially cities, where by 2050 more than 70% of the world population will live.

3. Infranomics – the universal discipline

The applied science and engineering community was seeking, in its own way, how to tackle the planetary scale problem of infrastructures design and control, viewing them as system of systems (SoS). The rapid growth of new types of infrastructures and high awareness of their role in sustainable development of modern society of risk, led to the concept of a new universal synthetic scientific discipline, namely, *infranomics (*or political economy of infrastructures*). Infranomics as a* term and its interpretation (*infranomics*= *infra*structures+eco*nomics)* was given in 2007 by Prof. A. Gheorghi of ODU (Norfolk, VA USA), to include *all the knowledge that ever may be required* to solve all the problems of critical infrastructures [4], but as yet, without reference to their relationship with each other (as it was done, in ancient times, by geography (a term, coined by Strabo),which included formalistically, almost all the knowledge accumulated by ancient Greeks, Egyptians and Assyrians. Gradually, it spun off over the centuries, such related to it disciplines as geodesy, geology, geo-physics, geo-chemistry, geo-ecology, geo-hygiene, geo-cancerology, and even geopolitics (see Fig. 2). Pursuing this line, it is appropriate to say that infranomics as a term is alike geography, a new universal umbrella science. Since none of the existing disciplines can provide the necessary decisions, *infranomics is meant to serve as the* discipline of disciplines, gathering all the knowledge required to solve critical infrastructures' problems.

According to [4], currently infranomics is a loose set of theories, assumptions, models, methods, and associated scientific and technical tools which are required for studying concepts, design, development, application, operation, management, maintenance, supply, resilience and survivability of meta-systems. Infranomics is meant to support analysis and decision-making process with respect to the meta-system (i.e., the entire plethora of technical components, stakeholders (those interested in financial success of the enterprise), ways of thinking, legal restrictions, etc.

4. Infranetics – an off-shoot of infranomics

Using the above geography-analogy, a new convergence science/technology should be shaped using the NBIC(S)-type concept in relation to the problems of the 21-st century second Nature. In order to achieve this, several issues should be first considered.

Infrastructures are the backbone of the world economy, the means of sustainable GDP growth, extension of the average life expectancy of a healthy population and an indicator of a region's strength. Neither the regional authorities nor industry nor academic institutions can afford to ignore the observed onset of the new combination of possibilities and risks that accompanies the emergence of a new generation of infrastructures [4]. The shape of modern society in the nearest future will be defined by the characteristics and services provided by the next generation of infrastructures. It seems necessary to develop the concept of convergent technologies that would best fit the solution of infrastructure problems. The world science community as represented in the UN concluded that the solution of the central problem of safe innovative development of a region/territory/country can be found by creating a *methodology of harmonized regulation of regional risk that amounts to optimal governance*. This methodology should be based on optimal management of systems of interdependent critical infrastructures (ICI) that play crucial role in the sustainable development of a modern society of risk. Research in this area is conducted by intersecting and cross-pollinating many disciplines, including previously unrelated bodies of research from different branches of sciences and engineering, to holistically solve the problem of optimal functioning of the planet's second nature. To achieve this goal, in this paper following working hypothesis is formulated and used: T*erritorial risk management /governance can be boiled down to managing/controlling the system of systems (SoS) of interdependent critical infrastructures (ICI), including fully the human factor component.*

The basis for the convergence of engineering science and technology in the general and cross-cutting issues of managing systems of interdependent critical infrastructures, are to varying degrees based on 1) digital mechanics (of solids, hydro-and air-) including assessing the likelihood of the events that initiate the accident and are responsible for the consequences of infrastructure failures; 2) theory of control (cybernetics) of large stochastic man-machine-environment systems with uncertainties, and 3) human factor analysis on the personal and social level. The outlined above allows offering [5], as an option, the following abbreviation for the convergent technologies/sciences that aim to holistically and quantitatively solve the main current and future problems of optimal functioning of the planet's second nature: **MAICS--** Digital Optimal Design and Stochastic **M**echanics, **A**rtificial Intelligence, **I**nformation Theory, **C**ognitive Science, **S**ocial Sciences**.** MAICS-- technologies may be spun off of infranomics and named *infranetics* (**infra**structures + cyber**netics).**

Infranetics as a convergent science emerged not as a whim, but as a logical outcome of efforts to overcome the intrinsic difficulties that are specific to the problem of managing SoS of critical infrastructures. This new discipline deals with: problem formulation; decomposition and downsizing of the multi-dimensional problem; indicating and developing methods for describing a new highly specific mega-object; choice of solution methods with accounting for their mutual compatibility; discovery and filling of information gaps. It also provides for the first time initial means for studying the *quantitative relation between the tangible and the intangible*: the quality of performance of SoS of CI with the quality of life and wellbeing of the society (see Fig. 3).

Figure 2. Analogy between ancient geography and current infranomics.

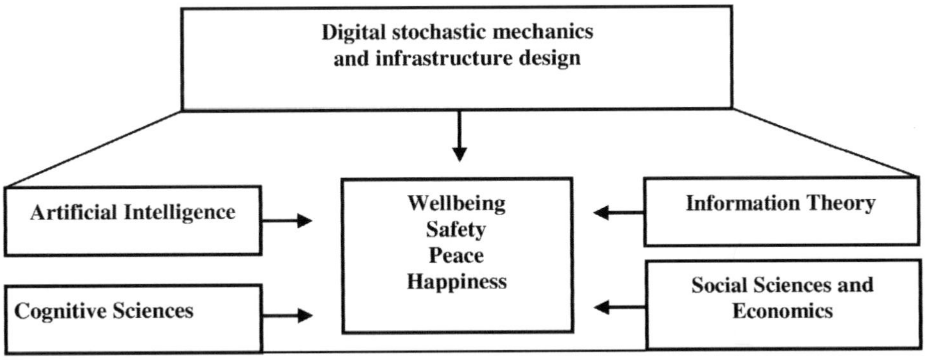

Figure 3. Framework of infranetics as a MAICS-convergent science.

5. Main Traits of Infranetics

The main features of infranetics as the research tool of modern infrastructure research and development could be formulated as follows. Infranetics uses goal-oriented approach according which a unified concept is formed that describes compatible ways and means for solving the problem in consideration. It manifests an *interdisciplinary approach* rather than narrow specialization, and accounts for *human factor* on each stage of the CI's life cycle. Infranetics transits from the analysis paradigm to the *synthesis paradigm* (especially, when using large amounts of raw data). It creates and uses *modern digital tools* for design (stochastic structural mechanics and mechanics of fracture, structural reliability and resilience), diagnostics, monitoring, maintenance/management and governance. It also transits to *nano-sizes* (in diagnostics, monitoring, protection against cyber-attacks).

6. Priority unsolved issues of Infranetics

From the stand-point of MAICS-technology following hot-spot issues are of first priority:

- Simulation of engineering and natural loads and impact on the elements, components and infrastructures as random functions/processes of time and coordinates, including global changes of climate;
- Methods of assessing the reliability and safety of multi-element systems subjected to combinations of random loads, physical fields and chemical processes;
- Diagnostics and monitoring methods of structural, functional and balance reliability and remaining lifetime of systems, structures and facilities;
- Damage assessment methods for infrastructures considering all the related uncertainties;
- Principles and methods of predictive maintenance of facilities, structures and systems;
- Methods and devices for systems condition and operation control and management;
- Theory of assessing complex risk of operating infrastructures;
- Optimization of selecting the devices and measures reducing the PoF and operating risk, under condition of restricted means;
- Practical methods of assessing *resilience and strategic preparedness* of critical infrastructures;
- Governance of SoS of infrastructures that connects their resilience with regional healthy life expectancy (HLE);
- Optimal governance theory of interdependent critical infrastructures.

7. Governance through infranetics

Safety protection is an important, but not the only purpose of the society. The society can allocate a certain share of resources for safety protection that must be continuously specified, since the society has other needs, such as clean air and water, healthy food, housing, health care, provision of pensions, education and other social services, which also increase healthy life time and its quality.

Management of a territory / region and its human conglomerates in the form of metropolis, cities, municipalities, companies, businesses, etc., is carried out by elected or appointed decision makers (DMs), each of which is expected to (quasi-) independently make decisions within her (his) competences and on their specific level. This group of DMs, together, collectively, but not in a tightly organized way, *de facto* controls (governs) such entities.

Governance as a term refers to all of the processes of state and social governing, whether undertaken by a regional government (city administration, directors of enterprises), market or network, whether over a family, tribe, formal or informal organization or territory and whether through the laws, norms, power or language [6,7]. Governance refers to the processes of interaction and decision-making among the authorized persons involved in a collective problem that lead to the creation, reinforcement, or reproduction of social norms and institutions.

Actually, current governance is about (quasi)collective management of territorial or regional risk in conditions of uncertainty of the modern world. In this general theoretical definition, we focus on the aspects related to management of SICI embedded into certain regional industrial or social networks, using confirmation or reproduction of national technical and social standards.

The practical conclusion from this concept is the need to orient theoreticians and applied scientists – experts in the field of risk – to meet the challenges of the tasks generated by the regional DMs in the daily management of the operation of the ICIs and ISCIs.

For this category of leaders it is necessary to create practical harmonizing decision-making support tools related to the effective functioning of regional enterprises and institutions, in the context of providing social needs and life options to individuals and regional society as a whole, according to the laws, ethical and aesthetic norms, historically formed in different regions of the country. This broader approach to governance allows selecting the best criteria for the design and long term operation management of ISCI that provides for sustainable development.

Key element in considering complex territorial risk is assessment of operation of interdependent critical infrastructures under the impact of natural, technogenic, mixed and all types of deliberate accidents and catastrophes. Safety and resilience of the existing and newly established ISICIs has a direct impact on the rate of innovations and adoption of new technologies, future industrial capacity and growth of the gross domestic and regional domestic product (GDP/RDP). The current stage of development of regional infrastructures and their capacity to create principally innovative infrastructures (i.e., smart power grids with built-in diagnostics, monitoring and protection systems, "green" buildings with solar power supply, motor roads generating electricity from moving cars, etc.) determines the future investment attractiveness of the municipality/territory.

8. Regional governance criteria

The conditions of *regional governance,* when regional-scale DMs make decisions based on their corporative or territorial criteria, requires having *harmonizing* benchmarks for all the regional entities (actors/players).

This paper offers from interdisciplinary positions an essentially novel approach to proactive predictive management of operational regional risk of SoS of critical infrastructures using the maximum social benefit criterion. It involves designing a methodology of assessing and managing regional risk, based on following four novel generalized indicators: regional healthy life expectancy (HLE); regional (municipal) life quality index (RLQI); SICI resilience, and origination and degradation entropy of regional SICI.

Average life expectancy in good health. HLE is proposed as the first integral parameter of risk value because it is an imperative indicator of municipality/regional sustainable growth. Research demonstrates [8] that actual regional HLE, having a biological limit and certain fractal properties, is a solution of a system of differential equations and can be presented as a logistic function of time. HLE depends on the current value of territorial HLE and the society's development level, as well as on optimality of the annual distribution of regional/municipal domestic product (RDP) for accumulation, consumption and safety control systems (protecting SICI, its staff and population against industrial accidents/catastrophes).The latter value indicates which means (a part of budget, RDP) may be used for reducing lethal accidents of ICI components and, consequently, the rate of decreasing or increasing death/ mutilation in a given area, and, consequently, its quantitative effects on the region's HLE.

Regional Life Quality Index (RLQI) is an integral social index that fuses the structural-functional reliability and safety of the elements, facilities and structures of ICI with economic indicators of their operation (regional domestic product) and social aspects of an area's sustainable growth (HLE and employment of population).Using RLQI allows the regional decision-makers (DMs) to pursue a balanced policy of managing the state, municipal and private property located in the given area (i.e. governance). RLQI is a non-additive parameter which provides seamless linking of indicators of mechanical, technological reliability and safety of the elements, facilities and structures of ICI with economic indicators of their operation and social aspects of sustainable growth of the area, including the generalized task of territorial safety and risk management.

Regional resilience is a very complex concept, which can be described as an *emergent* property of the territorial SoS, including the regional community. In a nutshell, regional resilience manifests itself in the quality of the reaction on a disaster or catastrophe: how fast and at what cost it rebounds to the pre-catastrophe state. The procedure of assessing this trait using infranetics tools is shown below.

Regional entropy. Modern physics supplies a useful tool – entropy principle - to generalize all the engineering and scientific parameters of an infrastructure problem. The principal idea of using entropic principle in interdisciplinary assessment of ICI efficiency is that the problem of the curse of dimensionality is removed due to reducing a multi-dimensionality task into (quasi) one-dimensional problem, since the dimensions of all the physics-engineering tasks under study are reduced to a single non-dimensional one – entropy. Simultaneously its correlation with probability of an accident or catastrophe (risk) is established.

9. Show-cases of implementation of the infranetics principles and techniques

The volume of the paper doesn't allow for any deployed examples of infranetics usage. So we restrict ourselves to two generalized cases. The first case shows how to parse an interdisciplinary problem into a solution-aimed chain of sub-problems which, in their totality, give the needed answer. In the second case the problem of regional governance is formulated, showing which disciplines, and who from the community of the municipality/region should be involved to solve the problem.

9.1. Case #1.Risk analysis of a main oil pipeline.

In order to conduct this kind of analysis it is necessary to verbally describe the whole chain of possible events that lead to the oil spill-catastrophe (Fig. 4). This will open the possibility to define/select the disciplines that could quantitatively describe each of the events (Fig. 5).

Then the problem is parsed into a chain of separate sub-problems, but in a way that the solution (output) of the first link in the chain serves as the input of the second link (problem) in the row, and so forth, until the output of the last problem emerges as the new-sought knowledge and gives the general solution of the problem (Figs.1, 4). Means and ways of parsing the considered convergent problem into a chain of consecutive problems, each of which is interconnected with the *previous* and the *consecutive* problem in the considered case are dictated by the risk scenario. Formalization of the choice of the mathematical tools to be used for solving each of the problems of the chain in order to achieve compatibility (in form and in dimension) of the pairs of the 'i-th output with the i+1-th input" in this case is also rather simple, as it flows from the physical property of each considered event. The details of this case may be found in [8].

When solving such problems researcher often discovers some yet missing theoretical and/or calculus components that need to be immediately solved in order to obtain the needed result in due time. Such info-gaps usually occur, as a rule, at the cross-sections of the disciplines involved in the research.

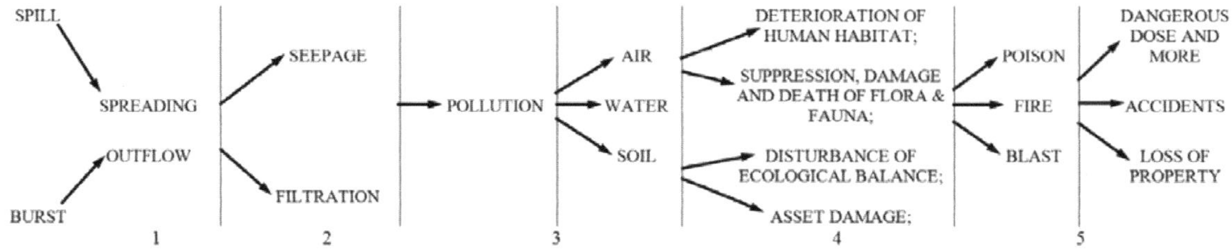

Figure 4. Main possible scenarios of a main oil pipeline spill.

STRENGTH OF MATERIALS, STOCHASTIC FRACTURE MECHANICS OF LARGE SYSTEMS, THEORY OF PLASTICITY AND STABILITY, STATISTICAL DYNAMICS, RELIABILITY THEORY OF LARGE SYSTEMS	HYDRODYNAMICS THEORY OF FILTRATION	AERODYNAMICS HYDRODYNAMICS THEORY OF FILTRATION	APPLIED BIOLOGY, ECOLOGY, THEORY OF ECOSYSTEMS, INDUSTRIAL MEDICINE	GAS DYNAMICS, COMBUSTION THEORY APPLIED AND BURNED MEDICINE, EXPERIMENTAL METHODS, COMPUTER MODELLING OF PROCESSES AND PHENOMENA
1	2	3	4	5

Figure 5. Main methods for describing stages of main oil pipeline critical failure (spill).

9.2. Case #2. Designing the regional risk management /governance strategy using infranetics concepts

The goal of the considered case is to build a strategy of optimal governance for a regional set of ICIs, using resilience as its main management tool.

This case is much more complicated than the previous one, because it deals with a set of interdependent critical structures, whereas case #1 considered just one critical infrastructure. The key element in considering complex territorial risk is in the assessment of resilience of interdependent critical infrastructures under the impact of natural, industrial/technological, mixed and all types of deliberate accidents and catastrophes, and its influence on the wellbeing of the community that works and lives inside the ICI cluster.

In Fig.6 the flow chart of constructing such strategy is presented. It can be seen that creation of such strategy involves a heterogeneous set of initial data and knowledge from multiple fundamental and applied, and engineering sciences. When solving this problem for several regions of the Russian Federation, author came across several info-gaps in macro-economics [9], demography, reliability theory of large systems [10], to name a few, which needed immediate attention, as they were road-blocks preventing reaching the assigned goal in due time. It seems that in this particular case infranetics concept threw new insight into the research problem; advanced research in several fundamental and applied disciplines; greatly enhanced and expedited the research.

Accumulating results of this kind of interdisciplinary research permits understanding what methods from what disciplines belong to infranetics. The results, obtained by the above scheme, make it possible to effectively manage the SICI in different types of municipalities and territories. Some details of this case may be found in [11-15].

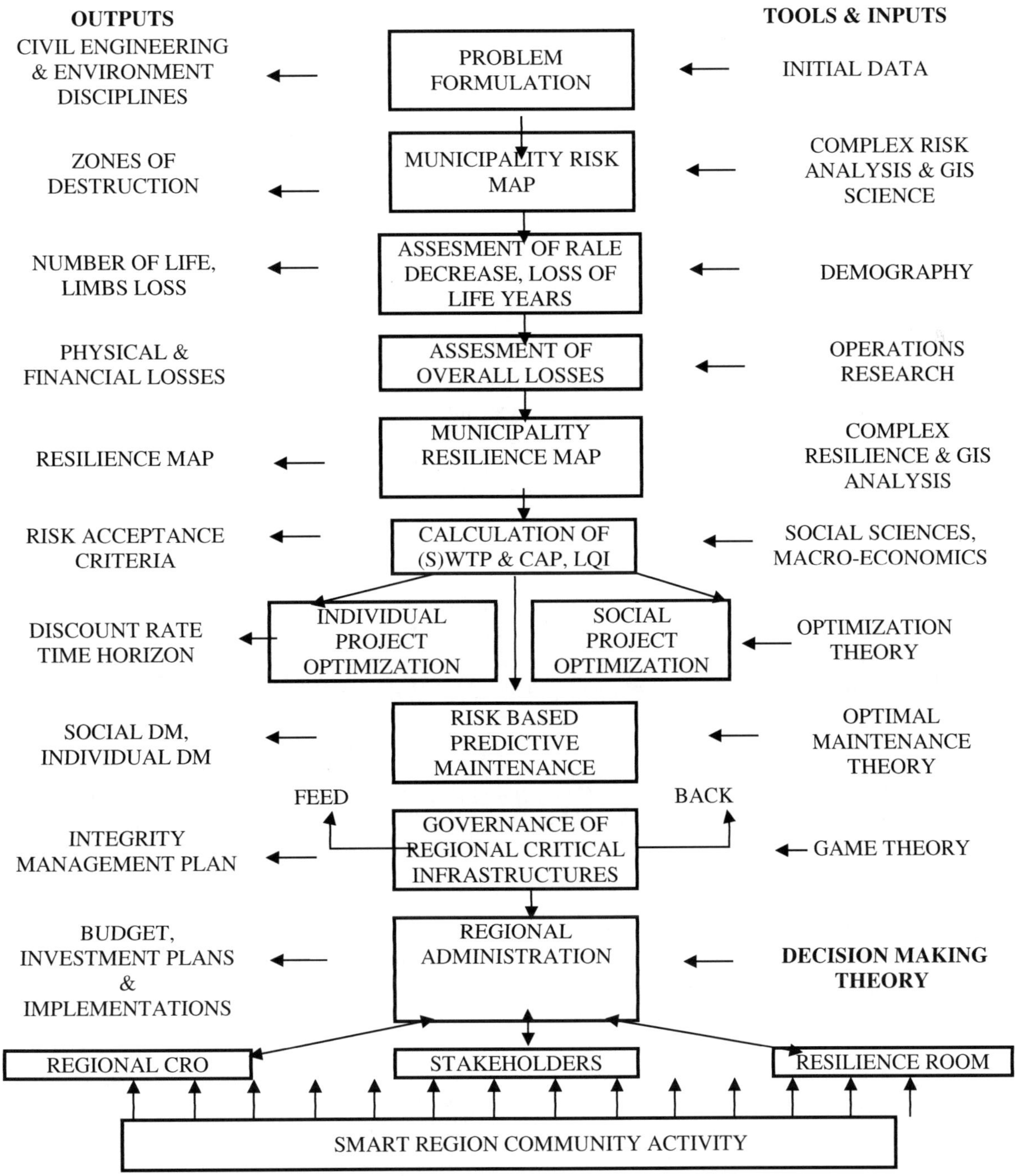

Figure 6. Regional governance strategy via regional resilience using infranetics principles and tools.

10. Conclusion

Infranetics manifests itself as a cross-pollinating complex-sciences discipline. It permits accurate-enough prognosis of the behavior of complex socio-technological and economic systems being designed or already performing, in usual and/or catastrophic situations. Infranetics uses goal-oriented approach according which a unified concept is formed that describes ways and means for solving the problem in consideration. If properly developed, infranetics could become the crucial discipline for governing municipalities and regions in the 21st century.

Infranetics creates practical harmonizing decision-making support tools related to governing of regional enterprises in the context of social needs and life options of individuals and regional society as a whole, according to the laws and historically formed ethical and aesthetic norms that exist in their region.

Infranetics makes possible effective regional governance via resilience of SICI in different moments of crisis, and to support decision-making in the daily operation of the CI, ICI and ISCI. The territorial and municipal level DMs receive a powerful tool for: monitoring and conducting predictive maintenance of CI; evaluating and minimizing territorial losses by *specific protection means*; supporting their intuitive and off-the-cuff decisions, especially in cases where the effects of these decisions will become apparent only in the distant future. DMs will also be able to monitor how their decisions affect sustainability of the quality of life, well-being and happiness of their constituents or employees in the long term.

11. References

[1] 2016 The Concept of the Strategy of Developing Convergent Technologies (Moscow: Kurchatov Institute)

[2] Pride V and Medvedev D A 2016 The NBIC-convergence phenomenon *Reality and Expectations. Philosophical Sciences* **1** pp 97-117.

[3] Timashev S A 2016 Infrastructures. Part 1. Reliability and Longevity (Yekaterinburg: Ural Branch, Russ.Ac. of Sci.)

[4] Gheorghe A V and Masera M and Katina P F 2013 *Infranomics: Sustainability, Engineering Design and Governance* (Switzerland: Springer) 438 p

[5] Timashev S A 2016 Infranetics-A New Convergent Science of the 21-st Century *Nauka Urala* **13-14**

[6] Bevir M 2013 *Governance: A very short introduction* (UK: Oxford University Press)

[7] Hufty M 2011 Investigating Policy Processes: The Governance Analytical Framework (GAF) *Research for Sustainable Development: Foundations, Experiences, and Perspectives* ed Wiesmann U and Hurni H et az (Bern: Geographica Bernensia) pp 403-424.

[8] Timashev S A and Bushinskaya A V and Malyukova M G and Poluyan L V 2016 Reliability and Integrity of Pipeline Systems (Yekaterinburg: Ural Branch, Russ.Ac. of Sci.)

[9] Timashev S A and Tyrsin A N and Kozlova O A and Makarova M N and Voronina L N 2014 Methodical Approaches to Assessing the Influence of Budget Expenditures on Average Life Expectancy *Scientific Review* **7-1**.

[10] Burukhina O S and Bushinskaya A V and Maltceva I N and Timashev S A 2018 Mechanical system reliability analysis using reliability matrix method *IOP Conference Series: Materials Science and Engineering* **365**

[11] Timashev S A 2011 Resilience and Strategic Preparedness of Critical Infrastructures *Proc. of the First Int. Conf. on Vulnerability and Risk Analysis and Management and the Fifth International Symposium on Uncertainty Modeling and Analysis* pp 764-771

[12] Timashev S A 2013 Unified quantitative criteria for managing regional risk *Proc. of the 11th International Conference on Structural Safety & Reliability* (NY: Columbia University New York)

[13] Guryev E S and Poluyan L V and Timashev S A 2014 Construction of dynamic risk maps for large metropolitan areas. *J. of Risk Analysis and Crisis Response* **4-2** (Paris: Atlantis Press)

pp 72–76
[14] Timashev S A 2014 Average life expectancy as a criterion for regional risk management *J. of Risk Analysis and Crisis Response* **4-1** (Paris: Atlantis Press) pp 10–19
[15] Timashev S A 2016 Infrastructure Resilience: Definition, Calculation, Application *Proc. of the WEEF, IGIP/ICL Conference*

IOP Conf. Series: Materials Science and Engineering **481** (2019) 012024 doi:10.1088/1757-899X/481/1/012024

Principles of Renovation the Territory of Residential Buildings Dating from 1960s to 1970s: Coping with Modern Housing Crisis

O Ryzhova[1], A Khrichenkov[2]

[1]Institute of Civil Engineering and Architecture, Ural Federal University, Mira str. 19, Yekaterinburg 620002, Russia
[2]Institute of Civil Engineering and Architecture, Ural Federal University, Mira str. 19, Yekaterinburg 620002, Russia

E-mail: av.khrichenkov@urfu.ru

Abstract. Large-panel residential construction was conducted in many countries worldwide; its peak was in 1960-1980s. By 2020, the majority of residential buildings built in the period from 1950 to 1970 will be more than 50 years old and their lifetime will be exhausted. In Russia the total living space of the buildings built during the considered period makes about 10% of all housing stock in the country. Housing that was built in the considered period, has become outdated morally and physically, it generates a housing crisis. In the majority of the European countries, residential territory renovation has already begun. In most cases, projects did not provide full demolition of residential buildings. In Russia, the only project that began to be realized is the renovation started in 2017 in Moscow. During the project full demolition of the existing building and new construction is provided. In other cities of the country realization of similar strategy is impossible and inexpedient for several reasons (lack of federal financing, the disinterest of investors). The purpose of the study is research of positive experience in coping with a housing crisis in Germany, France, Hungary and in other European countries. Methods: systematization, typology, comparative analysis, statistical data analysis. A result of a research: housing estate renovation principles are formulated; their classification and methods of realization are offered. These principles can be applied in the building renovation projects without full demolition and landscape modification and also in the complex sustainable development of the built-up territories.

1. Introduction

The beginning of panel housing construction in the USSR is considered to be 1945, when in Berezovsky (10 km from Yekaterinburg) a one-story panel house was built. Concrete panels for this house were produced in a factory that still functions in the town [1].

However the era of mass construction began when engineers of many countries, including Germany, the Netherlands and the USSR started looking into this mode of construction in the 1920s. In November 1955 the Resolution of the Central Committee of the CPSU, № 1871 "About elimination of excesses in design and construction" was signed. This document was the trigger for conversion from Stalinist neoclassicism to ascetic, standard and cheap architecture. New living spaces in huge amounts were required for the country: Khrushchev's plans ensured a norm of 9 square meters per person [2]. Barracks, dugouts and communal flats were inhabited. The shortage of housing and unsatisfactory conditions of accommodation occurred because of the consequences of World War II, not only in Russia, but also in countries of Eastern Europe. Therefore architects and town-planners were directed to develop standard projects for buildings and micro-districts. In Russia the civil engineer Lagutenko [3] was the author of one of the first standard house series, K7. During the sixties the K7 series was the main unit of the micro-districts.

The peak of panel housing construction was from 1950 to 1970, depending on the country. For example, in Russia the panel houses called "Khrushchyovka" were built up to 1985. Depending on the series type (short-lived or long-lived) and the construction materials, panel houses were calculated to have a lifespan of 25 to 150 years [4,5]. The majority of panel blocks have a lifespan of 50 years, which will expire by 2020. The moral and physical wearing out of buildings was extremely high, which spoke to an urgent need for building renovation. Coping with the post-war housing crisis from 1950 to 1970 has generated the crisis of old housing stock today.

Research objective is studying the renovation experience in the countries in which mass housing construction was carried out. Approaches to renovation (full demolition and new construction, partial demolition, building consolidation and combinations thereof) are considered. Detailed study of renovation projects has allowed us to reveal and systematize renovation principles. Key principles responding to renovation without full demolition, which is recommended for residential districts where aggressive transformation (pointed building or building demolition) has not been conducted, are marked out.

2. Base materials and research methods

Maps, plans and satellite pictures of the cities where renovation was conducted are the base materials for our research. Research by Russian and foreign experts in the field are considered. Such methods as systematization, a tipologization and comparative analysis are applied. Statistical data on building in the USSR (in the territory of modern Russia), and in particular in the territory of Yekaterinburg (earlier Sverdlovsk), gleaned from Russian statistical yearbooks [6]. Information about the arrangement and number of the remaining residential buildings has been procured from open data of the 2GIS company [7].

3. The situation in Russia

From 1950 to 1970 city housing stock in the territory of modern Russia has grown by 597.7 million square meters. At the moment the percentage of buildings in this period in general across Russia is 10% [6].

Mass building by residential and public building standard series was conducted in all large cities of the country and also in the new cities that were created near large industrial enterprises. The percentage of five- and nine-story buildings (these make up most of the standard projects) in the million-plus cities of Russia for 2017 are presented in table 1 [7].

Table 1. Shares of five- and nine-story buildings in Russian cities

City	Population	Five-floor buildings, %	Nine-floor buildings, %
Moscow	12 380 664	30.3	16.5
St. Petersburg	5 281 579	33.3	11.8
Novosibirsk	1 602 915	30.3	14.2
Yekaterinburg	1 455 514	30.1	13.6
Nizhny Novgorod	1 261 666	26.4	18.9
Kazan	1 231 878	33.8	17.9
Chelyabinsk	1 198 858	33.2	10.9
Omsk	1 178 391	41.9	14.7
Samara	1 169 719	25.3	11.6

Rostov-on-Don	1 125 299	17.6	13.1
Ufa	1 115 560	22.8	20.6
Krasnoyarsk	1 082 933	36.9	11.7
Perm	1 048 005	54.1	18.4
Voronezh	1 039 801	25.4	15.3
Volgograd	1 015 586	31.0	17.5

In the research conducted in Yekaterinburg, more exact borders of the historical period were set from 1958 to 1975. In these years, standards of urban planning design radically changed. The dynamics of housing growth in Yekaterinburg are presented in table 2 [8]. Data on population are provided by the Federal Statistics Service [6].

Table 2. Dynamics of housing growth in Yekaterinburg

Year	Number of buildings	Total living space, м2	Population
1958	216	416 201,71	-
1959	314	621 304,30	778 602
1960	316	610 349,65	-
1961	339	663 381,85	-
1962	288	685 213,01	853 000
1963	234	652 045,87	-
1964	202	576 159,39	-
1965	185	584 552,05	-
1966	179	600 817,20	-
1967	168	615 549,30	961 000
1968	177	653 222,93	-
1969	164	602 778,12	-
1970	177	672 098,18	1 025 045
1971	148	627 801,80	-
1972	140	676 269,38	-
1973	136	724 127,72	1 099 000
1974	128	655 727,36	-
1975	117	694 532,07	1 163 000
In total	3 628	11 332 132	

The share of the buildings built from 1958 to 1975 in Yekaterinburg comprises 27,5% of the general housing stock of the city. The majority of buildings have physical wear of about 50%. Demolition and resettlement of inhabitants are impossible because the number of hazardous and shabby dwellings is very high and municipal or federal financing for realization of this strategy of renovation is not great enough. For this reason the question of the need for a renovation strategy, which is attractive to private investors and in which the interests of inhabitants and municipalities will also be considered, is particularly acute.

The pilot project of renovation began in Moscow in 2017. Full demolition of buildings, in most cases, and demolition of all micro-districts and construction of new buildings with a higher density and number of floors are considered. Residential districts will be transformed to quarters with the maximum density of the building, according to the project. During the first stage (2017-2019) 5,171 houses are included in the program, and 10,000 inhabitants will be moved from their apartments [9].

IOP Conf. Series: Materials Science and Engineering **481** (2019) 012024 doi:10.1088/1757-899X/481/1/012024

The renovation project has caused major controversy and discontent, both among experts and inhabitants. Many have negative thoughts about demolition and consolidation. Inhabitants are dissatisfied that they will be moved to more remote areas and to apartments on higher floors. Seminars and conferences on comprehending the opinion of the population at the implementation of the renovation program are being held [10].

4. Hungary

Destructions after World War II, collectivization and resettlement of inhabitants from villages to the cities were reasons for an acute housing shortage in the countries of Eastern Europe. In the early sixties Hungary investigated the sphere of large-panel construction in the Soviet Union [11]. In the late sixties Hungarian engineers created a local system of large-panel construction. Panel building in Hungary, as well as in the USSR by micro-districts, was carried out. [12].

The first residential districts were built in Dunauyvarosh, Pech and Debrecen from 1961to1963 [13]. The first plant making panels for mass construction was constructed in Obuda in 1965 [13]. Now, in Budapest, more than a third of the population lives in "Panelház" - the standard name of the Hungarian panel buildings. Only 8% of residential districts are located in the downtown area; the others are in suburbs [12]. A similar situation occurred in many Russian cities.

In the 2000s, Hungary's government started a nationwide program of renovation of buildings built between 1960 and 1970[14]. However, in the Timea Csaba research [12] it is noted that municipalities did not allocate funds for creation of a comfortable environment and public spaces in residential districts of panel construction. Attention was paid first to renovation of buildings. The author of a research paper suggests that we focus on advantages of micro-districts and increase their number. Such advantages include extensive free spaces between buildings, tall trees, existing playgrounds and a social service network.

The following principles of panel building territory renovation in Hungary are distinguished:

- creation of public spaces;
- revitalization and natural landscape using;
- improvement of buildings technical condition and their esthetic qualities;
- solution of the shortage of parking spaces problem;
- functional saturation of building and improvement of social service system.

In many projects of building renovation in Budapest there is a tendency toward intensive gardening on roofs, as well as creation of private terraces and public spaces on roofs of buildings.

5. France

Building renovation projects have been implemented in France since 1977 [15]. In 2003 the national program of upgrading cities (Projet de rénovation urbaine, PRU) which required demolition of about 250,000 buildings and renovation of 400,000 residential buildings, began [16]. Generally this program included the social building that appeared after World War II. In post-war years building in France began to be conducted via municipal funds [17].

The decision of the choice of renovation strategy for each district or quarter was accepted depending on the condition of buildings, structural materials, population density and other parameters. For example, brick buildings are almost always reconstructed, and not demolished. In the projects providing resettlement of inhabitants during construction or resettlement during demolition of residential buildings, equivalent housing in the same area was offered to inhabitants. In 2004 the national agency of reconstruction of the cities was created for upgrading 500 quarters of France's cities. By 2013 during renovation 250,000 square meters of living space were updated or constructed. Implementation of the national program of updating cities is calculated to continue until 2030. Renovation is implemented both at the expense of the state and with the involvement of private investors.

The basic principles of renovation of building of 1950-1970 considered in PRU:

- decrease in number of floors (buildings more than 10 floors were demolished, on their place five-floor buildings were built);
- improvement of quality and the number of gardening in inhabited territories;
- change of strategy of retail placement (creation of "live" streets);
- improvement of transport availability of the remote areas, local transport network reorganization;
- differentiation of housing (in areas of social building more high-class housing appears).
- increase in energy efficiency of building.

6. Germany

The first experiments with panel building in Germany began in 1926 when in the city of Lichtenberg the pilot area of panel buildings was built. After World War II historically developed quarter building in many Germany cities suffered. Both in East and West Germany residential districts called "New development area" began to be built. When after 1990 a mass outflow of the population from dysfunctional areas and shabby housing began, the federal authorities started the renovation program [18]. Unlike the USA and France, Germany is famous for renovation projects without demolition.

One of the biggest residential districts on the area, Märkisches Viertel, has appeared in the West Berlin. It was built between 1963 and 1974. It was designed for 50,000 inhabitants, but the planned infrastructure could not satisfy their needs. Therefore to avoid a rise in crime and an outflow of inhabitants, buildings and territories were modernized [19].

In recent years the SFA bureau [20] has realized several projects in the city of Leinefelde. A distinctive feature is the radical decrease in density of housing stock: parts of buildings and even whole floors have been demolished, which has created a more comfortable environment.

The renovation principles which were used in the realized projects in Germany are given below:

- regulation of urban planning parameters of building by decrease in density and demolition of parts of buildings;
- revitalization of a landscape, creation of a green framework of the area;
- providing inhabitants with necessary social infrastructure;
- creation of the barrier-free environment;
- formation of network of public spaces.

7. System of renovation principles

Besides the experience of the European countries mentioned, researches of Ivanenko [21], Dagdanova [22], Prusov [23], Daniel M. Muntean, Viorel Ungureanu [24] and Sille Sommer [25] are studied. In these studies renovation of large-panel building in the cities of Denmark, Czech Republic, Estonia, Romania and other countries is described. The most widespread principles are systematized and divided into five groups: urban planning, architectural, social, functional and economic. Classification is presented in table 3.

Table 3. Renovation principles classification (for building date from 1960 to 1970)

Urban planning	Architectural	Social	Functional	Economic
1. Creating of "live" streets	1. Improvement of buildings' technical condition (physical wear decrease)	1. Preservation and strengthening of social communications between inhabitants	1. Functional saturation of spaces	1. Increase in market value of housing

IOP Conf. Series: Materials Science and Engineering **481** (2019) 012024 doi:10.1088/1757-899X/481/1/012024

2. Creation of visual skyline and unique image of the district/block	2. Improvement of accommodation conditions (obsolescence decrease)	2. Territory differentiation on private and public spaces	2. Social service ensuring	2. Building consolidation
3. Exception of "impersonality"	3. Increase in esthetic quality of architecture	3. Decrease in criminality		
4. Improvement of urban environment quality	4. Creation of barrier-free access to buildings	4. Change of borders of the land plots and property rights		
5. Preservation and strengthening of the initial planning concept of the district	5. Increasing the energy efficiency of building			
6. Sustainable use of territories				
7. Urban-planning parameters regulation				
8. Modernization of transport service local system				
9. Revitalization of landscape				

8. Strategies of renovation for regions

As was mentioned earlier, in the majority of Russia cities, realization of a renovation strategy with full demolition and new building similar to what has been realized in Moscow is impossible. The lack of financing, unwillingness of investors to develop the built-up territories and the problem of utilization of garbage from demolition of buildings are the main reasons. Therefore, it is expedient to consider alternative strategies.

1. Renovation without demolition, which includes sanitation and reconstruction/repair of buildings without vast landscape transformations. This strategy can be applied in small cities in which there is no need for building consolidation and only for buildings in good conditions.

2. Renovation with partial demolition, for projects with less density in building and with buildings that were physically outdated. This strategy can be applied to territories with adverse ecological and

social situations, such as in the freed territories. It would be possible to organize green spaces and public spaces of various functionalities.

3. Renovation with partial demolition and new construction. This strategy can be applied in territories in which building consolidation will take place. It can include full or partial transformation of the local transport service system and a rather drastic change of the landscape. More detailed study is necessary for consideration of the applicability of this strategy to the idea of transformation of micro-districts into quarters.

9. Conclusion

In research, the analysis of buildings dating from 1950 to 1960 in Russia in general and in million-plus cities in particular is provided. The need for development of a building renovation strategy to be considered for Russian cities other than Moscow is evident. We studied the foreign experience of overcoming the crisis of moral and physical obsolescence of buildings. We also considered the principles of renovation that have been applied in various European countries. We offered classification of the most widespread and effective principles. The optimum renovation strategies for regions are based on our findings.

References

[1] Architectural council of Moscow 2014 History of industrial housing construction: experiments with a framework and the panel https://stroi.mos.ru/builder_science/istoriya-industrialnogo-domostroeniya-eksperimenty-s-karkasom-i-panelu

[2] Meerovich M G 2017 *Urban policy in the USSR 1917-1929: from the garden city to the departmental industrial settlement* (Moscow) p 352

[3] Ogorodnikova O A 2018 Mass housing construction in the history of the Soviet daily occurrence *Universum Social sciences* **3**

[4] Chazova O L 2015 A problem of the first mass series housing demolition in the complex housing estate reconstruction *Taurian scientific observer* **4** pp 95-97

[5] Departmental construction norms BCH 53 1986 Rules of assessment of residential buildings physical wear

[6] Russian statistical yearbook 2007 Federal State Statistics Service (Moscow: Rosstat)

[7] International cartographical company 2GIS data https://2gis.ru/

[8] Fund of assistance to housing and communal services reforming data. Reform of housing and public utilities https://www.reformagkh.ru/

[9] Complex of urban policy and city building of Moscow https://stroi.mos.ru/novaia-proghramma-rienovatsii-piatietazhiek

[10] Kosareva N B 2017 *Accounting of inhabitants' opinions about implementation of renovation projects in Moscow* (Moscow: City Economy Fund Institute)

[11] Gabor Preisich 1998 *History of urban construction in Budapest 1945-1990* (Budapest) pp 77-116

[12] Timea Csaba 2006 *How we really live in panel block* (Budapest) p 66

[13] Timea Denes 2000 *Reconstruction of large-panel buildings* (Budapest: University of Technology and Economics)

[14] Balazs Szabo 2013 Ten years of housing estate rehabilitation in Budapest *Hungarian Geographical Bulletin* **1** (Budapest:) pp 113-120

[15] The state of National Urban Policy in France 2017 City Policy (Paris: Politique de la ville)

[16] Christine Lelévrier 2008 Mobilities and residential trajectories of relocated households during urban renewal operations (Paris, Institut d'Urbanisme de Paris) p 46

[17] Jean-Claude Croizé 2017 A time when France chose to use prefabricated panel construction systems (Cambridge: University of Cambridge) pp 877-886

[18] Meerovich M G, Malko A V, Kozlova L V, Gladkova E A 2017 Renovation of panel building of the 1960-1980s in Germany *News of higher education institutions. Investments. Construction. Real estate* **7** (Moscow:) pp 111–119
[19] Maerkisches-viertel district data https://www.mein-maerkisches-viertel.de/maerkisches-viertel/.
[20] Stefan Forster Architekten bureau web-site https://www.sfa.de/home/
[21] Ivanenko L V, Fayzrakhianova Y I 2012 *Foreign experience of the built-up territories management* (Samara: Samara state architectural and construction university) pp 77-83
[22] Dagdanova I B 2014 Modern principles of the organization of the city quarter as a tendency of inhabited environment humanization (on the example of the cities of Western Europe) *News of higher education institutions. Investments. Construction. Real estate* **7** (Moscow:) pp 75-96
[23] Dmitriy Prusov 2017 Principles of reconstruction the urban territory in dense building and complex geology conditions *Space & Form* **30** pp 189-202
[24] Daniel M Muntean, Viorel Ungureanu 2017 Adaptive design of large prefabricated concrete panels collective housing *Engineering and Technology International Journal of Architectural and Environmental Engineering* **3** (World Academy of Science,) pp 299-308
[25] Sille Sommer 2012 Soviet Housing Construction in Tartu Estonia [Master thesis in human geography] (Tartu: University of Tartu) p 67

IOP Conf. Series: Materials Science and Engineering **481** (2019) 012025 doi:10.1088/1757-899X/481/1/012025

Reindustrialization, International Practices, and the Opportunities of Ekaterinburg (the case of the Uralmash district)

A V Alekseeva[1], S V Pridvizhkin [2], V G Veniaminov[3]

[1]Ural Federal University, 17 Mira street, Ekaterinburg, Russia, 620002
[2]Ural Federal University, 17 Mira street, Ekaterinburg, Russia, 620002
[3]Ministry of Construction of the Sverdlovsk Region

E-mail: toni.pereskokova@gmail.com

Abstract. The article deals with the international practices of industrial areas renovation in the US and Spain. The main positive trends of the countries' development are described. The article also presents the theory of Mazayev, G.V. regarding the ranking of technological paradigms and their interconnection with a modern city. The city of Ekaterinburg is reviewed as a multi-paradigm city that includes industrial areas of the III technological paradigm. Considering the renovation from the point of view of the international practices and Mazayev's theory, a conclusion regarding Russia, particularly, Ekaterinburg, the Uralmash district, is made. The optimal path of developing the area of Uralmashzavod, PJSC, is presented. The conclusions regarding Ekaterinburg and, particularly, the Uralmash district are made based on the foreign countries' experience: the best scenario for the area of Uralmashzavod, PJSC will be transformation into a technology park with the advantages stated herein.

1. Introduction

The development of the city of Ekaterinburg has led to the necessity to change the urban planning. One of the important aspects of urban planning is the issue of industrial areas located within the city. The necessity to reduce the adverse impact on the environment leads to several conclusions, the main ones being transformation of the industrial area in order to give it a different function; and create conditions for the industry's development.

The article reviews the international practices of industrial areas renovation illustrated by an example of the US and Spain and the possibility to apply these methods in Russia, in particular, in Ekaterinburg, in the area of the Uralmash district.

2. International Practices

2.1. *The US*

The 2008-2009 world financial crisis led to the following conclusions regarding the reindustrialization in the US: "the industry is a consumer and a source of innovations, as well as the source of the increase in labour productivity, so these achievements should be applied to other sectors of the economy, which is easier to perform having its own industry". The main reasons for returning productions are as follows:

- Labour costs;

IOP Conf. Series: Materials Science and Engineering **481** (2019) 012025 doi:10.1088/1757-899X/481/1/012025

- Proximity to customers;
- Product quality.

One of the drivers of reindustrialization is the relative cheapening of labour as compared to China alongside the lowest energy prices among the world industrial leaders. [2]

According to the report "The Reindustrialization of the United States" [5], the authors define the reasons of reindustrialization commencement as follows:

- in the US, the labour costs per unit produced are some of the lowest among the world's major economies;
- low energy prices due to large amounts of shale gas;
- quite low cost of capital due to the policy of the Federal Reserve System;
- competitive export in the majority of the US traditional export markets due to relative weakness of the US dollar;
- start of the slow growth of the real estate market.

2.2. Spain

Over the last decades, Barcelona has turned into an urbanistic experiment ground that has yielded positive results. The comprehensive approach to changing the city is the basic approach. Currently, the main task is to return to a productive city, creating a new economy using urban planning means, and revival of industrial areas. Thus, the reconstruction in the city is taking place at several stages, locally, which contributes to gradual transformation of the city. The Barcelona development concept provides the main idea that there is no need to rebuild the existing infrastructure completely, that new functions should be implemented gradually, and that the functions of the area should be changed if necessary.

Great examples of this approach are the urban renewal area 22@, reorganization of the Sants district, and reconstruction of the Plaça Cerdà area.

3. Theory of G.V. Mazayev

Cities created by the industrial revolution when production changed from the manufacturing system to the industrial one began to be built according to the "live near work" principle. Thus, cities should be classified by several aspects, one of the most important being the main function of a city [8].

Mazayev believes that the most important aspect if the main function of a city [8]. In the modern economics, the main concept that characterizes the development of the economy both on a country's level and on a region's level is the concept of a "technological paradigm" [9].

"Technological paradigm is a complex of technologies that are characteristics for a certain level of production development; due to the scientific advance, as well as the technical and technological advance, there is transition from lower-level paradigm to higher-level ones" [10].

A brief characteristics of each technological paradigm according to the theory of Mazayev is presented in Table 1 below.

Table 1. Technological Paradigm Ranking according to the theory of Mazayev, G.V.

Technological Paradigm	Technology	Urban Planning
First	Water energy	Cities and towns were located near water bodies that provided the energy component
Second	Steam energy	Near constant access to water
Third	Electricity energy (ferrous metallurgy, heavy engineering, electric engineering, chemical industry, non-ferrous metallurgy)	Industry is an integral part of a city
Fourth	Electricity energy (automobile manufacturing, heavy weapons'	Closed administrative-territorial formations appear

	production, aircraft engineering)	
Fifth	Development of microelectronics, informatics technologies, and gas extraction, processing and refinement)	Clean productions near residential areas
Sixth	Development of nano- and biotechnology, genetic engineering, clean energy industry, and high-speed transport systems	Appearance of superagglomerations (utopia)

According to the theory of Mazayev, G.V.: "the technological paradigms do nor replace each other but mutually ensure the stable existence of the whole economic system: the "lower" paradigms ensure functioning of the "higher" paradigms that cannot develop, while the "lower" technological paradigms can exist without the "higher" ones and existed for a long time before" [9]. The master plan of the city and the prevailing technological paradigm are interrelated. They form the complex planning system that includes residential, social, transport, and infrastructure subsystems. Changing one element of the system leads to issues in other elements that are inseparably associated with it [9].

The change of the production trend in several major economies is considered as a factor of transition to the post-industrial society. In Russia, however, reduction of the industry's share is mostly caused by degradation and destruction [11].

Russia cannot move its industrial base that corresponds to technological paradigms III and IV to third countries – that would mean a national catastrophe and loss of statehood. Preserving these technological paradigms means preserving the corresponding cities' planning systems.

Another trend of changing the industrial areas of technological paradigms III and IV in Ekaterinburg (which is only a project so far) is dividing large industrial areas of the city with transport networks. The city's master plan provides for dividing the territory of the Uralmash plant with two city highways from the North to the South that would improve the connection of the northern part of the city and the city centre.

In a multi-function city, such as Ekaterinburg, industrial areas of the technological paradigm III are formed on the basis of the areas of the technological paradigm II. The majority of such areas are transferred to the category of lands for constructing business centres and residential buildings. As a result, the city's master plan is changed which deteriorates the living standards: the time of access to workplaces, as well as load on the transport systems, is increased. Thus, substantial change in the occupational structure occurs: occupation in production decreases, while occupation in the sphere of commercial services provision, social sphere, and management sphere grows; an industrial city is transformed in a multi-function centre of providing services.

4. Application in Ekaterinburg

Ekaterinburg is one of Russian industrial and business centres that is transforming from an industrial city into a post-industrial one. Initially, the city was built as an industrial centre, the so-called "city-plant". According to the theory of Mazayev, G.V., Ekaterinburg corresponds to the third technological paradigm, that is, industry is the main factor of the city's development. Business and service spheres do not provide such economic development as production.

Uralmashzavod, PJSC can be considered a production base that provides immediate increase of the economic component of the city and the whole region. Anyway, an industrial area cannot be reviewed separately from the city cluster. According to Josep Acebillo, the Barcelona's architect: "Industrial cities were clearly divided into zones and transport systems that connected the residential and the industrial areas of the city. Such a system will not work in post-industrial cities since the conditions and the economy have changed substantially. Nowadays, we do not live in one place and work in another, and buildings and districts have a lot more intertwined functions than several decades ago". Thus, the plant should be considered together with the Uralmash district, being its integral part.

IOP Conf. Series: Materials Science and Engineering **481** (2019) 012025 doi:10.1088/1757-899X/481/1/012025

Initially, the Uralmash district was planned to be a separate cluster having its own infrastructure provided by Uralmashzavod, PJSC. This area started to grow from the territory of the plant and gradually became a complete district with its own historical and cultural heritage. Architectural features, dominant elements, and the points of attraction of the district were formed under the plant's influence, as well as the transport system that provided not only for transportation of the people but also for the transport accessibility of the plant.

Currently, when passing to a post-industrial city, one cannot disrupt the combination and hybridity of the district's functions. At this stage, everything is interrelated within the territory of the Uralmash district – road networks are located so as not to prevent simultaneous usage of the city's transport network and industrial area's network. All this has been forming for more than one decade. Now the plant can without any difficulty assemble some equipment within its territory and easily transport it to its place of destination without using the city's transport structure.

If one of the radical methods applied in the US were used, namely, transferring the area outside the city's boundaries, the crisis that erupted in the US would not be long in coming. This would also inflict damage on:

- the district's transport accessibility;
- people losing their jobs due to the impossibility to work during the period of re-establishment and moving;
- the Uralmash district would become a "ghetto" as the territory would not have the main point of attraction – the plant.

According to the opinion of Acebillo, one of the best urban planning specialists, "one should deal with transformation of the existing buildings, as with any other urban planning operation, with the greatest care – first one should define the future intended use of buildings and then perform partial reconstruction preserving their main architectural features. The advantage of this approach is that this provokes implementing new functions to the existing fabric of the city. No complete renovations".

By 2035, the rate of population increase in Europe will amount to 0 % and in Russia, less than 1 %. Thus, Josep Acebillo asks: "Do you need vast territories in the periphery, or you can use the existing infrastructure anew?"

At the current stage, there is a problem of re-functioning of Uralmashzavod, PJSC. It is assumed that the most rational step will be transforming this area into a so-called technology park. A technology park is a territorial, scientific, technological, and technical base for implementing innovative projects; a complex uniting R&D institutes, industrial facilities, business centres, exhibition areas, educational institutions, and service facilities (transport means, access roads, residential settlements, and security). The International Association of Science Parks and Areas provides its own definition of an object of innovative infrastructure.

Thus, this will provide the possibility for the territory to become an innovation centre having its comfortable infrastructure both inside the plant and within the district. This will contribute to the revival of the existing district and will not disrupt the transport infrastructure that is formed as direct passage from home to work; it will also contribute to gradual transition from the third technological paradigm to a higher level.

Two main advantages of technology parks are:

- Exemption from property tax.
- The profit tax is reduced from 20 % to 13.5 %.

The technology park based on Uralmashzavod, PJSC may have prospective young partners located within the territory of the future technology park.

5. Conclusion

Based on the international practices, the conclusions regarding Ekaterinburg and the Uralmash district can be made. The territory should be considered as a whole and one should presume the advantages and drawbacks that each method may have regarding the urban planning and economic component, as well as the impact on people living in this district. Thus, one can conclude the optimal path for the

IOP Conf. Series: Materials Science and Engineering **481** (2019) 012025 doi:10.1088/1757-899X/481/1/012025

territory of Uralmashzavod, PJSC, is transformation into a technology park with the above-stated advantages.

6. References

[1] Belokrylova O S 2017 Change of the technological paradigms or the Fourth Industrial Revolution?: Institutionalization of the industry 4.0 in the South of Russia [In Russ] *Vestnik ekspertnogo soveta* **3(10)** pp 3-10

[2] Pobyvayev S A and Tolkachev S A 2015 Reindustrialization of the US and the EU [In Russ] *Mir novoy ekonomiki* **2** pp 29-36

[3] Loshakova I M 2015 Assessing the innovation potential of enterprises within the context of forming high-technology territorial and industry-specific systems [In Russ] Extended abstract of Cand. Sc. (Econ.) dissertation (Voronezh)

[4] Ussoltseva M S and Volkova Yu V Renovation of industrial areas in Saint Petersburg [In Russ] *Constructing Unique Buildings and Structures. ISSN 2304-6295/ 2 (29)/ 2015/ 98-111* pp 98-111

[5] Euler Hermes Economic Research Department Special Report "The Reindustrialization of the United States" *http://www.eulerhermes. us/economic-research/economic-publications/ Documents/economic_outlook_1187.pdf*

[6] How Barcelona is built [In Russ] *http://www.archplatforma.ru*

[7] Direct speech – Josep Acebillo. The former Barcelona's Chief Architect – on transformation of the old centre and large cities 2012 [In Russ] *The Village*

[8] Barkhin M G 1986 City. The Structure and Composition [In Russ] (Moscow: Nauka)

[9] Mazayev G V 2017 Impact of the technological; paradigm on the city's planning structure [In Russ] *Akademicheskiy vestnik UralNIIproyekt RAASN* **3** pp 10-14

[10] Glazyev, S Yu 2005 Choosing the Future [In Russ] (Moscow: Algoritm)

[11] Animits, Ye G and Silin Ya P 2013 The Middle Urals on the way to new industrialization [In Russ] *Ekonomika regiona* **3**

[12] Mazayev G V and Verkhovykh E Yu 2017 Influence of ideas of "postindustrialism" on development of industrial cities of Urals [In Russ] *Akademicheskiy vestnik UralNIIproyekt RAASN* **4** pp 11-16

IOP Conf. Series: Materials Science and Engineering **481** (2019) 012026 doi:10.1088/1757-899X/481/1/012026

Big data in urban planning and territory management

E A Pentsev[1] and O A Makarova[2]

[1] Department of Urban Development, Ural Federal University, 17 Mira street, Yekaterinburg, 620000, Russia
[2] Department of Urban Development, Ural Federal University, 17 Mira street, Yekaterinburg, 62000, Russia

E-mail: e.a.pentsev@urfu.ru, olga.makarova@urfu.ru

Abstract. Modern technologies are constantly evolving, almost continuously various innovations aimed at improving life appear in various branches. In addition to the rapidly growing number of inventions, the amount of information in public access is also growing. To store and process large volumes of information, BIG DATA technologies are used, which help improve the quality of urban planning studies. The article discusses the possibilities of using BIG DATA in urban planning, lists possible areas for using geographic information systems in economics, medicine, transport, environmental management, construction, municipal and state administration, science, and investment activities. There is identified possible users of GIS at various levels of management. The authors define the role of the federal universities in the creation, management of GIS, and training in this area.

1. Introduction

Every moment a person comes across information, it comes from a variety of sources and is easily acceptable – today there is no need to travel long distances and spend time to gain access to various interesting data.

Information is a resource that will never be depleted. Many people can use one and the same information for infinite number of times. Information was important at all times, however nowadays its value has increased even more.

By now the data bank, accumulated by humanity, is very large. For example, DigitalGlobe, which operate satellites performing remote ultra high resolution sensing of the Earth, have collected about 100 petabytes of data for over 17 years of operation [1]. That is why not only systematization, structuring and storing information, but also the rapid retrieval of the necessary data became an important aspect. And here is where BIG DATA come in handy, allowing automating the whole process.

For those who do not know, the largest BIG DATA is geographic information systems (further referred to as GIS). It is here where large amounts of data are organized layerwise and where the associated task of accessing it via WEB publishing mechanism is being solved.

2. GIS technologies in the modern society

The modern society cannot do without GIS technologies. It is impossible to build economy or perform modern management without them. The worldwide trends are such, that it is necessary to manage a large base of spatial data in time, and doing this requires GIS. Until recently, this task was difficult to solve because the data bank was small and the access to spatial data on the Earth (satellite photographs) was limited. However, during the last few years the situation changed for the better, and with the advent of new technologies the GIS rise to a new level. This allows GIS to be introduced into new areas of social activity.

GIS technologies have been in place in the world for over 50 years. During that time these technologies became widely distributed and appeared not only in fields were people professionally deal with geography, but in many other fields - business, emergency response services, municipal services, etc. This is partially due to the fact that this discipline covers a very wide range of issues, from purely theoretical to purely practical, from geography to programming, from business presentations to database design.

The purpose of GIS is to provide a decision-making process for optimal resource management, arranging transport and retail operations, utilization of real estate assets, water, forest and spatial resources. [2].

V I Laykin highlights several types of GIS industry applications:

a) administrative and territorial management:
- urban planning and objects design;
- maintaining of engineering communications inventory, land inventory, urban planning inventory, plantation inventory;
- predicting emergencies of technogenic and environmental nature;
- managing traffic flows and public transport routes;
- building environmental monitoring networks;
- engineering and geological zoning of the city.

b) telecommunications:
- trunked radio and cellular communications, traditional networks;
- strategic planning of telecommunication networks;
- optimal positioning of antenna, repeaters etc.;
- determining cable routes;
- monitoring the status of networks;
- operational dispatch control.

c) engineering communications:
- assessment of needs for water and sewage networks;
- modeling of natural disasters effects on engineering communication systems;
- designing engineering networks;
- monitoring of engineering networks condition and preventing emergencies.

d) transport:
- road, railroad, water, pipeline and air transport;
- management of transport infrastructure and its development;
- vehicle fleet management and logistics;
- traffic control, route optimization and freight traffic analysis.

e) oil and gas complex:
- geologic exploration and field survey works;
- monitoring of oil and gas pipelines technological operation modes;
- designing of trunk pipelines;
- modeling and analysis of consequences of emergency situations.

f) defense and law enforcement agencies:
- fast response services, armed forces, police, fire services;
- rescue operations and protective measures planning;
- simulation of emergency situations;
- strategic and tactical planning of military operations;

- navigation of fast response services and other defense and law enforcement agencies.

g) ecology:

- assessment and monitoring of natural environment condition;
- modeling environmental disasters and analyzing their effects;
- planning environmental protection activities.

h) forestry:

- strategic forestry management;
- lumbering management, forestry approach planning and road design;
- maintaining forest inventories.

i) agriculture

- planning agricultural land processing;
- accounting of farmers and crop lands;
- optimization of agricultural products and mineral fertilizers transporting.

This list is constantly being complemented, because the GIS are increasingly integrated into our life, activities and business. [3]

GIS have many features that should be considered. One of the features is that geographic information systems are elements of social informatization. This feature takes into account the integration of the GIS into education, science, production and applying the obtained information in practice.

Previously the information obtained from research scattered and its application was localized within the boundaries of the project, where it was generated and obtained. It was difficult and almost impossible to find the existing facts, check their availability. No data exchange was established. The lack of information on spatial data, in return, led to multiple repetitive works, inappropriate waste of time and money, creating obstacles to free utilization of data.

In 2006, in accordance with the Concept of creation and development of spatial data infrastructure in the Russian Federation (hereinafter - the Concept), formation of spatial data infrastructure (SDI) was started at the discretion of the Russian Federation Government decree no. 1157-r dated August 21, 2006. [4]

As stated by Sergey Menyaylo, RF Presidental Envoy in Siberian Federal District, at the XIII Interexpo GEO-Siberia 2017 international exhibition and research congress, held in April 2017 in Novosibirsk: "Without modern geographic simulation systems and reliable spatial information it is impossible to plan territorial development, perform effective traffic flow management, predict and prevent emergencies and deal with other important issues. [5]

During 13 years of operation of the Urban Development Code of the Russian Federation and of the Concept, new GIS and geographic portals have emerged, which significantly increased the quality and efficiency of management decisions at various level, due to SDI implementation. Transparency in making such decisions and increased investment in territorial development were reached through an accessible procedure of spatial data analysis.

However, due to the following facts:

- less then 50% of federal subjects of the Russian Federation and less than 10% of municipal entities of the country have developed and are maintaining the information systems for urban planning and geographic portals;
- each federal subject of the Russian Federation or municipal entity maintain it in a form, convenient for themselves (sometimes only by keeping the electronic copies of the documents, but not the graphic version of the information contained in them), using their own classifiers, static two-dimensional maps instead of multi-dimensional information maps etc.
- it is necessary to provide privacy of some spatial data,

further development of SDI is difficult and almost impossible without significant investments, because the burden on the budget of many, small, medium and large municipal entities referring to creating and maintaining the information systems for urban planning is infinitely high.

According to the Russian TASS information agency, at the Krasnoyarsk economic forum, held on April 20-22, 2017, Kirill Kuzmenko, Secretary to Council of the National Prospectors Union association, voiced an opinion concerning creation of a unified information portal, which would become the center of geographic data infrastructure and unite numerous bases under single standard, format and quality, leading to adoption of certain state standards.

In many developed countries, about 70% of managerial decisions in business and government are based on spatial data, i.e. the digital economy is based on the Earth data. Nowadays a considerable geographic data market has established in Russia. Its monetary value is expected to increase to about five times by 2020. However, these data are not compiled and analyzed in a proper way, and this leads to low quality and ineffective decisions being made. An operator is needed to collect this information into a single data bank, because the creation of a unified structure, especially in construction, will save millions of rubles, because it would take twice less time to accumulate the initial data etc.

It is especially urgent to create a unified database for urban planning and territorial management, because these are the fields where it is important to analyze a fairly large amount of information in order to make the right decision. Most of the information that is owned by the executive authorities, refers to territories, objects and particular addresses, that is why it becomes clear that GIS technologies, as tools for the most powerful and effective information assurance in the field of urban planning, are applicable to works with such data. Users of the geodata include federal ministries and departments, commercial customers and local self-governing authorities.

Regardless of whether the issue of developing the urban planning documentation of any level is solved, or any management decision is made, it is necessary to analyze a considerable volume of information that is relevant and presented in a certain way. However, the reality is that every structural unit and every administrative department "protect" their information and do not share it eagerly with other agencies. That is why the information is usually duplicated in every department, and sometimes is distorted due to different approaches to obtaining, updating and analyzing information, i.e. it is difficult, if not impossible, to see the real state of things in the municipal entity or the entire region.

Basically data are stored in Excel spreadsheets or in databases, however, even in such form it can sometimes be quite difficult to analyze the whole volume of the information and present it in a user-friendly manner upon request of specialists or residents. It is even more difficult to speak about making a certain management decision on its basis, which will have a strategic focus. Thus, to present complete and reliable information on one and the same object, it is often necessary for multiple agencies to interact, and analyzing this information takes much time and financial resources. This is why errors and inaccuracies may appear in urban planning documentation; inflated operating costs of various infrastructures (improvements, transport support etc.); risks and additional costs for the investor both at the stage of land plot allocation and at the stage of architectural and engineering design, leading to overall increase of the investment project cost; errors of municipalities, leading to legal expenses and additional costs of error correction etc. Since there is no agreed scheme of interdepartmental interaction, there is a need to establish a common base containing all the information from all departments.

As an example, let us consider the development of a project of partly-urbanized territory planning: in order to determine the number of places needed in the newly designed pre-school institutions it was necessary to know the number of children of pre-school age, living in houses already built. The information was requested from the Department of education, Department of Federal Migration Service Directorate of the Russian Federation and Department of Health. Letters of reply were received. What is interesting, the number of pre-school age children in two of these documents was two times different. However, the number of children living in a certain territory is a definite value, which means there already was a certain amount of inaccuracy at the level of design project preparation. This inaccuracy clearly affects the comfort of living in this area, while the administration of the municipal entity does not actually have any reliable information.

It is therefore appropriate to conclude that every municipal entity is simply obliged, according to cl. 56 of the Urban Development Code of the Russian Federation [6], to maintain information systems for urban planning (hereinafter ISUP) based on GIS technologies. However, in view of limited financial and human resources of required qualification, and requirements for provision of information security of such systems, the vast majority of municipal entities either maintains such systems nominally (for reporting only) or does not maintain them at all. In other words, the federal authorities granted these powers to municipal entities without considering, how these powers would be exercised.

The goal of information assurance of urban planning is to increase the efficiency of urban planning solutions in municipal entities, provide publicity, create conditions for growth of investment potential of the territories.

Here it is very important to be able to analyze a huge array of data from related industries. That is why the information contained in information systems for urban planning, as well as the basis of this system itself, can be used in various fields (see table 1).

Table 1. GIS application fields.

Field of application	Possible users
The information system for urban planning ensures compliance with the requirements of cl. 4 Ch. 7 of the Civil Code of the Russian Federation. [6]	Relevant departments of municipal entities and the Government of the subject of the Russian Federation. Private individuals and legal entities - in particular, - documents of unlimited use.
Maintaining a standby topographic plan of territory of inhabited localities and subject of the Russian Federation of various dimensions (initially – in electronic version, further – in digital version). In the future – creating a 3D engineering model of inhabited localities.	Relevant departments of municipal entities and the Government of the subject of the Russian Federation. Service recipients: design organizations, asset management companies, resource-supplying companies, private individuals etc.
Maintaining a standby address plan for inhabited localities of the subject of the Russian Federation.	Relevant departments of municipal entities and the Government of the subject of the Russian Federation. Separate units of special, municipal and emergency services (Emercom, Ministry of Internal Affairs, emergency medical service, gas emergency service, migration control service etc.)
Maintaining a standby plan of engineering communications (tracing, engineering network specifications, ownership, hydraulic models of pressure networks)	Relevant departments of municipal entities and the Government of the subject of the Russian Federation. Resource-supplying companies. Service recipients: developers, design companies, asset management companies, private individuals etc.
Maintaining a standby register of buildings and facilities (data on year of construction, designers and developers, technical specifications of objects and facilities, relevant drawings of buildings (in the future – 3D models), data on conducted repairs, data on resource consumption meters (water, gas, heat, electricity consumption) etc.).	Relevant departments of municipal entities and the Government of the subject of the Russian Federation. Asset management companies. Technical inventory bureaus. Service recipients: developers, design companies, private individuals, etc.

IOP Conf. Series: Materials Science and Engineering **481** (2019) 012026 doi:10.1088/1757-899X/481/1/012026

Field of application	Possible users
Maintaining a standby register of plantation in inhabited localities (location, species, planting year, measures taken etc.)	Relevant units of municipal entities. Asset management companies. Service recipients: developers, design companies, private individuals, etc.
Maintaining a standby register of geological and hydrogeological surveys (location and survey data by holes with indication of survey year and attachment of report)	Relevant departments of municipal entities and the Government of the subject of the Russian Federation. Various scientific and technical organizations (Lower Ob Basin Water Management Board, Institute of Geology of the Ural Department of Russian Academy of Sciences., Uralgidroekspeditsiya, etc.). Service recipients: developers, design companies, private individuals, etc.
Maintaining a standby plan of mineral deposits (developed deposits, survey data etc.).	Relevant departments of municipal entities, Government of the subject of the Russian Federation, Government of the Russian Federation. Various scientific and economic organizations (Institute of Economics of the Ural Department of Russian Academy of Sciences). Service recipients: investment companies, financial and industrial groups.
A system for monitoring and predicting traffic flows (location and operation of traffic lights, location of road signs (with reasons for their installation), collection of data from CCTV systems, data processing, information on the intensity of traffic flows at various time intervals and directions, determination of the parameters of gas contamination and noise levels in residential areas).	Relevant departments of municipal entities, Government of the subject of the Russian Federation, Government of the Russian Federation (especially State Traffic Safety Inspectorate). Service recipients: developers, design companies, private individuals and legal entities etc.
Maintaining a standby register on population (divided into age groups and location-specific, in the future – with data on places of employment, - education).	Relevant departments of municipal entities and the Government of the subject of the Russian Federation. Office for National Statistics. Service recipients: developers, design companies, legal entities etc.
Maintaining a standby register on motor vehicle registration (with reference to graphic map materials).	Relevant departments of municipal entities and the Government of the subject of the Russian Federation.

Field of application	Possible users
Inhabitant morbidity monitoring and prediction system (broken down by illness, age, place of residence and employment).	Relevant departments of municipal entities and the Government of the subject of the Russian Federation (especially the Ministry of Health).
The Natural Resource Potential system (assessment of the natural resource potential of the territory in terms of administrative and municipal division).	Relevant units of the Government of the subject of the Russian Federation (especially the Ministry of Natural Resources, Ministry of Economic Affairs and Ministry of Investment).
Telecom Operators Coverage Zones system	Relevant units of the Government of the subject of the Russian Federation. Telecom companies. Communication operators.
Specially Protected Nature Conservation Areas system (apart from map materials it includes the information from the State Inventory of Specially Protected Nature Conservation Areas, accounting information and regulatory framework).	Relevant units of the Government of the subject of the Russian Federation (especially the Ministry of Natural Resources). Ecological and environmental protection organizations.
Historical and Cultural Property Sites system (location, description, specifications of the protected sites, description of the protected site, - documents on renovation, protective legislation, historical references).	Relevant departments of municipal entities and the - Government of the subject of the Russian Federation (especially the Ministry of Culture). Monument protection departments. Non-governmental organizations. Service recipients: developers, design companies, legal entities etc.
Cataclysms and Emergencies of Natural Origin system (location, description, flood, forest fires, earthquake, landslide, karst hole prevention measures etc.).	Relevant departments of municipal entities and the Government of the subject of the Russian Federation.
Cataclysms and Emergencies of Technogenic Origin system (location and description of possible sources, possible consequence coverage areas of technogenic accidents, measures for prevention, localization and neutralization of such events).	Relevant departments of municipal entities and the Government of the subject of the Russian Federation.
Accounting and Management of State and Municipal Property in the Territory of the Subject of the Russian Federation system	Relevant departments of municipal entities and the Government of the subject of the Russian Federation.

Field of application	Possible users
Forest Administration of the Subject of the Russian Federation system (location, boundaries, quantitative and qualitative characteristics of forests, information on allocated plots, lumbering permits, lumbering and processing enterprises etc.).	Relevant departments of municipal entities and the Government of the subject of the Russian Federation.
Ecology of the Subject of the Russian Federation system (information on sources of water, air - basins and soils pollution, monitoring data (including comparative time analysis), pollution abatement and elimination activities).	Relevant departments of municipal entities and the Government of the subject of the Russian Federation. Ecological and environmental protection organizations.
Agriculture of the Subject of the Russian Federation system (data on location, boundaries, characteristics of the allocated agricultural land plots, zoning of the specific territory by preferred agricultural crops, crop data, etc.)	Relevant units of municipal entities and Government of the Subject of the Russian Federation (especially Ministry of Agriculture, Ministry of Investments, Ministry of Economic Affairs). Research institutions in agricultural sector. Service recipients: Business community.
Investment Sites of the Subject of the Russian Federation system (data on deployment, boundaries, characteristics of natural and engineering resources, human potential, examples of projects with calculated economic indicators).	Relevant departments of municipal entities and the Government of the subject of the Russian Federation (especially the Ministry of Investments, Ministry of Natural Resources, Ministry of Social development). Service recipients: Business community.
Science and Education system (location of scientific and educational facilities, description of activities (staff training), spatial and statistical analysis of data on the activities of organizations in the field of science and education, etc.)	Relevant departments of municipal entities and the - Government of the subject of the Russian Federation (especially the Ministry of Education and Science). The Russian Academy of Sciences and its branches. Service recipients: private individuals and legal entities, investment companies, business community.
Public Passenger Transport system (data on urban, suburban, long-distance routes of the transport system indicating the rolling stock, the timetable, the actual location of the rolling stock at a particular point in time, the number of passengers transported (in the future, information on the distribution of passenger flows).	Relevant departments of municipal entities and the Government of the subject of the Russian Federation (especially the Ministry of Transport). Service recipients: private individuals and legal entities, transport companies, business community.

Field of application	Possible users
Housing and Public Utilities system (data on buildings, engineering communications, transport facilities, technical condition, ongoing and scheduled repairs, analysis of efficiency of energy conservation measures, etc.)	Relevant departments of municipal entities and the Government of the subject of the Russian Federation (especially the Ministry of Construction, Ministry of Housing and Public Utilities, Ministry of Economic Affairs). Service recipients: private individuals and legal entities, asset management companies and homeowners associations, business community.

As K Norts noted, GIS are a powerful tool for business analytics and interaction with other enterprise systems.

Here, location data can be a versatile tool for advanced modeling, analysis, and processing of information. The ability of GIS to perform analysis and to interpret different information is what is relevant nowadays for most professionals in every company, because there is no need in searching and analyzing hundreds of analytic articles. They simply need to use the analytical capabilities of GIS to help benefit from the amount of the data available to them. And to achieve this, it is important to be able to find patterns in information, to eliminate redundancy and to make the right decisions. This can be done with modern GIS analysis capabilities based on BIG DATA technologies [7].

3. Conclusion

In these circumstances, Federal Universities have gained particular importance, serving as the only host to interdisciplinary staff able to quickly understand the complexities of subject-specific GIS and their construction, and to train a sufficient number of people for completely new professions. That is why it is important to involve the federal universities in the process of informatization and building of the country's digital economy. Federal universities should become centers for accumulation, storage, processing and use of spatial data in the territory of the respective federal districts, and state authorities (federal and regional) and local self-governing authorities should obtain regulated access to these data to enable analysis and the adoption of sound management decisions.

References

[1] The DigitalGlobe company transfers remote sensing data to the cloud on trucks / GIS-Association Geoinformation Portal. URL: http://gisa.ru/117867.html.

[2] Samoylenko V M 2010 *Geographic Systems and Technologies: Textbook* (Kiev: Nika-Center) p 24

[3] Laykin V I and Uporov G A 2010 *Geoinformatics: Textbook* (Komsomolsk-on-Amur: ASUHP Publishing House) p 58

[4] Concept of Creation and Development of Spatial Data Infrastructure of the Russian Federation (approved by the Russian Federation Government decree no. 1157-r dated August 21, 2006).

[5] XIII International Interexpo GEO-Siberia 2017 Exhibition and Congress is underway in Novosibirsk GIS-Association Geoinformation Portal. URL: http://www.gisa.ru/117598.html.

[6] Urban Planning Code of the Russian Federation dated December 29, 2004 no. 190-FZ (approved by State Duma of the Federal Assembly of Russian Federation December 22, 2004, rev. dated March 07, 2017).

[7] Norts K 2017 Are we missing the boat on the true value of analysis in GIS? / *J. ArcReview* JAR1(2017)80.

Modern Models of Public-Communication spaces of General Education Establishments within the Process of Social Adaptation of Students

I N Maltceva[1], N N Kaganovich[1], A I Redinkina[1], K A Tkachuk[2]

[1]Department of Architecture, Ural Federal University, 19 Mira street, Yekaterinburg, 620002, Russia
[2]Technical University of Munich, 21, Arcis street, Munich, 80333, Germany

E-mail: i.n.maltceva@urfu.ru

Abstract. The article is devoted to the issues of formation of internal and external public-communication spaces in civil buildings, in particular, during the development of spatial planning decisions of general education establishments. The possibilities and conditions for the formation of the project potential using innovative engineering technologies that undoubtedly determine the life cycle of the building and its further typological development are considered. The main factors that determine the relevance of the problem of communication and coordination of information and cultural relations, especially in educational buildings, are given. These factors allow to evaluate the need for investigation of stated problems. The idea of a school building as a standard organizational structure with a set of core functional blocks tightly interconnected can be reconsidered and substantially supplemented. To a large extent this concerns issues of personal and public communications, which are doubtless important for all potential participants of the educational process and for the development of educational programs. The article presents typological analogs from the world design, architectural and construction practice, as well as the results of the research and project approbation of the model of the ecopositive school building on the criterias of sustainable development, as part of the residential area of the perspective construction of the village "Polevodstvo" in the Yekaterinburg (Russia).

Keywords: school building, eco-sustainable architectural environment, social and communication space, forum, interactive ladder, social adaptation.

1. Introduction

In the modern society, one of the main tasks in the field of architecture is the creation of a comfortable environment for people, in other words, a harmoniously organized eco-sustainable architectural environment, as the surrounding space, along with the functional component, affects a person's well-being, emotional state and his formation as a person. Particular importance in matters of social adaptation in the process of forming the architectural space of most public buildings is assigned to its important component - public spaces. "Public space is a space for communication and social activity, organized in accordance with the dominant function" [1,2,3]. At the same time, the public space is projected as elements of the general communication inside bonded framework system and by filling it with new meanings it transforms to public-communication space. The influence of social factor and the aspects of the formation of public spaces can not be underestimated - this is expressed in the phenomenon of reproduction of the public sphere as a whole, when people go beyond their usual social circle and expand it, they must correctly react to other types of behavior, preferences, ideas and traditions. There social interactions occur on the verbal and non-verbal levels [4, 5].

Within public spaces classification mostly sporting and entertainment facilities, museum-exhibition, cultural-leisure and commercial buildings, multifunctional complexes, including universities, are considered. In the buildings of general education establishment, spaces for communication are especially necessary, since students spend most of their day at the school. Today these spaces actively "participate" in the process of education and upbringing. Modern models of the organization of public spaces are quite diverse. It can be various internal and external spaces, including atriums, open courtyards-atriums. All of them with the right organization can become a recreation, a place for holding social events, a center for communication, contacts, meetings, birth of ideas.

2. Russian and world practice of organization of internal public spaces

Particular importance in the organization of public spaces in the educational building is given to qualitative characteristics: universality and ability to layout transformations. In the same way are organized forum spaces (forum - the area where public life was concentrated in the cities of ancient Rome) in educational buildings. In the educational institutions, the forum can be organized in the zone of a concourse group or in a general center of school, integrated with the foyer of the assembly hall, combined with a winter garden or an exhibition hall [6]. This is how the transformation and universal use of them for different scenarios of "school" life takes place. Often the forum becomes the core of the complex volumetric and planning structure of the school building, the center or one of the centers (branched compositional scheme) of communication and bond structure. It combines the main functional blocks, creates the formative accent in the constructive structure, in the interior and exterior appearance, thus defining its spatial morphology. One example of such public spaces is the courtyard, an open or enclosed atrium courtyard; the size, shape and design decisions of which depend on architectural and planning, constructive solutions, climatic and historical factors, as well as from the design assignment and the author's concept.

Figure 1. Atrium of the school in Erestad (http://boombob.ru/img/picture/Oct/15/650b9ee737cd84ce8bbf4a0b412e8be7/9.jpg).

An example of such a solution is the school of the architectural bureau Tate Snyder Kimsey "Silverland" in the state of Nevada, USA (Fig. 1). The courtyard under the open sky unites the dining hall, the media center with a reading room, a sports zone, a cinema under the open sky. The open multifunctional space of energy efficient and full light of the school becomes not only a recreation place for rest and communication, but also a site for various school and social events. Public spaces in the form of amphitheaters and open "interactive" staircases are widely used in the planning of a school building. Architects combine these forms of organization of space with various methods of natural lighting. In addition to all the known advantages of atriums from the point of view of creating physical comfort and planning organization of the internal environment, they meet the criteria for organizing a modern public space - "transparency" and openness, visual communication with the environment. In many countries, school buildings use the already widespread spatial planning approach: a large single space surrounded by game and studying zones (rooms) of various occupancy that are able to transform depending on the changes in processes or their participants, from local and group to public spaces. The innovative project of the architectural bureau 3XN for the Danish gymnasium in Erestad (the district of Copenhagen) is a large space in which study areas are located around the main atrium with a staircase. The concept of design embodies the idea of the unity of teamwork and own thinking (Fig. 2).

Figure 2. Atrium of the school in Erestad
(http://boombob.ru/img/picture/Oct/15/650b9ee737cd84ce8bbf4a0b412e8be7/9.jpg).

In 2016 in Irkutsk (Russia) the Danish company CEBRA developed the innovative project "Smart School". The round shape of the building symbolizes the "unifying community" (Fig. 3). The separately standing volumes are united in a ring under a common roof and are grouped around the open courtyard. The internal space in the open air is the central connecting point of intersection of spatial and functional connections. The peculiarity is that this is also a training area for classes [7–9]

Figure 3. The project "Smart School" in Irkutsk, a general view
(https://ru.wikipedia.org/wiki/Умная_Школа).

According to the authors of the project, the circle symbolizes the unity of people and involves all students in the learning process [10].

Particular attention is devoted today to the semantic organization and design of internal open staircases, which are not only constructive elements for uniting levels in a building, but also in a certain sense a "uniting" social space for its inhabitants, a sort of "vertical place" for all-round communication, exchange of information and ideas. Students will always prefer the stairs to elevator, since are located in an attractive atrium space of three or four storey school buildings. "The ladder of the future" should look like this - "active", "information", as a symbol: "a ladder to the heights of knowledge." This, of course, is about ladders not included in the calculation of evacuation routes in accordance with fire safety standards. Such stairs also require safety in case of fire or emergency. This is due to the use of modern technologies: the use of fire curtains, the design of structural elements of stairs with the necessary requirement of fire resistance, the use of modern systems of automatic smoke removal and fire alarm.

"Greening" of atrium spaces by creating internal landscapes in school buildings is becoming especially urgent today. Mini-greenhouses, winter gardens, green corners (Fig. 4) serve as recreations, biodiversity laboratories or aquaponics (technology of growing crops). Through these buffer zones an integration of natural components into the internal environment of the building is happening, a semantic and visual connection with nature or urban environment is born [8].

Figure 4. The project "Smart School" in Irkutsk, a general view (https://ru.wikipedia.org/wiki/Умная_Школа).

The above-mentioned planning elements under the general name - forums, besides the organization of functional and visual interrelations of the internal premises, play the role of social and recreational zones of public activity and social adaptation of all participants, and in a certain sense are an extracurricular continuation of the main educational processes in the school building. In other words, the internal space should be as open as possible, have many inputs and outputs, be "transparent and responsive", while functionally zoned and have a flexible zoning structure. Although vertical communications and the supporting framework are among the least variable elements (anchors) in a building, they, at the same time, play a key role in the proper formation of public spaces. These qualities to a large extent determine the *planning potential of the building* as a basis for transitional typological development, and meet the basic criteria of eco-stability and viability of the architectural object.

3. Methods of research

The authors of the article have investigated modern trends and problems of the formation of a school building in the context of an interconnected and integrated system: participants - processes -

architecture - eco-sustainability. Considering the school as a social institution, a modern strategy of research in this field was adopted, in which, together with constructive and technological reliability, are considered such important properties [11, 12] as:

- *Communicativity* – accessibility and "openness" of the building, including social, architectural and planning, technological;
- *Adaptability* – universality and ability to transform internal space without loss of integrity and bearing properties of the object; technological - to the development of an infrastructure network; psychological - by creating various premises from private zones to large public areas;
- *Interactivity* – the possibility of reorganization or development of the architectural volume, the ability to respond to changes in the needs of society and to the reforms of training programs, the opportunity to get acquainted with the latest technologies in the field of ecology and energy efficiency in theory and on practice.

It can be concluded that practically all system properties and modern criteria for the formation of a school building imply the need to include public spaces in the general structure, more specifically, taking into account their linkage functions – public-communication space, which will ensure its viability and ability for subsequent development.

4. Model of a school building in the Sverdlovsk Region

As an approbation of the results of the research, the authors have developed a conceptual project proposal "Model of the school building the residential area of the perspective construction of the village Polevodstvo (Sverdlovsk Region, Yekaterinburg)".In addition to solving basic typological problems, the project proposes the creation of an interconnected system of various internal and external public spaces for movement, recreation and communication of all participants. The main planning method and the core of the composition of the building is a semi-enclosed internal open courtyard with a main entrance and a playground for various school events. The yard is surrounded by glazed gallery-recreations, which act as buffer zones between the courtyard and the main premises of the school. In addition these galleries provides opportunities for natural lighting and passive ventilation (Fig. 5).

This solution gives the building a natural appearance, security and safety for its inhabitants, while preserving the openness and visual connection with the external space. This model is the starting point for the development of the system of "anchors" in the structure of the building.

Figure 5. The courtyard.

In this project "anchors" are internal public spaces for various purposes: recreation galleries, halls, lobbies, foyer, staircase-forum, atrium, greenhouse "Edible Yard", green maintained roof. The authors

IOP Conf. Series: Materials Science and Engineering **481** (2019) 012027 doi:10.1088/1757-899X/481/1/012027

consider the possibility of expanding the social functions of the educational building as a center of communication and cultural activity of the residential area due to the autonomous use of the gym, library, workshops and assembly hall during the time free from the main functional process (Fig. 6).

Figure 6. Layout of public-communication spaces.

5. Systems of ecocertification of school buildings

Already at the design stage, the authors considered modern voluntary eco-certification systems that fully take into account the "life cycle" of buildings, from the beginning of construction to reconstruction and the demolition of obsolete objects, taking into account the issues of green and sustainable development of the building (Figure 7). In particular, during the study of school buildings, it was concluded that the requirements for educational facilities should differ from other types of buildings, since it is necessary to take into account the specificity of the typological properties of the object: the number and age characteristics of the participants in the process, as well as the purpose and objectives of the functional processes, planning requirements. Therefore, we turned our attention to the assessment system for school buildings LEED for School Green Building Rating System [7, 13].

The LEED assessment system for school buildings recognizes the unique nature of designing and building schools. This system is designed to address issues that the authors considered in the context of the topic of this article:

- *the sustainability of the construction site* or how to make the school a more integrated part of the community, allowing the building and its playing fields to be used for extracurricular activities and other functions;
- *the use of regional materials,* that is, to increase the demand for construction materials and products that are produced and produced in the region, thereby supporting the use of local resources and reducing the environmental consequences associated with transportation;

- *daylight and visibility* and thereby ensure for the people in the building the connection between the inner rooms and the outside through the access to the natural daylight and its sufficient quantity in the rooms with the regular stay of people;
- *the school as a learning tool* or the integration of the sustainable functions of a school building into an educational program.

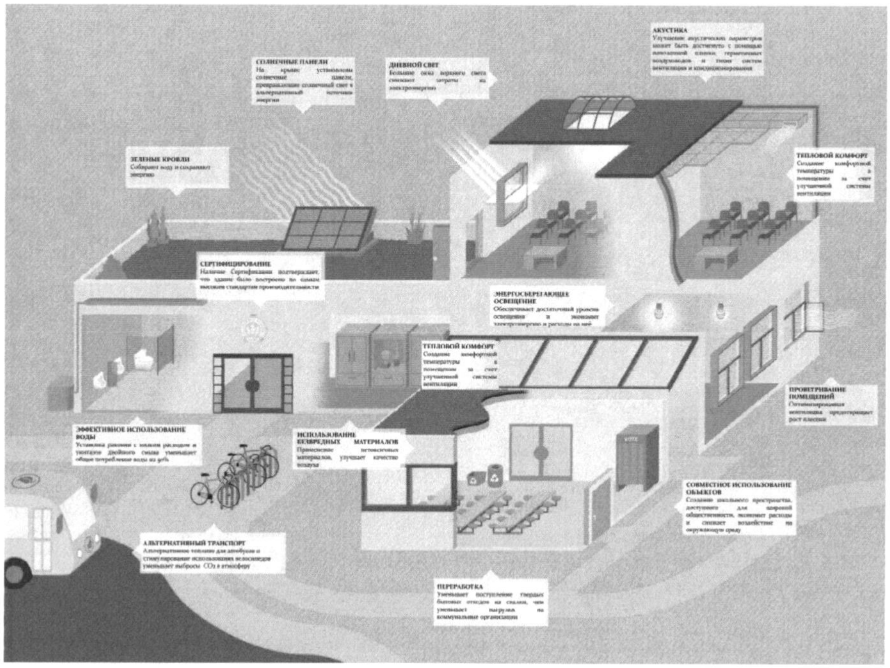

Figure 7. Certification of school buildings according to the LEED system (https://ehp.niehs.nih.gov/wp-content/uploads/2009/10/ehp.117-a448.g002.png).

Addressing the uniqueness of school premises and the health of children, the LEED for School Green Building Rating System provides a unique, comprehensive tool for schools built in accordance with green standards and sustainable development criteria, with measurable results. LEED for schools is a recognized standard for high performance schools that are safe for people's health, comfortable and cost-effective.

6. Conclusion

The problems of educational institutions are not only the development of educational programs and information technologies, qualified teachers and sufficient financing, but also the construction of modern school buildings that should become a full-fledged and comfortable architectural environment for learning and communication, connected with the surrounding society. In this environment, the diversity of forms of spatial organization must take into account the individuality of each student. The authors advocate the multivariate nature of the formation of school public spaces and agree with the words of M. Sartan, the General Director of Smart School, who says "... about the choice, about forks in the path of this choice, about the educational route, about the individualization of education and that everything bases on public spaces"[8]. These particular tasks the new architecture of school buildings must solve, in order to "participate" in the learning process, thereby becoming completely viable.

References

[1] Gelfond A 2015 *Public buildings and public space Relationship dualism* (Moscow: Russian Academy of Architecture and Construction Science Press vol 2) pp 18–32
[2] Moor M, Rowland J 2006 *Urban Design Futures* (London: Routledge) 216 p
[3] Gehl J 2010 *Cities for People* (Washigton: Island Press) 276 p
[4] Mastalerzh N 2013 Formation of the concept of public space as a structural element of the urban Environment *Russia. J. Architekton: news of universities* **9(43)**
[5] Thomas D 2016 *Placemaking: an Urban Design Methodology* (New York: Routledge) 159 p
[6] Bredikhin I 2011 To the discussion of notion and indication of the definition of a "public place" in administrative legislation (Russia, Chelyabinsk: South Ural State University Press vol **19**) pp 85-89
[7] Design Criteria of Sustainable School 2008 (Scotland: Construction Standarts for School vol **2**) pp 16–71
[8] Pashintceva T 2007 Center of gravity of Khodynskoe field. Experimental school for 1000 students *Russia. J. Architectural Vestnik.* AV02 (2007) 95
[9] Donkina M 2015 Smart School in Irkutsk Protest education conveyer *Russia. J. Ircity!ru* 16.09(2015)
[10] Grigiryeva E Smart School in Irkutsk 2015 *Russia. J. Project Baikal* **44**
[11] Anisimov V 2011 Problems of sustainable development of school buildings architecture *Russia. J. Architekton: news of universities* **2(34)**
[12] Anisimov V 2012 A systematic approach to the design of school buildings *Russia. J. Architekton: news of universities* **2(38)**
[13] Sorrell J 2005 *Design with distinction: the value of good design in higher education* (London: CABE) pp 11–16

IOP Conf. Series: Materials Science and Engineering **481** (2019) 012028 doi:10.1088/1757-899X/481/1/012028

Urban Planning Principals of Sustainable Neo-Industrial Territory Development

M V Pastukhova[1], L G Pastukhova[1]

[1]Ural Federal University, 19 Mira Street, Ekaterinburg 620002, Russia

E-mail: mony.past92@gmail.com

Abstract. Global trends in the innovative technologies' development are systematically entering our lives, introducing sociological and economic changes in the society. As a result of the introduction of innovative technologies, human life inevitably changes, and cities must respond to these changes in order to ensure the sustainable development of cities and society. The general goal of research is identifying the urban-planning principles for the sustainable development of settlement systems and cities, taking into account the neo-industrial scenario of the economic and political development. The research contains the analysis urban development strategies, description of the notion of neo-industrialization, disclosure of the concept of sustainable development, general principles and graphic models of urban development, sustainable development, neo-industrial development and neo-industrial sustainable urban development.

1. Introduction

Neo-industrial revolution, which had been started with the invention of microprocessor, makes whole society change drastically. Job forms modify and jobs distribution shifts.

Human activity qualitatively changes during the neo-industrialization: human devotes more time to learning and inventing automated machines. In turn the automated machines replace human in technical operations of manufacturing industry.

The more automated machines are improved, the more time human devotes to science and creation. Research area absorbs more and more people and science outgrows into the primary industrial force. The market economy is turning into knowledge-based economy.

The neo-industrial evolution of society will inevitably affect life and development of the city. So urban-planning strategies must consider factors of neo-industrial social conversions [1,11].

The general goal of research is identifying the urban-planning principles for the sustainable development of settlement systems and cities, considering the neo-industrial scenario of the economic and political development. Forming new urban planning principles of neo-industrial sustainable development will help to reveal the potential and form urban development strategy.

The principals will allow to form the criteria to estimate existing strategies relevance and to reveal the potential of neo-industrial territory development.

2. Methodology

Object of the article: urban planning principals of sustainable neo-industrial territory development.

IOP Conf. Series: Materials Science and Engineering **481** (2019) 012028 doi:10.1088/1757-899X/481/1/012028

Research methods (figure 1): Analysis of a complex notion allows to reveal several simple notions, that the complex notion consists of. Learning experience and a content analysis allows to reveal specific features: goals, priorities and principles. Generalization and synthesis of revealed principals allows to form new principles classify them.

Research methodology is illustrated as the methodology model (Figure 1).

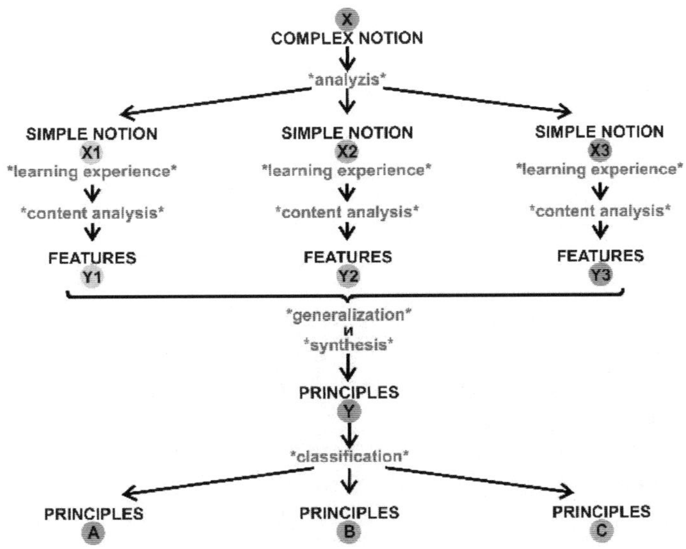

Figure 1. Methodology model.

The methodology model is a sample for the research model process. The research model (Figure 2) illustrates the process of the research

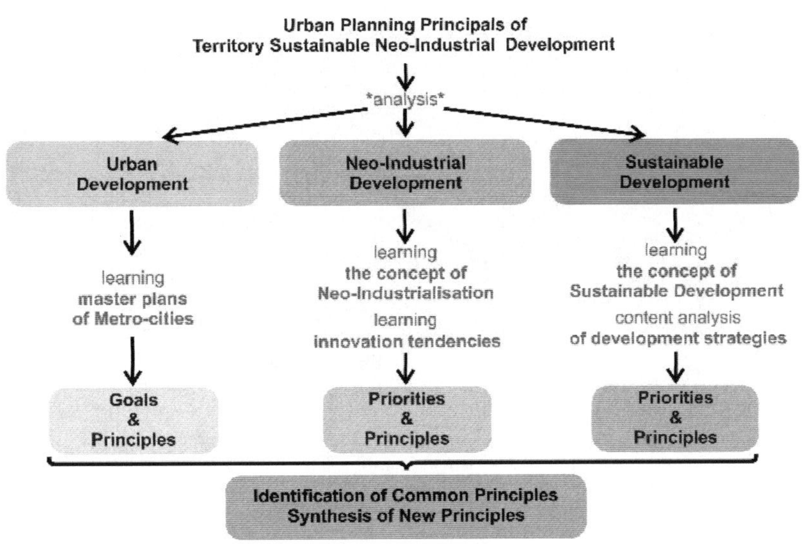

Figure 2. Research model.

3. Argumentation

The object of the research is a complex notion, that consists of simple notions: urban development, neo-industrial development and sustainable development. To define urban planning principals of sustainable neo-industrial territory development it is necessary to analyze urban development strategies, describe the notion of neo-industrialization, disclose the concept of sustainable development and to define general principles and graphic models of urban development, sustainable development, neo-industrial development.

3.1 Concept of sustainable urban development.

The concept of sustainable urban development is defined in the Un-Habitat "New agenda" [3].

New Urban agenda: making cities inclusive, safe, resilient and sustainable.

Description of the priorities of sustainable urban development according to the Un-Habitat "New agenda" is provided below [3]. The model of sustainable urban development is illustrated in Figure 3.

- Governance Structures: The NUA is anchored in participatory urban policies that mainstream sustainable urban and territorial development as part of integrated development strategies and plans, supported by institutional and regulatory frameworks linked to transparent and accountable finance mechanisms.
- Social Inclusion: Development must protect the planet and enable all inhabitants, whether living in formal or informal settlements, to lead decent, dignified, and rewarding lives, and to achieve their full human potential.
- Spatial Development: Balanced territorial development that accounts for different scales of cities and human settlements, strengthens their role in food security and nutrition systems, puts housing at the centre, builds infrastructure and services, facilitates trade, and connects farmers and fishers across value chains and markets
- Urban Prosperity: Inclusive and sustainable economic growth, with full and productive employment and decent work for all, is a key element of sustainable urban development where can people live healthy, productive, prosperous, and fulfilling lives.
- Environmentally Sustainable: Unsustainable consumption and production patterns, loss of biodiversity, pressure on ecosystems, pollution, natural and man-made disasters, and climate change and its related risks, undermine efforts to end poverty and to achieve sustainable development.

Figure 3. Model of sustainable urban development.

IOP Conf. Series: Materials Science and Engineering **481** (2019) 012028 doi:10.1088/1757-899X/481/1/012028

To define principles of urban development the review and content analysis of metro-cities spatial development strategies was held (Table 1).

Table 1. The result of content analysis.

No	Strategy title	Planning horizon	Scope of planning	Goals and principles defined in the strategy
1	Landesenwicklungs-plan Berlin-Brandenburg [4,5]	2008 - 2018	- spatial; - economical; - social; - transport; - ecological.	· creation of complex system of central areas; · development opportunity equality; · spatial interconnection of population growth and infrastructure development; · natural landscape protection; · transport infrastructure development according to pendulum migrations and international traffic flows; · restriction of building development on spare territories; · containment of the extensive development of small cities; · formation of a holistic ecological framework of the region.
2	Metro Vancouver 2040 Regional Growth Strategy [4,6]	2013 - 2040	- spatial; - economical; - social; - transport; - ecological.	· dense urban environment; · sustainable economics; · environment protection; · formation of integral areas; · sustainable transport system; · concentration of urban growth within the selected areas; · decentralization; · combination of densely populated urban areas and extensive natural landscapes; · prevalence of port logistics, industry and agriculture; · ethnic and cultural diversity as a recognition basis.
3	Copenhagen Finger Plan 2013 [4,7]	2013 - 2032	- spatial; - economical; - social; - transport; - ecological.	· maintaining competitiveness at the global level · dense of urbanized areas · limitation of new construction areas · priority of public transport and cycling · preservation of the natural landscape · maintaining the quality of the urban environment
4	The London Plan: Spatial Development Strategy for Greater London [4,8]	2014 - 2050	- economical; - social; - transport; - ecological.	· successful solvation of the problems of economic growth and population; · increasing competitiveness in the international arena; · variety; · development and accessibility of areas;

No	Strategy title	Planning horizon	Scope of planning	Goals and principles defined in the strategy
				· establishing leadership for environmental improvement; · a city in which everyone has easy, safe and quick access to jobs and services.
5	Plan Melbourne: Metropolitan Planning Strategy [4,9]	2011 - 2031	- spatial; - economical; - social; - transport; - ecological.	· high quality of life with increasing population; · limiting the growth of built-up area; · natural landscape protection; · promoting state growth and development; · efficient integration of the transport system.
6	CMAP, Go to 2040: Comprehensive Regional Plan (Chicago) [4,10]	2010 - 2040	- spatial; - economical; - social; - transport; - ecological; - financial.	· investment in the development of built-up areas; · maximum functional variety; · transit-oriented development; · informational clarity; · investments in public transport and the development of a multimodal transport system; · realistic plans and their binding to budget financing indicators; · region development as a center of the global economy and attraction of qualified personnel/

Defined goals and principles were generalized and, as a result, the key urban-planning principles were formed and illustrated as a model in Figure 4.

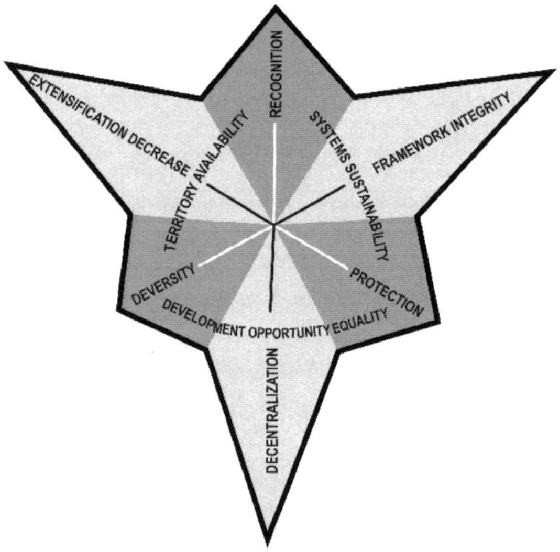

Figure 4. The model of urban development.

- Decentralisation;
- Development opportunity equality;
- Protection (of history and landscapes);
- Advance and maintenance cityscape quality;
- Extensification decrease, compact city;
- Framework integrety;
- Systems sustainability;
- Deversity;
- Area recognition;
- Territory availability.

3.2. Concept of Neo-Industrial development

To form the most relevant principles of territory development, the content analysis of innovation worldwide tendencies, that could affect the city development, was held: automatization, robotization, remote access, vertical farms, alternative energy.

The result of content analysis is introduced in Table 2.

Table 2. Content analysis of innovation worldwide tendencies.

No	Ten-dency	Loca-tion	Description	Result
1	Automa-tization	China 2017	Despite low labor price, the head of Changying Precision Technology Company in Dongguan, China, replaced 90% of labors with robots and automated systems [12].	- 250% production increase; - 80% defects reduce; - proceeds increase; - unemployment increase; - social protests emerge.
2	Robo-tization	Japan 2017	In 2015, ad agency McCann Japan's creative planner Shun Matsuzaka set himself a task he called the "creative genome project": he wanted to create the world's first AI creative director, capable of directing a TV commercial. The client was asked to fill out a form with all the elements they wanted to appear in the ad. The AI robot then scrambled the database for ideas (humans were required to actually produce the final creative). The two spots would then be thrown to a nationwide poll, where consumers could vote for which ad they preferred [13].	- 54% of public vote for the human created ad; - the most of executives chose the AI ad; - ad agency can use AI algorithms to create ads; - it appears, that not only labor can be replaced by robots, but also some of creative professions.
3	Robo-tization	USA 2017	BlackRock laid out an ambitious plan to consolidate a large number of actively managed mutual funds with peers that rely more on algorithms and models to pick stocks. As part of the restructuring, seven of BlackRock's 53 stock pickers are expected to step down from their funds. Several of the money managers will stay on as advisers. At least 36 employees connected to the funds are leaving the firm [14,16].	16% of total revenue represents machine-driven passive investing
4	Robo-tization	Germany 2017	In their study, researchers of the School of Business and Economics have shown that algorithms based on artificial intelligence are able to make profitable investment decisions [15].	Profitability decreased and even became negative at times. This decline was driven by the rising influence of artificial intelligence in modern trading - enabled by increasing computing power as well as by the popularization of machine learning.

No	Ten-dency	Loca-tion	Description	Result
5	Remote access	USA 2017	Automattic, the company behind the popular WordPress blogging website, is making the unusual decision to get rid of its San Francisco office. Its employees in the Bay Area will simply work full time from their homes, coffee shops, or wherever else they choose. Of the company's roughly 550+ employees, only about 30 of them live in the Bay Area. The rest of them are scattered across more than 50 countries [17].	Its employees in the Bay Area will simply work full time from their homes, coffee shops, or wherever else they choose. It gives employees a generous stipend to set up their home offices, meetings are held online, everything can be done via chat rooms, and the company pays for teams to travel anywhere in the world for when in-person time is needed
6	Remote access	Russia 2016	Russian company Beeline started the project called "BeeFree". According to the concept of the project 50-70% of employees will work remotely 1-5 days a week, that means all employees, except for those who work with special equipment [18].	- overhead decrease is expected.
7	Vertical farms	China 2017	The complex of vertical farms is now building in Shanghai. It allows to grow 350 times more production and consume 100 times less water than fields or greenhouses.	- agricultural territories reduce; - water consumption is reduced.
8	Vertical farms	USA 2017	Aerofarms in New Jersey (area 14 164 sq. m) produces more than 900 tons of plants a year. It is comparable to a simple farm with 140 000 sq. m area.	- agricultural territories reduce; - water consumption is reduced.
9	Vertical farms	France 2017	Vertical farm in the centre of Paris allows to produce 7 tons of berries a year, having area only 24 sq. m.	- agricultural territories reduce; - water consumption is reduced.
12	Alter energy	Germany 2017	Today in Germany 55% of consumed energy is produced by NPP. By 2022 it is expected to reduce consummation of NPP energy to 0%, going over to wind-generated energy.	In 2016 during lingering calm, the energy was provided to Germany from Switzerland.

As a result, the priorities can be defined and should be considered. Defined priorities and learning the notion of "neo-industrialization" [1,2] gives us an opportunity to define principles of neo-industrial development.

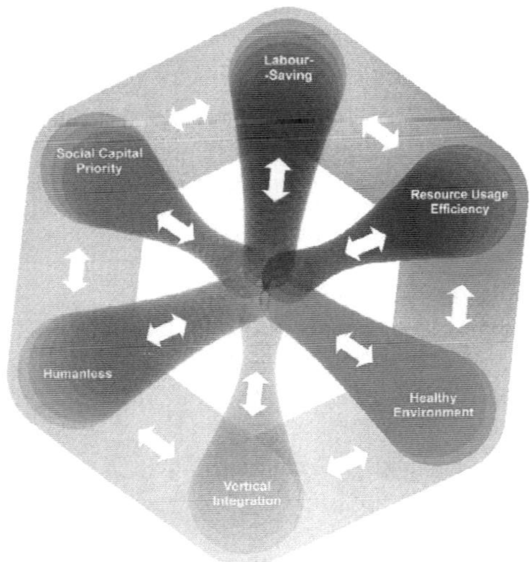

Figure 5. The model of neo-industrial development.

- Labour-saving;
- Vertical integration;
- Humanless;
- Resource usage efficiency;
- Healthy environment;
- Priority of social fund (capital).

4. Urban planning principals of sustainable neo-industrial territory development

As a result of generalization of different concepts principles, the general principles are revealed. Principles that cannot be generalized are synthesized as specific principles.

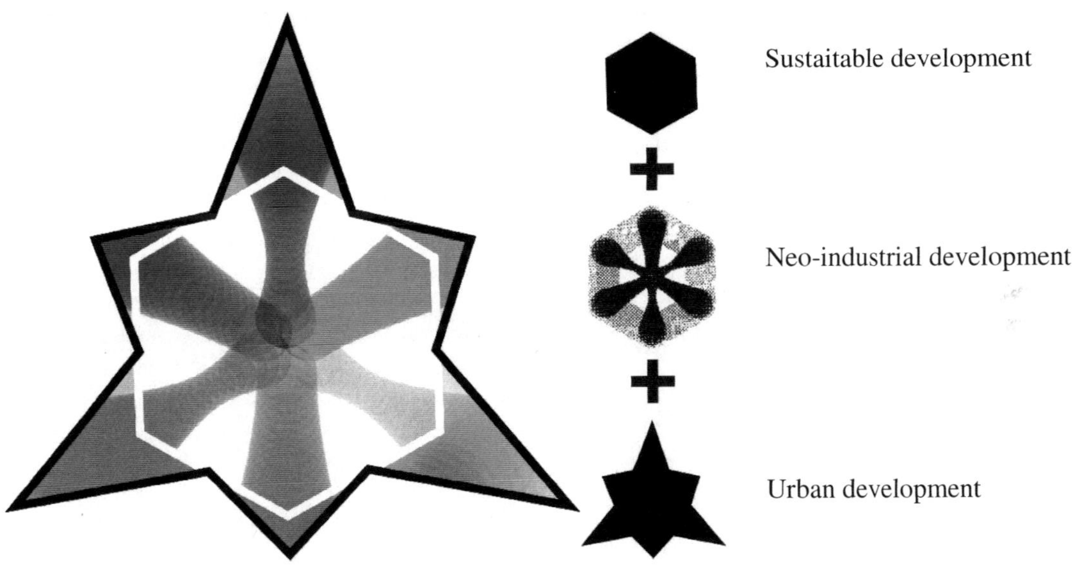

Figure 6. Model-symbol of Neo-industrial sustainable urban development.

Balanced complex development: Every element and every structure need to be developed. Development of one element must not threat the other elements development.

Priority of social fund: The innovation or the process can be integrated if it will not lead to social fund costs.

Areal independence and outlaying availability: People are free to choose place of residence and place of employment. Territorial availability of all places of employment and residential area. Integration of remote access concept in residential area development. Different areas can be connected with several different infrastructures.

Robotization as means of ecological and social security: it is necessary to define spectrum of processes that can be atomized with no threat for society and environment

Structural and processual territory integration: Maximum of safe processes must be integrated into the city contexture considering protection of existing structure and structure development. Integration can be reached by using innovative technologies.

Territorial clusters principle: Territorial clusters as basis for infrastructure, city and agglomeration development

Density and decompression balance: Urbanization and deurbanization balance, compact city concept, localization of anthropogenic affection, priority of intensive development.

Balanced vertical and horizontal integration: Scenarios of city development can be based on the concept of vertical integration of a company. If there is a strong company-developer, that is based in a city, the government must support the vertical integration of the company, and the company must develop city infrastructure by horizontal integration.

Diversity: Diversity of residential types, of spaces, of transport, of connections etc.

Hierarchy of functional connectors: Different areas can be connected in different ways: transport, energy, information etc. Every element of a system must have a general connection and accessory connections.

Originality: Originality and identity – the most important aspects when the concept of the area development is defined.

Joint innovation development: It is important that neighbored territory elements develop together, trying to consider interests of each other.

Resource usage efficiency: Efficiency usage of territory, people, financial, energetic, environment etc

Healthy robotization: Robotization is applied to reach people and environment security. Application of robots must be under control of government and must consider interests of citizens.

Decentralized hierarchy: Appointing several centers of development and defining several levels of city elements.

Framework adaptability: Any system must be based on existing elements. Liquidation of existing elements is allowable only if the element threats environment, people or the whole system.

Forming new urban planning principles of neo-industrial sustainable development will help to reveal the potential and form urban development strategy.

The principals will allow to form the criteria to estimate existing strategies relevance and to reveal the potential of neo-industrial territory development.

The principles, defined in this research need to be tested on different territory levels: agglomeration, city, district.

References

[1] Gubanov S and Musin M 2013 New Industrialization (Electronic jurnal Sverhnovaya realnost vol.**6)** (Moscow: Znanie – Vlast!) pp 20-27

[2] Pereslegin S. 2017, Jan. Eto ne prosto bezrabotitsa, a lisheniye chelovechestva printsipial'nogo smysla sushchestvovaniya https://www.znak.com/ [Electronic resource], https://www.znak.com/2017-01-09/znamenityy_futurolog_nyne_zhivuchie_pogibnut_v_adskoy_voyne_lyudey_i_kiborgov, (date of the application: 10.03.2017)

[3] The New Urban Agenda, Un Habitat III, 2016 http://habitat3.org/wp-content/uploads/Habitat-III-Zero-Draft-outcome-document-May-2016.pdf

[4] Muratov A 2014 *Strategicheskiy Masterplan: Instrument Upravleniya Budushchim* [Text]: [Study for the IV Moscow Urban Forum]. (Moscow: KB Strelka) 519 P

[5] März 2009 "Landesentwicklungsplan Berlin-Brandenburg" (LEP B-B), Gemeinsame Landesplanungsabteilung, http://gl.berlin-brandenburg.de/landesplanung/mdb-bb-gl-landesentwicklungsplanung-lep_bb_broschuere.pdf

[6] Lois E Jackson 2017 "Metro Vancouver 2040: Shaping Our Future" Regional Growth Strategy, Adopted by the Greater Vancouver Regional District Board on July 29, 2011, Updated to July 28, 2017 http://www.metrovancouver.org/services/regional-planning/ PlanningPublications/ RGSAdoptedbyGVRDBoard.pdf
[7] Ministry of the Environment of Denmark 2015 "The Finger Plan", A Strategy for the Development of the Greater Copenhagen Area https://danishbusinessauthority.dk/sites/default/files/fp-eng_31_13052015.pdf
[8] Greater London Authority March 2016 "The London Plan", The Spatial Development Strategy for London Consolidated with Alterations Since 2011 https://www.london.gov.uk/sites/default/files/the_london_plan_malp_final_for_web_0606_0.pdf
[9] Richard Wynne 2017 "PLANMELBOURNE 2017-2050", Metropolitan Planning Strategy, The State of Victoria Department of Environment, Land, Water and Planning, http://www.planmelbourne.vic.gov.au/__data/assets/pdf_file/0007/377206/Plan_Melbourne_2017-2050_Strategy_.pdf
[10] "CMAP: Go To 2040", Comprehensive Regional Plan, Chicago Metropolitan Agency for Planning, October 2010 http://www.browardmpo.org/images/WhatWeDo/LRTP/IL-CMAP-LRTP.pdf
[11] Kolyasnikov V A 2016 Yekaterinburgskaya aglomeratsiya: innovatsionnoye razvitiye i zhilishchnoye stroitel'stvo (Chelyabinsk: Izdatel'skiy tsentr YUUrGU)
[12] "The Chinese company replaced 90% of its employees with robots: productivity increased by 250%" (AIN) https://ain.ua/2017/02/05/kompaniya-zamenila-90-sotrudnikov-na-robotov
[13] Lara O'Reilly Mar. 12, 2017 "A Japanese ad agency invented an AI creative director — and ad execs preferred its ad to a human's" (Australia: Business Insider) https://www.businessinsider.com/mccann-japans-ai-creative-director-creates-better-ads-than-a-human-2017-3
[14] Laurence D. Fink March 28 2017 *Landon Thomas Jr.* "At BlackRock, Machines Are Rising Over Managers to Pick Stocks" (New York: New York Times) https://www.nytimes.com/2017/03/28/business/dealbook/blackrock-actively-managed-funds-computer-models.html?mtrref=undefined&_r=1
[15] Mar. 16 2017 *University of Erlangen-Nuremberg* "Deep learning and stock trading" (EurekAlert!) https://www.eurekalert.org/pub_releases/2017-03/uoe-dla031517.php
[16] Mar. 18, 2017 "Symbiotic intelligence learned to make money on the stock exchange" (Hightech) https://hightech.fm/2017/03/18/cindicator
[17] Julie Bort June 12, 2017 " $1 billion startup Automattic is closing its San Francisco office and having everyone work from home" (Australia: Business Insider) https://www.businessinsider.com/automattic-closes-san-francisco-office-to-let-everyone-work-remotely-2017-6
[18] Vereschagina Lena 14 March 2017 " How Beeline sent home thousands of employees" (The Village) https://www.the-village.ru/village/business/story/259078-beefree?utm_source=facebook.com&utm_medium=social&utm_campaign=letom-proshlogo-goda-bilayn-nachal-samyy&utm_content=5624337&fbclid=IwAR1DVAUWIb2T9m8hTdWjNzhjz_UdxCIiEPpoWP5_ocNflv2LUnP58czoDgQ

IOP Conf. Series: Materials Science and Engineering **481** (2019) 012029 doi:10.1088/1757-899X/481/1/012029

Improvement the scheduling based on the reducing of the dimensionality of the system

M G Dobrosotskikh[1],V Ya Mishchenko[2], M A Preobrazhenskii[3]

[1]Department of Technology, Organization of Construction, Examination and Management of the real estate, Voronezh State Technical University, 20-letija Oktjabrja, 84, Voronezh, 394006, Russia

[2]Department of Technology, Organization of Construction, Examination and Management of the real estate, Voronezh State Technical University, 20-letija Oktjabrja, 84, Voronezh, 394006, Russia

[3]Department of Physics, Voronezh State Technical University, 20-letija Oktjabrja, 84, Voronezh, 394006, Russia

E-mail: dobrmax@yandex.ru

Abstract. In this paper a full set of system parameters "building site + resources" has been determined. Methods of discrete description of the closed system dynamics and dynamics of the open system, interacting with external objects, have been formulated. In the four-dimensional phase-space the step function, quantitatively describing the construction production calendar plan, has been defined. It is shown that the stored resources do not determine the duration of construction project implementation, but rather limit it. On the other hand, the non-stock resources determine the possibility of construction project implementation only, when the upper boarder-line of the construction project duration is fixed. The method of reducing the dimension of the phase space by taking into account the spatial and technological relationships has been formulated. It is shown that the problem of reducing the dimension of the proposed algorithm can significantly extend the class of NP-solvable problem of optimizing scheduling and allow visualization of the algorithm to optimize construction production calendar-plan, enabling a significant expansion of the area of its practical application. The possibilities of the use of orthogonal sections of the graphical representation of construction calendar plan in three-dimensional configuration space in strategic and operational management of building production practice, has been demonstrated.

1. Introduction

Development of the methods of construction scheduling has a centuries-old history. However, at the first stages of development, they were presented in the form of empirical set of optima control rules of collective work [1]. Only at the turn of XIX and XX centuries, F. Taylor, in his classic work "Principles of Scientific Management", implemented the scientific formulation of the optimal manufacturing process management. Within this approach, L. Gant had developed the first scientific methods of schedule plan (SP) and formats of its visualization (the so-called Gantt charts/diagrams). From 1930-s, the study of the flow-line method of work organization, which allows avoiding the discrepancies, attributed to the linear and parallel methods of construction work implementation [2] has begun. The flow-line method can be characterized by the following features:

- splitting the scope and processes of work for creation of the most favorable conditions for individual workers, based on their specialization
- the maximum combination of work processes in time

The flow-line method of scheduling provides uniformity of resource consumption and the stable rhythm of production of finished products, creating favorable conditions for work of allied organizations: contractors, factories, and transport companies. However, the use of this method is limited by the need to combine equally rhythmic and multi rhythmic components of flows, which is rarely achievable in practice [3]. The practical use of such method is also limited by instability of the flow-line methods to external stochastic effects [4, 5]. Another significant disadvantage of the flow-line method, which does not allow performing the system optimization, is the iterative nature of the resource constraints accounted. These and other disadvantages of traditional methods of construction scheduling have stimulated the search for alternative methods of calendar planning [6].

New perspectives for SP improvement appeared in connection with the introduction of computer technology, thus a possibility to apply methods of operations research (linear and dynamic programming, multi-criteria optimization, game theory, and others) has arisen [7]. Application of these methods has allowed formulating and implementing a simple computer algorithm of description of SP of the construction project implementation, later called the Critical Path Method (CPM) [8].

Choice of critical paths is key moment of planning in this approach because it defines technological, financial and managerial risks [9]. However, the splitting of the set of operation sequences into critical and non-critical subsets is not uniquely determined [9]. As a consequence several critical paths are possible for one method of project implementation [10]. In addition, for large projects, The CPM analysis leads to NP-unsolvable problem and does not allow to make an unambiguous decision about optimal method of implementation [11].

The development of this approach allowed creating (Program (Project) Evaluation and Review Technique - PERT) [12], which is based on optimization of the process logical scheme. Recording of stochastic external influences, uncertainty and ambiguity of system dynamics is executed with the use of Graphical Evaluation and Review Technique - GERT) [13].

Currently used methods and algorithms of rational ordering in time and space of technological operations in the course of various technical and economic processes are very diverse [14]. These methods have been successfully applied in the planning of works at industrial factories [15]; for optimizing logistics processes [16]; for planning military operations [17]; for implementing state contracts with a large number of participants [18].

In domestic practice, computer technology for solving scheduling problems is used to visualize the solutions obtained by empirical methods as part of the standard software packages [19]. Meanwhile, the improvement of construction scheduling and optimization of construction project are possible with the use of deterministic finite automata [20]. One of the main components defining the deterministic automaton is a finite set of system conditions [21]. This paper is devoted to the formation of methods that can fully describe and visualize the conditions of the system "building site + resources".

2. Description of the dynamics of "building site + resources" system

Description of the dynamics of "building site + resources" system requires specification of sets of the discrete states of individual subsystems and dependence of the components of these sets from the time aspect. Building site can be quantitatively characterized by the following set of elements:

- W– types of work at the entire construction building site
- S – spatial elements (number of rooms, plots of land, etc.), in each of which all construction works are carried out with the workload volume $v_{k,j}$ ($k = 1,2\ldots,S; j = 1,2\ldots,W$)

In the equation (1) is shown summing up the volume of work in all the spatial elements, we get a full scope of work of types following:

$$V_j = \sum_{k=1}^{S} v_{k,j} \tag{1}$$

By this, let's assume that the capacity of non-stock resources and the specific consumption of the stored resource do not depend on the type of room.

For the construction project implementation the R_s types of stored and R_n non-stock types of resources are in need. At the beginning of construction project realization stored resources (financial, material) are described by their total volume R_i $(i = 1,2..., R_s)$. In contrast, the non-stock resources (machines, equipment, personnel) are described not by their quantity, but by the capacity $P_{j,i}$ $(i = 1,2..., R_n;\ j = 1,2...,W)$ for each type of work. For example, working personnel can have different qualifications and skills to perform various technological processes. Moreover, the productivity of machines and mechanisms is determined not only by their technical characteristics, but also by the work performed.

The system dynamics is determined by the time interval T (depending on the detailing level of construction scheduling - hours, shifts, days, weeks, etc., which are considered further to be the same, normalized to a single value). The duration of construction project measured by T, in contrast to the characteristics of the construction building site W, S, and to characteristics of its performers R_s, R_n, is determined dynamically during the planning process itself.

Dynamics of the closed system of the stored resource is described by the non-increasing function, because such resource can only be spent out or remain unused at certain stages of the construction project. In contrast, the non-stock resource may increase (i.e. workers can improve their skills and performance.

Dynamics of resources in the open system depends not only on the system properties, but also on its interaction with external objects. In particular, the financial resources can be replenished in the course of the project. It may also be necessary to lease or purchase equipment or hiring additional working personnel. The reverse processes are also possible (i.e. some machines and equipment are sold out, rented out, or got destroyed, some personnel is dismissed, etc.). The change in volume of stored resources in the T time period can be described by the function of $\Delta R_{j,t}$. For non-stock resource $\Delta P_{j,i,t}$ $(k = 1,2..., R_n;\ i = 1,2...,W)$ describes the change in capacity. These functions, depending on the sign, describe the increase or decrease in j-th power resource or T-th time interval. Therefore, accumulated due to external sources resources and capacity in the time interval $t_0 + 1$ can be define by the following equations:

$$R_{j,t_0+1} = R_j + \sum_{t=1}^{t_0} \Delta R_{j,t_0+1} \tag{2}$$

$$P_{j,i,t_0+1} = P_{j,t} + \sum_{t=1}^{t_0} \Delta P_{j,i,t} \tag{3}$$

For the determination of resources consumption, during the execution of construction works, it is necessary to set specific consumption costs of resources stored per unit of work performed $r_{j,i}$ and specific capacity of use of the non-stock resources $p_{j,i}$.

The task of the calendar planning of construction process consists of allocation of resources on the premises, according to the types of rooms, types of activities, and time intervals, with the fulfilling of condition of nonnegative balance of resources for any t. Quantitatively calendar plan is described by the step function $\delta_{t,k,i,j}$ that takes a value equal to one, if in the t time interval in the k room for the performance of i type of work the j resource is used. In all other cases, the function $\delta_{t,k,i,j}$ takes zero value.

The duration of the project realization T is limited only by the capacity of non-stock resource and determined by the condition of completion of all works in all rooms/premises, with full respect of technological conditions, which leads to the equalities:

$$\sum_{t=1}^{T_{ki}} \sum_{j=1}^{R_s} \delta_{t,k,i,j} P_{j,i,t} = \upsilon_{k,i} \tag{4}$$

$$T = \max(T_{k,i})\,;\, k = 1, 2, \ldots, S;\; i = 1, 2, \ldots, W$$

Equation (4) with any parameters has a formal solution, which might be unsatisfactory in practice according to the implementation duration criterion. Thus non-stock resources determine the feasibility of the project realization, only when upper border of the project duration is fixed.

On the other hand, storable resources do not determine the duration of completion, but only the possibility of its implementation. The total use of the stored resources should not exceed their availability at all times, that is quantitatively described by the condition

$$\sum_{t=1}^{t_0} \sum_{k=1}^{S} \delta_{t,k,i,j} P_{j,i} \leq P_{j,i,t_0} \tag{5}$$

In particular, for the closed system the left side of the equation (5) does not depend on the time value and equals to the original volume of stored resources. In this case, the condition (5) is simplified and takes the form of balance of stored resources during the implementation of the entire project:

$$\sum_{t=1}^{t_0} \sum_{k=1}^{S} \delta_{t,k,i,j} P_{j,i} \leq P_{j,i} \tag{6}$$

3. Visualization of the set of states "building site + resources"

Description of the dynamics of "building site + resources" system requires specification of sets of the four-dimensional character of the calendar-planning problem (room, work, resources, and time) does not allow displaying information about the dynamics and current state of the system in three-dimensional space in full. An even greater loss of information occurs, when the dynamics of the system is reflected on the plane. Therefore, any two-dimensional techniques of visualization of the calendar plan (such as, most commonly used in practice of planning, line diagrams, network diagrams, charts, organizational analysis, and others) inevitably reflect only part of the necessary information and require further explanations. These limitations are in act for any two-dimensional analytical building sites (matrices, tables, etc.).

The reduction of the system dimensionality by taking into account the spatial and technological relationships allows overcoming this contradiction, solving the problem by introducing the concept of planning unit, which combines the spatial and technological information about the building site's state. Let us call the planning unit (hereinafter - PU) the collection of information about the quantity of rooms at the building site premises, about the list of the work and workload volumes that need to be performed for entire project implementation. In such formulation, the problem of calendar-planning becomes three-dimensional and can be described by the step function $\tilde{\delta}_{t,n,j}$ that takes a value equal to one, if in t time interval in the nPU the j resource is used.

Since the capacity of the non-stock resource and the specific consumption of the stored resource do not depend on the type of room in the right sides of equations (4) - (6) , then the collection of indices i and k is simply replaced by an index n, taking values $n = 1, 2, \ldots, U$ in the interval (hereinafter U-

number of planning units building site). Therefore, in such problem formulation the equation, describing the calendar-planning scheduling for the time interval of project implementation with resource constraints take the following form:

$$\sum_{t=1}^{T_n}\sum_{j=1}^{R_S}\tilde{\delta}_{t,n,j}P_{j,i,t} = v_{k,n} \qquad T = \max(T_n)\,; n = 1, 2, \dots, U \tag{7}$$

$$\sum_{t=1}^{t_0}\tilde{\delta}_{t,n,j}P_{j,n} \leq P_{j,n,t_0} \tag{8}$$

$$\sum_{t=1}^{T}\sum_{n=1}^{U}\tilde{\delta}_{t,n,j}P_{j,n} \leq \sum_{n}^{U} P_{j,n} = P_j \tag{9}$$

In these equations P_j-is the total number of j storable resources.

Recording of spatial and technological schedule plan linkages within the PU allows to significantly expand the class of NP-solvable [22] problems of scheduling optimization. This is due to the fact that technologically impossible system states are excluded from consideration at the stage of task setting and do not contribute to a set of states of "building site + resources" system.

The main practical criterion making such problems belong to NP-solvable class is the possibility of creating the exact algorithm, which is not limited to a complete processing of all (or a priori to an undefined part) of the variants of studied system. For the NP-unsolvable problems, it was shown that even an exponential increase in the speed of computers does not significantly expand the class of practically solvable problems.

In addition, the decrease of the problem dimension can not only make the problems of optimization of more complex systems NP-solvable, but also determine the three-dimensional configuration space, the elements of which will fully describe the dynamics of the construction project. This dimension of the configuration space allows visualizing the algorithm of optimization scheduling that enables a significant expansion of the area of its practical application.

Lets define the discrete Cartesian configuration space, in which the x-axis represents the number of non-stock resource (j); the ordinate represents the number of the time interval (t); Z-axis represents the planning unit number PU(n). Thus, the elementary space unit is represented by a cube with the edges of unit length. In such a space, the function $\tilde{\delta}_{t,k,j}$ describes the position of j non-stock resource activated in the t time interval in n-unit of schedule plan. The example of graphical representation of three-dimensional schedule plan is shown in figure 1.

IOP Conf. Series: Materials Science and Engineering **481** (2019) 012029 doi:10.1088/1757-899X/481/1/012029

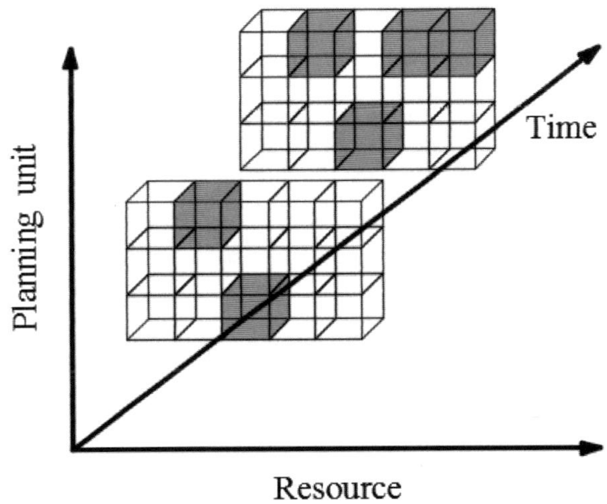

Figure 1. The three-dimensional graphical presentation of the part of schedule plan for the building site, described by the three PU.

The implementation of graphical presentation (figure 1) requires the use of five types of non-stock resources. To aid the visualization, two layers for time periods are shifted along the Y-axis. Highlighted are the elementary cells of configuration space, reflecting the PU, on which the work is done, along with time resources used (from the analytical point of view, the elementary cells of configuration space are those in which $\tilde{\delta}_{t,n,j}$). For example, in the first time interval t_1 only two types of resources are activated for the implementation of two PUs. In the second time interval $t_2$the scope of work has not changed and the same two PU are activated, but construction works are intensified and the number of resource types used has increased up to four.

Two-dimensional cross-section of three-dimensional graphical representation (GP) of schedule with planes perpendicular to the coordinate axes allows to get comfortable in the practice of the construction interface. Firstly, a set of front sections planes i =const (Figure 2) reflect PUs, in which in a fixed time interval t_i the construction work performed and non-stock resources are used. This information is a useful contribution to the practice of operational construction management, as it allows formulating the current parameters for each planning period (shift, day, week, etc.) the flow of vehicles and machinery, manpower and other resources.

The sections of the GP planes perpendicular to the PU axes (n = const) reflect the dynamics of the use of non-stock resource (figure 3). Analysis of this information allows to identify the dynamics of resources use and quickly redirect the non-stock resources (the use intensity of which in the analyzed period of time is small) to the other building sites. On the other hand, the high (close to 100% resource utilization) triggers the risks processes occurrence due to stochastic impact on the system. The analysis of sections with GP planes n = const allows the prediction of such risk processes. In the operational management practice, this information can also help to determine the times of the beginning of use and the complete release of each resource, which in its turn will optimize their use.

The sections with the GP planes perpendicular to the resources axis (j = const), reflect the dynamics of work in all areas of the building site (figure 4). The analysis of this information allows determining the intervals of high intensity work on the premises. This situation (as is by the utilization of high intensity of the recourses use) is unstable with respect to stochastic effects.

To aid the visualization, the figures reflect also non-activated PUs and resources, i.e. those for which the function $\tilde{\delta}_{t,n,j}$ takes a zero value for all non-fixed indexes.

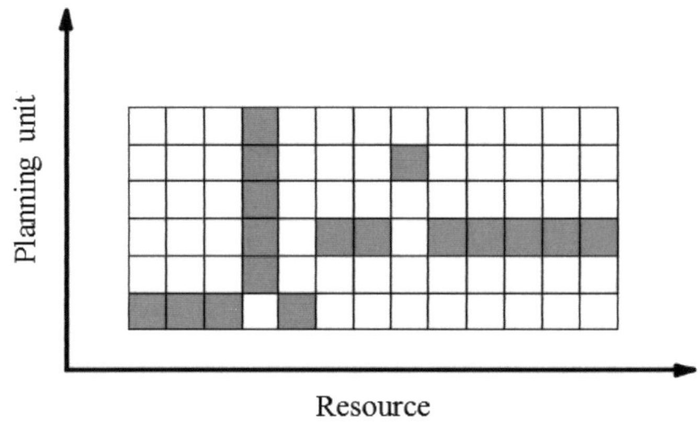

Figure 2. Example of GP (*i* =const) front section for the PU realization in the time period.

On the graphical presentation in figure 2 there are seven units of scheduling are implemented by using thirteen non-stock resources. Highlighted are the resources used by the implementation of PU. On this time axis the non-activated PUs and the resources are not displayed.

The figure 2 shows that the first PU uses the resources 1 - 3 and 5; the resource 4, described in the time period ,is used for the implementation of all PUs except of the first one; implementation of the third PU needs resources 6 - 13, with the exception of 8; resource 8 is only necessary for the implementation of the fifth PU.

In the figure 3 describes that at the first stage of realization only recourses № 1 – 4 are used. Therefore, at this stage, the remaining resources can be promptly forwarded to other construction sites. Resource number 4 is used very intensively and constantly needed. By the stochastic effects on the fourth resource the risk of default work schedule occurs. Therefore, it is appropriate to create the operational reserve of this resource. The same is true for the resource 8, which is not used only in the first scheduling stage. In contrast, the № 5 resource is not used and in the described period of time can be promptly redirected to the other construction sites; wherein the respective column section GP can be omitted. Maximum demand for non-stock resources can be observed in the third time phase, and the minimum - the second, and fourth, fifth and sixth.

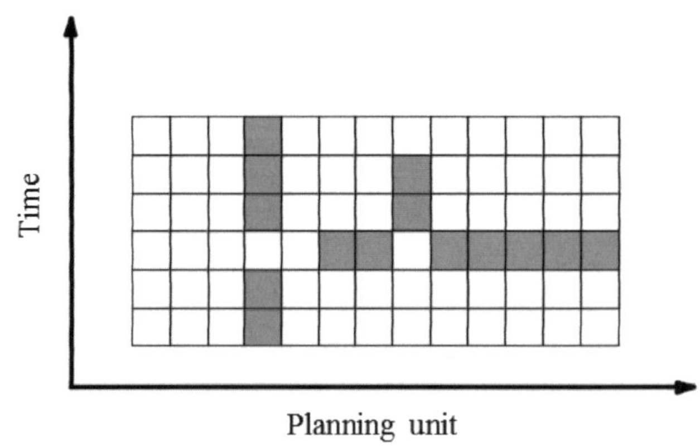

Figure 3. Example of GP section for the PU realization in six time periods, for which twelve resources are necessary.

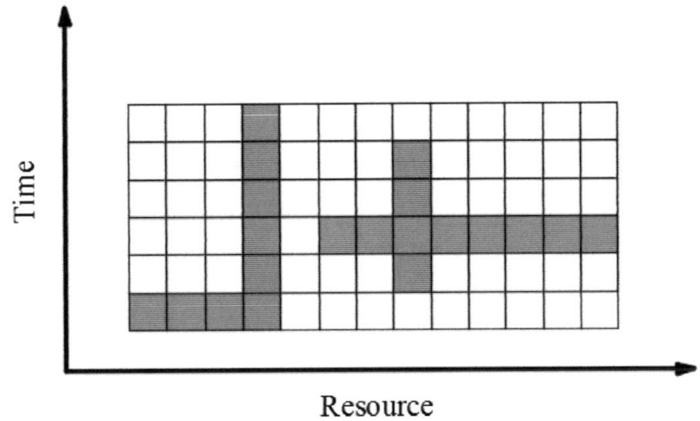

Figure 4. Example of GP section for the resource needed during the six time intervals of schedule planning for the implementation of the nine PUs.

In the figure 4 is shown, that four Planning Units 1 - 3 and 5 are not implemented. In contrast, the number PU 4 is implemented in all intervals except the third one. This situation requires an analysis of the possible risks of stochastic impacts. The fourth period of planning is very busy at this building site, in which the work is carried out at more than half of the PU of construction site. This situation also calls for a special analysis.

4. Conclusion

Thus, algorithms of schedule plan of construction production plan, based on the reducing of the dimensionality of the system described by taking into account the spatial and technological relationships allow us to solve important practical problems. First of all to construct the algorithm to improve NP-solvability of the scheduling problem. In addition, the reduction of dimension makes it possible to determine the three-dimensional configuration space, the elementary cell of which allows to completely lossless display the dynamics of the "building site + resources" system.

Orthogonal section of the configuration space allow us to formulate a method of visualization of schedule plan, which has a number of practical advantages in comparison with applicable previously (Gantt charts, diagrams of the PERT, network diagrams, and so on). It allows to graphically display information such as the dynamics of the project, including the information about the state of the building site and the use of resources at fixed intervals. In practice, such planning visualization allows us to formulate the current planning period for each flow of vehicles, machinery, manpower, and other non-stock resources. In addition the method makes it possible to predict the risk in relation to stochastic effects of the situation and plan preventive measures for their abolishment.

References

[1] Kustova T N, Kamakin O B 2001 *Economy history: Education guidance* (Rybinsk: RGATA)
[2] Dikman L G 2012 *The Construction Management* (Moscow: ASV Publishing House)
[3] Vasilyev V M 2001 *Management in construction* (Moscow: ASV Publishing House)
[4] Afanasyev A V 1982 *Spasmodic streams with continuous performance of peer works (in Improvement of the organization and management of construction)* (Leningrad: LISI) pp 13-22
[5] Gusakov A A 1994 *Organizational and engineering reliability of construction* (Moscow: SvR - Argus Publ)

[6] Larichev O, Sternin M 1992 *Knowledge-based approach for solving the multicriteria assignment problem* ed. Linster M (Arbeits papiere der GMD 630 Sisyphus Models of problem solving)

[7] Uskov V V 2016 *Innovations in construction: organization and management* (Moscow: Infra-Inzheneriya)

[8] Krüger, Wilfried 2006 Excellence in Change - Wege zur strategischen Erneuerung 3.Auflage, (Wiesbaden: Gabler Verlag) pp 212-213

[9] O'Brien J J 1965 *CPM in Construction Management and Scheduling by the Critical Path Method* (New York: McGraw-Hill)

[10] Shaffer L R, Ritter J B, Meyer W L 1965 *The Critical Path Method* (New York: McGraw-Hill)

[11] Smith S F, Fox M S, Ow P S 1986 Constructing and Maintaining Detailed Production Plans: Investigations into the Development of Knowledge-Based Factory Scheduling Systems *AI Magazine* vol 7 pp 45–61

[12] Project Management Institute 2013 A Guide to the Project Management Body of Knowledge 5th ed. Project Management Institute

[13] MacCrimmon K R, Ryavec C A 1964 An Analytical Study of the PERT Assumption vol **12** no 1 (in Opt. Res) pp 16–38

[14] Pinedo M 1995 *Scheduling: Theory, Algorithms and Systems* (New Jersey: Prentice Hall Englewood Cliffs)

[15] Muth J F, Thompson G L 1963 *Industrial Scheduling (*New York: Wiley) 577 p

[16] Meyer W 1992 *Geometrische Methoden zur L¨osung von Job-Shop Problemen und deren Verallgemeinerungen* PhD thesis (Universit¨at Osnabr¨uck Fachbereich Mathematik/Informatik)

[17] Baker R 1974 *Introduction to Sequencing and Scheduling (New York:* JohnWiley & Sons)

[18] Lawler E L 1982 Preemptive scheduling of precedence-constrained jobs on parallel machines ed. Dempster M A H, Lenstra J H, Rinnooy Kan A H G, Deterministic and stochastic scheduling (Proceedings of the NATO Advanced Study and Research Institute on Theoretical Approaches to Scheduling Problems held in Durham, July 6–17, 1981) vol 84 of NATO Advanced Study Institute Series C: Mathematical and Physical Sciences, Dordrecht (D. Reidel Publishing Co.) pp 101– 123

[19] Uskov V V 2011 *Computer Technologies in the Preparation and Management of Construction Projects* (Vologda: Infra-Inzheneriya Publ)

[20] John E Hopcroft, Rajeev Motwani, Jeffrey D Ullman 2006 *Introduction to Automata Theory, Languages, and Computation* 3 ed. (Massachusetts: Addison Wesley)

[21] Jacques Sakarovitch 2009 *Elements of automata theor*y (Cambridge: Cambridge University Press)

[22] Mishchenko V YA, Dobrosotskikh M G 2016 NP Solvable task of scheduling of construction, reconstruction and repair of objects *Proceedings of higher education institutions. Textile industry technology* vol. ***6*(366)** (Ivanovo: IVSPU) pp 13-20

IOP Conf. Series: Materials Science and Engineering **481** (2019) 012030 doi:10.1088/1757-899X/481/1/012030

Numerical simulation of thermal bridges in enclosure structures as an express-method for the expert assessment of thermal comfort in domestic premises

N D Korsun[1], D A Prostakishina[1], A S Novikova[1]

[1] Department of Building structures, Industrial University of Tyumen, 625000, Volodarsky str., 38, Tyumen, Russian Federation

E-mail: korsun1@ya.ru

Abstract. The paper presents the solution to the problem of numerical simulation of thermal comfort in domestic premises. Three thermal bridges have been selected as the key parameters for enclosure structures. They have been calculated on the basis of thermographic studies of a group of typical brick blocks of flats which are located in the city of Tyumen. The considered thermal bridges are as follows: the insulation connections in the outer wall openings; insulation in the outer wall corners; insulation in the outer wall-floor structure connections. The functional connections of the thermal comfort level versus the values of the thermal bridge parameters have been determined. The study has resulted in an express-method of the expert assessment of thermal comfort in domestic premises based on the total assessment of thermal bridges in enclosure structures. The method can be used for a complete structural survey of residential buildings, and in examination of building operation and heat-protective properties of enclosure structures.

1. Introduction

The indoor climate has a direct effect on humans and their life and represents the indoor climatic conditions embracing temperature, humidity and air velocity [1]. Observance of the necessary measures focused on comfortable indoor climatic conditions is one of the main tasks when designing and constructing residential buildings. Equally, reduction of energy consumption is currently one of the priorities for national development, and it is directly related to improvement of energy efficiency of buildings and their thermal protection.

2. Thematic justification

Vlasov O.E. [2], Bogoslovsky V.N. [3], Ilinsky V.M. [4], et.al paid attention to thermal engineering for construction and thermal capacity of buildings. They developed fundamentals of the methodology of thermal engineering for construction.

At present, energy efficiency of buildings and thermal comfort at minimum energy costs dominate [5, 6, 7, 8, 9, 10, 11, 12, 13]. The authors [6, 14] consider models of thermal comfort in homogeneous and heterogeneous thermal media, their advantages, limitations and range of application. The influence of air permeability of enclosure structures on energy efficiency of buildings is studied in [7, 8, 13, 15, 16]. The study is based on the results of surveys of residential buildings erected in Canada and Sweden over the last decades. The effect of the thermal protection level of a building on the amount of thermal energy losses is analyzed in [17]; operation and fuel-energy costs for a heating period of a block of flats are determined for a period of 5, 10 and 50 years of operation. The study on the influence of indoor air humidity on energy consumption of buildings, durability of building structures and human comfort, and a numerical method describing the moisture processes on the basis of experimental data are presented in [18].

Surveys of typical brick blocks of flats were conducted by experts of the Industrial University of Tyumen in the city of Tyumen from 2011 to 2017. They revealed a number of problems concerning thermal comfort of domestic premises in winter related to errors in design, construction and operation [19]. Based on the results of construction and technical expertise, when heat losses were measured by the thermal imaging control [20] using the FlirT335 thermal imager, conclusions were drawn on the most common defects related to the increased heat losses through the enclosure structures:

- defected insulation of connections in the outer wall openings as a result of poor air tightness of insulating glass units and poor junctions between window or door balcony blocks and the outer wall openings;
- defected insulation in the outer wall corners;
- defected insulation in the outer wall-floor structure connections; as a rule, this is due to the lack of thermal insulation in the end floor slab;
- empty vertical and horizontal mortar joints in brick outer walls; heterogeneous inclusions in outer walls.

These defects cause infiltration of outdoor air through the connections of outer and inner walls, connections of outer walls and floor structures or outer wall openings. This leads to a dramatic decrease of the thermal capacity of outer walls and condensate, fungi or mold formation, and hence, thermal comfort is disturbed.

3. Problem formulation and analysis

The revealed defects made it possible to formulate the problem in order to find the functional connections between the divergence of the parameters of thermal bridges in enclosure structures and the level of thermal comfort in domestic premises [21] on the basis of numerical simulation of heat flow processes in order to determine the acceptable limits.

The problem of numerical simulation consisted in a series of numerical experiments for each considered connection of the enclosure structure including thermal engineering irregularities which resulted in the values of the explanatory variable x_j $(j=\overline{1,\kappa})$ and the response of the system y_i $(i=\overline{1,n})$, where k is the number of factors, n – the number of experiments. Next, the functional connections and optimal assessments of the parameters b_0, b_j, b_{jl}, b_{jj} were determined by the regression analysis.

Thermal bridges in the connections of enclosure structures were considered as the objects of simulation: insulation of connections in the outer wall openings (Fig. 1); insulation in the outer wall corners (Fig. 2); insulation in the outer wall-floor structure connection (Fig. 3). The actual structural diagram and structural performance were taken into account in simulation of every connection.

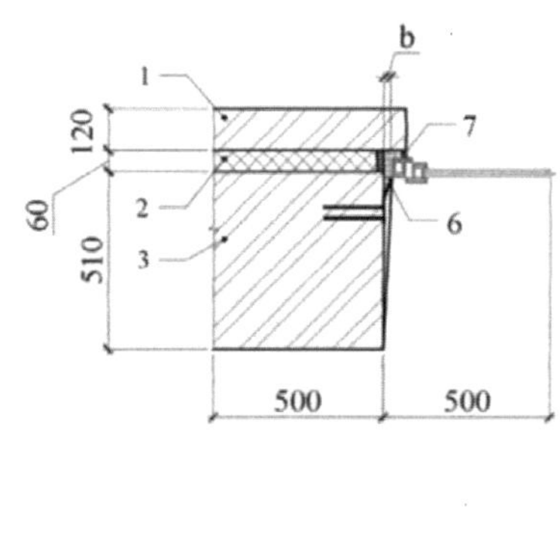

Figure 1 Connection in the outer wall opening: 1 – face layer, 2 – thermal insulation, 3 – bearing layer, 6 – assembly foam, 7 – insulation tape.

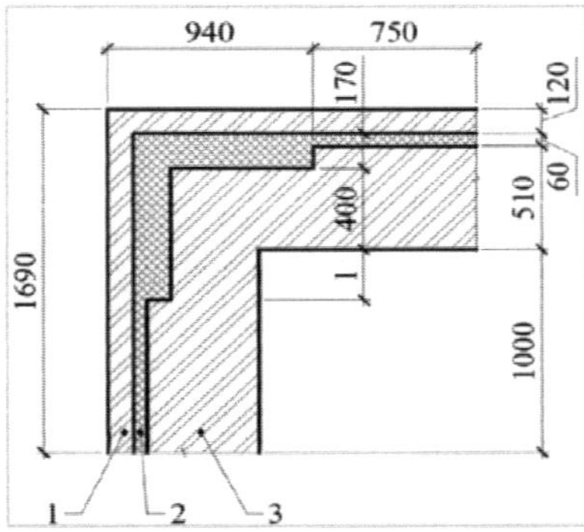

Figure 2 Outer walls connection: 1 – face layer, 2 – thermal insulation, 3 – bearing layer.

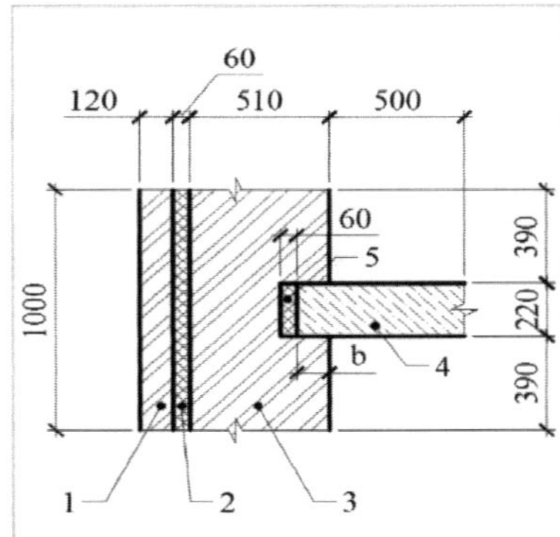

Figure 3 Outer wall – floor structure connection: 1 – face layer, 2 – thermal insulation, 3 – bearing layer, 4 – floor slab, 5 – thermal insulation in the end floor slab.

The temperature on the inner surface of the enclosure structure in the joint zone $\tau_{в}$,°C was taken as a target parameter in the whole of experiments. The level of thermal comfort was assessed in the following ranges:

- optimal level at $\tau_{в} > 16$°C;
- permissible level at 16°C $\leq \tau_{в} <$ 14°C;
- unacceptable level at 14°C $\leq \tau_{в} <$ 10,6°C;
- critical level at $\tau_{в} \leq$ 10,6°C.

The thermal physical properties of materials and geometric dimensions of the structural sections joined in the connections were taken as the variable parameters of the experiment; the range of the variable parameters was taken from the constructive considerations in accordance with the regulations and norms.

The considered variable parameters for the connection in the outer wall opening: the coefficient of thermal conductivity of the assembly foam x_1, $W{\cdot}(\text{m °C})^{-1}$ (in availability), $x_1 \in \{0;\ 0{,}7\}$; layer thickness of the assembly foam x_2, mm (in availability), $x_2 \in \{0;\ 60\}$ and the coefficient of thermal conductivity of the insulating self-expanding tape or mastic compound x_3, $W{\cdot}(\text{m °C})^{-1}$ (in availability), $x_3 \in \{0;\ 0{,}058\}$. The significance of the factor model parameters was determined by means of rough estimates using the Plackett-Berman method [22].

The considered variable parameters for the outer walls connection: the coefficient of thermal conductivity of the wall insulator, x_1, $W{\cdot}(\text{m °C})^{-1}$, $x_1 \in \{0{,}031;\ 0{,}110\}$; the length of the extra thermal insulation x_2, mm (in availability), $x_2 \in \{0;\ 500\}$ and the coefficient of thermal conductivity of the wall bearing layer, x_3, $W{\cdot}(\text{m °C})^{-1}$, $x_3 \in \{0{,}20;\ 0{,}81\}$.

The considered variable parameters for the outer wall - floor structure connection: the coefficient of thermal conductivity of the wall insulator, x_1, $W{\cdot}(\text{m °C})^{-1}$, $x_1 \in \{0{,}031;\ 0{,}110\}$; the depth of the floor slab support, x_2, mm, $x_2 \in \{120;\ 450\}$ and the coefficient of thermal conductivity of the insulator in the end floor slab, x_3, $W{\cdot}(\text{m °C})^{-1}$ (in availability), $x_3 \in \{0;\ 0{,}062\}$.

In order to describe the model, a second-degree polynomial-type function was taken, where the regression coefficients $\beta_0, \beta_j, \beta_{jl}, \beta_{jj}$ were replaced by the estimated coefficients b_0, b_j, b_{jl}, b_{jj}:

$$y = b_0 + \sum_{j=1}^{\kappa} b_j x_j + \sum_{j=1}^{\kappa-1} \sum_{l=j+1}^{\kappa} b_{jl} x_j x_l + \sum_{j=1}^{\kappa} b_{jj} x_j^2 \,. \tag{1}$$

In order to solve this problem, an active experiment was conducted using methods of the theory of planning of extreme experiments and saturated continuous D-optimal plans [22]. The ELCUT 6.1 software program was used to implement the experimental results [23].

4. Results

The experimental studies resulted in graphic presentations of the temperature fields with isotherms and heat flow vectors (Fig. 4, 5, 6), temperature values on the inner surface of the outer enclosure structure in the connection obtained for each experiment, and analytical dependencies of the objective function.

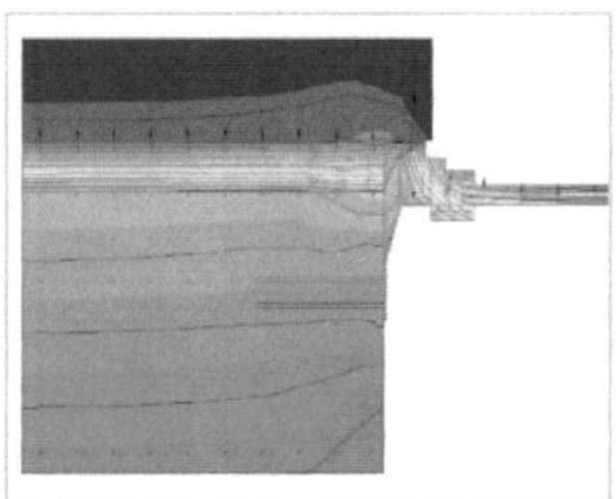

Figure 4 Temperature field of the connection in the outer wall opening.

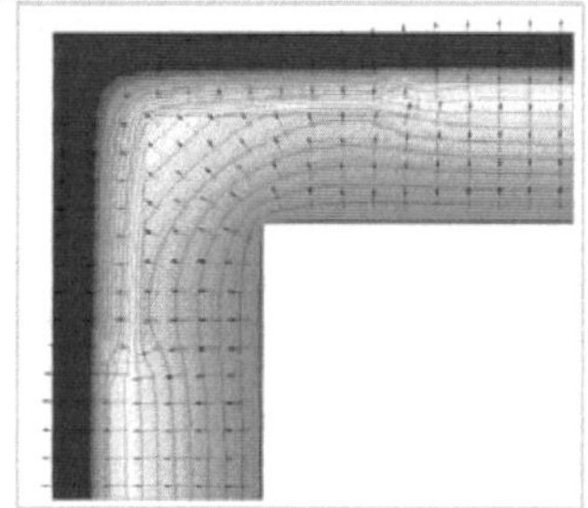

Figure 5 Temperature field of the outer walls connection.

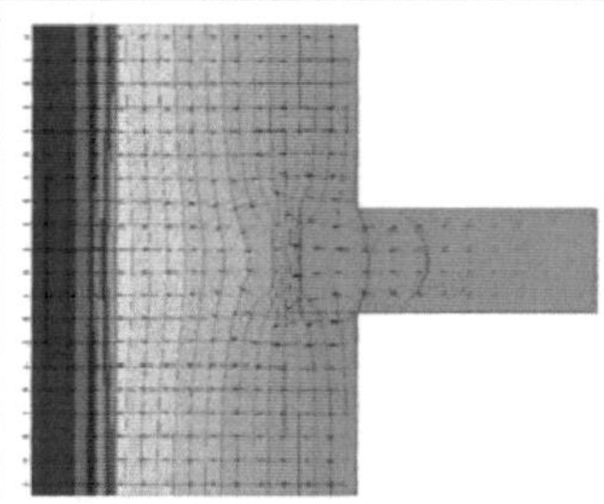

Figure 6 Temperature field of the outer wall - floor structure connection.

The analytical dependencies obtained are as follows:

- for the connection in the outer wall opening

$$y = 12,58 + 1,59 \cdot X_1 + 3,26 \cdot X_2 + 0,73 \cdot X_3; \quad (2)$$

- for the outer walls connection

$$y = 14,45 + 1,49 \cdot X_1 + 1,45 \cdot X_2 + 0,84 \cdot X_3; \quad (3)$$

- for the outer wall - floor structure connection

$$y = 15,12 + 3,70 \cdot X_1 + 0,94 \cdot X_2 + 0,93 \cdot X_3. \quad (4)$$

The analysis of the data obtained after the experimental plans (from 13 experiments of the saturated D-optimal plan for each connection) and the regression equations made it possible to draw the following conclusions:

1) for the connection in the outer wall opening, the whole of the studied parameters are significant; the best options for optimization of the connection are as follows: errors of closure insulated with the assembly foam with the coefficient of thermal conductivity to be 0.035 $W \cdot (m\ °C)^{-1}$, minimum error of closure - 20 mm and availability of the insulating self-expanding tape or mastic compound with the coefficient of thermal conductivity to be 0.029 $W \cdot (m\ °C)^{-1}$;

2) for the outer walls connection, the key parameters are as follows: the coefficient of thermal conductivity of the wall insulator (x_1, $W \cdot (m\ °C)^{-1}$) and the length of the extra thermal insulation (x_2, mm); the best options for optimization of the connection are as follows: thermal insulation of the outer wall with the coefficient of thermal conductivity to be 0.031 $W \cdot (m\ °C)^{-1}$, the length of the extra thermal insulation min. 500 mm and efficient brickwork with the coefficient of thermal conductivity to be 0,2 $W \cdot (m\ °C)^{-1}$;

3) for the outer wall - floor structure connection, the coefficient of thermal conductivity of the wall insulator (x_1, $W \cdot (m\ °C)^{-1}$) is the most significant parameter; the best options for optimization of the connection are as follows: thermal insulation of the outer wall with the coefficient of thermal conductivity to be 0.031 $W \cdot (m\ °C)^{-1}$, the depth of the floor slab support max 120 mm (min depth in

terms of design requirements), and availability of the insulator in the slab end with the coefficient of thermal conductivity to be 0.031 $W \cdot (m\ °C)^{-1}$.

5. Conclusions

Thus, it can be concluded that temperature conditions in domestic premises are the key factor for the winter period; in here, the key parameters are air temperature in the rooms and surface temperature of the enclosure structure in the zone of thermally-conductive inclusions. The challenges concerning thermal protection of buildings include: insulation in the outer wall - floor structure connection, insulation in the corners of the outer walls, insulation of connections in the outer wall openings.

A series of numerical experiments was performed for the thermal bridges which resulted in the obtained values of the target parameters and the given factors of the temperature conditions of the connections. Models of the selected indoor climate parameters were constructed using the regression analysis.

Based on the models built, the significant parameters influencing the indoor climate were determined:

- for the outer wall - floor structure connection - the coefficient of thermal conductivity of the wall insulator (x_1, $W \cdot (m\ °C)^{-1}$);
- for the outer walls connection - the coefficient of thermal conductivity of the wall insulator (x_1, $W \cdot (m\ °C)^{-1}$) and the length of the thermal insulation section (x_2, mm);
- for the connection in the outer wall opening - layer thickness of the assembly foam and the size of the error closure (x_2, mm).

The data obtained can be used to simplify the work of specialists and experts in a complete structural survey of residential buildings or in examination of building operation and heat-protective properties of enclosure structures in individual domestic premises, in particular, in the city of Tyumen. Since the regression equations are based on calculations of the temperature fields for particular brick walls that are widely used in residential construction in the city of Tyumen, their use will make it possible to exclude calculations of the temperature fields and conduct thermovision quality control of these structures in their expert assessment.

References

[1] Tekhnicheskij reglament o bezopasnosti zdanij i sooruzhenij [Technical regulations on safety of buildings and structures] (rus)

[2] Vlasov O E. Osnovy stroitel'noj teplotekhniki (k kursu otopleniya ventilyacii)/ O.E.Vlasov. M: Tip. VIA RKKA, 1938. - 94 s. [Vlasov O E. Basics of thermal engineering for construction (to the course of heating and ventilation)] (rus)

[3] Bogoslovskij V N. Teplovoj rezhim zdaniya. – M:Strojizdat, 1979. - 248s. [Bogoslovskij V N. Building thermal conditions] (rus)

[4] Il'inskij V M. Stroitel'naya teplofizika. - M:Vysshaya shkola, 1974. – 320 s. [Il'inskij V M. Building Thermophysics] (rus)

[5] Kuvshinov Yu Ya. Osnovy obespecheniya mikroklimata zdanij/Ucheb.dlya vuzov. – M: Izdatel'stvo Associacii stroitel'nyh vuzov 2012. – 200 s. [Kuvshinov Yu Ya. Basics of indoor climate for buildings / Manual for building universities] (rus)

[6] Cheng Y, Nin J and Gao N. Thermal comfort models: A review and numerical investigation *Building and Environment* 2012 vol **47** pp 13-22

[7] Orr H, Wang J, Fetsch D and Dumont R.Technical note: Airtightness of older-generation energy-efficient houses in Saskatoon *Journal of Building Physics* 2013 vol **36** pp 294-307

[8] Rohdin P, Molin A and Moshfegh B. Experiences from nine passive houses in Sweden – indoor thermal environment and energy use*Building and Environment* 2014 vol **71** pp 176-185

[9] Dodoo A, Gustavsson L and Sathre R. Building energy-efficiency standards in a life cycle primary energy perspective *Energy and Buildings* 2011 vol **43** Issue 7 pp 1589-1597

[10] Mlakar J and Štrancar J. Temperature and Humidity Profiles in Passive-house Building Blocks *Building and Environment* 2013 vol **60** pp 185-193
[11] Fountain M E, Xu T, Arens E and Oguro M. An investigation of thermal comfort at high humidities *ASHRAE Transactions* 1999 vol **105** Part 2 pp 94-103
[12] Zhang L Y, Jin L W, Wang N Z et. al. Effects of wall configuration on building energy performance subject to different climatic zones of Chia *Applied Energy* 2017 vol **185** part 2 pp 1565-1573
[13] Sassine E, Younsi Z, Cherif Y and Antczak E. Frequency domain regression method to predict thermal behavior of brick wall of existing buildings *Applied Thermal Engineering* 2017 vol **114** pp 24-35
[14] Friess W A, Rakhshan K and Davis M P. A global survey of adverse energetic effect of increased wall insulation in office buildings: degree day and climate zone indicator *Energy Efficiency* 2017 vol **10** Issue 1 pp 97-116
[15] Sobhy I, Brakez A and Benhamou B. Analysis for thermal behavior and energy savings of a semi-detached house with different insulation strategies in a hot semi-arid climate *Journal of Green Building* 2017 vol **12** Issue 1 pp 78-106
[16] Dos Santos G H and Mendes N. Heat, Air and Moisture Transfer through Hollow Porous Blocks *International Journal of Heat and Mass Transfer* 2009 vol **52**(9-10) pp 2390-2398
[17] Vatin N I, Nemova D V, Rymkevich P P and Gorshkov A S. Vliyanie urovnya teplovoj zashchity ograzhdayushchih konstrukcij na velichinu poter' teplovoj ehnergii v zdanii *Inzhenerno-stroitel'nyj zhurnal*. 2012. № **8** (34). S. 4—14. [Vatin N I, Nemova D V, Rymkevich P P and Gorshkov A S. Influence of a thermal protection level of enclosure structures on amount of thermal energy losses in buildings *Engineering and Construction Journal*] (rus)
[18] Teodosiu R. Integrated Moisture (Including Condensation)-Energy-Airflow Model within Enclosures *Building and Environment* 2013 vol **61** pp 197-209
[19] Novikova A S and Korsun N D. Obzor narushenij teplovoj zashchity na primere zhilyh ob"ektov goroda Tyumeni *Energosberezhenie i innovacionnye tekhnologii v toplivno-ehnergeticheskom komplekse*: materialy Mezhdunarodnoj nauchno-prakticheskoj konf. T 1 – Tyumen': TIU, 2016. – 438 s. - S. 237-241. [Novikova A S and Korsun N D. Survey of thermal protection defects as exemplified by residential buildings in the city of Tyumen *Energy saving and innovative technologies in the fuel and energy complex*: Collected papers of the International Scientific and Practical Conference. - Tyumen: IUT, 2016] (rus)
[20] GOST R 54852-2011. Metod teplovizionnogo kontrolya kachestva teploizolyacii ograzhdayushchih konstrukcij. - M.: Standartinform. – 19 s. [GOST R 54852-2011. Method of thermovision quality control of thermal insulation of enclosure structures] (rus)
[21] Novikova A S and Korsun N D. Ocenka teplovyh poter' na primere zhilyh ob"ektov goroda Tyumeni *Voprosy nauki i obrazovaniya: teoreticheskie i prakticheskie aspekty*: materialy Mezhdunarodnoj nauchno-prakticheskoj konferencii (CHekhiya, Praga, 16 maya 2017). – Vydavatel "Osviceni", Prague, ChechRepublic, 2017. - S. 127-131. [Novikova A S and Korsun N D. Assessment of heat losses as exemplified by residential buildings in the city of Tyumen *Issues of Science and Education: Theoretical and Practical Aspects*: Collected papers of the International Scientific and Practical Conference. – Prague, Czech Republic, 2017] (rus)
[22] Hartman K. *Planirovanie ehksperimenta v issledovanii tekhnologicheskih processov* / K. Hartman, EH. Leckij, V. SHefer i dr.; pod obshch. red. EH. K. Leckogo. – M.: Mir, 1987. – 522 s. [Hartman, K. Planning an experiment in studying technological processes] (rus)
[23] ELCUT [EHlektronnyj resurs].URL: http://elcut.ru/index.htm (data obrashcheniya: 13.11. 2015) [ELCUT [Electronic resource] .URL: http://elcut.ru/index.htm (reference date: November 13, 2015) (rus)

IOP Conf. Series: Materials Science and Engineering **481** (2019) 012031 doi:10.1088/1757-899X/481/1/012031

Increase of energy efficiency during overhaul of housing stock in Russian Federation

V Ya Mishchenko[1], S G Sheina[2], E P Gorbaneva[1]

[1]Department of Technology, Organization of Construction, Expertise and Property Management, Voronezh State Technical University, Str.20 letia oktyabrya, 84, Voronezh, Russian Federation
[2]Department of Urban Construction and Economy, Don State Technical University, Gagarin square, 1, Rostov-on-Don, Russian Federation

E-mail: mishenko@vgasu.vrn.ru, rgsu-gsh@mail.ru, elenagorbaneva@rambler.ru

Abstract. The construction sector covers various activities for the final consumption of energy resources, which have different consequences for energy consumption. Heating, cooling and lighting of premises for which most industrial energy consumption accounts for most of the energy consumption of buildings depend not only on the energy efficiency of the temperature and lighting control systems, but also on the efficiency of the buildings in which they are operating. The lack of generally accepted approaches to a comprehensive assessment of the effectiveness of the introduction of new energy efficient technologies, raw materials and equipment serves as a barrier to increasing the energy efficiency of domestic construction and overhaul of MAB. The article considers measures to increase energy efficiency during overhauls, which allow to reduce expenses for provision of housing and communal services and increase the market value of housing in the real estate market.

1. Introduction

The problem of energy efficiency is becoming critical in the 21st century, given the negative impact of energy technologies on the environment and the rising cost of energy resources. In Russia, energy efficiency is a resource for accelerating economic growth and a priority for transformational development of the economy. It should be noted that at present, Russia ranks 10th in the world in terms of energy consumption. In today's economy, no more than 1-2% of the potential for increasing energy efficiency is realized, 25% of the energy consumed by Russia is the residential sector, up to 15% of electricity and 50% of the heat energy does not reach the consumer.

The Russian Federation State Program "Energy Conservation and Energy Efficiency Improvement for the Period to 2020", developed in 2010, provides an assessment of Russia's energy saving potential (Table 1) [1].

It can be seen from the table that the construction industry has a high potential for energy saving. It is one of the main energy consumers, which poses the task of more rational use of energy resources at the stage of production of various building materials and operation of completed construction objects of various functional purposes.

Energy consumption in Russia for 2018 increased by 0.8% compared with 2017 [2]. As of January 2018, energy consumption increased by 0.3% in the Belgorod region's energy system, by 1.6% in the

Vologda region, by 0.2% in the Voronezh region, by 0.6% in the Kaluga region, as compared to the same period last year, the Lipetsk region by 0.7% [3].

Table 1. Estimation of the potential for increasing energy efficiency in selected sectors of the economy (million tons of equivalent fuel).

Energy consumption sector	Total	Including, by types of energy resources						
		Coal	Raw oil	Oil products	Gas	Fuel spills	Electric power	Heat
Agriculture and forestry	4.15	0.03	-	2.19	0.11	0.06-	1.04	0.72
Mining	1.59	-	-	0.20	-	2	0.53	0.86
Construction	0.73	-	-	0.29	0.01	0.01	0.36	0.06
Utilities	1.01	-	-	0.01	-	-	0.51	0.49
Heating	41.11	0.82	-	0.26	3.68	0.27	0.28	35.8
Hot water supply	23.11	-	-	-	3.60	-	0.25	19.26
Lighting (including transport infrastructure)	3.15	-	-	-	-	-	3.15	-

Energy consumption in Russia for 2018 increased by 0.8% compared with 2017 [2]. As of January 2018, energy consumption increased by 0.3% in the Belgorod region's energy system, by 1.6% in the Vologda region, by 0.2% in the Voronezh region, by 0.6% in the Kaluga region, as compared to the same period last year. , The Lipetsk region by 0.7% [3].

The problem of saving energy is undoubtedly urgent and vital. Exhaustion of natural resources, a negative impact on the environment and a high cost of energy are indicators to reduce energy consumption and production. Housing and communal services (utilities), industrial enterprises and mechanical engineering are the main consumers of energy [4].

2. Development of energy-efficient overhaul in the Russian Federation

Socio- economic problem of the housing and communal services sector in the territory of the Russian Federation is the provision of services for overhaul of buildings and structures [5-7]. Most buildings over-spend energy that goes to heating and cooling. The experience of other countries shows that overhaul takes into account the targets of the energy efficiency of modernization, the end result of which is the reduction of energy consumption while maintaining convenience and reliability.

In multi-apartment buildings (MAB), heat meters are recorded by counters due to low insulation of enclosing structures and the deterioration of in-house engineering systems. It should be noted that 40% of heat is lost through the walls, 18% through the roof and windows, 14% in ventilation, 10% in the basement. With an integrated approach to energy conservation, heat losses are minimized [8]. According to the federal law No. 261-FZ, energy saving is "the implementation of measures aimed at reducing the volume of energy resources used while maintaining the appropriate beneficial effect from their use" [1].

According to the current housing legislation, owners of residential and non-residential areas take part in the financing of the capital repair of MAB primarily through monthly contributions. In case of partial and untimely payment of contributions for the overhaul of the debt, they are levied compulsorily, taking into account the accumulated penalties and court costs of the regional operator. The collection rate in the Voronezh region was 90% for January-August 2018. The indicator exceeded the average level of Russia in payment of contributions.

Before each region, the task is to conduct a qualitative overhaul with the use of modern building materials. In the Voronezh region, a multi-level system of departmental and public supervision operates over the course of overhauls of multi- apartment buildings. A list of qualified contractors has been compiled and regularly updated.

However, measures to improve the energy efficiency of buildings are not included in the list of mandatory works for overhaul. The residents themselves can make this decision. By agreeing among themselves, they choose additional measures from the list developed by the Ministry of Construction of the Russian Federation for energy-efficient overhaul. The selected options require additional attachments. If the building energy consumption is reduced by 10%, tenants will receive financing up to 50% or subsidize the interest rate on the borrowed loan from the Fund for Assistance to Housing Reform. The support has been in effect since January 2017 under the Decree of the Government of the Russian Federation on the financial support of energy-efficient overhaul.

In 2017 in the city of Voronezh major repairs of two houses with the use of energy-saving technologies were made. Residents have cut utility costs by at least 10%.

3. Measures for energy saving in houses during major overhaul

According to the "Practical manual on increasing the energy efficiency of apartment buildings during major repairs" [9], developed by the Fund for the Promotion of Housing Reform, packages of energy-efficient measures have been formed for selective overhaul of the MAB and ensuring compliance with regulatory requirements for energy efficiency. They are combined in accordance with the principle of ensuring the achievement of normative indicators of energy efficiency and specific energy consumption corresponding to the period of putting the MAB into operation after repairs. Among the methods of energy saving, there are 4 directions [10-12]:

- ensuring effective thermal insulation;
- reduction of heat losses in the ventilation system;
- ensuring the tightness of window and door openings;
- reducing the power consumption by introducing modern economical devices.

With increasing energy efficiency in the construction sector, you can often encounter such barriers:

- lack of reliable and veracious information on energy efficiency and the costs and benefits of improving efficiency;
- budget constraints of local self-government;
- high transaction costs due to small individual investments;
- unattractive financial profit;
- the risk to unfamiliar materials, methods and equipment, or uncertain results.

Based on these data, it is necessary to take measures to overcome these barriers and improve existing methods of increasing energy efficiency. The policy of increasing energy efficiency is mainly focused on the "shell" of buildings, rather than on heating and ventilation equipment, although if you focus on equipment, you can significantly increase the energy-saving potential. There is a great potential for achieving greater energy savings by establishing and improving the regulatory framework of the Russian Federation [13]. Increasing efficiency by 10-20% is possible in most countries thanks to only the instruments, equipment and lighting products that can be purchased today. Therefore, the search for ways to optimize energy consumption in the construction sector is an urgent and extremely important direction of the country's development [14,15].

To carry out major repairs of multi-apartment buildings in order to increase energy efficiency, the following activities are necessary:

1) an increase in the degree of thermal protection of external walls to the standards of 2016-2020. Maintenance does not require additional costs. The life span corresponds to the life span of the building.

2) an increase in the degree of thermal protection of windows and balcony doors to the standards of 2016-2020 g. Operating costs - 0.5% of the capital annual maintenance costs. Life span is 15 years.

3) sealing of external entrance doors in the entrances with the installation of door closers (ensuring automatic closing of doors). Operating costs - 5% of the capital annual maintenance costs. Life span - 6 years.
4) the device of the radiator reflux screens (Fig. 1). Maintenance does not require additional costs. Life span is 15 years.
5) additional sectional entrance tambours (Fig. 2). Maintenance does not require additional costs. The life span corresponds to the life span of the building.
6) installation of common house heat metering devices. Operating costs - 1% of the capital annual maintenance costs. Life span is 10 years.
7) installation of a common house hot water meter. Operating costs - 1% of the capital annual maintenance costs. Life span is 10 years.
8) installation of an automated control unit for the heating system. Operating costs - 4% of the capital annual maintenance costs. Life span - 20 years.
9) installation of balancing valves (ventils) on the vertical risers of the heating system (Fig. 3). Operating costs - 1% of the capital annual maintenance costs. Life span is 10 years.
10) mounting of thermostatic valves (thermostats) on the heating devices (Fig. 4). Operating costs - 1% of the capital annual maintenance costs. Life span - 10 years.

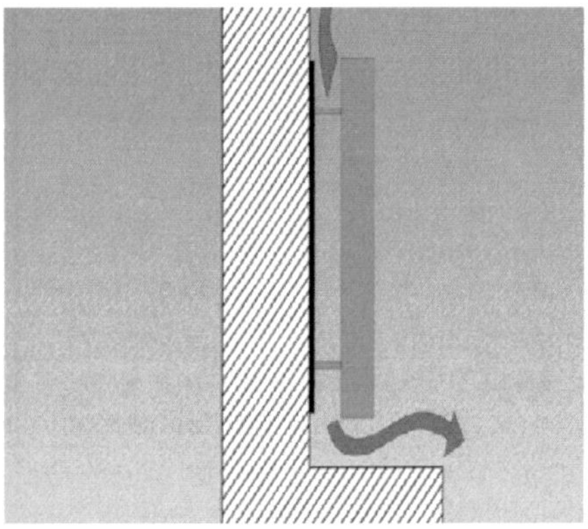

Figure 1. Schematic illustration of the device of the radiator reflux screens [9].

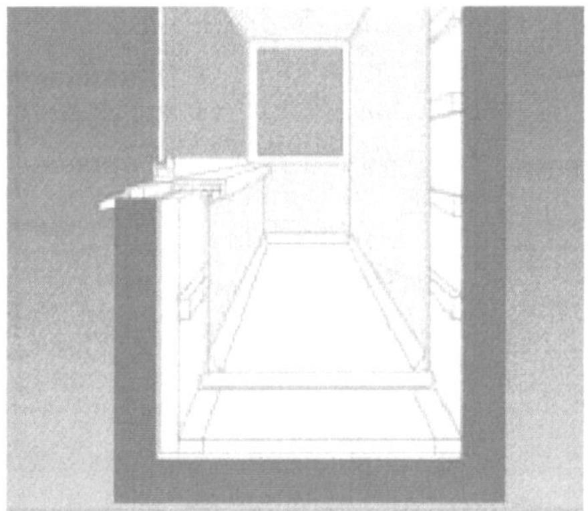

Figure 2. Schematic illustration of the additional sectional entrance tambours [9].

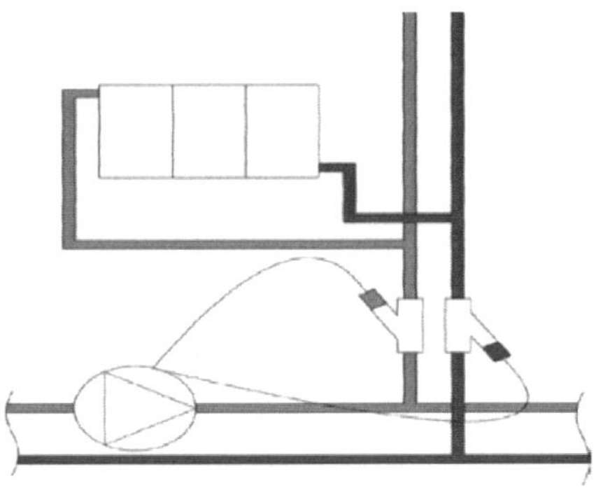

Figure 3. Schematic illustration of the installation of balancing valves on the vertical risers of the heating system [9].

Figure 4. Schematic illustration of the installation of thermostats on radiators [9].

4. Conclusion

Studies [18-22] have identified the need for the short term to take the following measures to improve energy efficiency:

- the adoption of policy measures that promote the efficient use of resources;
- expansion, recycling and informing consumers about possible options for reducing the use of natural resources;
- development and introduction of economic models into practice.

All of the above energy efficiency measures will be aimed at reducing the amount of energy consumed, while maintaining the quality of services provided in buildings.

References

[1] Federal Law No. 261-FZ of 23.11.2009 On energy saving and on improving energy efficiency and on introducing amendments to certain legislative acts of the Russian Federation [Electronic resource] URL: http://www.consultant.ru/document/cons_doc_LAW_93978/ (access date 22.09.2018)

[2] Ministry of Energy of the Russian Federation [Electronic resource access date 20.09.2018] URL: https://minenergo.gov.ru/node/10277

[3] Energy-efficient Russia [Electronic resource access date 20.09.2018] URL:http://www.energy2020.ru/general_news/news10875.php

[4] Queen A D 2014 Energy efficiency of a residential building Modern high technology **5** 171-2

[5] Mishchenko V Ya, Gorbaneva E P, Manukovsky A Yu and Safonov A O 2014 Genetic algorithms in solving multicriteria problems of optimization of resource allocation in the planning of energy-saving measures Scientific Bulletin of VGASU: Construction and architecture **3(35)** 77-82

[6] Mishchenko V Ya, Gorbaneva E P, Manukovsky A Yu and Safonov A O 2014 Increase of energy efficiency in the budgetary sphere of the Voronezh Region Scientific Bulletin of the VGASU: Construction and architecture **3(35)** 71-6

[7] Gorbaneva E P and Sheikina V V 2016 Analysis and methods for determining the risks of the investment-construction project Modern problems and prospects for the development of construction, the operation of real estate (Voronezh:VSTU) 189-197

[8] Bykov S A 2011 Aspects of energy conservation and energy efficiency in the overhaul of real estate in the Far East Russian Entrepreneurship **5(2)** 197-202

[9] Practical manual on improving the energy efficiency of multi-apartment buildings (MAB) during major overhaul 2015 Volume III Fund to promote housing and utilities reform p.249

[10] Mishchenko V Y and Gorbaneva E P 2004 The Role of Reconstruction and Modernization in the System of Preservation and Reproduction of Real Estate Objects Scientific Bulletin of the VGASU: Road and transport construction **3** 122-7

[11] Gorbaneva E P and Babeshko E A 2017 Problems of organization of construction production of energy-efficient real estate Current trends in construction and operation of real estate: Sat scientific 164-9

[12] Mishchenko V Ya, Barinov V N, Gorbaneva E P and Nazarov A N 2012 Organization of energy audit of socially significant objects in Voronezh Scientific journal:Engineering systems and facilities **3(8)** 115-123

[13] Aloyan R M, Fedosov S V and Oparina L A 2016 Energy efficient buildings - the state, problems and solutions (Ivanovo) p 276

[14] Tupikina A A and Chernov S S 2016 Energy efficiency as a priority direction of sustainable development: organizational and financial aspects Baikal Research Journal **7** [Electronic resource access date 21.09.2018] URL:http://brj-bguep.ru/reader/article.aspx?id=20585

[15] Oparina L A 2014 Fundamentals of resource and energy conservation in construction: Textbook. allowance (Ivanovo) p 256

[16] Gorbaneva E P, Dobrosotskikh M G, Kalinina E G 2016 Determination of the technical condition of buildings and structures using a monitoring system Modern problems and prospects for the development of construction, the operation of real estate (Voronezh:VSTU) 208-213

[17] Gorbaneva E P, Kolesnik S A and Tyurina P A 2016 Formation of an effective portfolio of investment and construction projects *Modern problems and prospects for the development of construction, the operation of real estate* (Voronezh:VSTU) 171-7

[18] Petrichenko K, Aden N and Tsakiris A 2016 *Tools for Energy Efficiency in Buildings. A guide for policy-makers and experts* (Copenhagen Center on Energy Efficiency (C2E2) p40

[19] Filushina K E, Minaev N N, Gusakova N V, Dobrynina O I, Zharova E A, Merkuleva Yu A and Runkova A S 2016 Regional Programs on Energy Saving and Energy Efficiency: The National Construction Sector *International Journal of Economics and Financial Issues* **6** **(S5)** 85-90.

[20] Mishchenko V Ya, Gorbaneva E P, Archakova S Yu and Dobrosotskikh M G 2017 Modeling of the implementation by teams of technological processes in organizational and technological design *FES: Innovation Economics* **6** 37-43

[21] Gorbaneva E P and Sevryukova K S 2017 The contribution of information and communication technologies to improving the energy efficiency of the construction sector *Information technology in modeling and management: approaches, methods, solutions* (Togliatti) 67-74

Technical indexes - building energy in tropical climate

Tran Ngoc Chan[1], Bui Quang Trung[1]

[1]Vietnam Association of Construction Environmental Engineering -VACEE

Abstract: In tropical climate countries including Vietnam, external heat absorbed into buildings in summer time is mainly depended on solar radiation transmitted through glass windows. This article aims to illustrate data representing the relation among SHGC, OTTV, and energy costs of air conditioning system in buildings.

1. Introduction

Abbreviation:
OTTV: Overall Thermal Transfer Value;
SHGC: Solar Heat Gain Coefficient of window glass

Building external structures play an important role to ensure comfortable environment for people working and living inside. Heat insulation requirement of external structures in winter has been regulated in TCVN [1]. For instance, external structure made by refined clay brick of 220 mm thickness secures a heat resistance R_o= 0.584-0.625 m^2.K/W, which is qualified to prevent condensation on the inner surface of the external structure, and prevent cold for human bodies due to heat irradiance from body to wall surface.

Meanwhile, in summer, the heat transmission into air-conditioned buildings due to temperature difference and solar radiation, which transferred through glass windows will be high. This leads to high energy expense for building air-conditioning system. To mitigate such expenses, Vietnamese QCVN 09:2017 sets thresholds for $OTTV_T$ – for walls, and $OTTV_M$ – for roof not exceeded 60 W/m^2 and 25 W/m^2, respectively [2].

Currently, trends of constructing high rise buildings with large glass windows has been common in Vietnam and worldwide, where Window-Wall Ratio (WWR) may be as high as 100%. In such cases, it is necessary to choose glass window with low SHGC to secure the regulated OTTV thresholds.

Apparently, glass with low SHGC is expensive, thus, resulting in investment surge. Nonetheless, in return, the energy expenses for air-conditioning system during the operation of the building would be reduced. To that end, it is crucial to compare different glass using plans with different SHGC to not only secure the threshold under QCVN 09:2017, but also reduce the energy expenses for the air-conditioning system in particular, and the whole building as well.

2. Correlation between SHGC, OTTV and energy expenses

OTTV calculation has been presented in the article [3]. OTTV is mainly dependent on the SHGC of the glass window, especially in the case of high WWR. In practice, when WWR $\geq$ 40%, heat transmitted through glass window accounts for 85-95% of the total heat absorbed into the building. As such, window glass serves as key element in saving energy consumed by air-conditioning.

To save air-conditioning energy, on one hand, WWR must be lower down, but on the other hand, low SHGC glass window must be chosen, and applying external structure to cover sunlight outside the window [4].

With the purpose of clarifying the correlation between OTTV and SHGC of window glass, calculations applied to different buildings with different functions, shapes, scales and orientations with four different SHGC have been implemented by computer software [5]. Results are presented in Table 1. Data in Table 1 is presented in Figure 1.

Table 1. Values of OTTVT , W/m^2 in correlation with SHGC of buildings with different WWR.

	Building charateristics and WWR	Plans of SHGC for window glass Plan 1: SHGC=0.85	Plan 2: SHGC=0.65	Plan 3: SHGC=0.45	Plan 4: SHGC=0.30
1	5-storey commercial-service centers, $\sum F_{floor.}$=18,436 m^2, North orientation, WWR=0.705	144.66	113.65	82.54	59.21
2	21-storey office building, $\sum F_{floor}$.=36,800 m^2, South orientation, WWR=0.688	137.03	107.88	78.72	56.86
3	23-storey office building,, $\sum F_{floor}$ =37,800 m^2, East orientation, WWR=0.392	87.65	70.15	52.64	39.51
4	14-storey apartment building, $\sum F_{floor}$.=18,755 m^2, South West orientation, WWR=0.372	86.32	68.76	50.27	36.76
5	14-storey office building, $\sum F_{floor}$.=16,628 m^2, South East orientation, WWR=0.298	69.35	54.92	40.49	29.67

Note: All buildings are located in Hanoi.

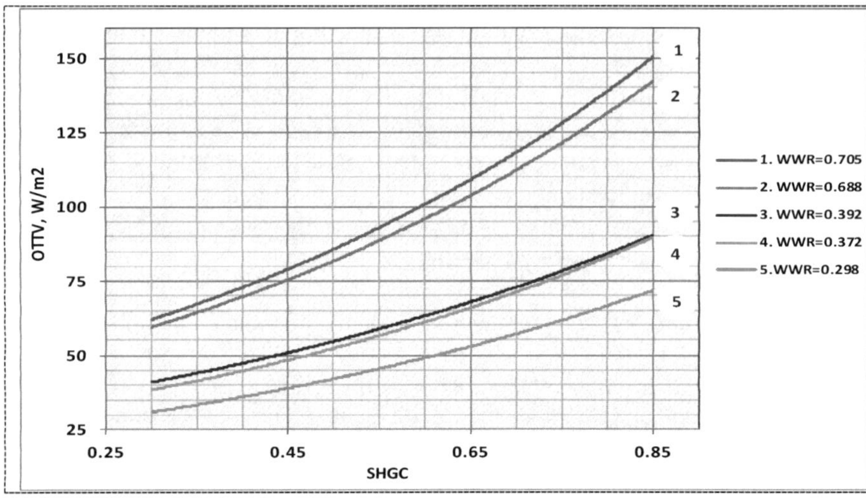

Figure 1. Variability of $OTTV_T$ in correlation with SHGC of buildings with different WWR.

The curves in Figure 1 demonstrate that regardless of difference in shape, scale, and orientation, the curves of the two buildings (Plan 1 and 2 in Table 1) with similar WWR (WWR = 0.705 and 0.688) are close to each other. Similar outcome is found in the Curve 3 and Curve 4 (Plan 3 and Plan 4 in

Table 1) with similar WWR = 0.392 and 0.372, respectively. To that end, $OTTV_T$ of the building is not affected by shape, scale, and orientation of the building, but the WWR.

3. Determination of technical-economic indexes for Commercial-Service Centers

For 5-storey Commercial-Service Centers (Plan 1 in Table 1), detailed calculation is conducted: it is necessary to identify OTTV, energy expenses for different electric usage demands, and investment capital for the window glasses of different SHGC. The Commercial-Service Centers is located in 5-storey podium with 18436 m^2 of a multi-storey complex located in Cau Giay district, Hanoi. The total external wall area is 7172.6 m^2 and total window area is 5059.1 m^2 (WWR = 0.705). Calculated results are presented in Table 2.

Table 2. Technical-economic indexes for Commercial-Service Centers in Cau Giay district, Hanoi.

Technical-Economic Index and its unit	Plans with SHGC of window glass			
	Plan 1: SHGC=0.85	**Plan 2: SHGC=0.65**	**Plan 3: SHGC=0.45**	**Plan 4: SHGC=0.30**
$OTTV_T$, W/m^2	144.66	113.65	82.54	59.21
Power expense for air conditioning system, kWh/year	1042,678.5	910,617.3	778,630.9	680,478.6
API, kWh	516.3	450.9	385.5	336.9
AEI, kWh/(m^2.year)	56.6	49.4	42.2	36.9
BEI, kWh/(m^2.year)	100.1	93.0	85.8	80.5
Investment capital for window glass (millionVND)	3642.55	5059.10	6070.92	7588.65

Note:
- API: Air Conditioning Power Index;
- AEI: Air Conditioning Energy Index – energy expense for 1 square metre floor in a year;
- BEI: Building Energy Index – Energy use index for whole building - – annual energy consumption for 1 m^2 of building gross floor are;
- Among the 4 plans, plan with SHGC = 0.3 secures $OTTV_T$ threshold as set forth in QCVN 09:2017;
- Investment capital for window glass for plans 1 to 4: 720,000; 1,000,000; 1,200,000 and 1,500,000 VN Dong/m^2.

Power price applicable to Cau giay -Hanoi area for usage of 401 kWh and higher is VND 2535 for each kWh. Annual economic profit by reducing power expenses is (1,042,678.5 – 680,478 .6)×2,535 = VND 918,176,746.5.

Period for investment return is 3,946,100,000/918,176,746.5 = 4.3 years. Data in Table 2 (Plan 2 and 6) are illustrated in Figure 2.

To detail the correlation between $OTTV_T$ and energy expenses for air conditioning for 1m^2 of the floor in AEI, kWh/(m^2.year), supplementary calculations of other buildings have been conducted and presented in Table 3 and Figure 3

As presented in Figure 3, there is difference between AEI = f ($OTTV_T$) function, and $OTTV_T$ = f (SHGC) function in Figure 1, which is a linear correlation, represented by straight lines (R^2 is approximately equivalent to 1). This facilitates the calculation in designing process to choose optimal $OTTV_T$ for energy saving of air conditioning system.

IOP Conf. Series: Materials Science and Engineering **481** (2019) 012032 doi:10.1088/1757-899X/481/1/012032

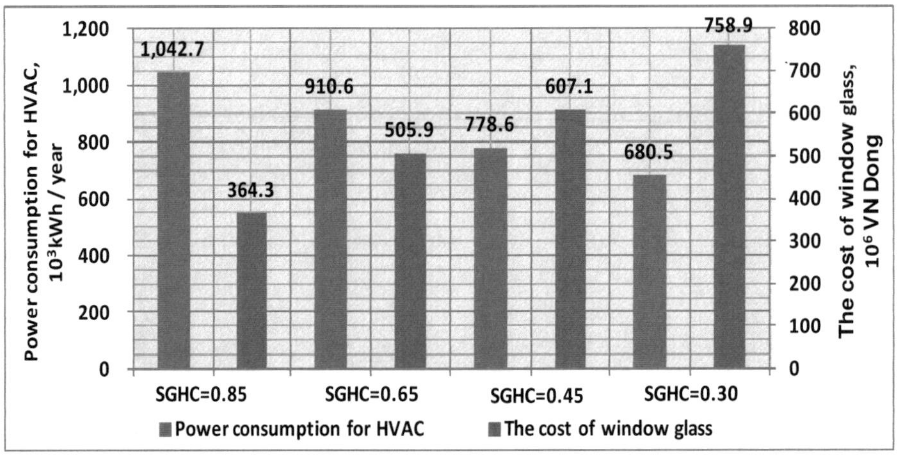

Figure 2. Power expenses and investment capital for window glass in 4 plans of different SHGC for the Commercial-Service Centers in Cau Giay district, Hanoi.

Table 3. $OTTV_T$ and AEI of different buildings of different SHGC in Hanoi.

	Building	Index	SHGC				
			SHGC= 0.85	SHGC= 0.65	SHGC= 0.45	SHGC= 0.30	SHGC= 0.17
1	5-storey commercial-service centers, $\sum F_{floor.}$=18,436 m^2, North orientation, WWR=0.705	$OTTV_T$, W/m^2	144.66	113.65	82.54	59.21	39.0
		AEI, kWh/(m^2.year)	56.6	49.4	42.2	36.9	32.41
2	23-storey office building,, $\sum F_{floor}$ =37,800 m^2, South orientation, WWR=0.392	$OTTV_T$, W/m^2	87.65	70.15	52.64	39.51	28.14
		AEI, kWh/(m^2.year)	57.6	51.53	45.5	41.0	37.26
3	14-storey office building, $\sum F_{floor}$ =10,182 m^2, East orientation, WWR=0.355	$OTTV_T$, W/m^2	96.02	78.35	60.68	47.42	35.94
		AEI, kWh/(m^2.year)	56.7	50.5	44.34	39.73	35.8

4. Conclusion

1. Technical-economic indexes in correlation with OTTV and power expenses in buildings of different functions, shapes, scales, and orientations, located in Hanoi, have been identified. Calculations show that building OTTV is mainly depended on WWR and SHGC of window glass (Figure 1);

2. To complete the calculation, a number of computer software have been programed, of which OTTV calculation software will be applied for implementing QCVN 09:2017;

3. Technical-Economic calculation of the correlation between energy expenses for air-conditioning system and investment capital for window glass of the Commercial-Service Center in Cau giay reveals that to comply with QCVN 09:2017, window glass of SHGC = 0.3 must be installed instead of window glass of SHGC = 0.85. In such cases, investment capital increases by VND 3946.1 million, and would be returned in 4.3 years by energy saving of air-conditioning system.

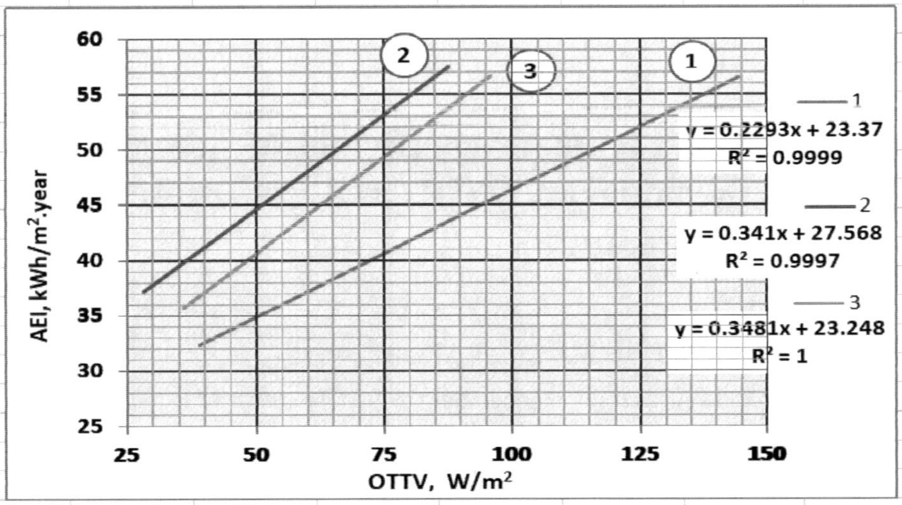

Figure 3. Dependency of AEI in $OTTV_T$ in buildings in Hanoi.
Note: Lines in the Figures are numbered by the order as presented in Table 3.

Reference

[1] Technical Standard TCVN 4605:1988 "Technical issues of covering structure – Designing Standards" 1998 (Hanoi: Construction Publisher) [in Vietnamese]

[2] Technical Regulation QCVN 09.2017/BXD "Buildings of efficient energy usage" 2017 (Hanoi: Construction Publisher) [in Vietnamese]

[3] Tran Ngoc Chan 2014 Limit heat transmitted through covering structure to reduce energy expenses for air-conditioning system *Scientific research journal "Air Quality in Building and Surrounding Environment"* (Haifa, [in Russian].

[4] Tran Ngoc Chan, Pham Van Luong 2015 Energy saving and efficiency in construction structures *Heat Energy Magazine* **125** (Hanoi) [in Vietnamese].

[5] Tran Ngoc Chan and Partners 2017 "Guidance for OTTV Calculation" supplemented into QCVN 09:2017/BXD. Ministerial-level Scienctific Research of Ministry of Construction, (Hanoi) [in Vietnamese].

IOP Conf. Series: Materials Science and Engineering **481** (2019) 012033 doi:10.1088/1757-899X/481/1/012033

Investigation of heat transfer coefficient of spherical element using infrared thermography (IR) and gas - water droplets (mist) as working medium

A H Abed[1, 2], S E Shcheklein[1], V M Pakhaluev[1]

[1] Ural Federal University named after the first President of Russia B. N. Yeltsin, 19 Mira St., Yekaterinburg 620002, Russia
[2] University of Technology, Iraq

E-mail: akraaam82@yahoo.com

Abstract. The changes that do occur in the air ambient properties with increasing concentration of water mist (fog) have a significant influence on heat exchange process between the building and the surrounding medium. In the present paper, the average and local heat transfer coefficient was estimated from surface temperature data obtained using infrared (IR) thermography. In particular, the experiments were performed on steady-state conditions under constant heat flux for a single sphere suspended in cylindrical channel using air as well as air / water droplets as working fluid. The influence of the different factors such as Re numbers and water flux density on heat transfer behaviors are examined. Five cases are tested under range of water flux density (0 -111.68 kg m^{-2} hr^{-1}). The experimental results confirmed that the heat transfer coefficient significantly increased with increase in water flux density. The heat transfer coefficients are respectively 1%, 19.7%, 90.2% and 134% higher than those in air-cooling. The results also revealed that infrared (IR) thermography it has proven to be very efficient in measuring the surface temperature distribution. The results obtained by infrared thermography (IR) are compared with calibrated thermocouples and the temperature distribution is found to be in close agreement with 2.94% average error.

1. Introduction

The changes that do occur in the air ambient properties with increasing concentration of water mist (fog) have a significant influence on heat exchange process between the building and the surrounding medium. The presence of water mist in a flow of gas provides an excellent technique of heat transfer enhancement for many industrial applications [1-2]. The heat transfer enhancement mechanisms of gas-water mist flow include both, the effects of water mist momentum on gas main flow and evaporation of water droplets. This technique offers extremely important increase of heat and mass transfer rate between the surface and main flow due to the following features: absorbs a large amount of energy in the form of latent heat during the evaporation process of water droplets, increase specific heat of mixture and increase the turbulence into main flow and inside the boundary layer [3-4]. Numerous experimental and numerical investigations on heat transfer enhancement of gas-droplets two phase flow have been conducted. Most investigations are found deal with topics of heat transfer such as flat plate heated wall, turbine blades and heat exchangers. As was shown by literature review on the experimental and theoretical research of gas-droplets flow [1-10], the single-phase heat transfer performance can be enhanced up to 70% by dispersed water mist in gas flow. The effect was manifested in a decrease in the surface temperature at simultaneous increasing of water mist concentration in the main flow. The literature survey points that all the experimental studies till date are based on thermocouples as one of the conventional devices that most frequently used to measure the temperature profile. Although that thermocouples can be accurately determine the temperature, but it is only a single/local point measurement, i.e., it is not capable of mapping temperature distribution at object surface. Also conventional measurement devices are not applicable when the electromagnetic heating induction is used, which creates eddies that lead to heat the temperature measurement devices itself. This will lead to a signal that does not precisely depict the temperature field being measured. Obviously, the direct method is hard to meet the requirements of engineering application. So, in the present paper, a non-invasive technique has been proposed to measure the surface temperature distribution and evaluate average and local heat transfer coefficient of spherical element using Infrared thermographic technique that can be used in practice. In recent year, Infrared thermographic temperature measurement has gradually become the most widely used measurement technique in industry, medicine, scientific research and military technology with many clear advantage of contactless measurement, remotely monitored, flexible, visual measurements in image, accurate and wide temperature range [11-14]. In addition, direct surface temperatures were recorded using K - type thermocouples for comparison to the Infrared (IR) thermography. Herein, the copper calorimetric sphere suspended in cylindrical channel was used as heated surface. The influence of water flux density (fog concentration) on heat transfer coefficient was stuided at range of Reynolds number under constant heat flux. by using infrared thermography (IR) measurement system, the non-uniform local surface temperature and heat transfer coefficient were estimated.

2. Experimental Set-up and Procedure

The schematic diagram of the experimental set-up and details drawing of heat transfer test model are presented in figure 1 and figure 2. The experimental facility consisting of air subsystem, test object, droplets liquid subsystem, Infrared (IR) thermography digital camera and data acquisition system. The air subsystem consists of transparent Plexiglas channel having a circular cross section of 50 mm diameter; 3 mm thickness wall is set vertically. An electrical air blower was used with variac voltage regulator to control airflow velocity. On the same axis, a pitot tube flow meter (compact type) with digital manometer was used to measure the mean flow velocity. The measured object is a copper calorimetric sphere 34mm in diameter inserted in flow main stream and it is rear supported by a low thermal conductivity textolite rod (k= 0.023 W/m °C, 4mm-diameter) which span across the Plexiglas channel. The test sphere was independently heated using 100W - high density cartridge heater having 8mm-diameter and 31mm-length. The surface temperature was directly measured using two calibrated thermocouples (K- type) were inserted inside the copper sphere in positions indicated by the numbers 1, 2 as shown in figure 2. All thermocouples are connected to multichannel data acquisition system that consists of analog signal input module OWEN MV110-8A and MSD200 data logging unit. The droplets liquid system was a 1.7 MHz ultrasonic atomization has the same structure as the one reported in ref. [15]. The droplets liquid generator consists of a piezoelectric transducers, transparent vessel, water tank and blower fan. The water mass flow rate was measured by the mass of water passing a droplets generator subsystem per unit time. To blend the liquid droplets with a controlled amount of air, a mixing chamber was installed and used as the blender. For contactless measurement, we used an infrared (IR) camera with a 320 x 240 pixel image dimension, and a 0.1m standard lens. Under constant temperature condition in the laboratory, the position and the focal length of the infrared (IR) camera are adjusted and aimed at the test object inside channel as shown in figure 1.

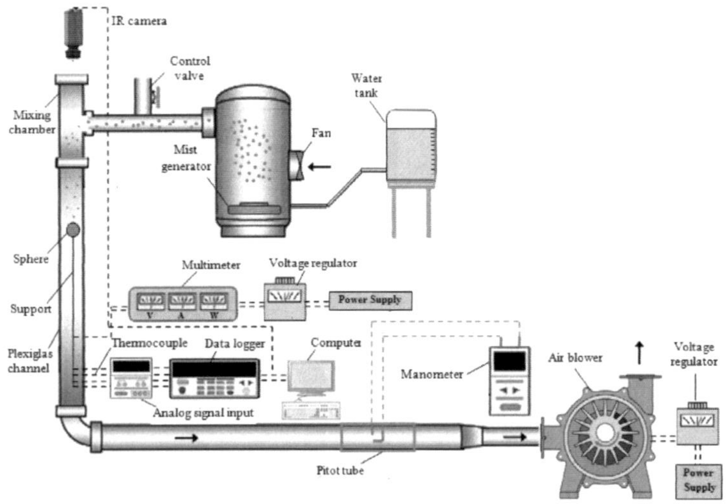

Figure 1. Schematic diagram of the experimental test bench.

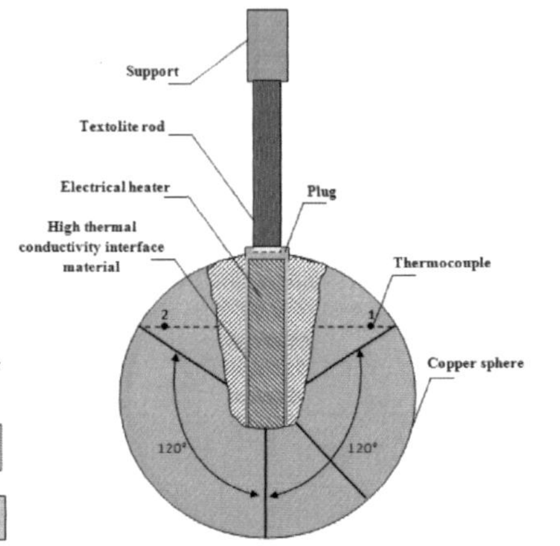

Figure 2. Schematic drawing showing the construction of heat transfer models.

Infrared (IR) measurement system gives a temperature distribution signature on the surface of sphere being measured as shown in figure 3. This will extensively help in understanding the influence of water droplets evaporation on heat transfer behavior.

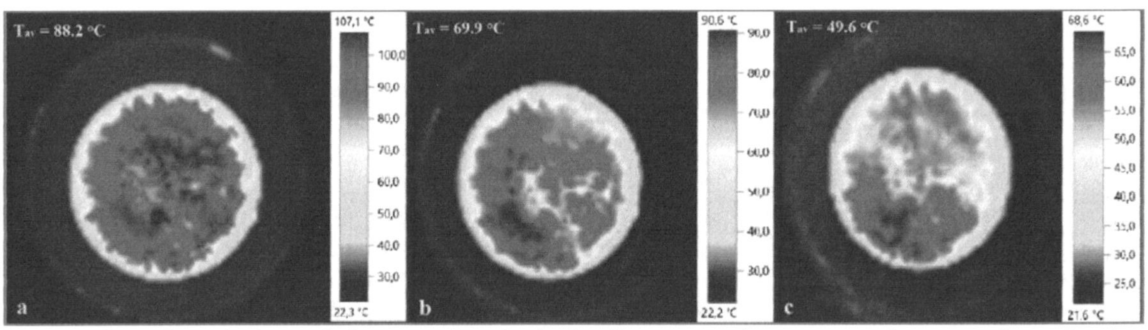

Figure 3. Infrared thermal image of spherical element under range of water flux density (Re = 25000): (a) only air; (b) 46.79 kg m^{-2} hr^{-1}; (c) 88.31 kg m^{-2} hr^{-1}.

In steady-state condition, the average surface temperature module as shown in figure 4 is recorded by the infrared (IR) measurement method to compare with the data obtained by thermocouples. For the purpose of validating the measurements made by infrared (IR) measurement method, experiments are carried out to estimate the surface temperature and heat transfer rate using contact-thermocouples method. Repeating the above process in the measurement of other cases of water flux and Re numbers, the final results are shown in figure 4. Indeed, if we regarded that thermocouple readings as true and reference temperature of sphere being measured, then Accordingly, the experimental data: the average error of infrared (IR) temperature measurement method is 2.94%.

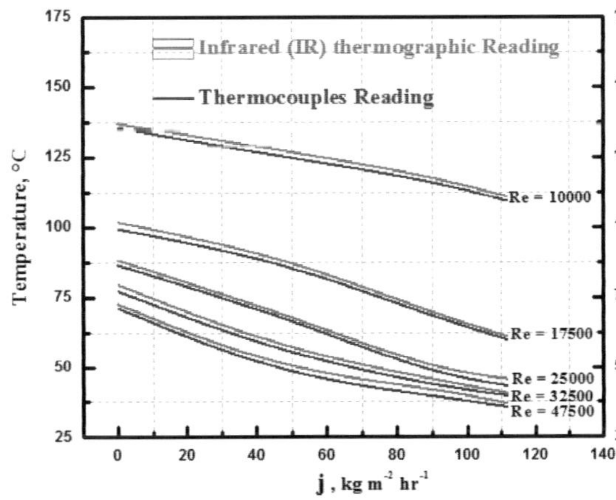

Figure 4. Comparison of surface temperature measured by infrared (IR) thermography and the surface temperature measured from thermocouples during verification tests

3. Mathematical formulation

This paper focuses on determination the average and local heat transfer rate in smooth sphere using infrared (IR) thermography with air as well as air / water droplets as working fluids. The uniform heat flux can be calculated by [16].

$$q = \frac{Q_{el.} - Q_{los.}}{A_s} \tag{1}$$

where $Q_{el.}$ – the overall heat generated by the electrical cartridge heater, $Q_{los.}$ – the heat loss obtained by measuring the average surface and the surrounding temperatures and estimated as 2.5% of the overall heat supplied. The local values of the heat transfer coefficient were obtained from [17].

$$h_x = \frac{q}{\left(T_{x.fs} - T_{ai}\right)} \tag{2}$$

where $T_{x.fs}$ – the local sphere temperature recorded by infrared thermal imager. For a single phase - air coolant, the average heat transfer coefficient can be estimated as [17]:

$$\bar{h} = \frac{q}{\left(T_{av.ts} - T_{ai}\right)} \tag{3}$$

where $T_{av.ts}$ – the average surface temperature obtained by the thermocouples readings indicated by the numbers 1, 2 as shown in figure 2. The average heat transfer coefficient for the air/water droplets flow can be estimated by using the wet bulb temperature at the inlet of the test section as the reference temperature that is [17]

$$\bar{h} = \frac{q}{\left(T_{av.ts} - T_{mi}\right)} \tag{4}$$

$$T_{av.ts} = \sum T_{x.ts} / n \tag{5}$$

Based on the capillary wave mechanism, the droplet diameter can be calculated using Lang equation [18].

$$d_p = 0.34\left(\frac{8\pi\sigma}{\rho F^2}\right)^{1/3} \quad (6)$$

The Reynolds number was obtained from [19]

$$\mathrm{Re} = \frac{\dot{m}_a D}{\mu A_t} \quad (7)$$

The heat transfer performance can be defined as the ratio of the heat transfer coefficient of air/water droplets flow to that of a single phase flow at constant pumping power. It can be written as [20]

$$\eta = \frac{h_{mist}}{h_{air}} \quad (8)$$

4. Experimental results and discussion

4.1. Infrared experimental results of air/water droplets flow

In this paper, two modules have been mainly studied for air/water droplets flow using Infrared (IR) temperature measurement system, the surface temperature measurement module and the heat transfer estimation module. The surface temperature plays an important role on heat transfer process. The distribution of local surface temperature over the surface of sphere as a function of diameter under different water flux is depicted in figures 5. It can be seen that the surface temperature tends to decrease gradually as water flux increases for constant surface heat flux. The surface temperature decreases about 5.5%, 20.2%, 37.6% and 49.4% compared with air - cooling under different water flux. It can be expressed by the evaporation phenomena on the sphere surface that leads to absorb a large amount of energy in the form of latent heat during the droplets evaporation process.

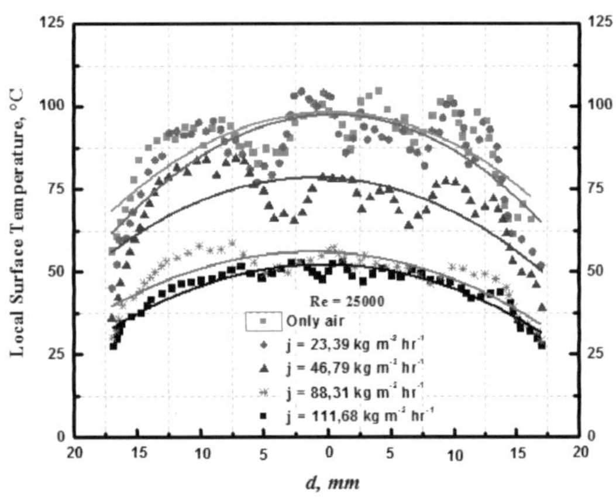

Figure 5. Distribution of the local surface temperature over the surface of sphere as a function of the diameter. (Re = 25000)

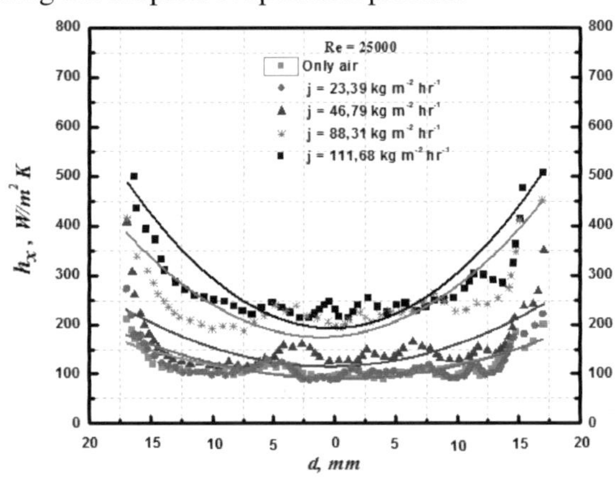

Figure 6. Distribution of the local heat transfer coefficient over the surface of sphere as a function of the diameter. (Re = 25000)

In the figures 6 the local heat transfer coefficient was related as a function of sphere diameter under range of water flux density. It can be observed that the effect of water flux density on heat transfer rate is significant for range of Re numbers. Using air/water droplets technique results in an enhancement of the heat transfer rate over air alone. The heat transfer mechanism may include three important physical effects: the convective heat transfer based on a decline temperature gradient, water mist evaporation on the heating surface and sensible heat of impinging water mist.

4.2. Direct measurement experimental results of air/water droplets flow

In the figure 7, the average surface temperature was related as a function of Re numbers under range of water flux density. It can be seen that the surface temperature tends to decrease gradually with increasing Re numbers. Also, it can be observed that the surface temperature decreases as water flux density increases. The surface temperature decreases about 4%, 17%, 37% and 47% compared with air – cooling under different water flux. In the figure 8, the dimensionless heat transfer performance was related as a function of Re numbers under range of water flux density. It can be seen that the heat transfer performance for all cases are generally more than unity and that indicate the effect of suspending water droplets into main flow on heat transfer behavior. At range of Re numbers (Re = 2500 - 10000), the heat transfer performance were in a range 1.0–1.26 for all water flux cases. This can be expressed that, when the surface temperature is hot (Ts >140), the water droplets may be completely evaporated before arriving to the heated surface due to the force of evaporation without wetting the heated surface. The heat transfer performance rapidly increases with increasing of water flux density for all cases of Re numbers. At Re numbers (Re = 47500), enhancement factor is 2.7 times of that with only air-cooling for water flux density (111.68 kg m^{-2} hr^{-1}).

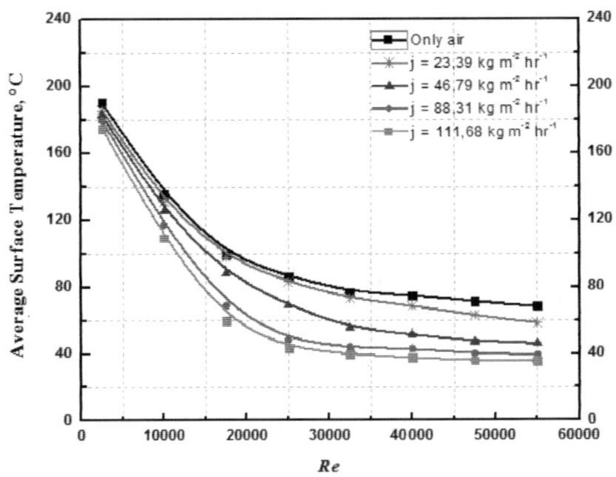

Figure 7. Variations of average surface temperature of sphere with Reynolds number under various water flux density.

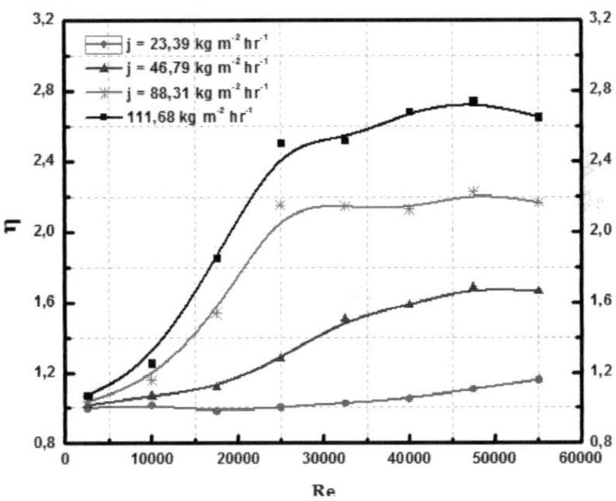

Figure 8. Variations of heat transfer performance with Reynolds number under range of water flux density.

5. Conclusions

A steady heat transfer and temperature profile measurements method for spherical element based on infrared (IR) measurement system is proposed in this paper. Based on the experimental results, the main conclusions can be summarized as follows:

- The results revealed the need to take into account changes of moisture content in air ambient in the development of refined methods to calculation of heat supply systems of buildings and structures.
- The infrared thermography (IR) it has proven to be very efficient in measuring the surface temperature distribution. The results obtained by infrared thermography (IR) are compared with a calibrated thermocouples and the temperature distribution is found to be in close agreement with 2.94% average error.
- The air/water mist cooling it has proven to be very efficient technique in removing heat from heated sphere. The surface temperature decreases about 4%, 17%, 37% and 47% compared with air-coolant by using different water flux density.
- The heat transfer performance for all cases are generally more than unity indicating that the presence of water droplets with different water flux given more efficient heat transfer enhancement than the application of air-cooling.

IOP Conf. Series: Materials Science and Engineering **481** (2019) 012033 doi:10.1088/1757-899X/481/1/012033

References

1] Tan X, Zhang J, Liu B, and Zhu X 2013 Experimental investigation on heat transfer enhancement of mist/air impingement jet *Science China Technological Sciences* **56** (**10**) 2456 2464

[2] Li X, Gaddis J and Wang T 2001 Modeling of Heat Transfer in a Mist/Steam Impinging Jet *J. Heat Transf.* **123(6)** 1086

[3] Wang T and Dhanasekaran T 2010 Calibration of a Computational Model to Predict Mist/Steam Impinging Jets Cooling With an Application to Gas Turbine Blades *J. Heat Transf.* **132(12)** 122201

[4] Yuen M and Chen L 1978 Heat-transfer measurements of evaporating liquid droplets International *Int. J. Heat Mass Transf.* **21(5)** 537-542

[5] Wataru N, Heikichi K and Shigeki H 1988 Heat transfer from tube banks to air/water mist flow *Int. J. Heat Mass Transf.* **31(2)** 449-460.

[6] Issa R 2010 Heat transfer optimization for air-mist cooling between a stack of parallel plates *Journal of Thermal Science* **19(3)** 253-260

[7] Li X and Wang T 2008 Two-Phase Flow Simulation of Mist Film Cooling on Turbine Blades with Conjugate Internal Cooling *J. Heat Transf.* **130(10)** 102901

[8] Kumari N, Bahadur V, Hodes M, Salamon T, Kolodner P, Lyons A and Garimella S 2010 Analysis of evaporating mist flow for enhanced convective heat transfer *Int. J. Heat Mass Transf.* **53(15-16)** 3346-3356

[9] Pakhomov M and Terekhov V 2010 Enhancement of an impingement heat transfer between turbulent mist jet and flat surface *Int. J. Heat Mass Transf.* **53(15-16)** 3156-3165

[10] Torfeh S and Kouhikamali R 2015 Numerical investigation of mist flow regime in a vertical tube *International Journal of Thermal Sciences* **95** 1-8

[11] Y H Li, X G Sun, G B Yuan 2007 Accurate measuring temperature with infrared thermal imager *Opt. Prec. Eng.* **15(9)** 1336–1341

[12] G H Li, L X Wu, M Wu and G X Qu 2004 Current status and applications of infrared thermography *Infrared Laser Eng.* **33(3)** 227–230

[13] Zhang Y, Wei B and Fu X 2014 An unsteady temperature field measurement method for large hot cylindrical shell forging based on infrared spectrum *Measurement* **58** 12-20

[14] Liu T and Pan C 2016 Infrared thermography measurement of two-phase boiling flow heat transfer in a microchannel *Appl. Therm. Eng.* **94** 568-578

[15] Kudo T Sekiguchi K, Sankoda K, Namiki N and Nii S 2017 Effect of ultrasonic frequency on size distributions of nanosized mist generated by ultrasonic atomization *Ultrasonics Sonochemistry* **37** 16-22

[16] Jiang G, Shi X, Chen G and Gao J 2018 Study on flow and heat transfer characteristics of the mist/steam two-phase flow in rectangular channels with 60 deg. ribs *Int. J. Heat Mass Transf.* **120** 1101-1117

[17] Wang T, Gaddis J and Li X 2005 Mist/steam heat transfer of multiple rows of impinging jets *Int. J. Heat Mass Transf.* **48(25-26)** 5179-5191

[18] R J Lang 1962 Ultrasonic atomization of liquid *J. Acoust. Soc. Am.* **34** 6–8

[19] Abed A and Shcheklein, S 2018 Investigation of hydrodynamic characteristics of laminar flow condition around sphere using PIV system *Journal of Physics: Conference Series* **1015** 032001

[20] Bian Q, Wang J, Chen Y, Wang Q and Zeng M 2017 Numerical investigation of mist/air impingement cooling on ribbed blade leading-edge surface *Journal of Environmental Management* **203** 1062-1071

IOP Conf. Series: Materials Science and Engineering **481** (2019) 012034 doi:10.1088/1757-899X/481/1/012034

The analysis of the possibility of use of nanofluid as the heat carrier for increase in efficiency of heat supply systems

V V Girfanova[1], A G Gevorgyan[1], V I Velkin[1]

[1]Department of Nuclear Power Plants and Renewable Energy, Ural Federal University, Mira Street 19, Yekaterinburg 620002, Russia

E-mail: gghayk@gmail.com

Abstract. The article is about possibility of applying nanofluids in heat supply systems as a coolant. The most effective nanofluids were selected by analyzing the thermal conductivity, concentration, size, mass, and velocity of nanoparticles. It is presented the dependences of the heat transfer coefficient of a nanofluid on the Reynolds number and the heat transfer coefficient on the volume concentration. It has been calculated, that with the addition of 1% vol. of nanotubes to the coolant path that gives heat in the evaporating section of the heat pump, the heat transfer coefficient of a nanofluid increases by 100%, and with adding 2% vol. or 3% vol. by 200% and 300%, respectively.

1. Introduction

The use of nanofluids in heating and ventilation systems can give a significant increase in heat transfer.

For the use of nanofluids as a coolant it is necessary to have the knowledge of their thermophysical properties. For this purpose we have worked on analysis of nanoliquids researches.

In works [1, 3, 4] on the study of heat transfer in nanofluids it is demonstrated heat conductivity of suspensions of ultradisperse oxides of aluminum, silicon and titanium in water at volume concentration about several percent, exceeds the thermal conductivity of pure liquid by tens of percent. Results of experiment with nanoparticles of various size show that heat conductivity of liquid on the basis of larger particles is rather well de-scribed by theory of Maxwell [2].

First of all, the obtained data were correlated with theoretical models, for describing the thermal conductivity of coarse-dispersed suspensions[9]. The first model of this type was created by Maxwell [5], who obtained a relation between the coefficients of thermal conductivity of the suspension λ and the carrier fluid λ_o.

2. The experimental data

The experimental data showed a duality: the thermal con-ductivity of nanofluids depend not only on the concentration and thermal conductivity of the nanoparticles, but also on the material and the size of the nanoparticles. Considerable increase of heat conductivity of nanoliquids is possible even in small concentration of particles[10]. At the same time, extent of this increase significantly depends on the mass of nanoparticles and the square dependence of growth of heat conductivity on the mass of nanoparticles shows the square of dependence on density: $\Delta \lambda \approx (p_d / p_m)^2$, where p_d – density of material of nanoparticles.

On the Setaram Sensys EVO TG-DSC installation the specific heat of the nanopowders SiO_2 and SiC was determined by experimental data [6] in the range of temperatures from 30°C to 80°C. With increase of temperature specific heat increases as well.

According to the experimental data [7], the heat capacity of the nanofluid is well described by the expression (1):

$$Cp = \frac{\lambda}{\alpha \times \rho} \tag{1}$$

where λ is the thermal conductivity of the sample, W/(m×K); α – thermal diffusivity, m^2/s; ρ – density of the sample, kg/m^3;

The results of calculating the heat capacity of the samples are given in table 1 [7].

Table 1. Thermal capacity(*Cp, J/(kg×K)*) of the studied heat carriers depending on concentration of nanoadditives.

Concentration, %	Name of coolant: Water+graphite	Water + carbon nanotubes	Water solution ethylene glycol + carbon nanotubes
0.2	3706.8	5204.4	2076.9
0.3	3748.2	5338.9	2037.8
0.35	3753.5	5444.8	1993.0
0.4	3774.2	5465.4	1950.5

The obtained results show that the heat capacity of the nanofluid exceeds the heat capacity of pure water and the growth dependence increases with increasing nanoparticle concentration.

The analysis of results [8] and table 1 showed that the use of nanofluid with carbon nanotubes is most effective.

Due to the theory of similarity we will calculate liquid thermolysis coefficient – waters and nanoliquids - water + carbon nanotubes. Let us consider the set-in turbulent flow of liquid with constant physical properties in a round cylindrical pipe. Initial data: d = 40 mm; T_{zh} = 105°C; T_c = 95°C; l = 4.4 m. at ϑ = 0.3 m / s; Re = 42253; $Nu_{zh,d}$= 131.20.

The heat transfer coefficient is determined by Nusselt's number, at the same time heat conductivity of liquid is calculated at temperature equal to average of temperatures on an entrance and an exit. When $Nu_{zh,d}$=131,2 and the concentration of carbon nanotubes is 1% vol., the heat transfer coefficient, according to formula (2) and (3) are:

$$\alpha_{zh} = \frac{Nu_{zh,d} \times \lambda_{ж}}{d} = \frac{131{,}2 \times 0{,}55}{0{,}04} = 1804{,}0 \frac{Вт}{m^2 \times c} \tag{2}$$

$$\alpha_{н} = \frac{Nu_{н,d} \times \lambda_{н,}}{d} = \frac{131{,}2{,}\times 1{,}1}{0{,}04} = 3608{,}0 \frac{Вт}{m^2 \times c} \tag{3}$$

Thus, the heat transfer coefficient of a nanofluid increases by 100%, in addition of 2% vol. or 3% vol. by 200 and 300%, respectively.

According to the calculations, in the figure 1 the dependence of the relative heat transfer on the volume concentration of nanotubes is presented.

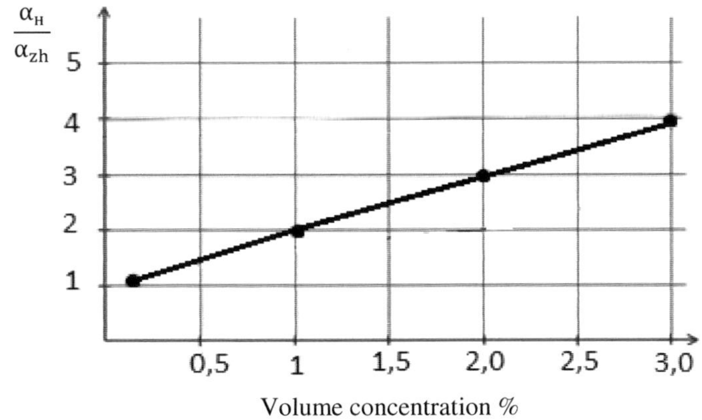

Figure 1. Dependence of the relative heat transfer on the volume concentration of nanotubes in the heat carrier.

It can be seen from Figure 1, when the concentration of nanoparticles increases, the heat transfer coefficient increases as well in comparison with pure water. It means that it is possible to reduce the surface area of heating of heat-gas supply systems' heat exchange devices.

The dependency graph of the heat transfer coefficient of nanofluid water/carbon nanotubes on Reynolds number is submitted in the figure 2.

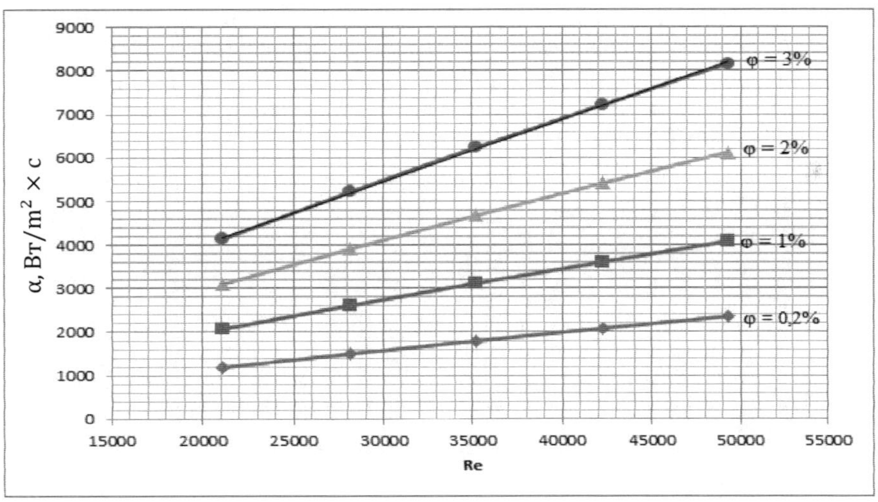

Figure 2. Dependence of the heat transfer coefficient of a nanofluid on the Reynolds number.

3. Conclusion

The obtained dependences show that a significant increase in the heat transfer coefficient is possible with a slight increase in the Reynolds number. At the same time, the degree of this increase essentially depends on the concentration of nanoparticles and their velocity.

On the basis of the study, it can be assumed that the use of nanofluids has good prospects for improving heat exchange devices and improving the energy efficiency of heat supply systems. The most effective is the use of a nanofluid with carbon nanotubes.

IOP Conf. Series: Materials Science and Engineering **481** (2019) 012034 doi:10.1088/1757-899X/481/1/012034

References

[1] Eastman J A, Choi S U S, Li S, Thompson L J and Lee S 1998 Enhanced thermal conductivity through the development of nanofluids Materials reearch society (Boston: FallMeeting) pp 3–11

[2] Choi E S, Brooks J S, Eaton D L, Al-Haik M S, Hussaini MY, Garmestani H, Li D and Dahmen K 2003 Enhancement of thermal and electrical properties of carbon nanotube polymer composites by magnetic field processing *Journal of Applied Physics* **94** pp 6034–6039

[3] Wang X, Xu X and Choi S U S 1999 Thermal conductivity of nanoparticlefluid mixture *J. Thermophys. heat trans.***13** pp 474–48

[4] Maxwell J C 1881 *A treatise on electricity and magnetism* vol. 2 (Oxford: Clarendon Press) p 1435

[5] Rudyak V Ya 2005 Statisticheskaya aerogidromekhanika gomogennykh i geterogennykh sred. T. 2.Gidromekhanika [Statistical aero hydromechanics of homogeneous and heterogeneous environments, vol. 2: Hydromechanics] (Novosibirsk: NGASU edition) p 468

[6] Gul'bin V N 2011 Razrabotka kompozitsionnykh materialov, modifitsirovannykh nanoporoshkami, dlya radiatsionnoy zashchity v atomnoy energetike[Development of composite materials modified with nanopowders for radiation protection in nuclear power] *Yadernaya fizika i inzhiniring[Nuclear Physics and Engineering]* **2** pp 272–286

[7] Anakulov M M 2014 liyaniye uglerodnykh nanotrubok na izmeneniye teplofiziche-skikh i elektrofizicheskikh svoystv vodnogo rastvora etilenglikolya 65 (antifriz) i vody [Influence of carbon nanotubes on the change of thermophysical and electrophysical properties of an aqueous solution of ethylene glycol 65 (antifreeze) and water] (Moscow) p 74

[8] Trubitsyna G N, Frolikova V S and Barzenkova V V 2016 Otsenka vozmozhnosti ispol'zovaniya nanozhidkostey v sistemakh teplosnabzheniya i ventilyatsii[Evaluation of the possibility of using nanofluids in heat supply and ventilation systems] *Mezhdunarodnaya nauchno-prakticheskaya konferentsiya «Molodoy uchenyy - vyzov i perspektivy»[International Scientific and Practical Conference "Young Scientist: Challenge and Prospects"]* **8** (Moscow) p 206

[9] Danilov V S, Korzhavin S A, Shscheklein S E and Velkin V I 2012 experimental research of efficiency of the combined system of solar heat *International Scientific Journal for Alternative Energy and Ecology* **3** p 77

[10] Velkin V I, Shkolniy A V, Kirillov M P, Achkeev M V, 2007 *Patent for the useful model RUS 69942* 16.07.2007

IOP Conf. Series: Materials Science and Engineering **481** (2019) 012035 doi:10.1088/1757-899X/481/1/012035

Optimization of Work of Ventilation System in the Building with "Warm Attic"

A D Sukhov[1], M G Ushakov[1]

[1]Department of Heat and Gas Supply and Ventilation, Ural Federal University, Yekaterinburg, 620002, Russia

E-mail: suhov-ad@bk.ru, paritet.m@mail.ru

Abstract. The article is devoted to the optimization of the ventilation system of a residential house with a warm attic. A new technical solution was proposed for the installation of an industrial fan in the space of a warm attic. The operation of such a system is provided by the phenomenon of ejection. The article presents the proposed method for calculating the ejection stream of an industrial fan. The advantages of this ventilation system were listed on the basis of the analysis. There is a table comparing the centrifugal exhaust fan with individual exhaust fans in the conclusion of the article.

1. Introduction

The paper analyzes the functioning of the ventilation system in a building with a "warm attic", reveals the shortcomings of its operation, propose and calculate new technical solutions for optimizing its operation, resource and energy saving.

The ventilation of residential buildings is provided in a natural way, without the use of machines, due to the gravitational pressure due to the difference in the temperatures of the external and internal air, and also under the influence of wind.

The inflow of outside air occurs through special valves installed in walls or window frames, and removal is carried out through the air intake devices installed in the rooms and further into the duct system, into the warm attic and the exhaust shaft (figure 1) [1].

At present, the use of a warm attic is widespread in construction. Also, since the introduction of this structural element in 1980, millions of square meters of housing have been built throughout the Russian Federation.

2. Relevance and main task

Warm attic as an energy-saving component is effective. It allows to exclude heat loss through the overlap of the last floor. But during the operation of buildings with a warm attic problems are identified, the solution of which is an important task.

Normative and recommendatory documents with the receipt and generalization of experience in designing and operating a warm attic have been amended and supplemented. This shows the urgency of solving problems with this constructive element.

The main task of the research task is to find technical solutions that will help solve the standard problems of buildings with a warm attic and optimize the operation of the ventilation system of these buildings.

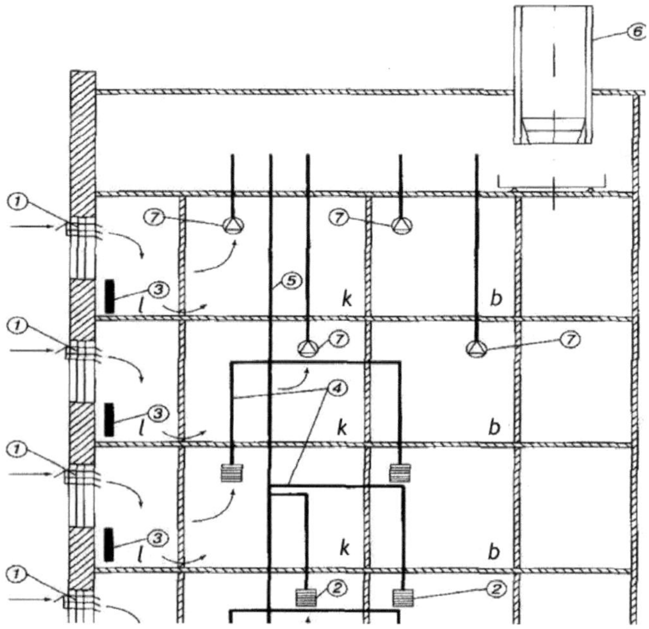

Figure 1. Scheme of the natural ventilation system with the separate and common prefabricated exhaust duct in a building with a warm attic.
Notation: *l* – living quarters; *k* – kitchen; *b* – bathroom; 1 – inflow unit; 2 – exhaust device; 3 – Heater; 4 – exhaust ducts; 5 – prefabricated exhaust duct; 6 – exhausts shaft; 7 – exhausts fan (individual).

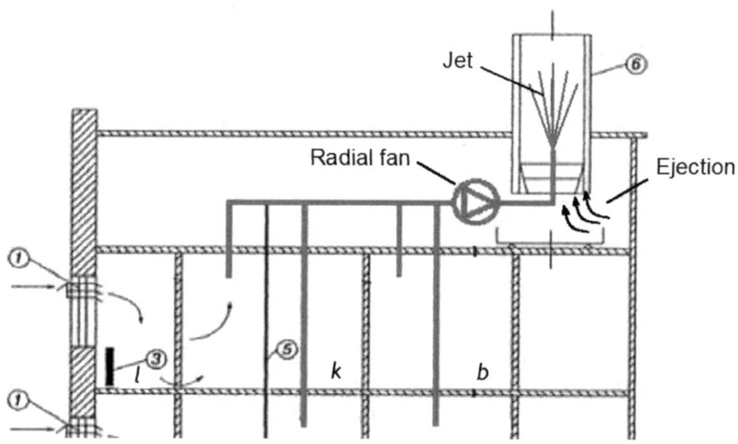

Figure 2. Scheme natural ventilation system with the ejector.

3. Technical solutions

There are many ways to optimize the operation of building ventilation systems with a warm attic. The following technical solutions were adopted in this work:

- The applied exhaust channels of GOST 17079-88 have a large cross-sectional area. Aerodynamic calculation shows the possibility of reducing the cross-sections of the channel as a consequence of the higher available pressure compared to the pressure losses in the ventilation system.
- Air removed from the premises of the last floors of buildings using household fans will be removed by an industrial fan that also removes air from these rooms, but delivers it to the exhaust shaft. The jet ejects the air, which is removed by natural ventilation from the total volume of the attic and leaves the exhaust shaft (figure 2). This system allows a large amount of air to be ejected with a small flow rate, which can be controlled by changing the diameter of the nozzle creating the jet. Calculating the ejection jet is a difficult task. There are basic provisions for calculating this jet further in the article.

IOP Conf. Series: Materials Science and Engineering **481** (2019) 012035 doi:10.1088/1757-899X/481/1/012035

4. The procedure for calculating the jet

There are the following items for calculating the ejection jet:

a) For the proper operation of such a system, it is necessary to determine the critical sections of the jet X_1, X_2 according to formula [2]:

$$X = \overline{X} \cdot m \cdot \sqrt{F_n}\,, \tag{1}$$

$\overline{X}$, the relative distance from the air distributor to the considered section of the jet;

F_n, the cross-sectional area of the shaft, m^2.

The distance from the nozzle to the cutoff of the exhaust shaft must be at least the section X_1 and no more X_2.

b) Speed factor m for a compact jet is determined by the formula [2]:

$$m = \frac{0.67}{\tan \alpha_{0.5V}}, \tag{2}$$

$\tan \alpha_{0.5V}$, the angle between the axis of the jet and the line joining the centre of the initial section of the jet to the point at which $V=0.5V_x$.

c) Let us determine the air flow rate L_x in the sections of the jet X_1, X_2 according to formula [2]:

$$L_x = \frac{2 \cdot x}{m\sqrt{F_0}}. \tag{3}$$

F_0, the cross-sectional area of the air distributor, m^2.

d) Ejected air flow $L_{Ejection}$ determined by the formula:

$$L_{Ejection} = L_x - L_0\,. \tag{4}$$

5. Advantages of the adopted technical solution

In the scale of mass building optimization of the sections, of the channel, of the exhaust shaft and the use of an ejecting jet allows:

- to increase the usable area of the premises;
- to reduce the amount of materials for construction (concrete and reinforcement);
- to reduce transportation costs for the delivery of reinforced concrete products;
- to reduce the labor costs for the manufacture and installation of this design;
- to work natural ventilation without disturbance during blowing exhausts shaft;
- not to fall the precipitation into the "warm attic" through the exhausts shaft, which reduces the amount of moisture in it;
- to save energy at the work of ventilation equipment;
- not to create, because of the fan installed in the attic, resistance to the operation of natural ventilation (In contrast to household);
- to work natural ventilation during the warm period of the year;
- to increase available pressure due to air ejection;
- "Overturning traction" due to wind pressure does not occur.

Table 1. Comparative table of energy consumption of ventilation equipment.

Fan Type	Apartments	Qty. for 1 apartment pcs.	Total	Flow rate (m^3/h)	Power 1 unit[a] (W)	Total power (W)
Household	4 fl. x 4 ap.	3	48	2560	8.8	422
Centrifugal	4 fl. x 4 ap.	-	1	2560	120	120

[a] Household and centrifugal fans with minimal energy consumption.

The technology with ejector jets can be used not only in new construction, but also in the reconstruction of constructed buildings, with a minimum capital cost that pays off with energy savings in the operation of ventilation equipment (Consumption is less than a minimum of 3.5 times, table 1).

6. Conclusion

A detailed analysis of the proposed method for optimizing the operation of the ventilation system of a residential building showed that the one is practically meaningful. The installation of an industrial centrifugal fan in the space of a warm attic allows solving many problems of the natural system of ventilation, as well as significantly reducing the consumed electric power.

In addition, this technical solution will allow you to select the optimal section of the ventilation unit. This will give a significant economic effect due to the increase in the useful area of the apartment and reducing the cost of ventilation units.

The resulted algorithm of calculation of an ejection stream will allow to make exact selection of the ventilating equipment for the accepted technical decision.

Reference

[1] Ministry of Regional Development of Russia 2012 TR AVOK-5.2-2012 Technical recommendations on the organization of air exchange in apartments of residential buildings (Moscow: AVOK) p 26 [In Russ]

[2] Shumilov R N 2000 *Theoretical foundations of ventilation Aerodynamics: Tutorial 2nd ed Revised and add* (Ekaterinburg: USTU) p 92 [In Russ]

[3] GOST 17079-88 Reinforced concrete ventilation blocks Technical conditions (Moscow: Standardinform) p 8 [In Russ]

[4] Ministry of Regional Development of Russia 2012 SP 60.13330.2012 Heating, ventilation and air conditioning (Moscow: FCS) p 76 [In Russ]

[5] Recommendations for the design of reinforced concrete roofs with a warm attic for residential buildings of different heights. Approved prot. OJSC "TsNIIEP Housing" dated October 25 1979 No 34 (Moscow: TsNIIEP) p 13 [In Russ]

[6] GOST 17079-88 Residential and public buildings Indoor microclimate parameters (Moscow: Gosstroy of Russia, GUP TsPP) p 14 [In Russ]

[7] Ministry of Regional Development of Russia 2001 SANPiN 2.1.2.1002-00 Sanitary and epidemiological requirements for residential buildings and premises (Moscow: Gosstroy of Russia) p 15 [In Russ]

[8] Grimitlin M I 2004 *Indoor air distribution* (SPb .: ABOK North-West) p 37 [In Russ]

Resource-saving in the operation of a building with a correctly executed waterproofing

A V Vasilev[1], V I Yamov[1]

[1]Department of construction production and expertise of real estate, Ural Federal University named after the first President of Russia B.N. Yeltsin, Mira Street 19, Ekaterinburg, 620002, Russian Federation

E-mail: a.vasilev@gydromix.ru

Abstract: The article is devoted to the issues of waterproofing of expansion and technological joints. The necessity of using reliable waterproofing methods during construction is identified and justified. Particular attention is drawn to the shortcomings of existing methods, namely bentonite and hydrophilic waterproofing cords related to installation problems and other technical issues. System installation waterstop for technological joint mating base plate-wall and the repair scheme of the one with using the polyurethane resin are described. Comparison is given of the cost of repairing the technological joint and the installation of a waterstop in new construction. The influence of the construction quality on the further cost operation of the building is revealed.

1. Introduction

The main feature of the current situation in the world is urbanization, which is an increase in the role of cities in the development of society. Cities act as areas of concentration of huge production capacities, concentration of information, innovation and cultural potential, which provide the main part of the growth of national wealth.

The cornerstone in this development is the quality construction of new facilities (factories, houses, warehouses, roads). At performance of each stage of construction according to the technological regulations there will be an increase of durability and reliability of a design.

The most important component of construction is high-quality building materials used in compliance with the technologies and regulations prescribed by the manufacturer. Improving the quality of construction will entail the saving of finance in the future operation.

The quality of products is a consequence of many factors: the quality of personnel, the quality of production organization, the quality of machinery and technology, and the quality of management.

One of the main directions for improving the quality of construction is the waterproofing of buildings and structures.

2. Materials for waterproofing

In structures exposed to water, expansion and construction joints, and cracks should be given increased attention, since more than 99% of all leaks take place through these ones that occupies less than 1% from the entire surface area of concrete. That entails expensive repairs. If you look at the market of building, you can conclude that the main emphasis is on hard waterproofing, which does not withstand cracking in the body of concrete and roll-bitumen waterproofing, which also has its problems: careful surface preparation, observance of surface moisture, priming.

Technological joints are time gap in concreting, they are created when the construction of a building requires a technological pause in the production of structures. It can be said that this is the combination of the old concrete layer with the new one, which results in the formation of a technological weak point through which leaks can occur in the future.

At the moment, the most common type of waterproofing of technological joint is swelling cords made of hydrophilic rubber and bentonite cords made from modified natural sodium bentonite clay and unvulcanized polymer (butyl rubber). Bentonite and hydrophilic waterproofing cords have the property of swelling and increasing in size when moisture contacts in them, which in theory ensures waterproofing and sealing, even under hydrostatic pressure. In the project, these cords look like a reliable structural element providing reliable waterproofing, but in practice there are many problems such as:

- Surface Unevenness – Profile Setup Problems;
- Fastening of swelling cords data – use of additional elements for fastening: anchors and sealants;
- Problems associated with leaching of bentonite cord with active water inflow (Fig. 1);
- Problems related to the preservation of the geometric dimensions of the swelling cord, made from hydrophilic rubber prior to pouring concrete: dependence on rainfall - when water contacts the cord, swelling occurs, and when pouring concrete on the swollen profile, the latter will return to its initial geometric state after some time and it will leave an empty space, due to this leaks may start.

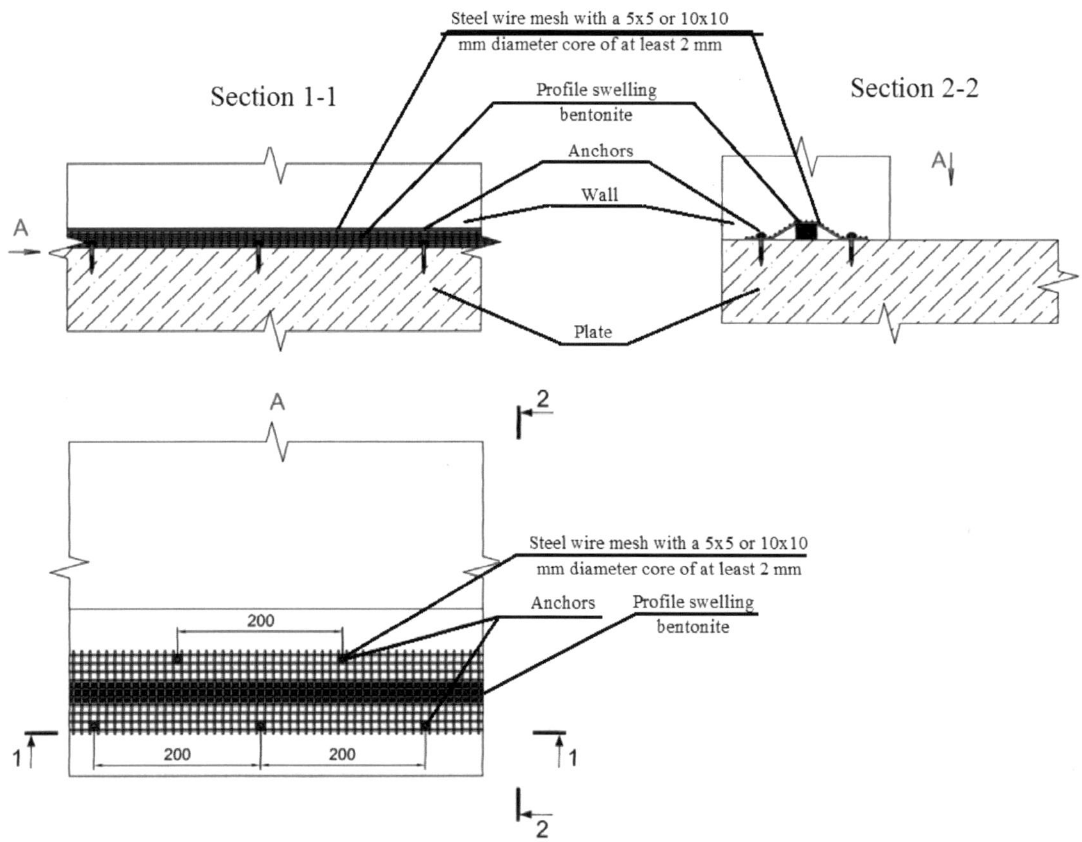

Figure 1. Installation of bentonite cord.

It is also necessary to take into account the most important element of the waterproofing of buildings and structures, namely the sealing of expansion joints. Expansion joint - is designed to reduce the load on structural elements in the places of possible deformations arising from fluctuations in air temperature, seismic phenomena, uneven sedimentation of soil and other influences that can cause dangerous own loads, which reduce the bearing capacity.

Since the cavity of the expansion joint does not require watertightness to the filler, lately the filling material is a sheet of expanded polystyrene which is embedded in the seam when it is formed. Such filling allows free compression and opening of the seam without any stresses of the conjugated elements. At the same time, sealant are used as fillers of the cavity for compacting expansion joints of small sizes up to 30 mm and pasting of roll-bituminous materials with a compensator on the expansion joints. Regularly for contour sealing of all types of joints waterproofing tapes on a polymer basis are used.

A significant problem in the waterproofing of expansion joints is its movement. For example, as the temperature decreases, the joint gap increases, and tensile stresses arise in the sealant material, which must be perceived by this one.

3. Waterstops

To date, there are special materials for technological and expansion joints, namely waterstops. These materials are made of plasticized PVC, can be joined into the system and welded, both with each other - to obtain the required length, and with other ones. Welding is carried out with a thermal knife or a construction hair dryer at a temperature of 185-195°C.

Waterstops when concreting is embedded in the body of each slab. All the waterstops have anchors (or anchor ribs) (Fig.2). Due to these protrusions, the waterstop is firmly retained in the body of the concrete elements. These concrete elements are interconnected through an elastic material - a waterstop, which plays the role of a barrier and prevents the penetration of water into the structure.

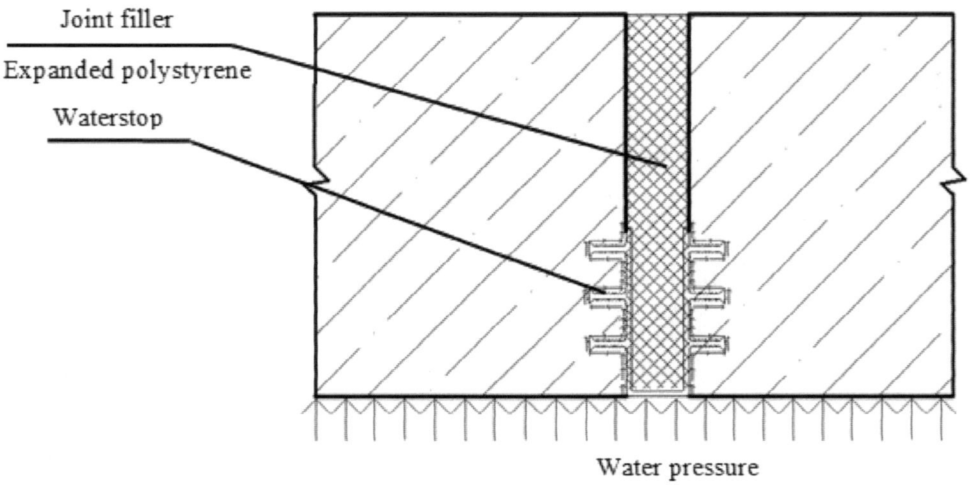

Figure 2. Waterstop in the concrete.

Joint movements are compensated:

- for technological joints - due to the elasticity of the material of the waterstop;
- for expansion joints - due to deformation of the central element of the waterstops (compensator) and the inherent elasticity of the material.

The principle of operation is very simple: due to the different anchors and ribs, these materials lengthen the path of water.

4. Comparison of cost

Now consider the practical part related to finance for waterproofing buildings and structures.

At the beginning examining at the cost of waterproofing of technological joint mating base plate-wall with using a waterstop (Fig. 3).

This waterstop is mounted using special staples (3 pieces per 1 meter), which are attached with a binding wire to the transverse reinforcing bars of the upper belt of the reinforcing cage of the foundation plate or slab. Staples prevent the movement of the waterstop from the design position. The reinforcement frame in the upper part of the outer walls should allow unobstructed placement of the waterstop.

Table 1. Cost of material and installation of waterstops for new construction.

Name	Cost of material	Cost of works	Total
Waterstop for waterproofing technological joint - mating foundation plate-wall	6$	4$	10$

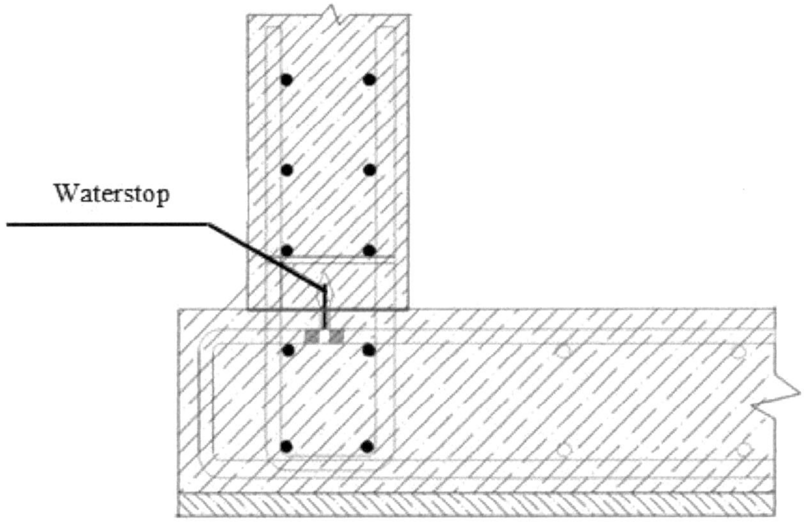

Figure 3. Waterproofing of technological joint mating base plate-wall.

Now consider the cost of repair with polyurethane resins (Fig 4).

The following water penetration technology is used:

1. Break the joint. The dimensions are 30x30 mm.
2. To put the repair material with the formation of fillets. With active water inflow, using fast-setting repair composition.
3. Drill the holes at a distance of 50-70 mm higher in one row parallel to the joint at an angle of 45 under the injection packer. The distance between the holes is 200-250 mm. Depth of drilling 2/3 of the thickness of the wall.
4. Clean the holes from cement dust, dirt and foreign elements.
5. Install injection packer.
6. Pressurize the hydroactive elastic polyurethane foaming resin with low viscosity.
7. Within 15 minutes after the injection of foam is injected with elastic polyurethane resin
8. Remove injection packer.
9. Fill the drill holes with the repair composition.

10. Apply an elastic waterproofing in two layers through a glass net for a width of 300-400 mm along the joint.

Table 2. Cost of repairs, taking into account works and materials.

Name	Cost of material	Cost of works	Total
Injection of a technological joint with active leaks using polyurethanes - mating a foundation slab-wall	44$	51$	95$

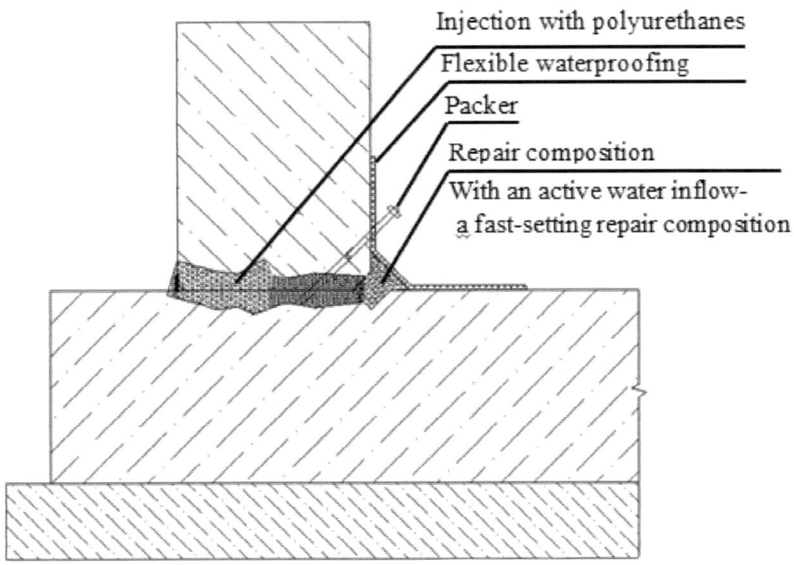

Figure 4. Repair of technological joint mating base plate-wall

The cost is calculated taking into account the exchange rate of the Central Bank of the Russian Federation as of 18.09.2018.

This cost of materials and works is calculated from the average cost for the city of Ekaterinburg (Russian Federation) and does not include overheads and estimated profits.

5. Conclusions

From the foregoing, it follows that the cost of the original waterproofing with waterstop is 9.5 times cheaper than the subsequent repair of this interface without the use of similar elements. These indicators show with certainty that with properly executed waterproofing, resource saving is occurring in the construction industry.

6. References

[1] Lyapidevsky B V, Lander A F 2008 Technological regulations for the installation of waterstops AQUASTOP for the installation and restoration of waterproofing of expansion and technological joints in reinforced concrete structures of underground and buried structures (Moscow: Typography of the construction industry of the city of Moscow) p 56

[2] Shilin A A, Zaitsev M V, Zolotarev I A, Lyapidevskaya O B 2003 *Waterproofing of underground and buried structures during construction and repair* (Tver: Russian trade mark) p 399

[3] Popchenko S N 1981 *Waterproofing of structures and buildings* (Leningrad: Stroyizdat) p 304

IOP Conf. Series: Materials Science and Engineering **481** (2019) 012037 doi:10.1088/1757-899X/481/1/012037

Review of effective tools for monitoring the performance of membrane bioreactors

A S Ilinykh[1], V A Ilinykh[1], E V Migalatiy[1]

[1]Department of water treatment, Ural Federal University, Ekaterinburg, 62002, Russia

E-mail: a.s.ilinykh@urfu.ru

Abstract. The production of high quality wastewater after treatment plants using a membrane bioreactor (MBR) probably requires less control than traditional processes of activated sludge. However, the use of MBR leads to additional problems associated with the activated sludge filtration, the formation of a sediment layer on the surface of the membrane and membrane fouling. Operation engineers require simple tools for controlling and monitoring the operation of treatment plants that can provide timely information about the biological state of the activated sludge and the filterability of the waste fluid, as well as the risks of membrane contamination. To this end, effective means of monitoring MBR and methods for determining biological activity are considered. For a broader characterization of the wastewater composition, the use of total nitrogen and total carbon analyzers is considered, which have significant advantages over the analysis of chemical and biological oxygen consumption (COD, BOD).These tools can be used regularly to ensure timely intervention and more stable monitoring of the operation of MBR systems.

Key words: membrane bioreactor, sewage, biochemical wastewater treatment, treatment plants, analysis of operation MBR.

1. Introduction

As is known, the main water users are industrial enterprises, agriculture, household needs of the urban population. Considering the fact that 75% of industrial enterprises and 72% of Russia's population are located in the European part of the country, where only 16% of surface waters are located, and the problem of ensuring the proper quality of sewage discharged into water bodies becomes more acute.

The huge volume of sewage produced by industrial enterprises and agricultural organizations represents, on the one hand, a potential danger, on the other hand is a vital resource for mankind and the environment. In Russia, as in most developed countries, the bulk of this wastewater is treated in centralized treatment facilities, where the content of pathogenic microorganisms and nutrients decreases to the required value before the sewage is discharged into water bodies. Or disinfection of wastewater for subsequent reuse in industrial plants is carried out.

The process of cleaning with the use of activated sludge, which has already become traditional, has been used for more than a hundred years for treating both domestic and industrial wastewater [11, 15]. Probably, this biological process has the longest history of application in industry. The active sludge process is used to treat wastewater of various origins in different environmental conditions.

Purification processes using activated sludge can be disrupted from various biological phenomena, such as swelling and reduced sedimentation in sedimentation tanks, which lead to the formation of a poor runoff that contains biomass particles [6]. The increase in the number of membrane bioreactors (MBRs) over the past 15 years is due to the desire to reduce pollutant emissions, increasing demands on the quality of discharged wastewater, the need for reuse of water, and new developments in the field of membrane technology [8].

MBR are a recognized technology that is currently used in construction treatment facilities [10]. Although the MBR includes the use of an activated sludge process, problems with the removal of particles from the secondary settler are mitigated by membrane separation. However, MBR are unable to process the required volume of wastewater, with critical contamination of membranes or low biomass filtration [8]. In addition, because of the need for constant membrane mixing and chemical purification to remove contaminants, the capital and operational costs of owning MBR are higher than for active sludge processes [12].

A significant part of the literature on MBR examines the influence of various properties of waste water on the process of membrane separation and the tendency to contaminate membranes [9, 12, 16]. The results of the studies convincingly indicate that higher concentrations of extracellular polymers and / or soluble microbial polymers are the keys to explaining the low biomass filterability and, as a result, increase the tendency to contamination [14]. The literature is complex and marred by a change in extraction methods and analytical methods. Given the uncertainty regarding operational issues affecting biomass filtration and the costs of owning MBR, the question of how best to monitor the operation of MBR and their biomass becomes more urgent. This review analyzes a range of analytical methods (tools) available to engineers and operators for monitoring biomass status and cleaning efficiency in MBR.

In this review, methods of chemical analysis for determination of ion concentration, ion sums of pollutants and methods for volumetric evaluation of solids are considered.

Analytical instruments were evaluated from the practical point of view, by the cost of equipment and reagent, ease of use, repeatability and time spent on the analysis, which allows you to quickly respond to changes (on the same day). Although this review is largely focused on the operation of MBR, many methods are transferred to the operation of activated sludge plants.

2. Evaluation Tools

2.1. The cost of reagents and capital equipment costs

Initially, it is worth to estimate the costs of the analysis, which include two categories. First, the cost of consumable reagents, and, secondly, the amount of capital investment. The lower these costs, the more likely that the tests should be included in the daily analysis. These costs can vary significantly depending on the manufacturer, the amount of purchase and the place of purchase.

2.2. Location

From the point of view of obtaining data that can be used on the day of sampling. Methods that can be used on-site are, of course, the most profitable.

2.3. Time

The time it takes to perform each analysis or prepare it for automated processing is another determining factor of suitability for daily use.

2.4. Training

An analysis that requires considerable scientific knowledge to perform is less desirable than one that is easy to use. To this end, technological processes are required, which are simple and require less preparation. These processes deserve more appreciation.

2.5. Accuracy

This category is intended to distinguish between tests with a low accuracy of the operational value and with uncertainty of the value reading.

3. Evaluation results

An established set of analytical chemical tests is usually used to monitor both the initial runoff and the filtrate in order to monitor the progress of the necessary chemical reactions and the removal of suspended particles. The data obtained from this analysis can help the operator understand that there are no serious process failures and that that the wastewater meets the regulatory requirements for discharge without studying the internal processes. Typically, the initial purified runoff analysis involves the determination of various ions (for example, NH_4^+, PO_4^{2-}) and total measurement (COD and TOC).

4. Ionic analyzes

The determination of the concentrations of nitrogen and phosphate ions is often carried out by colorimetric analysis using commercially available test kits. Although these kits are a quick and easy option, large treatment plants or complex processes may require testing of several parameters on many samples. In this case, the economic cost of using the test suite can quickly increase, taking into account the costs and time associated with ordering, delivery, storage, actual use and final disposal.

A recommended alternative for high throughput is ionic analysis using ion chromatography. Ion chromatography offers lower detection limits and the possibility of analyzing without interference for high-color or sulphide-containing waters, for which colorimetric determination is often unsuitable. A full range of base cations (Li_2, Na_2, NH_4, K_2, Ca_2O и Mg_2) and anions (Fl_2, Cl_2, Br_2, NO_2, NO_3, PO_4^{2-}; SO_4^{2-}) can also be quantified. It can also be used to detect transition metals such as iron and manganese.

Although the capital costs of ion chromatography are significant, one should not overlook the advantages of integrated ion analysis conducted on site, which is valuable for membrane processes that have a risk of formation (inorganic) sediment. The ability to quickly check the Lanzhelle saturation index or calcium carbonate precipitation potential, and adjust or initiate acid dosage, can save many tens of thousands of dollars for cleaning chemicals, extend the life of the membrane and shorten the time for the regeneration process.

5. Total amount of analyzes

A summary general analysis is tests that are aimed at giving an overall assessment of the sample. They usually include BOD5, COD and total organic carbon (TOC). They are often used to assess the effectiveness of organic loading or removal processes. [3] BOD5 limitations are often characterized by regulatory requirements [5].

5.1. BOD5

Biological oxygen demand (BOD5) is the amount of oxygen expended on aerobic biochemical oxidation under the action of microorganisms and the decomposition of unstable organic compounds contained in the water under study. Thus, the indicator is a relatively raw indicator of degradable material in the sample and depends on the viability of the microbial population. BOD5 has a long history of application [7] and is valuable for quantifying the potential biodegradability of a sample. For the daily operation of the station, the use of BOD5 is impractical due to a five-day incubation [4]. An obstacle to an accurate analysis in BOD can be iron ions, sulphides or reduced nitrogen compounds [3]. In the latter case, "carbon BOD" can be established by suppressing nitrogenous bacteria with nitrification inhibitors.

Despite the shortcomings of the BOD indicator, it is important to periodically compare the COD: BOD 5 ratio. This may indicate a change in the purification ability or the toxicity of the incoming substance. The cost of the reagents for this analysis is minimal, the equipment requirements include incubators and dissolved oxygen or pressure measuring devices. To date, the search for a technology that will allow the measurement of BOD5 in a short time, or can be used as an online sensor, continues.

IOP Conf. Series: Materials Science and Engineering **481** (2019) 012037 doi:10.1088/1757-899X/481/1/012037

5.2. COD
Chemical oxygen demand (COD) is the amount of oxygen consumed during the chemical oxidation of organic and inorganic substances contained in water under the action of various oxidants. This test gives results in less than 3 hours. Reagents are usually cheap, but disposal costs can be in the same order. COD as an organic load measurement can be artificially high in the presence of reactive inorganic elements, such as Fe^{2+} [3]. Despite this, COD is the most commonly used method for estimating the consumption of oxygen spent on oxidation [2]. Although interference caused by oxidation of inorganic compounds can not be a significant problem for domestic sewage, industrial wastewater can contain high concentrations of inorganic compounds in the reduced state (e.g. H_2S). In the aerobic MBR process, H_2S will be oxidized to form SO_4^{2-} ions. Therefore, the organic balance of the mass, based on COD, will require measurement of the influence and flux of various types of sulfur. Despite these drawbacks, COD is the most commonly used method for estimating oxygen demand [2]. The necessary equipment is a heating unit and a spectrophotometer, and it can be performed even in basic laboratory rooms.

5.3. TOC
Given the shortcomings in the measurement of BOD5 and COD, there is a growing trend towards direct analysis of total organic carbon (TOC) compounds in wastewater [1]. The method used for the broadest range of TOC analysis is high-temperature / infrared, but is applicable only to pure water (i.e., MBR filtrate), the UV / persulfate method has lower detection limits.

TOC analyzers are available that include simultaneous measurement of the total "bound" nitrogen, which makes it a very attractive primary tool for large laboratories. It should be noted that not all TOC analyzers are accurate enough for processing samples with solid particles [17], therefore the samples to be analyzed need to be filtered using a 0.45 mm filter prior to TOC analysis. A more appropriate name for the analysis results will be the total amount of dissolved organic carbon (KRU), and not the total amount of organic carbon of TOC. More reliable analyzers have the ability to determine both TOC and CRU and allow a more complete picture of the movement of carbon compounds.
The accuracy and usefulness of analyzers of this type is high, which could not but affect the magnitude of the capital costs that are associated with the high cost of the device. However, the cost of consumables is the minimum amount of high purity oxygen. Laboratory requirements are the desktop space, access to the oxygen of the bottle and clean water of the required quality. The analysis can be carried out in fully automatic mode, without the need for operator intervention.

6. Analysis of suspended solid waste water particles
In the MDB system, the analysis of the undiluted water (solids in which it is suspended) can provide important information either about the removal of nutrients or about the propensity for contamination. For example, the level of ammonia, nitrate and orthophosphate in supernatant water taken from different zones can help determine the causes of nitrification / denitrification problems or the removal of phosphates. Normally, the supernatant water is separated from the biomass by centrifugation and then filtered (0.45 mm) before analysis by analysis [13]. Supernatant water can significantly differ from the filtrate of the MBR process in terms of the content of organic substances and colloidal materials. This is due to the preservation of all materials that are not soluble enough to pass through the MBR membrane. Sampling and analysis of the water in the water require more caution than simply measuring the filtrate. The foam present on the top of the MBR reactors can contaminate the samples. Secondly, centrifugation and filtration should occur as soon as possible after sampling [6], or the results will depend on the continuation of microbial removal of substrates. For example, phosphate accumulating organisms (PAO) can release phosphate in conditions of low dissolved oxygen.

7. Conclusions
Timely and reliable analyzes are of vital importance for the stable and economical operation of MBR. Various technical solutions were considered and evaluated to create a standard set of timely and

reliable analytical tools for monitoring and regulating the operation of MBR. The choice of technologies for monitoring the operation of MBR and their biomass should be carefully analyzed. For less equipped laboratories, colorimetric analysis (including COD) will remain the basis of the laboratory operation. It is recommended to invest large treatment plants of industrial enterprises in their own local laboratory systems that are capable of measuring nutrient elements, and also helped in monitoring sedimentation on the surface of membranes. The combination of such indicators as total organic carbon, total nitrogen yield results that are faster and more accurate than BOD, and have a lower environmental load than COD. The introduction of the latter into local laboratories will prevent violations in the operation of treatment plants, as well as avoid a complete violation of the cleaning process, which will ensure the exclusion of emergency discharges of waste water into water bodies.

8. References

[1] Aziz J A and Tebbutt *et al* 1980 Significance of COD, BOD and TOC correlations in kinetic models of biological oxidation *Water Resources* **14** pp 319-324

[2] da Silva *et al.*, 2011. Optimization of the determination of chemical oxygen demand in wastewaters *Analytica chimica acta* **699** pp 161-169

[3] Frimmel F H *et al* G 2011 3.01-Sum Parameters: Potential and Limitations (Treatise on Water Science. P. Wilderer/Elsevier, Oxford) pp 3-29

[4] Henze M Y *et al* 2008 *Wastewater Characterisation. Biological Wastewater Treatment: Principles, Modeling and Design* (London: IWA Publishing)

[5] Higgins J *et al* 2004 A Review of Water Quality Criteria in Australian Reclaimed Water Guidelines and Sewage Effluent Discharge Licences (Environmental Protection Agency)

[6] Jenkins D *et al* 2004 *Manual on the Causes and Control of Activated Sludge Bulking and Foaming and Other Solids Separation Probelms* (Lewis Publishers)

[7] Jouanneau S *et al* 2014 Methods for assessing biochemical oxygen demand (BOD): a review *Water Resources* **49** pp 62-82

[8] Judd S and Judd C 2011 *The MBR Book In: Principles and Applications of Membrane Bioreactors for Water and Wastewater Treatment* second ed. (Butterworth-Heinemann, Oxford) pp 55-207

[9] Kiristaev A V 2008 Ochistka stochnykh vod v membrannom bioreaktore [Wastewater treatment in a membrane bioreactor] *Synopsis of a thesis for a degree of Candidate of Technical Sciences "Water supply, sewage, construction systems for water resources conservation* (Moscow:) p 27

[10] Kraemer J T *et al* 2012 A practitioner's perspective on the application and research needs of membrane bioreactors for municipal wastewater treatment *Bioresource Technology* **122** pp 2-10

[11] Lofrano G and Brown J 2010 Wastewater management through the ages: a history of mankind *Science of the Total Environment* **408** pp 5254-5264

[12] Rosenberger S *et al* 2005 The importance of liquid phase analyses to understand fouling in membrane assisted activated sludge processesdsix case studies of different European research groups *The Journal of Membrane Science* **263** pp 113-126

[13] Rosenberger S and Kraume M 2002 *Filterability of Activated Sludge in Membrane Bioreactors* (Amsterdam: Elsevier Science B.V., P.O. Box 211)

[14] Sheng G-P *et al* 2010 Extracellular polymeric substances (EPS) of microbial aggregates in biological wastewater treatment systems: a review *Biotechnology Advances* **28** pp 882-894

[15] Tilley D F 2011 *Aerobic Wastewater Treatment Processes; History and Development* (London: IWA publishing)

[16] Trunov P V 2010 Osobennosti protsessa ochistki stochnykh vod v pogruzhnykh membrannykh reaktorakh *National University of Urban Economy* **93** (Kharkiv) pp 133-137

[17] Vanrolleghem P A and Lee D S 2003 On-line monitoring equipment for wastewater treatment processes: state of the art *Water Science and Technology* **47** pp 1-34

IOP Conf. Series: Materials Science and Engineering **481** (2019) 012038 doi:10.1088/1757-899X/481/1/012038

Climatized module as the basic unit of high-tech medicine. Problems of development and application

V A Tomilin[1], M Yu Ananin[1], N D Perfilieva[1]

[1]Department of Architecture, Ural Federal University, Yekaterinburg, 620002, Russia

E-mail: vtxceasarx@gmail.com, m.y.ananin@urfu.ru, nataly_pnd@rambler.ru

Abstract. In this article a study of climatized module application as a product for clean rooms complex formation in different technological schemes was made. The problems of clean rooms modular formation as well as its all advantages and disadvantages are considered. The clean rooms classification was analyzed. The scheme of climatized module classification has been derived. The problems of modern rules and regulations in sphere of clean rooms organization in Russian Federation was revealed and assumptions for improving of this situation were made. The following issues are considered: ensuring a high level of clean rooms equipping and execution; development of modular construction in medicine, pharmaceutical production, production of microelectronics and other fields on the climatized module basis; the possibility of upgrading obsolete technological solutions; creation of a structural classification of the climatized module for both medical technology and for various areas of possible application. The conclusions on the research were made.

Key words: clean rooms, complex, prefabricated building, medical climatized module, particle concentration, hermetically sealed.

1. Introduction

In the development of modern high-tech production clean rooms (hereinafter CR) construction and application occupies an important place [1].

In the usual premises of industrial or medical processes air purity is estimated by the mass concentration of pollutions in the air. Unlike them, CR are characterized by a countable concentration of particles, that means their number per air volume unit. The main factor in CR concept determining is that these rooms are characterized precisely by the countable concentration of particles, i.e. the number of particles per air volume unit as well as the size of these particles (0.1, 0.3, 0.5 microns and etc.). Therefore the peculiarities of maintenance and determination of purity indicators, specific requirements for control devices, air particle counters, interior finishings, general climate support system, etc. 2].

The purpose of this study is to improve the medical services level through the competent climatized module (hereinafter MK) use in the clean room complex (hereinafter CRC) organization and the creation of better conditions for high-tech industries (pharmaceuticals, microelectronics, military industry, etc.).

This goal determines the following objectives: 1) ensuring a high level of equipment and execution of CR; 2) development of modular construction on the CM basis; 3) the possibility of obsolete technological solutions upgrading; 4) the creation of CM structural classification for both medical technology and various areas of possible application. The study was carried out by the method of morphological analysis of information and its synthesis.

2. State of the field

The idea of using CM in high-tech medicine is not the new one (fig. 1). Throughout the entire world, the creation of hermetic, multi-profile, antibacterial reusable rooms with the possibility of high factory availability has proved its superiority over traditional interior finishes. In several countries the CRC development is standardized by regulatory documents. For example, in USA special rules for CR interior finishing by definite kinds of materials, that have proved themselves well enough over the past 30 years, and certain norms for air exchange and equipping such premises has been adopted. In Russia CM implementation for CRC is insufficiently explored theme and it has limited application because of inadequate compliance of regulatory documents with modern realities, as well as the high cost materials for CM. However, CM development is changing and some companies actively engaged in CM development and application for CRC aims [3].

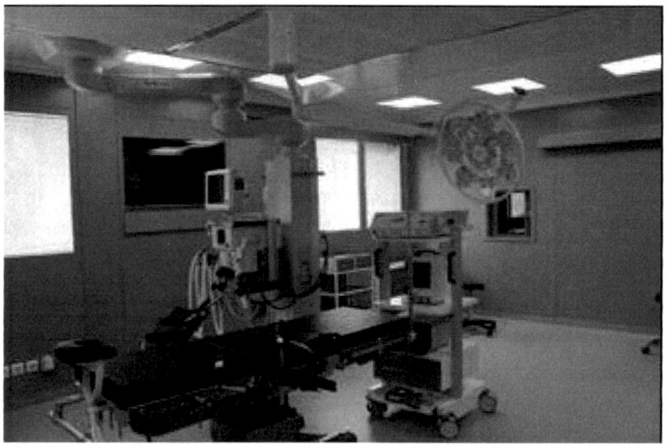

Figure 1. Application example of CR on CM basis for operating room.

3. CM design features

CM is a separately designed, multifunctional and reusable module for the organization of special cleanliness classes rooms. CM is a fast-erectable construction (prefabricated building), which has a huge number of advantages over the traditional finishing in the field of CRC formation in pharmaceutical production, micro-electronics production, the formation of special cleanliness classes premises in medical technology, etc.

Its main features are followings: 1) premises air-tightness and high-tech antibacterial finishing with increased wear resistance; 2) CM is a full-fledged integrated solution that includes engineering systems solutions such as ventilation, automation engineering, water treatment and water disposal, electrical engineering, equipping with necessary technologies, etc.; 3) increased factory readiness and an individual project for further CRC servicing and reorganization development; 4) the possibility of multiple use of the CM (dismountable technology) as a prefabricated building (hereinafter PFB); 5) the possibility of complex technological units formation, when using CM as PFB or stand-alone self-supporting modules [4].

The CM development in our country and all around the world has made it possible to reduce the undesirable consequences of the post-operation period, as well as to form full-fledged technological processes without violating the purity of air exchange, cleanliness classes, and the room mismatch in the process performed in this unit. It should be noted that the execution and development of the CM

IOP Conf. Series: Materials Science and Engineering **481** (2019) 012038 doi:10.1088/1757-899X/481/1/012038

must be carried out by high-quality specialists with extensive experience in this field, understanding of technological processes, the ability to differentiate technologies (for example, medicine from pharmaceutical production and etc.), and also be able to verify the products quality standards [5, 6].

However, despite all the advantages of CM for CRC, they have a number of disadvantages that limit the CM development in Russia and their mass use. Such shortcomings include: 1) the lack of a strict regulatory and legal framework for the design, use and implementation of CM in CR. Nowadays, there are a number of regulatory documents that describe the principle of constructions and implementing of the CRC in Russia. They consider the air cleanliness classes, finishing materials quality, air-tightness, chemical treatment and resistance to it and other aspects. But at the same time, in that documents there is no clear explanation of how the CRC should look in the general constructive and technological execution. Nowhere is it said about the possibility of CM use for the CRC organization as medical equipment. This disadvantage leads to unfounded freedom of choice, which encourages irrational solutions, as well as the design decisions complication; 2) the high cost of the full complex for the CR organization on the CM basis (from the project development to the CM commissioning). It is worth to mention the high price is determined by final products high quality characteristics, as well as by the uniqueness of each manufactured unit. Perhaps the creation of a universal modules series, as well as design solutions that can be laid out as separate parts in modern technology, with all the previously worked out and calculated engineering loads and equipment, will lead to a price reduction in both in production and in designing. After all it is necessary to remember that the CRC organization with the help of CM is the development of high-tech medicine, and, consequently, the improvement of the public health quality [7, 8].

4. CM specific features

CM is a module that is directly related to PFB [9]. According to its structural base it is a metal frame, which can be executed in several forms [10]. The most used and popular ones are: 1) stand-alone modules inside a common room. These modules have a frame, which is covered with enclosing panels. All necessary engineering systems are located under the floor and (or) in the thickness of the wall structures. Air purity maintenance systems (air exchange and air conditioning systems) are located on floors and places specially designated for them, or on the module itself. 2) The module that is installed in the finished premises. This module is a metal frame inside the room with hanging panels. All engineering systems are behind the panels. 3) The module, which is the PBF itself. This stand-alone module can be used to form buildings by combining several elements. A good example of this use of CM may be feldsher-midwife station (hereinafter FMS), which are built in hard-to-reach areas or in areas with different emergencies [11].

Enclosure structure is a composite panels made of various materials with special covering, which must withstand a large number of chemical treatment cycles. Such panels can be plaster-metal, panels made of HPL plastics, stainless steel filled with gypsum wall board and etc. The ceiling is made using the clip-in system, for the possibility of constant control over the ceiling space and engineering systems [12].

The main feature of the whole CM is that it is hermetically sealed. Floor covering, ceiling structures hairline joints, joints of wall fences, doors, glass blocks - all these elements are carried out hermetically with the help of sealing rubber, shapes, sealants and etc [13-15]. For example, in the company LLC «NPO StroyMedService» a unique system of fixing and sealing HPL panels in the CM was developed. Their method allows fixing and sealing the panels with the possibility of their further «hot» replacement, or completing reorganization.

The special design, the structure air tightness, the sustainable development, the modern technologies use and the multifunctional performance are defined CM as modern medical equipment, which should and can improve the level of CRC execution, the level of public health, and increase the implementation level of various production facilities where CR is needed.

IOP Conf. Series: Materials Science and Engineering **481** (2019) 012038 doi:10.1088/1757-899X/481/1/012038

5. CR classification

As it was mentioned earlier, the CM for CRC has a number of design features that depend on the CR classification.

An important CR characteristic is the definition of its class. The CR is characterized by a classification number that determines the allowable concentration of micro-particles in 1 cube meter of air. In the CR, several clean zones can be organized. Each clean zone should be in a laminar airflow, which allows to create local clean zones, such as a laminar flow unit. The nature of the air flow in the CR largely depends on the features of the construction of both the CM and the climate systems.

In table. 1 the CR classification according with interstate standard GOST ISO 14644-1 «Cleanrooms and associated controlled environments – Part 1: Classification of air cleanliness by particle concentration» is illustrated.

Table 1. Cleanliness classes according to GOST ISO 14644-1.

Cleanliness class	The maximum permissible number of particles in 1 cube meter of air with dimensions equal to or greater than, μm					
	0,1	0,2	0,3	0,5	1,0	5,0
1 ISO	10	2				
2 ISO	100	24	10	4		
3 ISO	1000	237	102	35	8	
4 ISO	10000	2370	1020	352	83	
5 ISO	100000	23700	10200	3520	832	29
6 ISO	1000000	237000	102000	35200	8320	293
7 ISO				352000	83200	2930
8 ISO				3252000	832000	29300
9 ISO				35200000	8320000	293000

In table 2 there is a more detailed CR classification according with different standards.

Table 2. CR classification according with different standards.

GOST ISO 14644-1	GOST R 50766-95	USA standard 209 E	USA standard 209 D
1 ISO	R1	-	-
2 ISO	R2	-	-
3 ISO	R3 (1)	M 1.5	1
4 ISO	R4 (10)	M 2.5	10
5 ISO	R5 (100)	M 3.5	100
6 ISO	R6 (1000)	M 4.5	1000
7 ISO	R7 (10000)	M 5.5	10000
8 ISO	R8 (100000)	M 6.5	100000
9 ISO	R9 (1000000)	-	-

The air purity requirements during CRC design process are usually determined by technical conditions, technological assignment, technological regulations (processes), etc.

IOP Conf. Series: Materials Science and Engineering **481** (2019) 012038 doi:10.1088/1757-899X/481/1/012038

6. Micro-pollutions sources

The ambient air contains a huge amount of micro-particles, which directly or indirectly affect the process in the CRC. Micro-pollution of air is allocated by personnel, enclosing structures, equipment, penetrate into the CR from the environment. In the clean rooms 70 80% of pollution comes from human activities, 15-20% from the work of specialized equipment, 5-10% of pollution comes from the environment [16].

These indicators show that the CM for the CPC must be hermetically sealed and equipped with an air exchange system set.

A clean room is an artificially created environment. There are no such conditions of purity on the Earth. For example, ISO class 5 corresponds to air purity in the atmosphere at an altitude of more than four kilometers. Maintaining a given cleanliness class is a constant struggle with possible pollution sources, maintaining a permanent barrier between them and an external natural and polluted environment.

NASA conducted a study on the relationship between the particles number and the number of microorganisms in the air. This was necessary to simplify classifications and calculations, as well as standardization and licensing of CR. Calculation of microorganisms in the air is a long and costly research, and the micro-particles calculation is carried out quickly and accurately enough. The NASA standard NHB 5340 was developed, which established a clear dependence between the number of particles and microorganisms. This dependence is illustrated in Table. 3.

Table 3. Dependence between the particles number and the number of microorganisms according with NASA standard NHB 5340.

Cleanliness class according USA standard 209 D	Particles		Microorganisms	
	Diameter, μm	Quantity in 1 ft^3 (L.)	Suspended in 1 ft^3 (L.)	Precipitated per 1 ft^2 per week (1 sq. m. per week)
100	> 0,5	<100 (<3,5)	<0,1 (<0,0035)	1200 (12900)
10000	> 0,5	<10000 (<350)	<0,5 (<0,0176)	6000 (64600)
	> 5,0	<65 (<2,3)		
100000	> 0,5	<100000 (<3500)	<2,5 (<0,0884)	30000 (323000)
	> 5,0	<700 (<25)		

The development of that standard promoted the development of various classifications, including the Russian one. GOST ISO 14644-1 was created with reference to the American scientists developments and researchers of scientists from around the world.

7. CM classification for CRC

Based on the above facts, a morphological analysis of the information was performed and an intrinsic model of the CM classification was determined.

The main idea of classification is to designate CM as equipment that allows producing better CR quality. The use of CM will allow reducing the time and resources for the CRC technology development. The classifier creation will allow the most accurate and fast formation of technical requirements for CM equipping.

From the research, the following main parameters when choosing CM can be determined: 1) the main point of the classifier is the premises purpose. This parameter means its technological belonging. From this point it is possible to get an extensive premises list (medicine, pharmaceutical production, microelectronics, military industry, laboratories of different directions, etc.). Having determined the premises purpose, it becomes possible to determine the area of these premises, their overall dimensions, as well as necessary equipment. This item will allow to immediately get a necessary equipment list, its load on all engineering systems and the frame as a whole; 2) The second important point is the cleanliness class (Table 1). Taking into account this classification, it is possible to distinguish the cleanliness class parameter in the future room. Moreover, based on this characteristic, it is possible to single out air distribution and climate systems. This is the equipment selection for capacity, dimensions, power consumption, the class of filter elements, etc. The equipment data must correspond to the cleanliness class (Table 1). Having obtained this data, it becomes possible to immediately determine the necessary characteristics of the equipment; 3) The next item is chosen according with points 1 and 2 of the classification/ This is the engineering systems choice. In this paragraph, the need for water and sewerage, their location in the module is determined. The necessity in providing with electricity, its costs an demanding capacities for equipment and the connection points number are also determined. Based on point 1, important secondary engineering systems are identified (for example, medical gas supply systems); 4) CM design features. At this point the construction scheme is selected: the frame (depending on the equipment weight, its attachment, etc.), enclosing structures (choice of wall panels, ceiling panels, door types, etc.), choice of lighting, choice of floor covering; 5) the last item of the classifier - additional systems. This point is the most extensive, because it includes systems that, based on item 1, differ for each technology. These are such systems as automation and dispatching system, smart lighting system, emergency prevention systems, air quality control systems, etc.

If to develop this classifier and take into account all aspects and problems in the formation of tasks for the CM production, it will be possible to shorten the project production time, and also allow the customer to get the most extensive product view and find out more about the necessary product characteristics.

The main problem for today is the lack of a full understanding of all the CM advantages over the traditional finishing, as well as the lack of strict control of the CM compliance and equipment with all the necessary technological processes characteristics in which this CM will be used. The development of the classifier will allow to avoid annoying and sometimes critical errors in the CRC formation, as well as additional costs that, in the case of organizing high-tech production, can be quite critical [17].

8. Conclusion

After conducting this research, conclusions were drawn on the following issues: 1) ensuring a high level of equipment and execution of clean rooms. The use of CM is the logical and most rational way to organize a clean room complex in different technological schemes. CM is a high-tech product that allows to reduce the problems that arise in the formation of complex technological processes; 2) development of modular construction on the basis of CM. Use CM in modular construction is the most rational technological scheme. Simplicity of constructive solutions in combination with high technologies, as well as increased factory readiness allow to install the building ready for operation in the shortest possible time, without huge labor costs; 3) the possibility of upgrading obsolete technological solutions. CM is a flexible unit that has huge potential for application in the finished architecture and can be located in the working technological scheme, which will allow increasing the level of technological production, as well as reducing the costs of reconstruction and major repairs; 4) creation of CM structural classification. Criteria analysis for classification will create a full-fledged and voluminous CM classifier that will simplify the development elements and decision-making on the introduction of CM into the high-tech production technological scheme.

This study made it possible to determine all the advantageous CM characteristics, to define the classification of these modules and to designate the MC as an integral part of the high-tech production progress.

References

[1] Nadolenko V 2011 *Modular clean rooms as a simple and economical solution* (Saint-Petersburg: Technology in the electronics industry) pp 59–61

[2] Unno K 1985 Clean rooms *Current state and trends of the environment monitoring in semiconductor manufacturing* **4** p 11

[3] Kuznetsov V I, Mchedlishvili B V, Sisakyan A N, Fursov B I and Shestakov V D 2005 *Clean room* (Moscow: Ecological systems and devices) pp 36–37

[4] Tomilin V A, Ananin M Yu, Gubanov A N and Zimberg E O 2018 Use of medical climatized module for organizing high-tech medical technology in remote and hard-to-reach areas *I International Conference on Advances in Science and Technology* (Ufa: Press) pp 69–75

[5] Adam F M 2001 Features of prefabricated buildings installation *Installation and special works in construction* **2** (Saint-Petersburg:) pp 12–16

[6] Markin A V 2017 Features of the concept of "multiplicity of air-exchange" in relation to clean rooms for production purposes *The electronic technician. Series 3: Microelectronics* (Moscow:) pp 70–78

[7] Kholodova L P, Ananyin M Yu, Fedorova M S 2013 Architectural Analysis of Planning Decisions of the First Factory Hospitals of the Urals *Architecton: News of Universities* **4** (Yekaterinburg) pp 15–25

[8] Fedorova M S, Kholodova L P 2014 Key milestones for the development of standards for the design of military hospitals *Architecton: News of Universities* **3** (Yekaterinburg) pp 10–20

[9] Adam F M 2001 Full-scale construction of modular prefabricated low-rise buildings (*Proceedings of the scientific and practical conference "Post-Soviet urban planning"* (St. Petersburg: Gosstroy of Russia GU Institute of Urban Development) pp 119–121

[10] Tamplon F F 1976 *Metal enclosing structures* (Sverdlovsk)

[11] Asaul A N, Kazakov Yu N, Bykov V L, Prince I P and Erofeeev P Yu 2004 *Theory and practice of using prefabricated buildings in normal conditions and emergencies in Russia and abroad* (Saint-Petersburg)

[12] Rybakov V A 2011 *Fundamentals of structural mechanics of light steel thin-walled structures* (Saint-Petersburg)

[13] Kuznetsov V I, Mchedlishvili B V, Sisakyan A N, Fursov B I and Shestakov V D 2004 Clean room (Russian patent № 2224182)

[14] White V 2004 *Designing clean rooms* (Moscow)

[15] Fedotov A E 2012 *Foundations of GMP* (Moscow)

[16] Sineglazov V M, Fedosenko V N and Radko E N 2014 Mathematical models of pollution dynamics of clean rooms *Electronics and control systems* (Kiev: National Aviation University) pp 114–120

[17] Tomilin V A, Ananin M Yu, Gubanov A N, Zimberg E O 2017 Prospects for the use of fast-moving buildings for medical technology *Collected Articles on the Materials of the IV International Scientific and Practical Conference "A New Word in Science and Practice"* **3** (Ufa) pp 23–31

IOP Conf. Series: Materials Science and Engineering **481** (2019) 012039 doi:10.1088/1757-899X/481/1/012039

Reducing the risks of water and territory pollution by toxic metal-containing effluents of machine-building industry

G B Brayalovsky[1], O B Naschetnikova[1], E V Migalatiy[1]

[1]Ural Federal University, 17, Mira Street, Ekaterinburg 620002, Russia,

E-mail: o.b.naschetnikova@urfu.ru

Abstract: The article is devoted to the baromembrane method for cleaning washing waters of degreasing sites and for the bright nickel plating of galvanic productions. The authors analyzed specific features of technologies for part metallization in terms of the formation and separation of washing wastewater flows for their separate cleaning. Studies were carried out on the membrane extraction of nickel sulfate and sodium phosphate compounds on two types of membranes: nanofiltration OPMN-P and reverse osmosis Hydranautics ESPA-1. High selectivity (98-99,9%) and specific productivity of OPMN-P membranes (25 l/m^2 per hour) were proved during the extraction of nickel sulfate and sodium phosphate at concentration ranging from 50 to 200 mg/l specific to washing water of the processes of degreasing and bright nickel plating. The advantage of using nanofiltration membranes OPMN-P over reverse osmosis Hydranautics ESPA-1 membranes was substantiated. A technological scheme to create a closed water supply system at these sites was developed.

1. Introduction

Machine-building industry, the most important technological stage of which is the processing of metal products by applying various kinds of metal coatings, can be considered a critical infrastructure as it is extremely water-intensive and environmentally hazardous.

Technological lines of part metallization are linked to chemical and electrochemical treatment of metal surfaces in a water medium. Therefore, the formed wastewater contains toxic compounds of surfactants, acids, alkalis and transition metals ranging from tens of milligrams to hundreds of grams per 1 dm^3 [1, 2]. Soluble complex compounds of heavy non-ferrous metals (copper, nickel, chromium, cadmium) are especially dangerous, since they are difficult to neutralize in traditional treatment facilities and pose a threat to the pollution of water bodies [3]. These toxic substances can cause diseases of the central nervous system, blood vessels, heart, liver and other organs in humans. In addition, they have a mutagenic effect, both for humans and aquatic biota.

To this end, the task of improving the technology of purification of metal-containing wastewater in order to create closed technological cycles and reduce the discharge of polluted effluents into bodies of water is highly relevant.

2. Specific features of the part metallization technologies

The process of applying metal coatings usually consists of two stages that is preparation of the product surface for coating (mechanically or chemically) and application of coating [4, 5]. Because of these processes, two main types of effluents like waste electrolytes and various washing waters are formed.

As a rule, concentrated effluents are regenerated and processed individually in each specific metallization line.

After each of the preparatory and basic technological operations the product is washed in cold or hot water. Washing products after the electrolyte bath helps to maintain the chemical composition and purity of the solution in a subsequent processing bath. The washing is usually carried out in the flow-through baths.

This method of washing uses a large amount of water (up to 2 m^3/m^2 of coating), and a significant amount of electrolyte enters the wastewater. Reduction of the amount of washing water and decrease in the loss of metals can be achieved through the installation of recovery baths, the use of advanced washing methods, etc. [4, 6].

All waste waters from the chemical and electrochemical metal processing are divided into three main groups based on the content of contaminants: cyanide-containing (contain highly toxic hydrocyanic acid compounds); chromium-containing (contain highly toxic compounds of chromium (VI)); acid-alkaline (contain free mineral acids and alkalis, salts of heavy, alkali metals and ammonium, organic compounds, etc.) [5].

The joint disposal of cyanide-containing and acid-alkaline or chromium-containing wastewater is unacceptable because of the possibility of formation and emission of highly toxic gas - hydrogen cyanide - into the production facilities. The joint disposal of chromium-containing and acid-alkaline wastewater and their subsequent neutralization inevitably results in a significant overexpenditure of reagents.

Therefore, cyanide-containing and chromium-containing effluents are processed separately and the solutions of the recovery baths and washing waters in the areas of nickel plating, copper plating, galvanizing, degreasing and other operations are mixed in acid-alkaline waste and sent to a neutralization station. Neutralization of ions of heavy non-ferrous metals is carried out through adding water-soluble alkaline reagents to wastewater. As a result, metal ions are converted into sparingly soluble hydroxides that precipitate. Flocculant polyacrylamide is used for better and more complete and rapid coagulation of hydroxides. The clarification time of neutralized water is 40 - 60 minutes; the sediment volume is up to 10% of the total volume of the liquid [1, 7].

The reagent method is the most widespread in the Russian practice of neutralizing the wastewaters of galvanic productions. Its main advantage is an extremely low sensitivity to the initial content of contaminants. The main disadvantages are high residual salinity of purified water and high humidity of the sediment. This calls for additional purification of clarified water, as well as construction of facilities for sediment dewatering [8].

The amount of many components in the treated wastewater exceeds the maximum permissible discharge values for both urban sewerage and water bodies. The existing approach - neutralization of the general wastewater - leads to a presence of sediment with a high content of heavy metals, which cannot be extracted due to its complexity. The sediment is currently subject to landfill. The amount of sediments is constantly increasing. Today, they are the main pollutants of the environment on the territory of cities with machine-building industries.

3. Experimental results

To solve this problem, the authors conducted research at one of the machine-building enterprises in Ekaterinburg. The research focused on the possibility of processing the nickel sulfate and sodium phosphate electrolytes using membrane plants equipped with the OPMN-P nanofiltration membrane.

The choice of this type of membrane is due to the effective parameters of membrane separation, i.e. high specific permeate productivity and selectivity in many components (including various electrolytes in a soluble form). In addition, affordability and availability of domestic production of these membranes make them extremely attractive for use in treating wastewaters from ions of heavy non-ferrous metals.

The authors carried out experimental studies on membrane purification of washing waters of the bright nickel-plating process with a concentration of nickel sulfate in the 10-50 mg/l range, as well as of washing waters of the degreasing process with a concentration of sodium phosphate in the 5-100 mg/l range. Moreover, in order to study the membrane characteristics at various concentrations of electrolytes, the authors conducted experiments on the concentration of washing waters to 10 g/l for nickel sulfate and sodium phosphate respectively.

The experiments were carried out on a roll-type laboratory membrane installation using various types of membranes: nanofiltration OPMN-P and reverse osmosis Hydranautics ESPA-1.

The aggregated results of the experiments are shown in Figures 1 and 2.

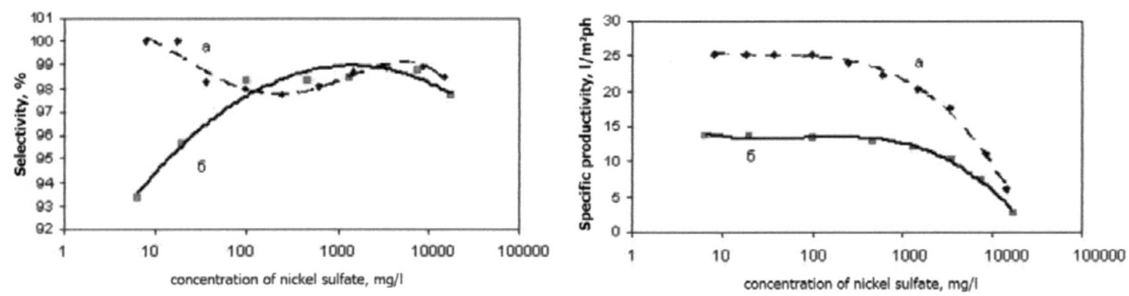

Figure 1. Dependence of selectivity and specific productivity of membranes on the concentration of nickel sulfate; operating pressure 0,4 mPa, solution temperature 22°C
membranes: a – OPMN-P, b – Hydranautics ESPA-1.

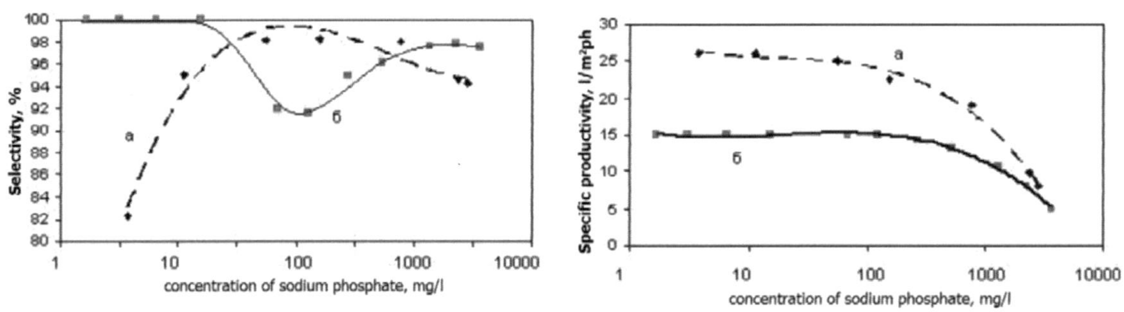

Figure 2. Dependence of selectivity and specific productivity of membranes on the concentration of sodium phosphate; operating pressure 1,0 mPa, solution temperature 22°C
membranes: a – OPMN-P, b – Hydranautics ESPA-1.

Figure 1 shows that the nickel sulfate selectivity of the OPMN-P membrane at the concentrations typical for the washing waters of the nickel-plating process (ranging from 10 to 50 mg/l) has a maximum value of about 99.9%. A further increase in the concentration to 1000 mg/l leads to a slight decrease in selectivity to a level of 97.5– 98%. High selectivity of this type of membrane is due to the presence of a negative charge on its surface. In this case, even the ions, which are smaller than the diameter of the membrane pores, are electrostatically repelled from its surface and remain in the concentrate stream.

In this concentration range, the specific productivity of the OPMN-P membrane remains stable – 25 l/m^2 per hour. However, further concentration leads to a sharp decrease in the specific productivity of the membrane, which is unacceptable.

In contrast, the membrane separation parameters of the reverse osmosis membrane Hydranautics ESPA-1 are much lower in given concentration range: selectivity is only 93-95% and specific productivity is 15 l/m^2 per hour, which is 1.7 times lower than that of the OPMN-P membrane.

A series of experiments on the membrane extraction of sodium phosphate, the results of which are shown in Figure 2, demonstrate that specific productivity of Hydranautics ESPA-1 membrane is significantly lower than that of the nanofiltration membrane at close values of sodium phosphate membrane selectivity at a concentration of 50 to 80 mg/l (approximately 98.4 99.9%), and it decreases with increasing solution concentrations.

Thus, the possibility of use and effectiveness of Russian nanofiltration membranes OPMN-P in the development of technologies for cleaning washing wastewater from galvanic productions has been proved.

Based on the research results, the technologies of separate processing of individual solutions of galvanic lines have been developed. The basis of the implemented technological schemes is the membrane concentration of the recovery baths solutions in degreasing and bright nickel-plating sites. These technologies made it possible to reduce significantly the removal of nickel and phosphate ions to general wastewater treatment plants. However, they did not solve the problem of a large consumption of washing water, which is irrationally discharged into the city sewage system.

Thus, a new technological scheme has been developed (Figure 3) that allows comprehensively processing washing solutions of degreasing and bright nickel-plating sites, preventing their discharge into acid-alkaline wastewater.

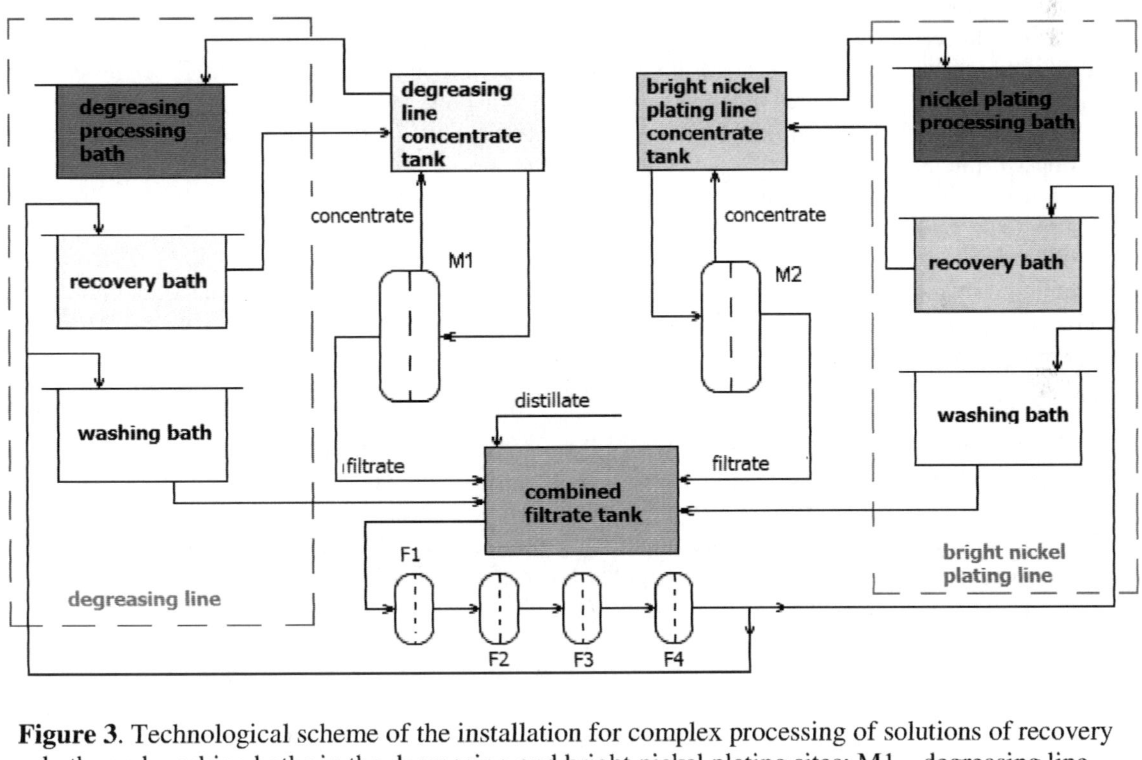

Figure 3. Technological scheme of the installation for complex processing of solutions of recovery baths and washing baths in the degreasing and bright nickel plating sites; M1 – degreasing line membrane unit; M2 – bright nickel plating line membrane unit; F1 – mechanical cleaning filter; F2 – sorption filter; F3 – H-cation filter; F4 - OH-anion filter.

Basic components of the degreasing solution are sodium phosphate and caustic soda. The salt composition of the technological solution of bright nickel plating is determined by nickel sulfate.

According to the technological scheme, the solution of the degreasing line recovery bath is periodically discharged into the concentrate tank of the degreasing line. This solution is concentrated as follows: the solution from the concentrate tank of the degreasing line is directed to the membrane unit. Two streams are formed at the exit from the membrane unit:

- concentrate (with increased salt content);
- filtrate (permeate with low salt content).

Concentrate is returned to the concentrate tank of the degreasing line, and permeate is returned to the combined filtrate tank. Thus, a solution with a low salt content leaves the concentrate tank and the concentration occurs. The concentrated solution, if necessary, is sent to compensate for the losses in the processing bath of the degreasing line.

The solution of the recovery bath of bright nickel-plating line is pumped to the concentrate tank of the bright nickel-plating line. This solution is concentrated similarly to the concentrate of the degreasing line.

The filtrate is sent to the combined filtrate tank, and the concentrated solution is sent to compensate for the losses in the processing bath of the bright nickel-plating line.

The filtrates are mixed in a combined filtrate tank, in which the following reactions occur:

$$NiSO_4 + Na_3PO_4 \rightarrow Ni_3(PO_4)_2\downarrow + Na_2SO_4$$

$$NiSO_4 + 2NaOH \rightarrow Ni(OH)_2\downarrow + Na_2SO_4$$

The formed sediment of nickel phosphate is retained on the mechanical filter (F1). Sulfate ion, chloride ion and excess phosphate ion are retained on the anion exchange filter (F4). Sodium ion is retained on the cation exchange filter (F3). Organic substances and surfactants, which could have entered washing baths on the surface of the parts, are retained on the sorption filter (F2). The resulting desalted water is sent to the recovery and flow-through washing baths.

The scheme implements the circulation of water from the flow-through washing baths through a combined filtrate tank.

It should be noted that for continuous operation of ion-exchange filters, it is necessary to adjust the technological process in such a way that the solutions of both recovery baths are treated simultaneously.

The issue of regeneration of ion-exchange filters was not considered, since the salt content of the combined filtrate is less than 20 mg/l, and simple replacement of ion-exchange resins or cartridges is the most rational option.

4. Conclusions

Based on the results of the experiment, OPMN-P nanofiltration membrane having higher productivity compared to reverse osmosis membranes was chosen for complex solutions processing of recovery and washing baths in the degreasing and bright nickel-plating sites. Thus, during the concentration of a nickel sulfate solution on the OCMN-P membrane from 0.5 g/dm^3 to 1 g/dm^3, electricity consumption is 34% lower than during a similar process on the Hydranautics ESPA-1 membrane.

The developed technological scheme allows to:

- return valuable components to processing baths;
- completely stop the discharge of solutions of the recovery baths and washing water into the acid-alkaline stream;
- create a circulating water supply system in the degreasing and bright nickel plating sites.
- In the future, the introduction of this technology will reduce the pollution of water bodies with metal-containing wastewater and improve the ecological situation in Ekaterinburg.

Acknowledgments

The most pleasurable thanks to Pyrkova Tamara, Foreign Languages Department, the Institute of Social and Political Sciences, Ural Federal University named after the first President of Russia B. N. Yeltsin.

References

[1] Yakovlev S, Karelin Y, Laskov Y and Voronov Y 1985 *Industrial Wastewater Treatment* (Moscow: Stroyizdat) 335 p
[2] *Handbook on Flexible Automated Galvanizing Lines* 1989 ed Zubchenko V (Moscow: Machinostroenie) 672 p
[3] Vinogradov S 1998 *Environmentally Sound Electroplating Industry* ed Kudryavtseva V (Moscow: Globus) 302 p
[4] Krivoshein D, Kukin V, Lapin V et al 2003 *Engineering Protection of Surface Water from Industrial Effluents* (Moscow: Vysshaya Shkola) 344 p
[5] Korotin A 1984 *Electroplating Technique* (Moscow: Vysshaya Shkola) 200 p
[6] Aksenov V, Anikin Y, Migalatiy E et al 1988 *Wastewater Treatment in Electroplating Plants and Printed Circuit Board Manufacturing* (Sverdlovsk: UPI) 53 p
[7] Getmantsev S, Nechaev I and Gandurina L 2008 *Industrial Effluent Water Treatment Using Coagulants and Flocculants* (Moscow: ASB) 272 p
[8] Yakovlev S, Volkov L, Voronov Y and Volkov V 1999 *Industrial Effluent Water Treatment and Utilization* (Moscow: Khimiya) 448 p

IOP Conf. Series: Materials Science and Engineering **481** (2019) 012040 doi:10.1088/1757-899X/481/1/012040

Modern Tendencies in Design of Public Spaces in term of Sustainable Development

O S Burukhina[1], S A Serbin[1], S V Vartanyan[1], I N Maltceva[1]

[1]Department of Architecture, Ural Federal University, 19 Mira street, Ekaterinburg, 620002, Russia

E-mail: olgaburuhina@mail.ru

Abstract. The Flight of Fantasy is a multifunctional complex located on the territory of perspective development in Dubai, United Arab Emirates. The project was designed for the International Student Competition " MultiComfort House Students Contest" hold by Saint-Gobain ISOVER. The project was developed taking into account the standards of energy-efficient construction and sustainable development and meets all the requirements of the competition. This paper describes the main conceptual solutions adopted by the authors. Since Dubai has a tropical desert climate, it was necessary to protect all spaces from overheating in summer and to ensure the possibility of comfortable movement between apartments and public spaces. Also it was necessary to offer an optimal solution for residential spaces that meet all the needs of residents. The architectural concept should fit the modern trends of the city development and use innovative materials.

1. Introduction

Dubai is the largest and most populous city in the United Arab Emirates (UAE). Dubai has a tropical desert climate, Köppen classification Bwh, because of its location within the Northern desert belt. Summers are extremely hot and humid. The weather between December and March is warm and considered to be the most comfortable climatic conditions of the year. The Emirate of Dubai has experienced a rapid transformation. From a desert city with an estimated population of 10,000 in 1900, today it is a mega-city with a population of over 2.1 million people and between 8-10 million annual visitors.

Based on these conditions and world experience [1,2,3,4,5], we have developed an architectural concept and construction solution. In addition, the project involves the infrastructure of the complex, building services systems and technical equipment. The main characteristics meet the requirements of energy efficiency and multicomfort for newly built buildings. Compliance with these requirements was the main complexity of the project. In a hot climate it is necessary to solve other tasks, unlike widespread energy-saving techniques for buildings in a temperate climate. In these climate conditions, about 60% of the total energy consumed by the building spends on cooling. Therefore, it is necessary to take into consideration using renewable energy sources, and avoid overheating of the premises.

2. Architectural concept

Historically, the symbol of the UAE was the image of a falcon, on whose breast a dhow schooner was depicted [6]. The dhow is a traditional swimming device for pearl hunters. It was sails and waves that we were inspired of. The idea of waves pushed us to associations with a number of physical

IOP Conf. Series: Materials Science and Engineering **481** (2019) 012040 doi:10.1088/1757-899X/481/1/012040

phenomena. Also we tried to make the architecture fit in the surrounding of the site and maximize the views to the waterfront and architectural dominant. Special focus was given on developing the sustainability dimensions as well as the comfort ones. As a result, we got the following semantic chain: waves - physical interaction of waves - interference of coherent waves (Figure 1).

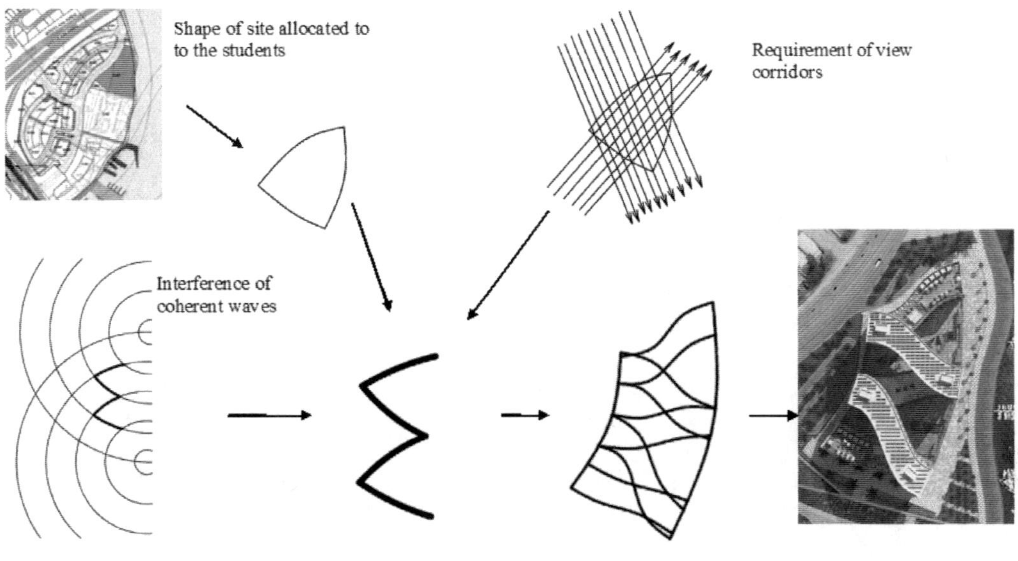

Figure 1. Architectural concept.

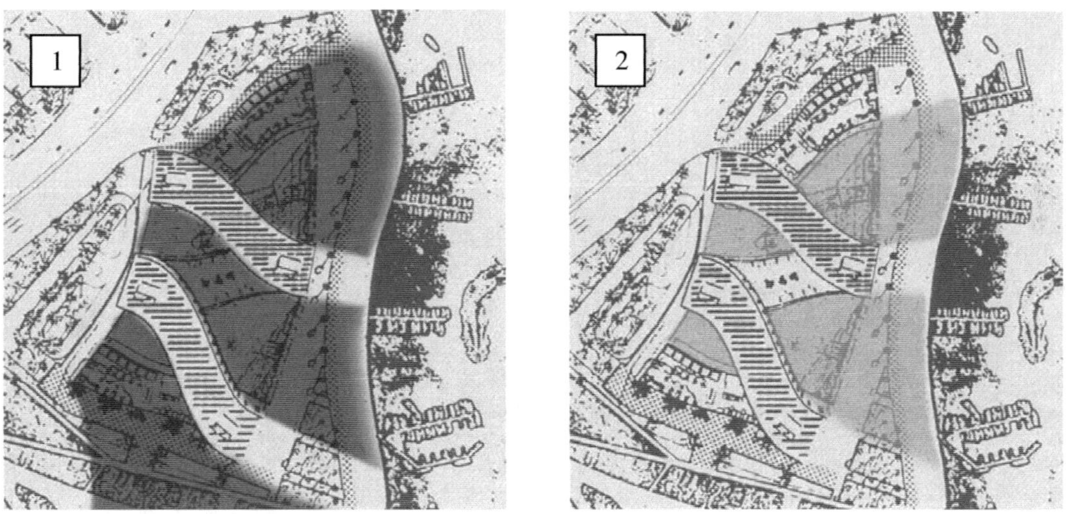

Figure 2. Object visibility: 1 - above 5 floors (shown in purple); 2 - at the height from 1 to 5 floors (green).

The following architectural features of the object can be distinguished. The object consist from five five-storey blocks disposed in special order to compose ten-storey building. The first floor is of commercial premises, the remaining floors are residential. The apartments are designed for different segments of the population: young specialists, foreigners of different nationalities living and working in the country. About 10% of the apartments are designed for long-term rentals of up to 6 months for people involved in cultural events. The area of apartments is set to 50 - 70 sq.m. per person (according to the requirements of the development region). There are two machine rooms in the underground part of the building: they are tidal energy turbine control center (TCC) and a cooling system engine compartment, as well as a two-storey underground parking for 500 cars. All traffic infrastructure is brought underground.

Figure 3. Visualization: 1 – the view of a complex from the water; 2 – the view of a complex from the city.

The dominant element of the complex is a glazed air-conditioned atrium, in which public, cultural and commercial premises and walking paths and recreation area, as well as walking paths and recreation area are located. On the roofs of the first and fifth sections there are a swimming pool, recreation area and a restaurant with the views to the waterfront, where the residents of the complex can spend their free time at any season. We provided for a spacious multicomfort atrium that unites the entire complex of five buildings into a single one following the modern methods of effective planning of smart functional and public spaces [7,8,9], as well as the requirements for creating a comfortable microclimate of development.

Figure 4. The trafficable roof.

3. Construction solution

The main construction material of the object is reinforced concrete with fibre-reinforced polymer. The concrete was chosen self-sealing with a self-cleaning surface. Self-compacting concrete prepared according to the method supposed by Hiro Okamura (or analogs) [10,11], has a high plasticity, good adhesion to the reinforcement and does not require vibration. This effect is achieved, in particular, by the addition of a polycarboxylate. A self-cleaning surface that does not require finishing is achieved by the addition of titanium dioxide nanoparticles. Such concrete has high aesthetic qualities, and also triggers the decomposition reaction of many harmful bacteria on the concrete surface.

We use carbon fiber reinforced polymer as a fibre-reinforced plastic. Carbon fiber was chosen because of having the best strength properties [12]. Due to high average annual temperatures in the UAE, the reinforcement should be selected according to the expansion coefficient. Carbon fiber has value of this coefficient closer to concrete than steel, which provides the best grip.

4. Fire security

All reinforced concrete structures of the complex have a fire resistance limit of REI 120, which is provided by a protective layer and composition of concrete, as well as protective coatings. In our project, there are five special shelter rooms (one for each building). They are protected by layers of reinforced concrete and sheet refractory coatings. Their walls are able to withstand the onslaught of fire for two hours. Each safe area is equipped with an autonomous ventilation system. Sealed fireproof doors prevent the penetration of smoke. In safe areas, people can hide until the emergency services take control of the situation or their turn for evacuation approaches. The fire lift is designed to move 5,500 kg (50-60 people).

The early warning system works continuously, 24 hours a day. As soon as a smoke detector, a thermal sensor or an automatic water extinguisher is triggered, a powerful fan system is activated. They pour clean cool air into the building through refractory ducts. Fresh air displaces smoke from the stairwells, ensuring the safety of evacuation routes.

5. Insolation

Due to the location "from north to south," the walls with windows are located on the west and east sides, which provides the required insolation. Also we propose to use special automated shutters on all window openings to protect the premises from overheating.

6. Solar energy generaion

The roof surface of the upper five-story sections is equipped with BenQ SunForte 333 PM096B00 solar panels. They are performed by reverse-contact technology, which allowed to obtain an output power of 333 watts with a confirmed efficiency of 20.4%. A total of 600 panels are supposed to be installed. The dimensions of one panel are 1559x1046x46 mm. The average generation of solar cells was calculated in special program, it is about 250 kWh / day. Monocrystalline silicon is used as the main material of the panels.

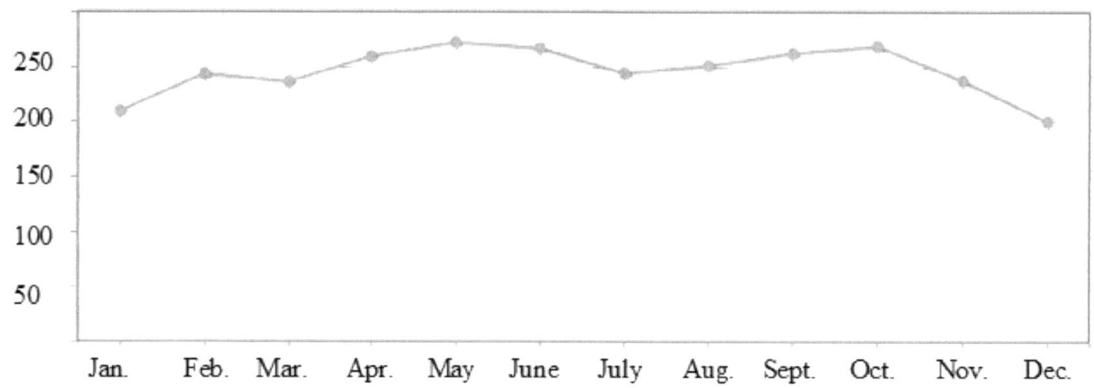

Figure 5. Annual solar energy generation.

Window are equipped with blinds also consisting of solar panels. Solar energy, collected during the day, is supplied to the electrolyuminous foil in the dark at night and provides lighting for the premises. This device allows the use of a large surface to collect sunlight. The smart house system allows to adjust the inclination of the collecting surface depending on the position of the sun, and thus improve the efficiency of the device.

7. Tidal energy

In the underground part of the building we arranged a reduced version of the Turbine Control Center, developed by OPENHYDRO and awarded several awards. CCT is designed to fix and control the energy output from tidal massifs and export it with the greatest efficiency to the network. It simplifies the architecture of the array, which requires only one submarine cable to exit: creating a transformation in the economy of tidal massif development, which increases the reliability and predictability of the entire energy supply system.

8. Artificial park

We propose to equip a part of the territory around the complex with artificial trees, in which there are solar batteries instead of leaves. This device is called "Sun Tree". The structure of the location of solar cells repeat the regularity of the leaves one discovered by Aidan Dwyer, which allows increasing the energy output in the summer by 20%, and in winter by 50%. Also, this design allows trees to take up less space and be more weather resistant than flat panel constructions on the roof.

Along with the "solar trees" it is proposed to establish Wind Tree. Recently several generators of this type applies in the form of a tree about 7 meters in height apply in north-west France. People who have already seen the device in action agree that Wind Tree has a certain artistic value. The creators of the wind generator were inspired by the leaves waving in the wind. The device uses tiny blades that are dispersed inside the artificial leaves. The blades rotate independently of the direction of the wind, while the wind farm is capable of generating electricity even at a minimum wind speed (about 2 meters per second).

1

2

Figure 6. The artificial trees: 1 – Sun Tree; 2 – Wind Tree.

9. Air-conditioning system

The air quality and the ventilation rate are considerable in common comfort. In our project we propose to use the cooling system of Alfa Laval, called SeCool [13] (with the corresponding additional products, including Maxi Cooling) using seawater. Sea water directly cools the plate heat exchanger of the chiller. Circulating fresh water is cooled in two VETs with gaskets and titanium plates and then used to cool the condensers of air conditioners.

The advantages of this system are that it does not consume fresh water, which is especially important in the zone of the Northern Belt of the Desert, where Dubai is located. Also demineralized water doesn't lead to corrosion. The plant equipment is in contact only with non-aggressive and uncontaminated fresh water. Aggressive and clogged sea water is present only in the dismountable heat exchangers of the SECOOL circuit.

The only requirement to use such a system is the presence of a source of sea water nearby, which is performed on our site. The above mentioned system will be connected to the geothermal cooling network in the ground, which will reduce the maintenance costs of the complex.

9.1. Air-conditioning of underground premises

For the ventilation of underground premises, the Jet Trans Fans system is installed. It allows to remove exhaust gases at an increased rate and to control the operation of each fan individually. The number of cars in the parking lot is determined automatically using a CO_2 sensor. The system allows you to regulate the load and thrust of specific fans, reducing energy consumption in general, and can operate in a completely reversible mode.

10. Acoustic isolation

The most simple and effective way to protect against noise is to use of multi-layer frame sound-proof structures. Such designs provide the same protection from noise, as well as an additional built-up brick wall 10 times larger.

A structure in which there are two or more layers provides better sound insulation than one thick layer. We propose multilayered frame partitions consist of several alternating layers of rigid (dense) and soft (light) building materials.

11. Elevators

Ensuring the availability for low-mobility groups and creating a barrier-free environment is an imprescriptible feature of multicomfort. Our complex is equipped with OTIS Gen2 Life lift with a lifting capacity of 1020 kg, the maximum speed of which reaches 1.6 m/s. The components of the Gen2 Life compact elevator are installed inside the shaft and eliminate the need for a machine room. This saves construction costs and frees the area. When the elevator is operated, there is no need for harmful substances, due to so-called zero lubrication.

IOP Conf. Series: Materials Science and Engineering **481** (2019) 012040 doi:10.1088/1757-899X/481/1/012040

The optional Gen2 Switch technology provides up to 100 runs at the time of power failure. When powered from a single-phase 230 V network or from renewable energy sources, the system requires less installed power than a household microwave oven.

12. Calculation result at MCD

To calculate the amount of energy spent on the operation of the building we chose the southernmost section of the complex. The calculation was held out in a specialized program MultiComfortDesigner. The results obtained fully satisfy the energy efficiency criteria (Figure 7). Annual heating costs in the cold period are 13.24 kWh / m2, and conditioning costs in the warm period are 5.68 kWh / m2. Quite low indicators were achieved due to the fencing of the internal courtyard with a glass dome.

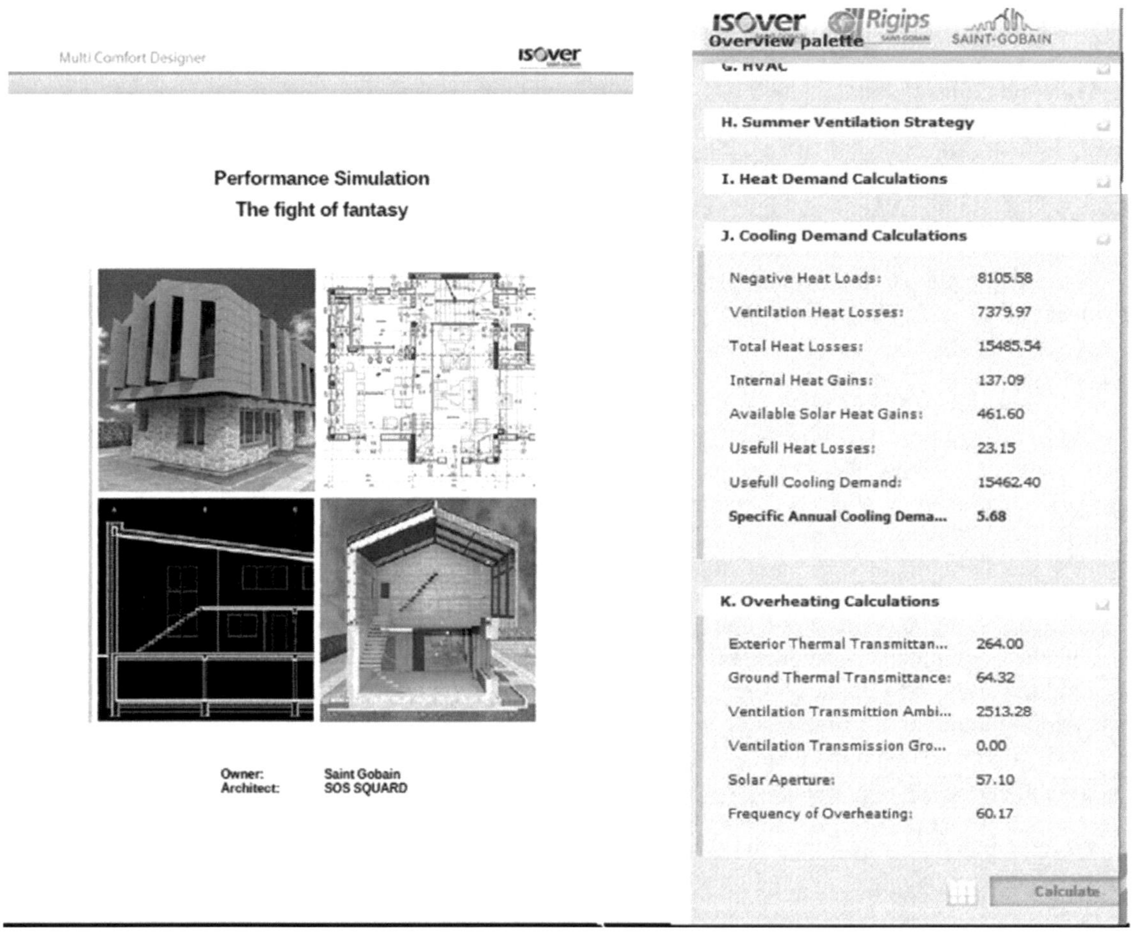

Figure 7. The artificial trees: 1 – Sun Tree; 2 – Wind Tree

13. Conclusion

The project of a multifunctional international complex, designed at the Ural Federal University, meets both the requirements for creating a microclimate in hot operating conditions with a significant reduction in energy consumption for these purposes, and preserving the traditions and modern development of such a multinational community as the city of Dubai and the United Arab Emirates.

14. References

[1] Nabil A 2017 principles of architectural decisions of energy efficient buildings in hot climates *Architectural Exploring* **9** pp 31-43

[2] Maltceva I, Baturin A, Panteleeva A and Serbin S 2016 Methods of energopastive construction - project "Red yard" in Brest, Belarus *Safety Problems of Civil Engineering Critical Infrastructures* ed V Alekhin (Ekaterinburg: Ural Federal University) pp 140-7

[3] Tkachuk K, Burukhina O and Maltceva I 2017 Renovation of residential houses and the adjacent territory in Madrid with the applying of modern energy efficient technologies Safety problems of civil engineering critical infrastructures *Safety Problems of Civil Engineering Critical Infrastructures* ed V Alekhin (Ekaterinburg: Ural Federal University) pp 434-41

[4] Tsitsin K 2013 Energy-saving technologies – future of housing construction *Strategic Decisions and Risk Managemen*t **77** pp 50-1

[5] Chen L and Bilyushova T 2012 Modern energyefficient buildings *Vologdinskiye Reading* **80** pp 80-81

[6] Nezhinsky K 2008 *World Blazons* (Moscow: EKSMO)

[7] Zalejska-Jonsson A, Lind H and Hintze S 2013 Energy-efficient technologies and the building's saleable floor area: bust or boost for highly-efficient green construction buildings *Buildings* **3** pp 570-87

[8] Rodriguez S, Voss K and Todorovic M 2014 Energy efficiency evaluation of zero energy houses *Energy and Buildings* **83** pp 23-35

[9] Mitterer C, Künzel H M, Herkel S and Holm A 2012 Optimizing energy efficiency and occupant comfort with climate specific design of the building *Frontiers of Architectural Research* **1** pp229-35

[10] Okamura H 1995 Mix-design for self-compacting concrete *Concrete Library of JSCE* **25** pp 107–20

[11] Shesternin A, Korovkin M and Eroshkina N 2015 Basics of self-compacting concrete technology *Young Scientist* **86** pp 226-8

[12] Stepanova V, Stepanov A and Zhirkov E 2013 *Fibre-Reinforced Polimer Rebar* (Moscow: Paper airplane)

[13] Sviridiva E 2014 Thermal insulation is an important element of energy-efficient construction *Energosovet* **37** pp 72-4

IOP Conf. Series: Materials Science and Engineering **481** (2019) 012041 doi:10.1088/1757-899X/481/1/012041

Numerical and experimental studies of gas cleaning in multi-cyclone elements with filter inserts

G I Belyaeva[1], A T Zamalieva[2], M G Ziganshin[3]

[1,2] Gazprom transgaz Kazan LLC, 420073, Kazan, Russia
[3] Thermal power engineering, Gas supply and Ventilation Department, Kazan State University of Architecture and Engineering, 420043, Kazan, Russia

E-mail: gulnazкa16@mail.ru[1] albina-0587@rambler.ru[2] mjihan@mail.ru[3]

Abstract. As a result of the tightening of standards for the content of suspended particles with a particle size of less than 10 μm in the atmospheric air, the development of production structures in the urban environment often becomes closely linked to the problem of deep cleaning of heterogeneous emissions, that requires, at a minimum, two-stage cleaning. In this paper, we consider the effectiveness of using a multi-cyclone apparatus for such purposes, the elements of which combine both stages of purification. A bench test of such an element, which is a cyclone filter with a diameter of 250 mm on the basis of a serial cyclone TsN-11, was carried out. A numerical model of the multi-cyclone has been investigated, which has made it possible to establish the optimal arrangement of the cyclone elements with respect to the trajectories of the streams of trapped particles.

1. Introduction

Recently, regulations on the content of suspended particles in air with dimensions less than 10 μm have been tightened. Accordingly, the requirements for cleaning of heterogeneous emissions from highly dispersed components were increasing. One-step processing of technological emissions cannot provide the required degree of cleaning at acceptable energy costs. In order to efficiently processing production emissions containing of the $PM_{2,5}$, PM_{10} classes particles, a multi-stage purification system is needed with cleaning devices that optimally take into account the physicochemical properties of all emission components, both pollutants and its neutral part [1]. In this paper, the results of a study of the apparatus combining the stages of coarse and final cleaning of heterogeneous emissions with PM_{10} and $PM_{2,5}$ classes are presented. The efficiency of the multi-cyclones can be calculated by methods based on theoretical and experimental data [2]. Now, quite complete and reliable data on the efficiency of the multi-cyclones have been obtained in studies carried out both with natural (physical) and on numerical models [3, 4]. However, they relate only to the studied devices of a specific size and design, and the study of innovative concepts must be carried out fully.

2. Methods

2.1. The bench test of the multi-cyclonic filter element

The scheme of the experimental stand is shown in Fig. 1. Passage of dusty air through the cyclone 7 was provided by a fan 14. Chamotte dust passed through the measuring manifold 1 into the air duct 2 with a diameter of 140 mm. The dusty flow through the duct 2 through the inlet 6 of the cyclone 7 go into the ring space between its body and the exhaust pipe 8. The airflow at the inlet to the cyclone was measured with a bellmouth inlet 1 with an internal diameter of 140 mm and a micromanometer 3. The air flow rate in the ductwork varied in three modes, with the minimum open (first), intermediate (second) and fully open (third) throttle valve positions 15. To find the resistance of the cyclone, the pressure loss ΔP was measured by a micro-manometer 9 at points 5 and 10. The inlet velocity in the cyclone was measured prior to dusting by a thermal anemometer 19 (Model Testo 425) installed in the round access port 4. Because of the rotational motion of the flow, the particles of large and medium sizes clash with the inner wall of the cyclone and slid into the hopper 17 that has the hermetic shutter 16. On screw downward trajectory of the flow, fine dust fractions were deposited on the fabric filter 13. Then the flow turned in front of the hopper and return along an ascending screw path through the exhaust pipe 8. The flow velocity inside the cyclone element was measured with the Testo 425 thermo-anemometer. Measurements were made at two levels along the height of the element. The upper level was located 40 mm below the top of the cylindrical part, the lower one - directly under the exhaust pipe. The measuring points were along the diameter of the cyclone element. The scheme of their arrangement, is shown in Fig. 2.

Samples of the filter material were weighed before and after the test. In addition, after the experiment, tissue samples cut out from the beginning, middle and the end of filter to determine the countable concentration of the particles deposited. Its value was obtained by direct counting of particles on samples under a MIN-8 microscope with 400 times magnification, on which all settled particles with dimensions from 0,25 μm and higher were clearly visible [5]. The determination of a countable number of $PM_{2,5}$, PM_{10} classes particles is necessary to reliable compare the intensity of their settling from the flow with filter under different regimes.

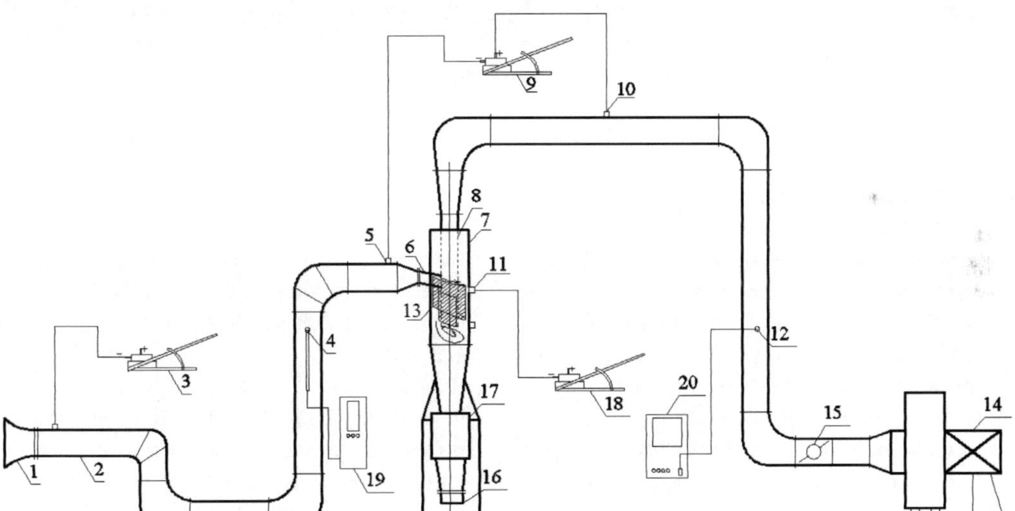

Figure 1. An experimental test bench for testing a filter element made of a standard cyclone TsN-11 with a diameter of 200 mm: 1 – bellmouth inlet, 2 – duct, 3,9,18 – micro-manometers, 4,5,10,11,12 – ports of access, 6 – input pipe, 7 – cyclone, 8 – exhaust pipe, 13 – fabric filter, 14 – fan, 15 – throttle valve, 16 – hermetic shutter, 17 – hopper, 19 – thermo-anemometer, 20 – impactor.

In this paper, a laboratory microscope MIN-8 with additional illumination was used to study opaque objects in reflected light. A digital camera (resolution 12 megapixels, a lens with an aperture of F/1.8 absorbs 50% more light to increase the contrast and improve the quality of the pictures in poor light conditions) was installed on the inclined attachment of the microscope, after which have been achieved the focusing with a microscope. To create a clear image of the dusty surface of the filter, illumination was arranged on the right and on top in addition to the usual illumination of the stage.

In addition to bench tests, by the methods of computational hydrodynamics, the tangential components of velocities and the static pressures in several cross sections of the cyclone were calculated [6, 7]. The obtained results of numerical simulation are confirmed by results of empirical tests and are comparable with the results of external researchers, presented, for example, in [8, 9, 10].

2.2. The choice of the location of filtering cyclone elements in a multi-cyclone on the basis of numerical studies of the motion of a dispersed flow

Since the work is aimed to increasing the efficiency of cleaning the dispersed flow with the particle classes PM_{10}, $PM_{2,5}$, that is, dust of medium size and fine, including sub-micron dust, when choosing the arrangement of cyclone elements, it is necessary, as far as possible, individually to take into account the particle trajectories in flow. Previously, using the Computational Fluid Dynamics (CFD) method with the RANS (Reynolds-Averaged Navier-Stokes equation) turbulence model, a 2-d model of the serial CB-11 multi-cyclone was created, consisting of 11 cyclone elements with a diameter of 245 mm and a semi-tangential gas inlet. Two versions of the battery cyclone model, differing in the arrangement of the cyclone elements, are constructed (Figs 5, 6). At the Gambit preprocessor created an unstructured finite element mesh based on triangular elements. It allows more accurately reflect the geometry of region, although for three-dimensional numerical studies of channels with an obstacle, a structured grid that requires a smaller computational resource would be preferable. Before the problem solving beginning, the following boundary conditions were established: "Velocity Inlet" - uniform distribution of the velocity at the inlet of the channel (0.5 m/s); "Pressure Outlet" - atmospheric pressure at the outlet of the channel; "Wall" is the boundary condition for the walls. The formulation of the boundary conditions had the following singularity. The impermeable surfaces of the cyclone elements perform a trapping function for particles, and the walls of the cyclone body cary a function of their bouncing. Therefore, the "Wall" condition for adjusting the boundary conditions of a discrete phase model was created for both - obstructions and channel walls.

3. Results and Discussion

Results of bench tests of a filtering cyclone element

Figure 2 shows the results of measurements of the flow velocity inside the cyclone at the positions of the throttle valve 15 at the lower point V1, V2, V3 and in the upper point V4. It can be seen that the flow velocity at the cyclone center under the exhaust pipe (at distances of about 0.12 m) is 21-25 m/s for all regimes. The results of the tests also showed that the velocity in the upper part of the cyclone decreases to 17.5 m/s, and asymmetry of the velocity diagrams is observed throughout the whole height of the exhaust pipe that has the boundaries 0.06–0.14 m. This indicates the preservation of the flow precession in the exhaust pipe after passing the filter and serves as an indirect confirmation of the qualitative conservation of the hydraulic flow regime in the cyclone element after the installation of the filter insert.

Figure 3 shows the results of measuring the flow pressure inside the cyclone at various concentrations of supplied dust.

The results show that the vacuum inside the cyclone with a fully open throttle valve with supply 0,145 and 0.141 kg of dust under the ring space (distances from 0,01 to 0,04 and 0,16 to 0.19 m) is in the range of 400-600 Pa, and the pressure diagram under the exhaust pipe (distances from 0,04 to 0,16 m) is also asymmetric.

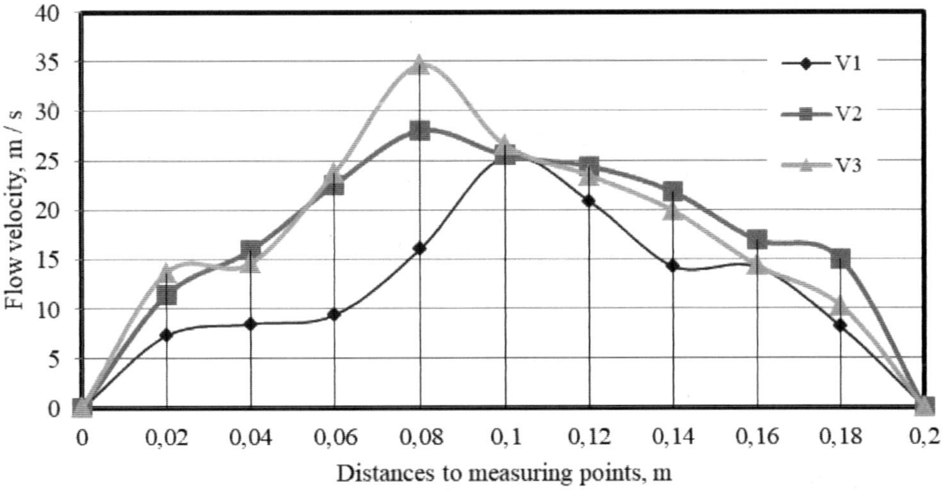

Figure 2. The velocities of not dusty air flow inside the cyclone under the exhaust pipe, V1, V2, V3, m / s in the first, second and third positions of the throttle valve 15, respectively.

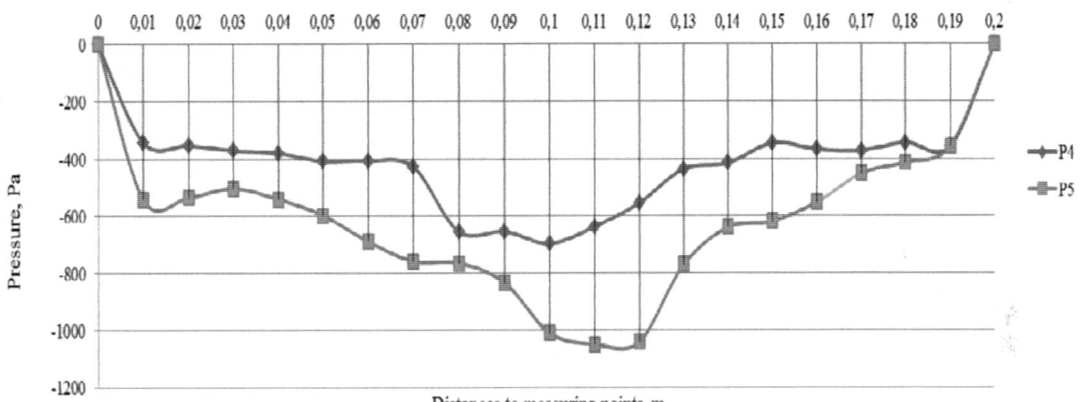

Figure 3. Underpressure P4 and P5, Pa, at the flow inside the cyclone in its lower part at the fully open throttle valve, with supply 0.145 and 0.141 kg of dust, respectively.

3.1. Numerical studies of the motion of a dispersed flow in a multi-cyclone

Numerical calculations of this multi-cyclone construction carried out by the statistical two-parameter model of turbulence k-ε, whose main advantage is the availability of a computational resource for problems close to real conditions. The inlet velocity of dusty gas flow into the multi-cyclone is assumed to be 4 m / s. In Fig. 4, Fig. 5 as an example, the velocities and static pressures in the horizontal section of the battery cyclone are presented. Calculations showed that pressure peaks occur on the frontal parts of the elements, not only in the first, but also in a subsequent rows. This allows to optimize the arrangement of the tangential inlets of the cyclonic elements.

The highest velocities are observed between the cyclonic elements, which is explained by the narrowing of the flow between them.

IOP Conf. Series: Materials Science and Engineering **481** (2019) 012041 doi:10.1088/1757-899X/481/1/012041

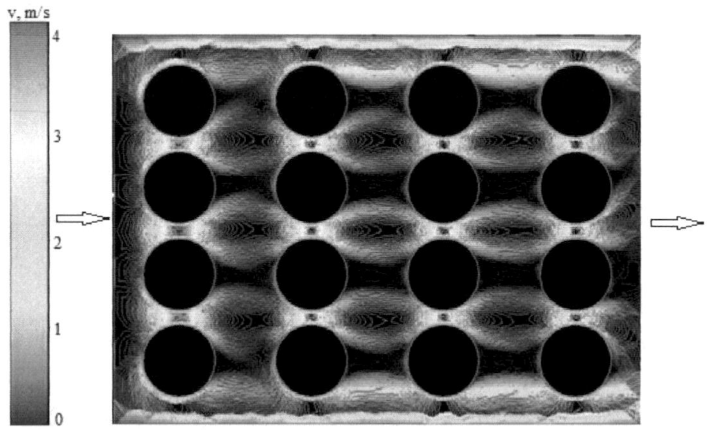

Figure 4. The velocity diagrams in the first model of the multi-cyclone.

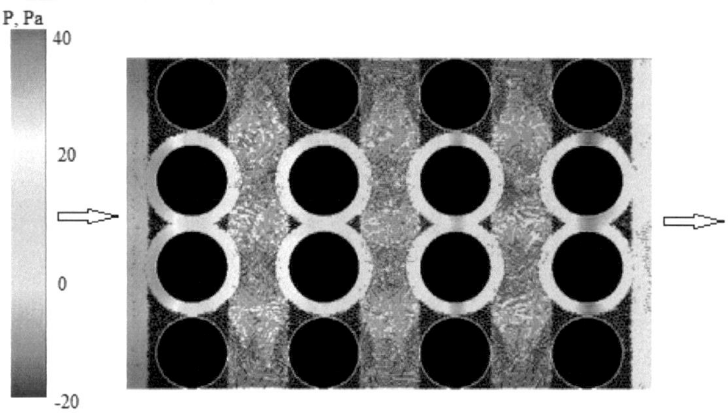

Figure 5. Distribution of static pressure in the second model of the multi-cyclone.

The results of numerical studies show that in the working space of second model between the elements, both the values of the velocity and the pressure are higher, although the flow rate characteristics at the inlet of the elements body are the same for both models. This is because the flow in the second model passes only in the spaces between the cyclone elements. In the first model, a significant part of the flow passes in the zone near the walls of multi-cyclone body that creating less resistance than the areas between the elements. Calculations clearly show that these zones represent local resistance such as one-sided sudden contraction and sudden expansion. The created numerical model of multi-cyclone allowed to estimate the influence of various factors on the efficiency of dust trapping in cyclones.

The results of the investigations made possible to arrange the inlet branches of the cyclone elements in the multi-cyclone, according to the first and second models, taking into account the hydraulic resistance of the apparatus. Then solved the problem of finding the most advantageous location of the elements inlets, taking into account the particle trajectories in the apparatus. A 2d-model of a multi-cyclone is constructed. An example of a geometric model of part of the multi-cyclone with dimensions of 1325x445 mm with 4 cyclonic elements with a diameter of 245 mm is shown in Fig 6. The number of cells of the grid is more than 150 thousand. According to the previous models (Fig. 4, 5) it could be seen that the elements of the multi-cyclone for the incoming flow represent an array of obstacles. A numerical experiment was carried out with the injection of particles with size ranging from 5 to 45 μm into the air flow in front of such an array. The problem of the turbulent flow motion equations closure solving by including the k-ε model. The flow with the particles reached a velocity of 0.5 m / s to the obstacles and flowed around them. At the same time, part of the particles, together with air, escaped from the channel, but some part of particles was kept on the obstacles. On the basis of fraction of the total mass flow of particles passing through the channel, the penetration and the degree of precipitation are determined, and also the trajectories of particle motion are detected. The study was carried out under standard conditions T = 20°C, P = 101325 Pa. The density and dynamic viscosity of the flow are assumed to be constant: $\rho = 1.205$ kg / m^3, $\eta = 18.1 \cdot 10^{-6}$ Pa · s. The developed numerical model of the multi-cyclone made it possible to evaluate the influence of various factors on its hydrodynamic characteristics and to find the location of the cyclonic element inlets that best correspond to the trajectories of the particle flows in the multi-cyclone body.

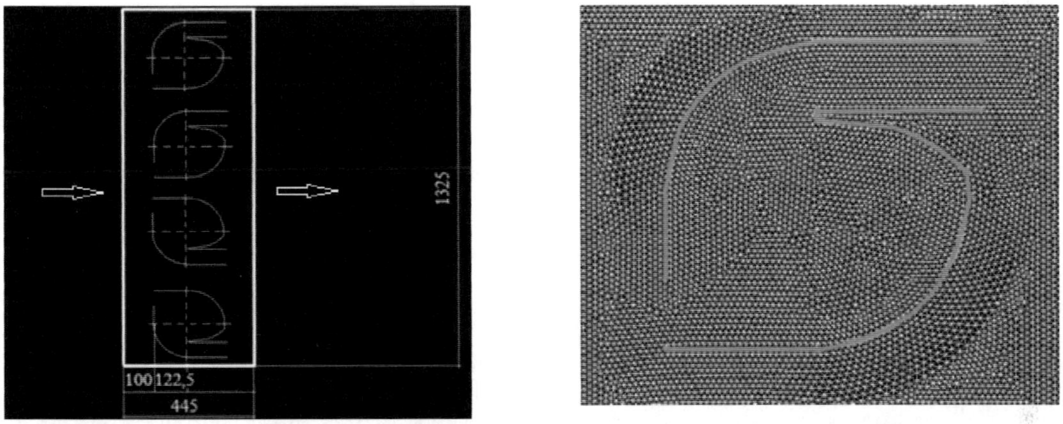

a) b)

Figure 6. Geometric model of a part of a multi-cyclone body with a configuration of cyclone elements: a - a calculation model; b - an example of a calculated unstructured finite element mesh with triangular cells around one of the cyclone elements.

4. Conclusion

Thus, the experimental study carried out showed the effectiveness of the multi-cyclone cell with a tangential inlet and a filter insert. In a numerical study, the hydrodynamic characteristics of dispersed flows in a multi-cyclone body were obtained and the best scheme for locating cyclone elements with the optimum localization of their inlets was determined.

References

[1] Zamaliyeva A T, Belyaeva G I 2017 *Ter.Neftegas* **6** pp 106-10
[2] Mazyan W I, Ahmadi A, Ahmed H, Hoorfar M 2017 *Journal of aerosol science* **110** pp 36-42
[3] Balestrin E, Decker R K, Noriler D, Bastos J C S C, Meier H F 2017 *Separation and Purification Technology* **184** pp 54-65
[4] Osama Hamdy, Magdy A Bassily, Hesham M. El-Batsh, Tarek A. Mekhail 2017 **46** pp 81-97
[5] Zamalieva A T, Ziganshin M G, Potapova L I 2017 *News of the KSUAE* **4** (42) pp 415-24
[6] Zamaliyeva A T, Belyaeva G I and Ziganshin M G 2018 *Ter.Neftegas* **6** pp 114-19
[7] Chu K W, Wang B, Yu A B and Vince A 2009 *Powd. Technology* **193** pp 235-47
[8] Chu K W, Wang B, Yu A B and Vince A 2012 *Chemical Engineering Science* **73** pp 123–39
[9] Zhou Z Y, Kuang SB, Chu KW and Yu AB 2010 *Jour. of Fl. Mech.* **661** pp 482-510
[10] Belyaeva G I and Zamaliyeva A T 2017 *Gas Ind.* **5** pp 107–11

IOP Conf. Series: Materials Science and Engineering **481** (2019) 012042 doi:10.1088/1757-899X/481/1/012042

BIM technologies in customer service business processes. Commercial pricing classifier

S V Pridvizhkin[1], N S Davydov*[1], A V Belkevich[2] and M K Kozhevnikova[1]

[1]Ural Federal University, Ekaterinburg, 620002, Russia
[2]BIM-Cluster, LLC, Ekaterinburg, Russia

E-mail: nikitadavydov14@icloud.com

Abstract. In the construction industry, implementing the information modeling technologies directly in the business processes of the Customer's Department will allow improving the following procedures on a deeper scale: designing buildings and structures with the parameters that enable to receive consolidated cost estimates in an automated mode at various stages; monitoring and control over construction and installation works at any stage of the process; and creating a federated model.

Keywords: construction company, construction, BIM technologies, effectiveness, implementation, pricing, calendar planning.

1. Introduction

Currently the BIM technologies are used worldwide more and more, which, in its turn, allows to substantially increase the quality of construction project execution, shorten the time frame, and reduce the estimated construction cost that provides the customer with the opportunity to improve the effectiveness of a company's activity within the construction industry. In this paper, the authors describe the process of implementing the BIM technologies in the activity of the Customer's Department of a construction company and study the possibility of using the classifier with commercial rates for a qualitative transformation of the business processes using the BIM approach [1].

The subject of the study is a commercial pricing classifier. The purpose of the study is to decide on the estimated and engineering calculations for information models of buildings. The methodology of the study is:

- theoretical: a theoretical analysis of literary sources;
- experimental and theoretical: modelling, analysis, deduction.

2. Methods

The construction industry's problems of ensuring the effectiveness of capital investments in the Russian Federation are constantly discussed by the top-level state authorities, including the Main Department of State Expertise. Conferences and round tables are held, and the enormous amount of work is being done in order to solve the problem [2]. The key stages of this are the following two –

taking a decision on investments with detailed justification of the amount of the capital investments in a project and the design activities that also include obtaining the project's expert review.

This study was conducted using methods:

- Theoretical: a theoretical analysis of literary sources;
- Experimental and theoretical: modelling, analysis, deduction.

On the basis of the available literary sources, the authors processed and analyzed research data on information technologies in construction. Simulated the work of customer service departments to identify the need to create a classifier of commercial rates. According to the results of the simulation, an analysis was made of the existing needs and suggestions for a specific solution to a particular task [3].

The objectives of the work is to resolve the difficulties of the customer service. This issue is being actively developed by Glavgosexpertise and various project committees [4].

Currently, the main mechanisms aimed at ensuring the effectiveness of capital investments in construction are the examination of the state investment construction projects carried out by the Ministry of Economic Development of the Russian Federation with regard to the effectiveness of using the federal budgetary funds, and the comprehensive technology and price audit of the largest investment construction projects and verification of the correct determination of the estimated construction cost carried out by experts. Actual mechanisms that are widely used nowadays have several drawbacks, which substantially reduce the possibility of taking weighed decisions in investing in construction of this or that object.

On the other hand, constant redistribution of powers among the competent authorities, the possibility to obtain a non-state expert review, and other approaches to evaluating the investment construction projects and their effectiveness do not undergo any significant changes. In practice, the increased focus is on the compliance assessment of the documents, while choosing essentially correct and profit-making decisions and their suitability for modern engineering solutions and technologies are left aside.

Besides, various experts note that the de facto project evaluation is usually started at the last preparation stage and the assignment is almost never evaluated in regard to the correctness of design and other input data. This is with the exception of such situations when the technology and price audit of the largest investment construction projects is performed. Currently, the obsolete base and index method is used for evaluating the cost of investment construction projects, as well as outdated itemized construction estimates [5]. The Main Department of State Expertise determines the project cost taking into account only the design and construction costs and does not consider further costs on operating the property, though sometimes such costs exceed the investments in project construction.

In order to ensure implementing the principle of effective usage and investment of budgetary funds, currently the regulatory documents of the construction industry are being amended [6].

"Increasing the effectiveness of investing budgetary funds is planned to be achieved by creating a database with similar projects, implementing the investment justification, fixing the economic effectiveness criteria and evaluating the planned marginal cost of construction taking into account the construction price norms and the cost of similar projects, updating the cost standards and using the input cost assessment method instead of the base and index method, and developing the consolidated construction cost standards" [7].

However, the transition to accounting the operation costs and costs on construction object demolition using the information modeling technologies is in progress. Experience has proven that quite often the substantial reduction of construction costs for this or that object by using available materials, products, structures, and works leads to the necessity of large investments in maintenance and minor repairs of a building or a structure in future [8]. The increased construction costs due to using high-quality materials, process, and engineering equipment, as well as performing works with better quality will allow reducing the operation costs and extend the service life of a construction object [5].

3. Results

Within the current business environment a comprehensive solution is needed that provides for creating a standard framework for developing the mechanisms of evaluating all costs and creating new methods, standards, and classifiers. It is important to pass over to a new level of organizing works and to actively use the existing information regarding the projects that have already passed the state expert review - that is, to pass over to informatization and creating new information environment where an information model of an object can be created. The information model should include all the data required to take a decision on capital investments, the materials and the information about the objects. [9]

Transforming the business processes using the BIM approach will allow designing buildings and structures with any characteristics, forming various types of costs estimates by various methods, monitoring and controlling the construction and installation works at any process stage, and creating a federated model [10].

In a narrower sense, this will lead to the following [11]:

- Eliminating discrepancies and errors in detailed design documents directly at the design stage;
- Creating a method to transfer a part of the works from the Detailed Design Documents stage to the Basic Design stage, as well as from the Basic Design stage to the Sketch Design stage;
- Expediting the generation of the planned cost of a construction object;
- Improving the accuracy of budget planning and control over spending of the funds allocated for construction;
- The possibility to monitor the performance of construction and installation works online;
- Drawing up and updating the as-built project documents.

Developing the information modeling technology for the above-mentioned business processes requires an item-by-item analysis of the requirements thereto.

The Design block in the created software allows:

- Facilitating the process of making any changes to the design documents;
- Computerizing the process of calculating volumes and areas of the designed objects;
- Eliminating discrepancies in the detailed design documents in a timely manner.

The following data sources are required for the operation of this system:

- The developed and adapted classifier of standard design solutions of a construction company;
- The technical assignment for a project and a model;
- The comprehensive classifier of all structural elements used and types of works performed.

According to the results of using the improved Design block a customer receives the following:

- Calculated data on the volumes and cost of materials used;
- The complete cost calculation for construction and installation works;
- Technical and economical parameters of an object and the effectiveness of project implementation;
- The design at the Sketch stage;
- The design at the Basic Design stage;
- The detailed design documents.

Further on, Classifier of Structural Elements and Cost Estimate Creation blocks allow to obtain the following:

- The possibility of prompt interaction of all departments involved;
- Computerized collection of volumes and rate setting;
- A works and materials reference book that is standardized for a construction company;
- All the data from the project's technical and economic parameters are used;
- The bill of quantities is drawn up;
- According to the results of the design works, it is possible to down load the bill of quantities taking into account the volume of the required materials;

- Automatic generation of the bill of quantities;
- Creating various templates of contracts with cost estimates.

The Construction and Installation Work Control and Federated Model block allows to carry out the following:

- Technical supervision and operative control over the construction process;
- Comparing the object model with actual built structures;
- Making prompt changes in the sections of detailed design documents and creating a federated model of the construction object;
- Preparing data for controlling the use of a building;
- Working model;
- The construction schedule;
- The current federated model and documents are created during the process in a real-time mode;
- The bill of quantities is constantly downloaded and the budget is reviewed in order to control the expenditure. [12]

The statutory regulation in the sphere of information modeling are developed quite actively. On March 6, 2018, Order No. 410 of the Federal Agency on Technical Regulation and Metrology established the Project Technical Committee PTC 705 “The Information Modeling Technologies at All the Stages of the Construction Object Life Cycle”, “pursuant to several provisions of the Action Plan to implement the evaluation of the economic effectiveness of investment justification and information modeling technologies at all the stages of the construction object “life cycle”, composed of many organizations, including the Main Department of State Expertise.

Currently, according to the proposal by the Project Technical Committee 705, the “National Standardization Program” for 2018-2019 includes 27 topics, among them the issue of developing of set of basic standards; proposals regarding the documents contributing to the development of the concept are accepted.

“The PTC 705 Medium-Term Standardization Program provides for the standards that ensure the information security of the information modeling technologies (ISO 19650-5), the standards that detail the requirements to an information model at all stages of the life cycle of buildings and structures, and the standards that establish the methodology of using the information modeling by specific participants of the construction process (“role” standards)” [13].

4. Discussion

The information model of a building includes a large amount of various information. It is quite important that all this information is classified correctly. Consequently, when developing information modeling software classifiers must be created correctly as well. Correct classification provides the possibility to use the information model of an object in other software packages.

Standards for the methodology of forming the information on construction resources as a whole are planned to be developed for creating the construction resources classifier.

Currently, the following areas of using the classifiers in information modeling are identified conventionally:

- Creating various specifications;
- Analysis of the information model in other programs.

There may be several applied tasks and programs that can work with the information model of an object, but for the transformation of business processes the following developing areas are a priority:

- The calculation of the construction cost of a building or a structure;
- Thermophysical and acoustic reports according to the information model of a building.

The calculation of the construction cost of a building or a structure is necessary for:

- Construction companies, contractor and subcontractor companies;
- Design institutes and organizations;

- Developer;
- Technical customer that is the top lieutenant of the developer.

The construction cost calculation within the information model can be divided into several stages as follows:

- All volumes are calculated by the information model
- The volumes are then linked to the rates and types of works, thus carrying out the item-to-item calculation;
- Further, the item-to-item calculation is recalculated as a cost estimate using the rates, volumes, and types of works to be performed;
- The item-to-item calculation is used to draw up a work schedule that details the types of works and can be broken down by floors, sections etc.

From the technical perspective, the object construction cost can be obtained at any stage of information modeling. Yet, as is proven by the experience, no exact calculations can be made at the Basic Design stage as the model's level of detail for the first calculations should be sufficient, that is, the model should include exact geometric reference, form, and orientation, the connection should be described, and the non-graphic information presented. Because of this, the construction cost calculation will be performed in a more accurate manner when the information model's level of detail is higher [14].

On the other hand, it is quite often that the cost calculation allows verifying the information model for the degree of information filling and the flaws. When making calculations, several errors can be noted that were made in creating the information model.

In order to make estimate and engineering calculations according to a building's information model the approved geometric parameters of a building are required, as well as the identification and classification of all structures. When obtaining the information on the composition of the object's structures the exact classification of the construction materials used is of great importance, as are the types of rooms and a type of works.

Adopting the above-mentioned complex measures provides the possibility to pass on to a new level of taking decisions on capital investments in construction, correct planning, coordination, and further use of a construction object [15].

In order to reform the construction industry and assist in carrying out business processes in a reasonable manner, unified information environment should be created, a takeoff in developing information systems should be performed, and the formats of electronic documents should be improved for all investment construction projects.

5. Conclusions

1. The introduction of classifiers gives effect to the company, an effective BIM-approach for planning and evaluating investment and construction projects;

2. This simple solution reduces the routine burden on employees;

3. The classifier of commercial rates is the property of the company and the cost of its creation will be disposable

References

[1] Mamaev A E et al 2016 Applied application of BIMbuilding model to control investment and construction project of a construction project *Vestnik SanktPeterburgskogo politekhnicheskogo universiteta im. Petra Velikogo* – Bulletin of St. Petersburg Polytechnic University. Peter the Great, **1** (**3**)

[2] Brainard S L 1997 An Empirical Assessment of the Proximity Concentration Tradeoff between Multinational Sales and Trade *American Economic Review* **87(4)**

[3] Astaf'eva N S, Kibireva Yu A, Vasil'eva I L 2018The advantages of using and problems of implementation of building *Construction of unique buildings and structures***8 (59)**

[4] Lipsits I V, Kosov V V 2014 *Investment project: methods of preparation and analysis* (Moscow: BEK Publ)

[5] Kozhevnikov K I, Pridvizhkin S V, Bazhenov O V, Kozhevnikova M K, Zasukhina Y A 2017 *Advances in Economics, Business and Management Research* **38**

[6] Zolotov V A and Semenov V A 2014 Perspektivnye sxemy prostranstvenno-vremennoj indeksacii dlya vizualnogo modelirovaniya masshtabnyx industrialnyx proektov *Trudy Instituta sistemnogo programmirovaniya RAN.* **26** (2)

[7] Lukashova A, Chumachev D 2013 London is a city of contrasts, rains and fogs *Respublikanskaya stroitel'naya gazeta – Republican construction newspaper* **12 (513)**

[8] Zakrevskaya L V, Lubin P A, Avdeev S N, Gandelsman I A, Filippov S V 2017*Zhurnal grazhdanskogo stroitelstva* **75** pp 123-128

[9] Voronin I A, Izatov V A 2014 About the possibility of determining the estimated cost of construction projects using the BIM design methods *Proc. of the All-Russia conf. "Economic, organizational and management problems of construction complex development in Russia"*

[10] Astashov A M and Oshkina L M 2014 Rol informacionnyx texnologij proektirovaniya v reformirovanii inzhenerno-stroitelnogo obrazovaniya *Integraciya obrazovaniya.* **18** no 4 (77)

[11] Talapov V V 2011 *Foundations of BIM, introduction to building information modeling* (Moscow, DMK Press Publ)

[12] Barabanova T A 2017 Using of BIMtechnology in the technical operation of buildings *Proc. of the international scientific conference "Integration, partnership and innovation in construction science and education"*

[13] Malakhova V V Cluster approach to development of region construction complex *Economics and management* **1**

[14] Drapkin I M, Mariyev O C, Chukavina K V 2015 A Quantitative Evaluation of Import And Export Potential of the Foreign Direct Investment in the Russian Economy with a Gravity Approach *Journal of New Economic Association* **4**

[15] Pridvizhkin S V, Bazhenov O V, Vyguzova M A 2017 *Materials Science and Engineering*, **262** 012165

IOP Conf. Series: Materials Science and Engineering **481** (2019) 012043 doi:10.1088/1757-899X/481/1/012043

Drying and shrinkage of cement paste for 3D printable concrete

G S Slavcheva[1]

[1]Department of Building Materials Technology and Structures, Voronezh State Technical University, 20-letiya Oktyabrya St., 84, Voronezh 394006, Russia

E-mail: gslavcheva@yandex.ru

Abstract. The article presents the results of the study of drying and shrinkage of three compositions of cement paste different in the content of microsilica and superplasticiser within the range of variation typical for 3D printable concrete compositions. Plate samples kept in the environment with RH = 20, 30, 50, 70 % were used to simulate different drying conditions of material in thin layers of 3D printing structures. Rates of mass loss and shrinkage of hardened cement paste are compared with criterial parameters of their structure. It is shown that increased dosage of microsilica and superplasticiser ensures the reduction of total volume of pores, increase of the content of nanopores, while simultaneously the area and energy activity of solid phase surface increases. These factors secure the ability of cement paste to better retain the water in the structure under all drying conditions. As a result, reduction of absolute values of mass loss by 3 - 4 times and shrinkage by 2 times in comparison with reference system is ensured for cement paste with maximally dense structure. At the same time, there is a twofold increase of relative deformability on each percent of mass loss, which may lead to significant tension in 3D printing structures due to moisture gradient.

1. Introduction

3D concrete printing (3DCP) is one of the emerging technologies that can minimize the supply chain in the construction process by automatically producing building components directly from digital model without human intervention and complex formwork. Potential advantages of this process include ability to get multicomponent materials of different functionality directly on the construction site, to print the freeform constructions without formwork, to reduce materials consumption and building work content [1 - 3]. One of the most complex problems in the implementation of the abilities of 3D printing in different fields of construction is the problem of creating mixtures with functional characteristics that correspond to their using. Active development of the research has allowed achieving quite a wide range of effective mixtures for 3D printable concrete. When it comes to optimisation of mixture compositions, researchers usually focus their attention on such issues as rheological behaviour, extrudability, buildability and structural build-up of concrete [4 - 18]. However, the problems of 3D printable concrete behaviour during working-time remain unexplored. The problem of drying shrinkage may be one of the most significant among them, since intensive drying of the material may take place in the thin layers of 3D printing structures and lead to formation of cracks in the structures.

As a rule, high-performance printing concrete is obtained from low water/cement ratio (0.2 - 0.3), efficient plasticisers, ultradisperse fillers (microsilica and nanosilica, metakaolin, etc.), which allow forming solid high-density structure of composites. The experience of studying similar high-performance cement-based materials has shown [19 - 26] that the process of their shrinkage is different from the development of shrinkage of ordinary cement-based materials. This article is dedicated to the study of processes of drying and shrinkage of high-performance cement paste in the conditions simulating drying in 3D printing structures and correlation of characteristics of these processes with parameters of material structure.

2. Theoretical framework

Drying shrinkage of cement-based materials is defined by joint impact of various types of water differing in connection strength with the structure on its development. According to the existing perceptions [27 - 30], it is supposed that at the initial stage of drying in the range of relative humidity of the environment RH = 80 – 95 % capillary water is removed from large pores r > 100 nm. Capillary pressure in these pores is relatively low, thus the values of shrinkage stress and deformations are small. When RH decreases in the range from 80 to 40 %, water is removed from the pores in radius of 20 nm < r < 100 nm, at the same time the value of capillary pressure grows and, correspondingly, drying shrinkage increases. After capillary water is removed from the pores, adsorption water is removed from the surface of solid phase (RH < 40 %). It is known that adsorption water compensates for a part of excess surface energy of solid phase. In the process of desorption solid phase particles begin to be affected by growing forces of surface tension, for which reason their volume reduces.

The value of drying shrinkage of the material is defined by the level of drying which depends on humidity of environment and structure's ability to retain water. The latter is determined by quantitative ratio of different types of water in the material. This ratio depends on the following parameters of cement-based materials structure:

- volume and structure of hard phase, as its surface area and surface energy determine the effect of contraction forces when adsorption water is removed;
- volume and size of pores defining the value of capillary forces.

High-performance printing concrete usually has high-density fine-dispersed structure different from the structure of ordinary cement-based materials in characteristics criterial for the development of shrinkage. Therefore, comprehensive study and comparison of the structure and parameters of moisture shrinkage of cement paste as a shrinkage carrier in cement structure is required for predicting operational deformative behaviour of high-performance printing concrete.

3. Experimental programme and methods

3.1. Materials and mix design

Portland cement CEM I 52.5, microsilica (specific surface 24500 m^2/kg, SiO_2 content ~ 70%); superplasticiser (SP) of trademark Sika ViscoCrete 20HE based on polycarboxylic ethers were used to produce the samples of cement paste.

Ordinary cement paste (Table 1) of reference content and high-performance cement pastes of MB-8 and MB-30 series were studied. The change of structure of high-performance cement pastes was ensured by variation of the dosage of microsilica and plasticiser. Consistence of cement paste was assumed as a fixed factor for the experiments (slump 180 mm for other compositions).

Table 1. Mix composition.

Material	Mass/ mass cement loss		
	Control	MB-8	MB-30
Cement	1.0	1.0	1.0
Microsilica	-	0.08	0.3
Plasticiser	-	0.005	0.018
Water/cement ratio	0.40	0.34	0.24

3.2. Experimental procedures

Sample plates 10 × 40 × 160 mm in size dry in the environments with different humidity were used in assessment of deformations and mass loss to simulate humidity conditions of material usage in thin layers of 3D printing structures. Series of cement paste samples (6 in each series) were produced for experimental research from one batch of each of the studied mixtures. Series of cement paste samples hardened during 24 hours in moulds. Then the samples were put in dessicators above saturated salt solutions with humidity RH = 20; 30; 50; 70 %. Maturing in dessicators was carried out with constant temperature (20 ± 2)^{0}C until stabilisation of their mass and linear dimensions. Deformations were determined by a comparator with indicating gage with division value of 0,01 mm. Mass and sizes were monitored within 1, 3, 7, 14, 28 days since the moment of placement into dessicators and then after every 14 days. Sample weighing error accounted for ± 0.1 g, error of sample length change accounted for ± 5 μm. Within-run variability ratio in assessment of deformations and moisture losses accounted for 8 - 10%.

Simultaneously, structural characteristics of cement paste were monitored. Assessment of specific surface, total volume of pores as well as the volume of pores in radius of less than 20 nm (this radius is considered to be the border of effect of capillary forces) of cement paste was carried out through four-point nitrogen adsorption BET method on SoftSorbi-II ver.1. Surface energy of hard phase was estimated by calorimetric method by heat of wetting the prepared sample of cement paste at t = 20°C in Calvet calorimeter C80 using mixing cell with circulation. Phase composition of cement paste was controlled by X-ray diffraction method (CuKα radiation, λ = 1.541788 Å, diffractometer ARL X'TRA). At the same time the degree of cement hydration in the content of alite phase was evaluated. Scanning probe microscope JEOL JSM-7001F was used to assess the morphology of cement paste structure.

4. Experimental results and discussion

4.1. Structural characteristics

Combined application of microsilica and superplasticiser additives results in the change of hardened cement paste structure by parameters the values of which are presented in table 2. The thickest structure is typical for MB-30 system. Total volume of pores in cement paste structure is characterised by minimal values of the volume of pores (0.24 m^3/m^3) with up to 60 % of pores in radius of less than 20 nm from their total volume; specific area of the surface reaches maximum value of 51.3 m^2/g for the studied systems. MB-8 system containing small amounts of microsilica and plasticiser is similar to control system in the values of density and porosity and to MB-30 system in the values of specific area of solid phase surface. Such change of parameters of solid phase and porosity is naturally caused by parameters of the cement substance of the studied systems. Morphology of the particles of the studied systems is evaluated according to the data of electronic microscopy (Figure 1).

Table 2. Structural characteristics of cement paste.

structural characteristics		control	MB-8	MB-30
Density, kg/m^3		1680	1850	1970
Total volume of pores, m^3/m^3		0.34	0.33	0.24
Content of pores in radius r, in % of total volume of pores	r < 20 nm	24	53	60
	r > 20 nm	76	47	40
Total specific surface of solid phase of the material, m^2/g		29.1	49.4	51.3
Heat of wetting of solid phase of the material, kJ/kg		15.2	27.1	31.2
Characteristic of intensity of diffraction maximums	d = 4.86 Å, for $Ca(OH)_2$	1265	746	496
	d = 1.76 Å, for C_3S	65	116	100
Hydration degree, %	82	78	64	66

IOP Conf. Series: Materials Science and Engineering **481** (2019) 012043 doi:10.1088/1757-899X/481/1/012043

a) control b) MB-8 c) MB-30

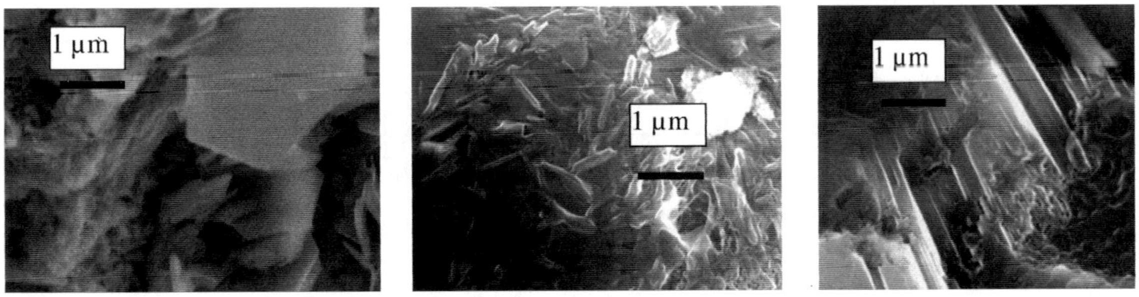

Figure 1. Typical photos of the structure of hardened cement paste samples.

In the formed structure of hardened cement paste there is mostly C-S-H gel of amorphocrystaline morphology. However, the size of the particles of new developing formations in systems MB-8 and MB-30 is significantly smaller than in the control system. The smallest size of particles of C-S-H gel and the densest packaging are typical for MB-30 system. A significant number of plate crystals of $Ca(OH)_2$ is recorded only in the control system. According to the results of X-ray phase analysis (Table 2), it has been established that in systems MB-8 and MB-30 the intensity of diffraction maximums typical for crystals of $Ca(OH)_2$ decreases by 3 - 4 times as related to the control system. The samples of MB-8 and MB-30 series are characterised by high density and extremely low free volume of pores for the development of structure formation processes. Therefore, for them the degree of cement hydration at 28-day age is characterised by lower values than in check samples. It was this transformation of structural characteristics of cement paste that determined the change of energy potential of the surface of hard phase and pore volume. Also, almost twice as higher energy activity of solid phase surface is typical for MB-8 and MB-30 systems. Specific heat of wetting reaches 27 - 31 kJ/kg as against 15.2 kJ/kg in the control system.

4.2. Drying and shrinkage

Our experiments have determined that drying and shrinkage process is directly related to characteristics of the composition and structure of the studied systems. Significant differences in density and porosity, in surface area and solid phase energy have shown that the value of mass loss for the cement paste of the studied series differs approximately by 3 - 5 times (Figure 2) while the value of deformations differs approximately by 2 - 3 times (Figure 3) under different drying conditions.

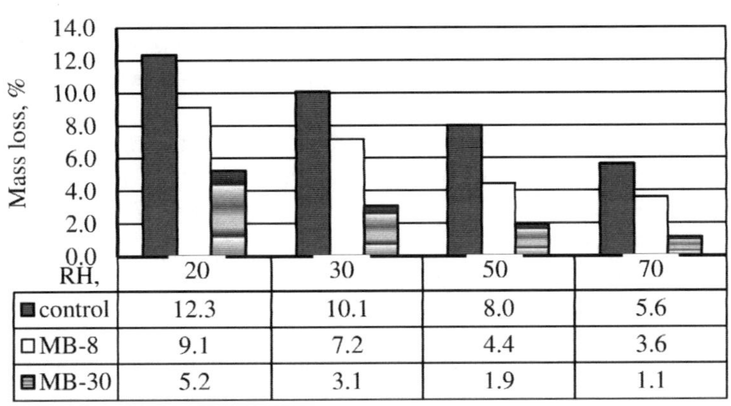

RH,	20	30	50	70
■control	12.3	10.1	8.0	5.6
□MB-8	9.1	7.2	4.4	3.6
▤MB-30	5.2	3.1	1.9	1.1

Figure 2. Mass loss of cement paste under different drying conditions

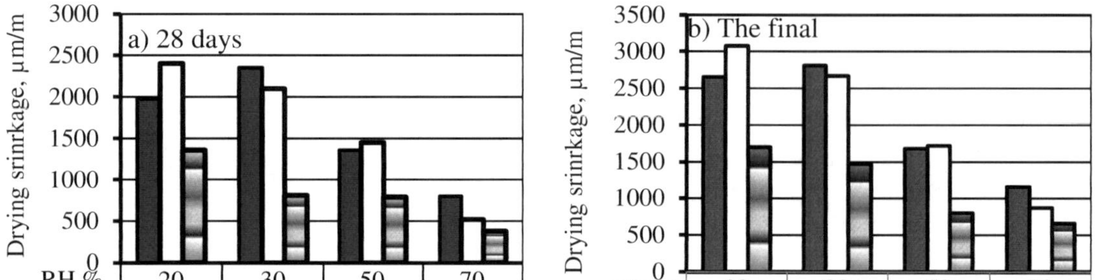

RH,%	20	30	50	70
control	1980	2350	1360	800
MB-8	2400	2100	1450	520
MB-30	1360	810	790	380

RH, %	20	30	50	70
control	2650	2810	1680	1160
MB-8	3075	2670	1720	870
MB-30	1700	1470	800	660

Figure 3. Drying shrinkage of cement paste under different drying conditions

On drying in the environments with RH = 5 0% and RH = 70 %, where adsorption water is almost not removed, the difference in the values of mass loss of ordinary (control) and high-performance (MB-8 and MB-30) cement paste is especially notable and can reach almost a fivefold value (Figure 3). For instance, moisture loss on dehydration of the control series samples in the environment with RH = 70% accounts for 5.6% by mass, while in the samples of MB-8 and MB-30 series the losses are 3.6% and 1.1 % correspondingly. The value of shrinkage of the control system in the environment with RH = 70 % where mostly capillary water is removed already reaches 800 μm/m at the early stage of drying. By the end of 28 days the deformation of control cement paste exceeded the shrinkage in systems MB-8 and MB-30 almost threefold, total shrinkage is twice different. The obtained data on the difference in the processes of drying and shrinkage of the samples of the studied series are explained by the difference in binding force of the material and water. Higher values of binding forces of the material and water which retain the water in the structure correspond to the higher values of specific surface and surface energy of solid phase as well as to the smaller radius of pores. As a result, a considerable part of liquid stage happens to be adsorption-related, and the transfer of moisture in high-performance cement paste can be mainly determined by diffusion of adsorbed molecules on the solid phase surface. So, under the drying conditions in thin layers with normal operating humidity the structure of high-performance cement paste can retain moisture better and resist the growth of deformations.

In the environments with RH = 20, 30 % when not only capillary, but also adsorption water is removed, deep dehydration is typical for the samples of MB-8 and MB-30 series. For these series the value of mass loss with RH = 20 % is 3 - 5 times higher than with dehydration of RH=70 %. The difference in mass loss when drying in the environments with RH = 20 % and RH = 70 % is only twice for the samples of the control series.

Minimal values of shrinkage are noticed for MB-30 system. This system with the largest content of microsilica and SP is characterized by the highest value of specific area and surface energy of solid phase, low porosity and predominance of nanopores r < 20 nm in the total volume of pores. As a result, this does not only prevent from removing water, but also facilitates the reduction of deformation value under all drying conditions. That said, under the change in humidity in the range of RH from 70 % to 20 % total shrinkage increases by 3.5 times for MB-8, by 2.3 times for the control composition, and by 2.6 times for MB-30 system.

For high-performance cement pastes of MB-8 series in the environments with RH =20, 30, 50 % the value of drying shrinkage almost corresponds to the deformations of cement paste of the control composition. Only in the environment with RH = 70 % the value of deformations in MB-8 system is significantly lower than the shrinkage of control series samples. Increased deformability of MB-8 system can be conditioned by the following. This system is characterised by the values of area and surface energy of solid phase and relative content of nanopores comparable with MB-30 system. This happens to be the predominant factor of the reduction of deformation value in the environment with RH = 70 % when capillary water is removed. At the same time, MB-8 system is similar to the control one by the value of pore volume, therefore the effect of non-compensated surface forces acts as a factor of deformation growth with removal of water from smaller and smaller pores when drying in the environments with RH = 20, 30, 50%.

So, parameters of the structure of high-performance cement pastes of MB-8 and MB-30 series predetermined the reduction of mass loss when drying by 3 - 4 times as compared to control ordinary cement paste and reduction of shrinkage by 2 times.

4.3. Coefficient of linear shrinkage

Coefficient of linear shrinkage (**β**) was used for quantitative criterion evaluation of the value of shrinkage of cement pastes. This relative value physically represents a ratio of material deformations $\boldsymbol{\varepsilon}$ (μm/μm) to the measurement unit of its moisture content $\boldsymbol{\Delta W}$ (g/g). Coefficient of linear shrinkage is graphically interpreted as slope ratio of shrinkage curve to the axis of its moisture content change (Figure 4). This index allows quantitatively evaluating the ratio of mass loss value to material shrinkage and can be used for predicting shrinkage stress in the cement volume with humidity gradients.

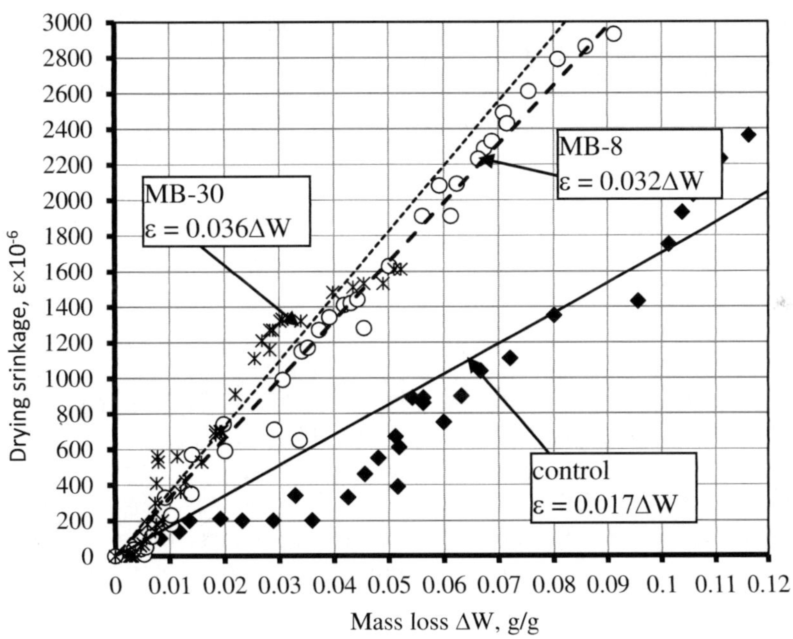

Figure 4. Dependence on drying shrinkage of cement paste on mass losses.

After summarising the aggregate data on the values of moisture shrinkage and moisture losses of cement paste of the studies series with dehydration in the environments with RH = 20, 30, 50,70 %, the design value of coefficient of linear shrinkage accounted for:

- $\beta_{control} = 0.017$,
- $\beta_{MB-8} = 0.032$,
- $\beta_{MB-30} = 0.036$.

So, relative deformability of high-performance cement pastes happened to be twice higher than in ordinary cement paste. This proves the proposed assumption of the increase of binding force of the structure with water for high-performance cement pastes in relation with the increase of content of nanopores $\boldsymbol{r} < 20$ nm, increased dispersion, area, and energy activity of the solid phase surface.

5. Conclusion

Joint employment of microsilica and superplasticiser typical for obtaining high-performance printing concrete is accompanied by significant change of structural characteristics criterial for moisture shrinkage. Total porosity is reduced almost by 1.5 times while the content of pores in radius of less than 20 nm is increased (up to 60 % of total volume); specific surface area and surface energy of solid phase increases almost twice.

Structural changes happen to be a growth factor for the binding force of structure with water. For high-performance cement paste as compared to ordinary cement paste, total reduction of moisture losses when drying reaches a three-fourfold value. And this is a positive factor for the hardening of material in thin layers of 3D printing structures as it helps to preserve moisture for hydration and hardening. If an optimal high-density structure is created, cement paste shrinkage may be reduced twice.

At the same time, quantitative criterial evaluation of deformability potential of cement paste by the coefficient of linear shrinkage $\boldsymbol{\beta}$ has shown that the value of shrinkage for high-performance cement paste under the change of moisture content by 1 % will be twice higher than shrinkage of ordinary cement paste. As a result, the level of internal stresses in thin layers of 3D printing structures may significantly rise even when operating moisture content is slightly changed. Therefore, when it comes to high-performance printing concrete, drying shrinkage can make considerable contribution to the development of stresses in constructions under cyclical moistening-drying during the whole period of operation.

Acknowledgements

This research work has been carried out in the frame of the 7.10781.2018/11.12 Project, financed by the Russian government in order to obtain initial scientific results broadening the participation of subordinate educational organisations in the implementation of the National Technological Initiative.

References

[1] Tay Y, Panda B, Paul S, Noor M, Tan M and Leong K 2017 3D printing trends in building and construction industry: a review *Virt. Phys. Prot.* **12(3)** 261–276

[2] Kothman I and Faber N 2016 How 3D printing technology changes the rules of the game: Insights from the construction sector *J. Manuf. Technol. Man.* **27(7)** 932–943

[3] Bos F, Wolfs R, Ahmed Zand Salet T 2016 Additive manufacturing of concrete in construction: potentials and challenges of 3D concrete printing *Virt. Phys. Prot.* **11(3)** 209–225

[4] Roussel N and Lanos C 2003 Plastic Fluid Flow Parameters Identification Using a Simple Squeezing Test *Appl. Rheol.* **13** 132–141

[5] Zahia T, Roussel N and Lanos C 2005 The squeezing test: a tool to identify firm cement-based material's rheological behaviour and evaluate their extrusion ability *Cem. Concr. Res.* **35** 1891–99

[6] Le T, Austin S, Lim S, Buswell R, Gibb A and Thorpe T 2012 Mix design and fresh properties for high-performance printing concrete *Mater. Struct.* **45(8)** 1221–32

[7] Le T, Austin S, Lim S, Buswell R, Law R, Gibb A and Thorpe T 2012 Hardened properties of high-performance printing concrete *Cem. Concr. Res.* **42** 558–566

[8] Perrot A, Lecompte T, Estell´e P and Amziane S 2013 Structural build-up of rigid fiber reinforced cement-based materials *Mater. Struct.* **46(9)** 1561–68

[9] Feng P, Menga X, Chenb J, Ye L 2015 Mechanical properties of structures 3D printed with cementitious powders *Constr. Build. Mater.***93** 486–497

[10] Malaeb Z, Hachem H, Tourbah A, Maalouf T and Hamzeh F 2015 3D concrete printing: machine and mix design International *IJCIET* **6(6)** 14 – 22

[11] Perrot A, Rangeard D and Pierre A 2016 Structural built-up of cement-based materials used for 3D- printing extrusion techniques *Mater. Struct.* **49**1213–20

[12] Lecompte T and Perrot A 2017 Non-linear modeling of yield stress increase due to SCC structural build-up at rest *Cem. Concr. Res.* **92** 92–97

[13] Ma G, Li Z and Wang L 2018 Printable properties of cementitious material containing copper tailings for extrusion based 3D printing *Constr. Build. Mater.* **162** 613–627

[14] Wenga Y, Li M, Tan M and Qian S 2018 Design 3D printing cementitious materials via Fuller Thompson theory and Marson-Percy model *Constr. Build. Mater.* **163** 600–610

[15] Kazemiana A, Yuan X, Cochran E and Khoshnevis B 2017 Cementitious materials for construction-scale 3D printing: Laboratory testing of fresh printing mixture *Constr. Build. Mater.* **145** 639–647

[16] Wolfs R, Bos F and Salet T 2018 Department Early age mechanical behaviour of 3D printed concrete: numerical modelling and experimental testing *Cem. Concr. Res.* **106** 103–116

[17] Qian Y, and Kawashima S 2018 Distinguishing dynamic and static yield stress of fresh cement mortars through thixotropy *Cem. Concr. Compos.* **86** 288–296

[18] Paul S, Tay Y, Panda B and Tan M 2018 Fresh and hardened properties of 3D printable cementitious materials for building and construction *Arch. Civ. Mech. Eng.* **18** 311–319

[19] Kaprielov S, Karpenko N, Sheinfeld A and E.Kouznetsov 2003 Influence of multicomponent modifier containing silica fume, fly ash, superplasticizer and air-entraining agent on structure and deformability of high-strength concrete *7th CANMET/ACI Int. Conf. on Superplasticizers and other chemi-cal admixtures in concrete* (October 10-13, Berlin: Germany) pp 99–107

[20] Kaprielov S, Sheynfeld A, Kardumian H and Dondukov V 2006 Characteristics of the structure and properties of high-strength concrete, containing multicomponent modifiers including silica fume, fly ash and metakaolin *16th Int. Baustofftagung (IBAUSIL)* (September 12-15, Weimar: Germany) **B2** pp 77–84

[21] Alsayed S 1998 Influence of superplasticizer, pasticizer, and silica fume on the drying shrinkage of high-strength concrete subjected to hot-dry field conditions *Cem. Concr. Res.* **28(10)** 1405 – 15

[22] Baroghel-Bouny V, Mainguy M, Lassabatere T and Coussy O. 1999 Characterization and identification of equilibrium and transfer moisture properties for ordinary and high-performance cementitious materials *Cem. Concr. Res.* **29** 1225–38.

[23] Baroghel-Bouny V, Mounanga P, Khelidj A, Loukili A and Rafai N 2006 Autogenous deformations of cement pastes: part II. W/C effects, micro–macro correlations, and threshold values *Cem. Concr. Res.* **36(1)** 123–136

[24] Yang Y, Sato R and Kawai K 2005 Autogenous shrinkage of high-strength concrete containing silica fume under drying at early ages *Cem. Concr. Res.* **35(3)** 449–456

[25] Jiang Z, Sun Z and Wang P 2005 Autogenous relative humidity change and autogenous shrinkage of high-performance cement pastes *Cem. Concr. Res.* **35(8)** 1539–45

[26] Soliman A, and Nehdi M 2011 Effect of drying conditions on autogenous shrinkage in ultra-high performance concrete at early-age *Mater. Struct.* **44** 879–899

[27] Wittmann F 1973 Interaction of hardened cement paste and water *J. Am. Ceram. Soc.* **56(8)** 409–415

[28] Ayano T and Wittmann F 2002 Drying, moisture distribution, and shrinkage of cement-based materials *Mater. Struct.* **35(3)** 134-140.

[29] Shimomura T and Maekawa K 1997 Analysis of the drying shrinkage behaviour of concrete using a micromechanical model based on the micropore structure of concrete Mag. Concr. Res. **49(181)** 303–322

[29] Hansen W 1987 Drying shrinkage mechanisms in Portland cement paste *J. Am. Ceram. Soc.* **70 (5)** 323–328

[30] Bazant Z and Kim J 1991 Consequences of diffusion theory for shrinkage of concrete *Mater. Struct.* **24** 323–326

Estimates of the influence of the solar irradiation spectrum energy on the intensity of heat treatment of solid waste with hexavalent chromium

M I Bilalov[1], M G Ziganshin[2]

[1]Thermal power plants Department, Kazan State Power Engineering University, Russia, 420066, Kazan, Russia
[2]Thermal power engineering, Gas supply and Ventilation Department, Kazan State University of Architecture and Engineering, Russia, 420043, Kazan, Russia

E-mail: mjihan@mail.ru

Abstract. Methods of processing chromium containing wastes are considered. Positive and negative aspects of widely used methods of transfer of hexavalent chromium to trivalent are indicated. It is proposed as the most effective way of thermal neutralizing of chromium-containing waste in a reducing environment. Field tests of samples containing hexavalent chromium and organic compounds, which showed the possibility of intensification of heat treatment processes using solar energy, were carried out. An estimation is obtained for the wavelength ranges of the near UV, the visible, and the near IR spectrum, within the limits of which photochemical reactions for the reduction of hexavalent chromium are possible. This will reduce the temperature of heat-treatment and, together with it, reduce energy consumption in heat treatment devices. The latter is also important for reducing greenhouse gas emissions.

Keywords: hexavalent chromium, solid waste, the electrical and electronic equipment, heat treatment, solar irradiation spectrum

1. Introduction

In 2002, the European Union introduced the original RoHS Directive (Restriction of Hazardous Substances), also known as Directive 2002/95 / EC, and in 2011, 2015, subsequent directives RoHS2, ROHS3 were published. After that, the use of hexavalent chromium (Cr VI) significantly decreased in the EU, as well as in many other countries. At the same time, we note that the directives impose restrictions on electrical and electronic products, while allowing the content of Cr VI in them to be 1000 ppm [1]. This is a significant number, as the stream of electronic wastes is growing rapidly in both developed and developing countries. For example, according to the Environment Directorate of the General European Commission ('DG Environment'), waste of electrical and electronic equipment (WEEE) such as computers, TV-sets, fridges and cell phones, is one, the fastest growing waste streams in the EU, with some 9 million tonnes generated in 2005, and are expected to grow to more than 12 million tonnes by 2020. Hence, such wastes can contain up to 20 thousand tonnes of Cr VI, which were used for anti-corrosion protection. It should also be borne in mind that hexavalent chromium compounds can be used in a number of other industries, for example, in tanning, metalworking, or in painting processes, which are widely used in various spheres of production, including construction. Therefore, the problem while doesn't lose acuity, and the challenges of finding the best ways to Cr VI-containing waste processing remains to this day.

One of the possible ways of neutralizing hexavalent chromium compounds is to convert it to a less toxic trivalent or bivalent state. It can be carried out by various methods, among which currently used are mainly reagent, electrochemical, membrane, sorption and thermal methods. Each of them has its advantages, as well as general and specific imperfections. They are widely represented in the literature (see, for example, [2-8]), so we only note here that all methods are energy-consuming, and reagent and sorption methods are also material-consuming methods. Exceptions are the methods of membrane and heat treatment. At the same time, the membrane method requires preliminary purification of waste water from oils, surfactants, solvents, organics, and is characterized by the high cost of membranes. Therefore, as an optimal method, the method for neutralizing chromium-containing waste by heat treatment in a reducing medium of carbon monoxide is considered. CO as a fuel burnout product should be oxidized to carbon dioxide CO2 with the reduction of Cr VI to Cr III and/or to Cr II.

However, a lot of fuel is spent for heat treatment, since it is necessary to heat up to a temperature of about 700 ° C and higher not only chromium, but the entire mass of waste. In addition, as we see, there is an inevitable output of carbon dioxide. Therefore, the creation of a heat treatment system that meets modern challenges, including the low-carbon requirements for the production and/or use of alternative energy sources, should aim to maximize the current reduction in energy consumption from carbon fuels. It is known that Cr VI in a mixture with organic compounds can be reduced at low temperatures if the mixture is exposed to solar or other radiation in the far and near ultraviolet (UV), and even in the visible part of the spectrum. So, for example, it is known that the canvases of artists of the 19th and 20th centuries, which used a lot of chrome yellow pigment in their works, under sunlight, and, perhaps, illuminated by daylight lamps, have now begun to acquire a greenish shade.

2. Methods

To determine the effect of photo-oxidation under solar irradiation on the efficiency of heat treatment of contained the Cr VI waste of coloring, field tests of the samples was carried out. Placement of samples at two experimental sites: in Kazan (latitude 55° 36 ', longitude 49° 19', altitude above sea level 128 m) and Naberezhnye Chelny (latitude 55° 42 ', longitude 52° 20', altitude above sea level 103 m), Russia, Republic of Tatarstan, is presented in Fig. 1, photo a, b, c, d, e. Geodetic heights at the location of samples in Kazan 102 m, in Naberezhnye Chelny 70 m.

The material for testing was $K_2Cr_2O_7$ potassium dichromate of "chemically pure" class. The choice of material and class of purity was dictated primarily by the desire to ensure unambiguity and convenience of observation in the field conditions after a change in the state of the sample, which in this case can easily be done due to change of the sample color by the course of the experiment. 14/07/18 in Kazan, samples 1, 4 were installed in opaque white PET-beaker at a level of +0.250 m above the surface, with an inclination of the axis to the south by approximately 55° (Figure 1, photo a), which made it possible to irradiate the bottom of the beaker better when the sun is at the zenith. Sample 2 were installed vertically in a transparent PET bottle. In Naberezhnye Chelny, the samples were installed in transparent double beakers of polypropylene (PP) at a level of +2,800 m above the surface vertically. In Kazan, sample 1 was covered with a magnifying glass (an x6 loup), sample 2 – with a PET-plate, sample 4 - with a transparent polyethylene film PELD. Samples 6, 7 were placed 9 days later (Figure 1, photo b). Sample 6 was coated with a transparent no color PELD film, No. 7 with a transparent PELD film of pastel green color (Pastel Green, RGB 119 221 119, CMYK 46 0 46 13). Samples 6, 7 are also installed with an inclination of the axis to the south by approximately 55°. After 5 days, the samples 3, 5 (Figure 1, photo c) were installed (also with an inclination of the axis to the south of approximately 55°). The latter was prepared with natural drying oil. Sample 3 is covered with PET-plate, sample 5 – with PELD-film, both coatings are transparent and colorless. In the city of Naberezhnye Chelny on July 15, 2018 at 12:00, 4 samples 10, 11, 12, 13, were installed, mixed on drying oil oxol (Figure 1, photo d). Samples 11, 12 are covered with a colorless polyethylene transparent film, sample 13 – with an opaque lid, sample 10 covered with a magnifying glass with a sixfold increase (× 6 loupe). To protect against accidental mechanical damage at a height of 0.5 m above the samples, a metal grid with 3 × 3 mm cells is installed (Figure 1, photo e). Due to the magnifying glass it was possible to increase the layer temperature by an average of 10°C. In Naberezhnye Chelny on clear July days the layer temperature under the magnifier reached 43°C, in Kazan - up to 46°C. The temperature of the other samples was approximately equal to the ambient temperature. In addition to heating, the loup played the role of a filter on the range of the radiation spectrum incident on the samples within 300-6000 nm [9]. The transparency of the used polymer materials in the UV part of the spectrum is also small. In the near UV region in the interval λ = 200-400 nm, the integrated light transmittance of a 100 μm PELD film is about 45%, in the visible part of the spectrum – 70% and in the IR part of the spectrum to λ = 2300 nm – 80%. Approximately the same transmittance spectra in the UV and IR regions have PP: from 250 nm for thin films, or from 370 nm for thick (from 1 mm thick and thicker) to 2300 nm ([10]). However, PP is more transparent than PELD and has a transmission in both the visible and in the IR parts of the spectrum to 90%. According to [11], PET has a transmissive power in the visible and in the IR part of the spectrum up to 1500 nm, close to 90%. However, unlike PELD and PP, even a thin PET film (10 μm thick) does not transmit radiation in the near UV region up to 320 nm, which is explained by the content in its molecular structure of the carbonyl group C = O, which is one of the types of chromophore groups [12]. Also, according to [11], the transmittance of PET drops to 10% in the far IR region at 1500 nm, and then has two transmission windows of 60-75% for the ranges 1500-3500 and 14000-19000 nm. Thus, the radiant energy incident on various samples was different in spectrum and intensity.

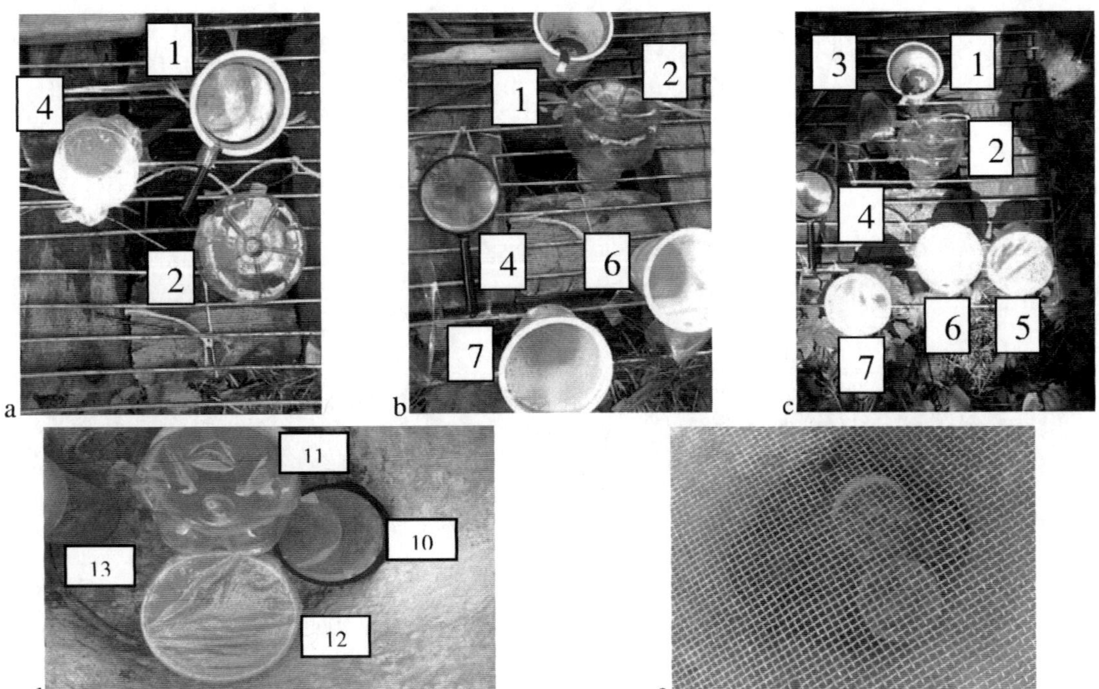

Figure 1. Installation and location of samples in Kazan (a, b, c) and Naberezhnye Chelny (d, e): a - 14/07/18; b - 23/07/18; c - 28/07/18; d, e - 15/07/18.

In the IR region of the spectrum, the irradiation conditions for samples 10, 11, 12, 13 were close to the irradiation of samples 5, 6 coated with a 100 μm PELD film, but 10, 11, 12, and 13 samples were irradiated 10% more intensively. For the PELD film from the transmission range of 200-400 nm in the

IOP Conf. Series: Materials Science and Engineering **481** (2019) 012044 doi:10.1088/1757-899X/481/1/012044

near-UV region, only the interval 345-400 nm is important for our tests, since the radiation in the 200-345 nm range is completely absorbed by the ozone layer of the atmosphere. Therefore, irradiation of samples 5, 6 under a colorless PELD film occurred in the range of 345-2300 nm, and a sample 7 under a 50 mm thick PELD film received the most intense irradiation in the visible spectrum at a wavelength of the order of 490 nm. Samples 2 and 3 under the PET plastic received energy comparable in intensity to PP in the range of 345-1500 nm (the interval 320-345 nm of the near UV region of the spectrum is absorbed by the ozone layer of the atmosphere).

3. Results and Discussion

Figure 2 shows the change of the outdoor temperature in Kazan. From these data it is clear that the daylight average temperature was lower of 20°C during 3 days, and the daylight maximum temperature 30°C and higher fixed during 10 among 32 of test days. The icons in Figure 2 allow you to distinguish between types of temperature during the day. Here the maximum and average temperature was prepared for darkness, daylight and a full day.

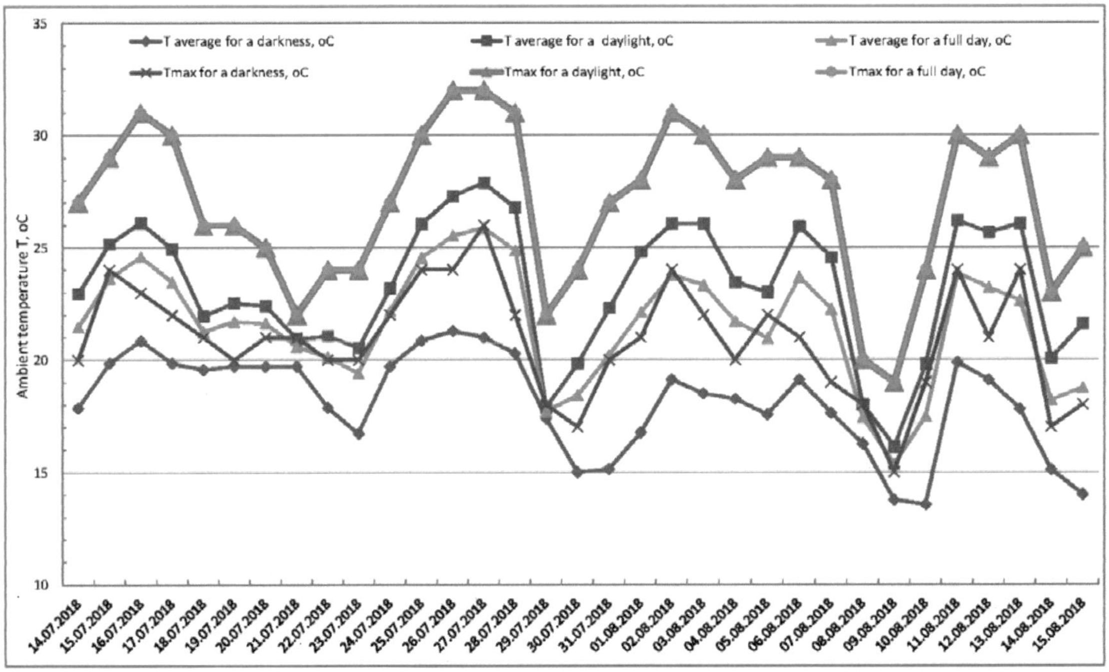

Figure 2. Graph of outdoor temperature for the period of field tests from 14/07/18 to 15/.08/18 in Kazan.

The photographs in Table 1 show the change in the state of samples 1-7 from 16/07/18 to 15/08/18. The results of the test showed that the photochemical effect is manifested on the surfaces of all the samples. The most intensive changes occurred in the samples 1, 4 on the drying oil "Oxol" and 3, 5 on natural drying oil, and least intensively - in sample 2 on the drying oil "Oxol". Sample 1 for 1 week was covered with a magnifying glass, and sample 4 - with a no color PELD film. The remaining time the sample 1 stood open, and PELD-coating of sample 4 was replaced by a magnifying glass. The increased transformation of the material surface in samples 1, 4 can be explained by an increased sample temperature, exposure to atmospheric moisture (after the formation of the film), and a longer exposure to solar radiation. Sample 3 was coated with a 400 µm thick PET-plate, and sample 5 - with a colorless PELD film 100 µm thick. The next day after installing the samples, a dark film of drying oil formed on their surfaces. The film of the material on the walls of the beaker acquired a light brown

IOP Conf. Series: Materials Science and Engineering **481** (2019) 012044 doi:10.1088/1757-899X/481/1/012044

color. In samples 3, 5 the material with the original color of potassium dichromate were not observed. The change in the state of samples 3, 5 indicates a photochemical effect not only and not so much on the pigment as on the film-forming material, with the acceleration of oxidative processes in it. The further extension of these processes into the depth of layer occurs by a chain mechanism. The high intensity of the process in samples 3, 5 also indicates a better proceeding of reactions in them when using natural drying oil, in comparison with the drying oil "Oxol".

The sample 2 in the PET bottle was installed vertically and had the poor geometric conditions of irradiation. Despite the length of testing, its material was transformed only in the form of separate blackening zones of the upper layer, but the underlying layers, after sampling of part of materials on 15/08/18, retained their original color. Consequently, the presence of irradiation in the range of 345-400 nm of the near UV spectral region in sample 2 did not provide the intensification of the process.

Table 1. Change in the state of some samples during the tests due to solar radiation.

Dates	Sample Numbers						
	1	2	3	4	5	6	7
16/07/18	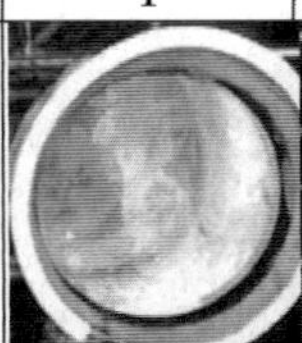	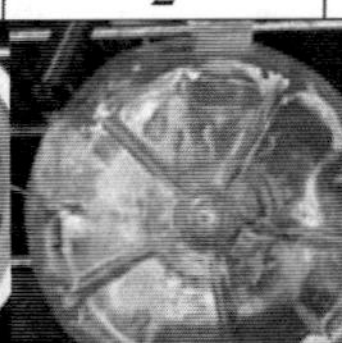		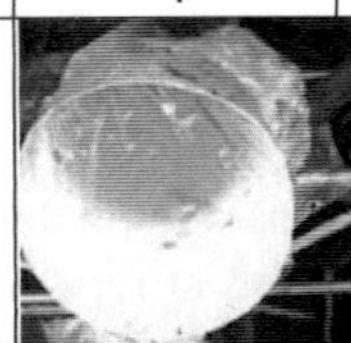			
23/07/18		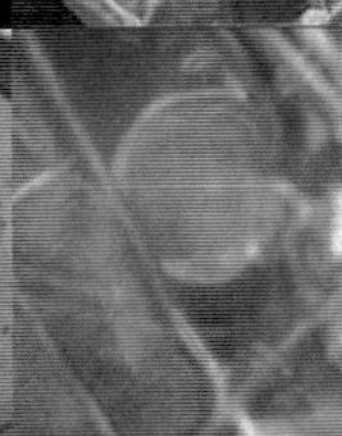		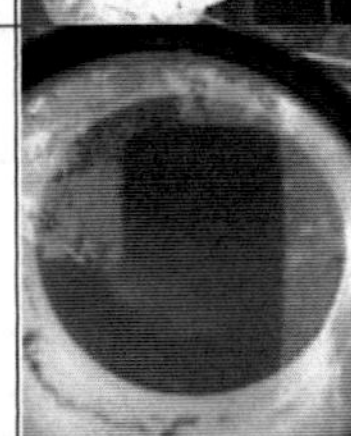		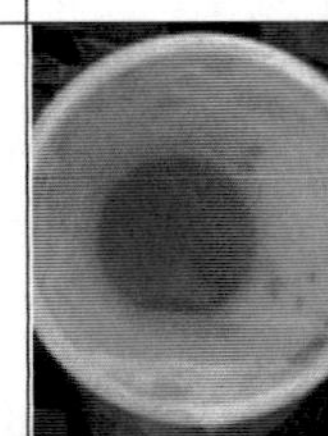	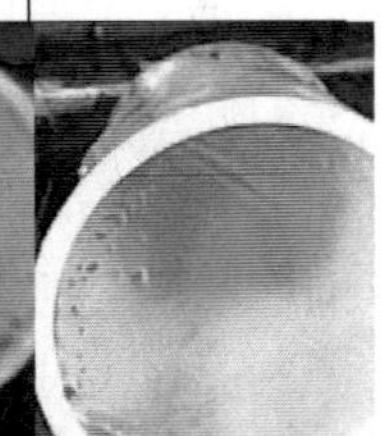
28/07/18	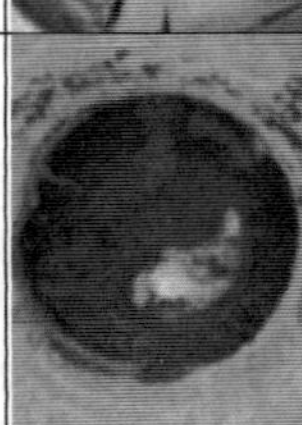	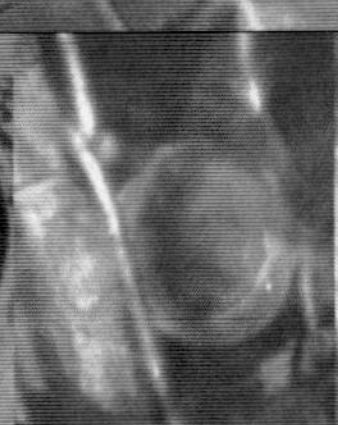	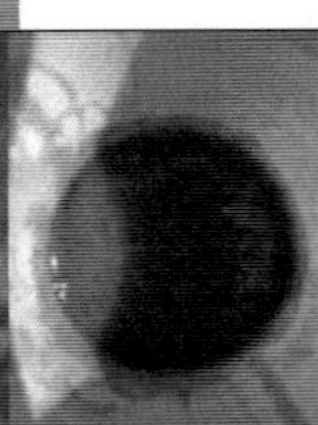	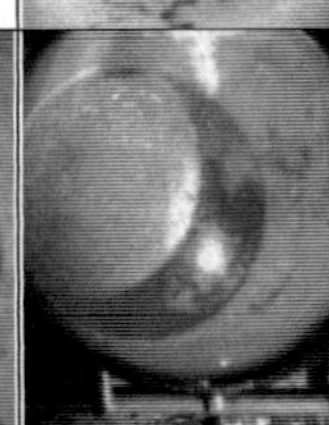	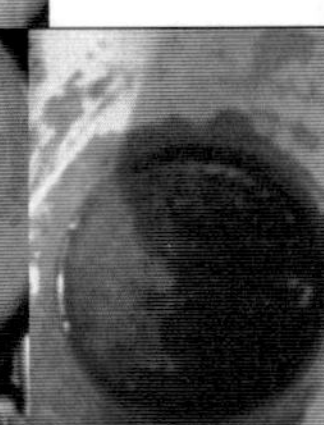	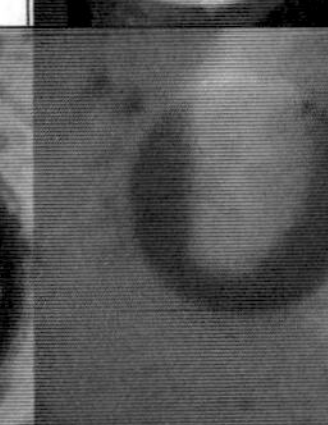	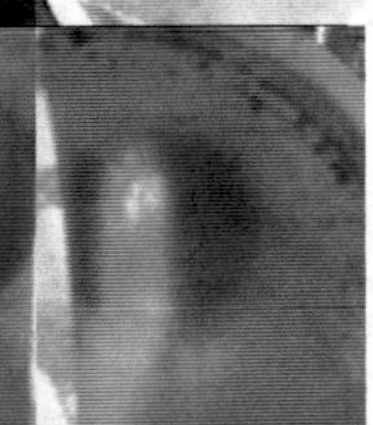
03/08/18	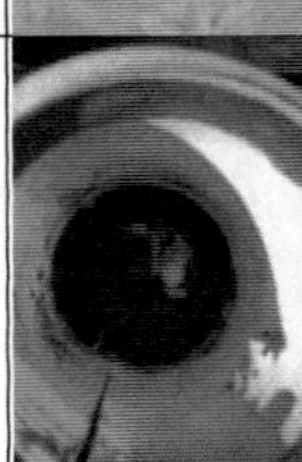	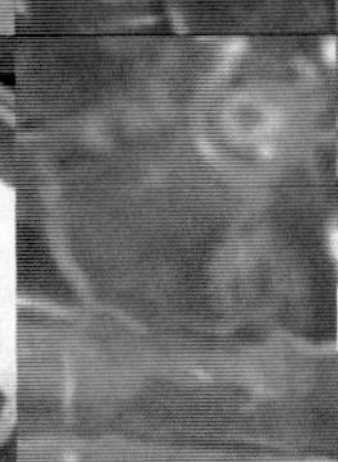	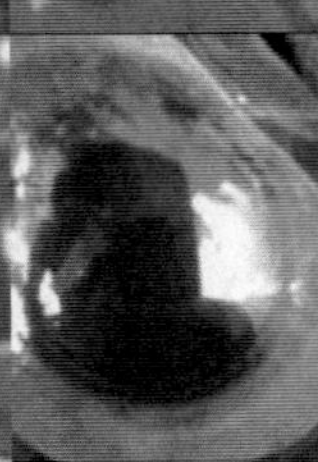	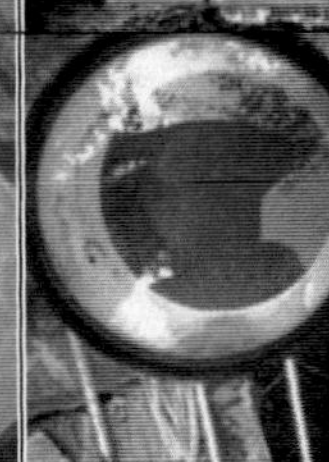	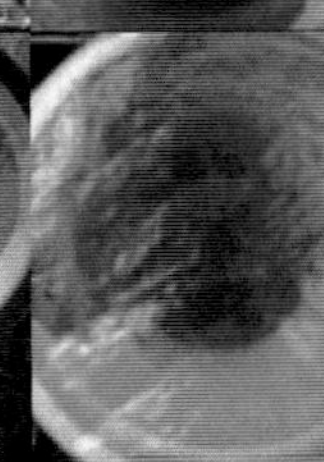	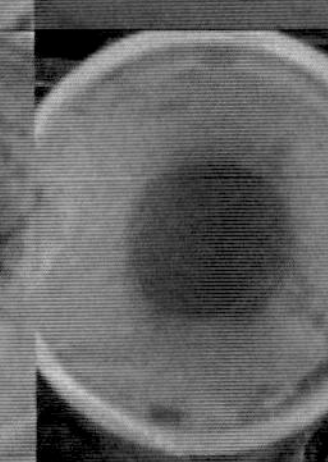	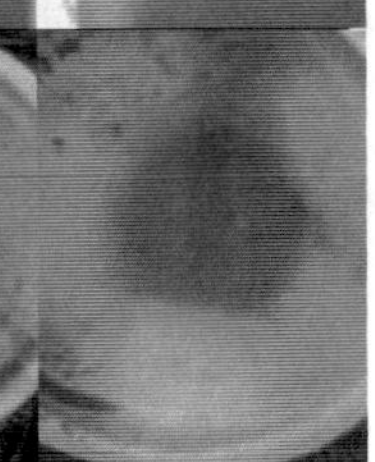
13/08/18	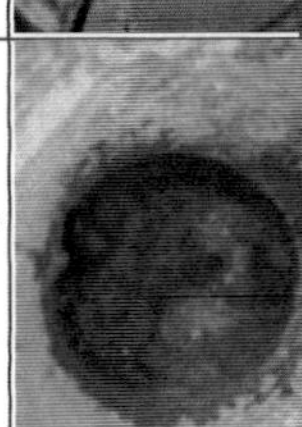	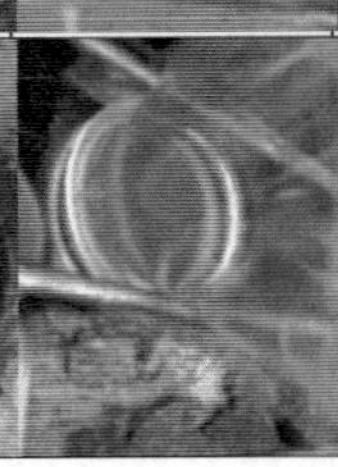	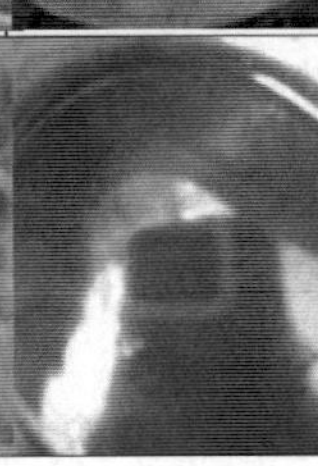		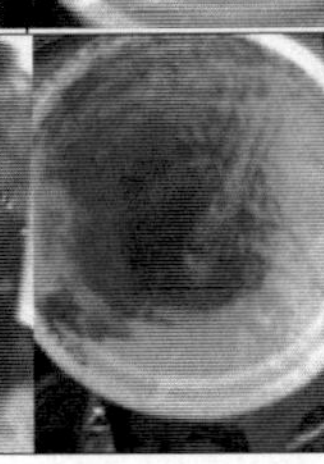	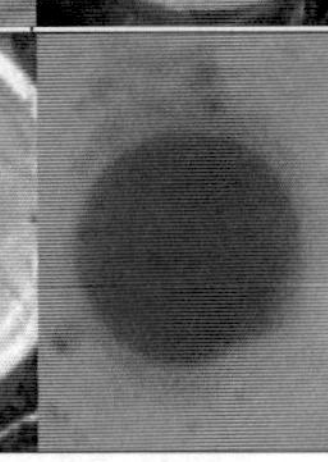	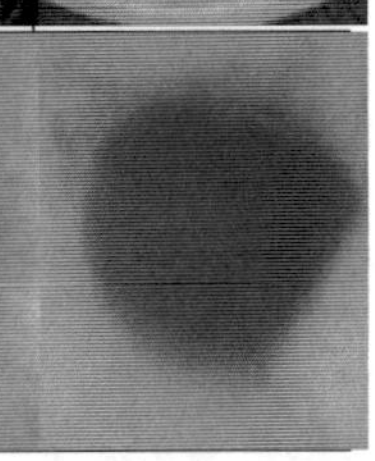

In samples 6 and 7, the changes occurred at an average rate and, in comparison with sample 2, were expressed in a much more (some of twice in the time) intensity formation of the blackening areas of the surface and the acquisition of a red - cherry color of the subsequent layer. At the same time, the lower layers, adjacent to the bottom of the beaker, retained the original color in both samples. During the tests the water didn't go into samples. We can also see that in sample 7, coated with a 50 micron PELD color film, the processes were more intensive than in sample 6 covered with a non-colored PELD film 100 microns thick. Hence, irradiation in the range of 345-400 nm in the Near-UV spectrum and in the interval of 700 -2000 nm in the Near-IR spectrum didn't give the advantage to sample 6.

A photometric analysis of samples at a wavelength of 540 nm was performed in accordance with GOST 31956-2012, method A, which corresponds to ISO 18412:2005, "Water quality – Determination of chromium (VI) – Photometric method for weakly contaminated water". The analysis was carried out with the UNICO SQ-2800 model spectrophotometer that permits the work operation in the wavelength range of 190 nm to 1100 nm, and is supplied of a dot matrix LCD display for photometric results. The correlation coefficient of the calibration graph was 0.9993.

The results of photometric analysis in the main corresponded to the results of the visual observations presented above. The greatest transition of hexavalent chromium to the trivalent and the divalent state occurred in samples 1 and 4–18% and 17%, respectively. Transformations in samples 3, 5 on natural drying oil – 16% and 4% are also noticeable. At the same time, in sample 7, in spite of significant visual changes, the hexavalent state of chromium was changed only by 1%. This fact can be explained by the fact that under the conditions of sample 7, an intensive conversion of dichromate to hexavalent chromate of potassium, having a black color, took place. The photometry also showed changes close to zero in sample 6. The latter does not contradict visual observations. At the same time, unexpectedly high result (about 35%) was fixed at sample 2 that visually almost has unchanged. The latter can only be explained by the error in sampling in the field conditions, in connection with which this result is recognized as not significant. It should also be noted that photometry with high accuracy, corresponding to the used model of the spectrophotometer, was hampered by the presence of organic substances and, possibly, a desiccant in the sample. Therefore, the analysis can be regarded as semi-quantitative.

4. Conclusions

Thus, the carried out investigations have shown the possibility of intensifying heat treatment of waste containing hexavalent chromium and organic compounds using solar energy. An estimate is obtained for the wavelength ranges of the near UV spectrum, the visible spectrum and the near IR spectrum, within the limits of which photochemical reactions for the reduction of hexavalent chromium are possible. This will reduce the temperature of the heat-setting process and, together with it, reduce energy consumption in heat treatment devices. The latter is also important for reducing greenhouse gas emissions.

References

[1] Waste Electrical & Electronic Equipment – European Commission ('DG Environment'). Available at: http://ec.europa.eu/environment/waste/weee/index_en.htm [Accessed 9 September 2018]

[2] Kumar R V, Basumatary A K, Ghoshal A K and Pugazhenthi G 2015 *RSC Adv* **5** (9) 6246-54

[3] Mondal S K and Saha P 2018 *Chemical Engineering Research & Design* **132** 564-83

[4] Yuxiang Zhao, Dejun Kang, Zhong Chen, Jiajun Zhan and Xiaoqi Wu 2018 *Int. J. Electrochem. Sci.*, **13** 1250 –59

[5] Malik D S, Jain C K, Yadav Anuj K 2017 *Appl Water Sci* **7** 2113–36

[6] Garg K K, Rawat P, Prasad B 2015 I*nt J Water and Wastewater Treatment* **1**(1) 1-9

[7] Owlad Mojdeh, Aroua Mohamed, Daud Wan, and Baroutian Saeid 2009 *Water, Air & Soil Pollution* **200** (1-4) 59-77

[8] Turanov D Yu 2010 *Uspekhi v oblasti khimii i tekhnologiy XXIV* **9** (114) 10-20

[9] Suleymanov M Zh 2006 *Spectral characteristics of transparent coatings of solar collectors In the XXI Int. Conf. collection "Equations of state of matter"* (Chernogolovka: IPCP RAS) 207-10

[10] Speranskaya T A 1976 *Optical properties of polymers* (Leningrad: Chemistry) 136
[11] Krynin A G and Khokhlov Yu A 2013 *Aviation Materials and Technologies* **4** 31-34
[12] Serova V N 2010 *Optical and other materials based on transparent polymers* (Kazan: KSTU) 540

IOP Conf. Series: Materials Science and Engineering **481** (2019) 012045 doi:10.1088/1757-899X/481/1/012045

Application of the laser cutting of wood-containing materials in construction

A T Gabdrakhmanov[1], A A Bobrishev[1], L N Shafigullin[1]

[1]Kazan Federal University, 423800, 68/19 Mira avenue, Naberezhnye Chelny, Russia

E-mail: ATGabdrahmanov@kpfu.ru

Abstract. One of the popular materials in the construction is plywood. Experimental studies of the laser cutting of plywood of various thicknesses (4, 6, 8 and 10 mm) under different cutting conditions are presented in this paper. In the experiments, the focal length, cutting speed and power were varied. The aim of the work is to study the samples and select the optimal cutting mode.

1. Introduction

Valuable building properties of wood determine areas for its effective use. The low density of dry wood with relatively high strength and rigidity (along of a fibers) makes it appropriate to use of wooden structures in the coverings of public, industrial and agricultural buildings; it is easiest to implement constructive measures against the decay with the most complete using of the best structural properties of dry wood. Ecological purity makes the wood particularly useful for the construction of housing, particularly of cottage type. In the enclosing parts of heated buildings well used small thermal conductivity of dry wood across of the fibers. The chemical resistance of dry wood justifies the predominant using of metal-free and especially glued wooden structures for coatings of chemical workshops and warehouses.

One of the popular materials in the construction is plywood. It is a wood-layered plate, which is made on the basis of thin sheets of veneer. Each layer is laid with different directions of fibers to give strength.

The laser cutting of wood and plywood is the most economical, fast and convenient way of cutting these materials, because it is made without mechanical contact with the material itself, with the absence of waste and consumables [1-18].

The cutting of plywood is possible on any, the most complex contours with a petty detail, by using sample instruments (saws and mills) easy not cope, because there is no mechanical contact.

Veneer sheets (from 0.6 mm) are also easy to laser-treat, allowing to cut out the most original subjects from different breeds, which gives great opportunities for decorators, furniture manufacturers and other interior items. When laser cutting veneer edges are not torn, are not break and are not burn. Due to the minimum thickness of the cut (the thickness of the laser beam), the gaps between the parts are very small, which makes it possible to interchange the cut elements while preserving valuable material.

Experimental studies of the laser cutting of plywood GOST 3916.1-96 of various thicknesses (4, 6, 8 and 10 mm) under different cutting conditions are presented in this paper. In the experiments, the focal length, cutting speed (V) and power were varied (P). The aim of the work is to study the samples and select the optimal cutting mode.

2. Experimental studies

Specific energy of cutting, i.e. what energy it is necessary to spend on the destruction of a unit of mass of a substance is determined by the formula:

$$S_p = c \cdot T_p,$$

where c is thermal capacity; T_p is the temperature of cut material.

For plywood, the value of $S_p = 5.4 \cdot 10^{-6}$ J/kg.

Optical parameters, such as focusing of laser radiation - the size of the focusing spot, the focal length of the optical focusing system, the displacement of the focal plane relative to the surface the processed materials, the degree of the polarization radiation and distribution of power density over the beam section have a significant effect on the profile of the cut, its depth and width [19-20].

The technological laser AUTOM 150 was used for cutting the plywood. AUTOM Laser 150 is a portal-type laser cutting and engraving machine. It is effective in the processing of non-metals - for cutting sheet material: leather, cloth, paper, cardboard, rubber, wood, organic glass, plastic. Technical characteristics of this laser technological complex are presented in table 1.

Table 1. Characteristics of laser AVTOM 150.

Parameter	Value
The size of treated surface	1300x2600 mm
Laser power	150 W
Laser emitter	CO_2 glass tube, water cooling
Engraving, speed	1000 mm / sec
Engraving depth	0-25mm (depending on material)
Maximum speed the movement of machine	25000 mm / sec
Positioning accuracy	<0.01 mm
Graphic format	DST, PLT, BMP, DXF
Support software	LASERCUT 5.x software
Power supply	220V ± 10% / 50 Hz

The results of experimental studies are presented in figures 1-6.

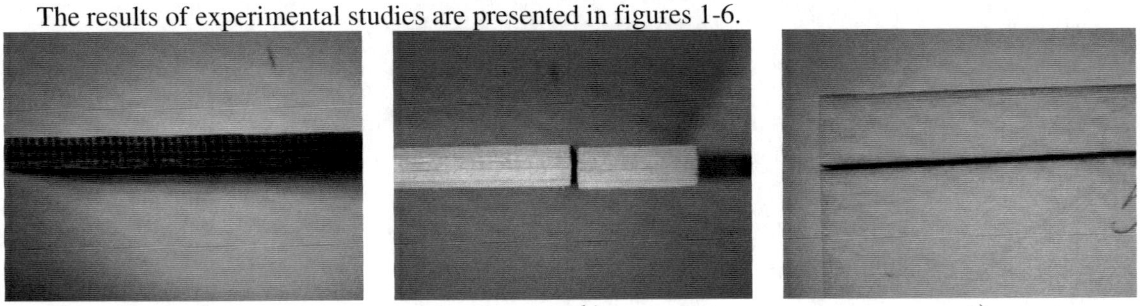

a) b) c)

Figure 1. P = 25W; V = 0.6 m/min; gas flow rate 5 m^3/h (air), h = 4 mm; a) type of cut; b) type of cut in the profile; c) type of cut from above.

Figure 1 shows the most effective cutting mode for plywood thickness of 4 mm, cutting is carried out to the full depth, it can be cut parts of almost any shape. When the cutting speed is increased by a factor of 3 (Fig. 2), it is necessary to increase the laser radiation power by 4.8 times, and in this mode it is not impossible to cut the details of the complex shape. When it is necessary to cut out small parts or curvature less than 4 mm, it is need to reduce the speed by 2 times (Figure 3).

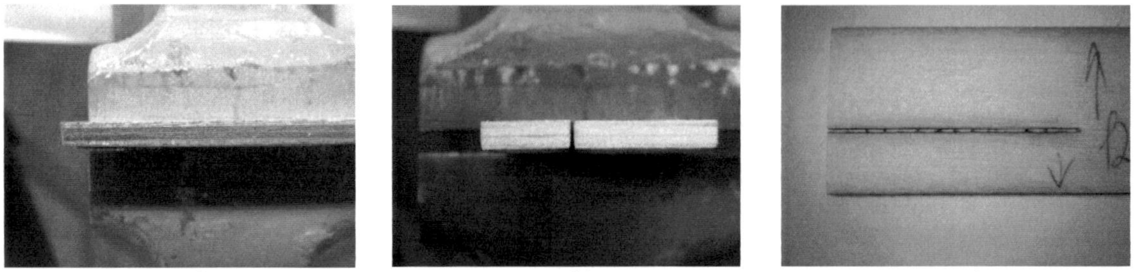

a) b) c)

Figure 2. P = 120W; V = 1.8 m/min; gas flow rate 5 m^3/h (air), h = 4 mm; a) type of cut; b) type of cut in the profile; c) type of cut from above.

a) b) c)

Figure 3. P = 25W; V = 0.3 m/min; gas flow rate 5 m^3/h (air), h = 4 mm; a) type of cut; b) type of cut in the profile; c) type of cut from above.

Figures 4-6 show photos of the cut in the most effective modes for plywood thickness of 6, 8, 10 mm.

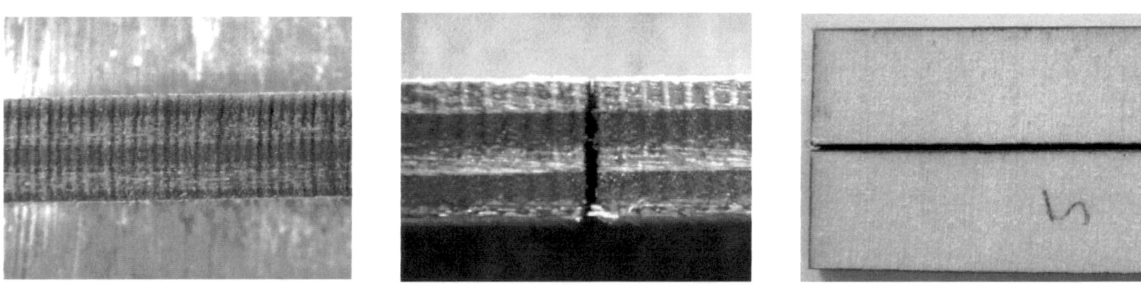

a) b) c)

Figure 4. P = 30W; V = 0.6 m/min; gas flow rate 5m^3/h (air), h = 6 mm; a) type of cut; b) type of cut in the profile; c) type of cut from above.

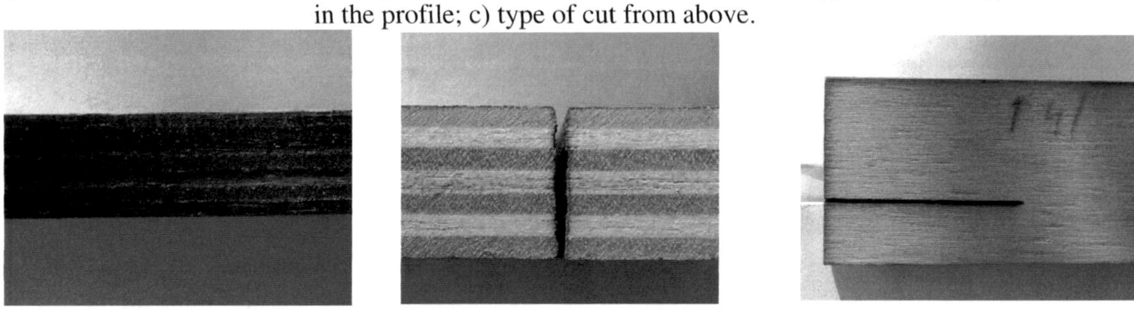

a) b) c)

Figure 5. P = 35W; V = 0.6 m/min; gas flow rate 5m^3/hour (air), h = 8 mm; a) type of cut; b) type of cut in the profile; c) type of cut from above.

a) b) c)

Figure 6. P = 40W; V = 0.6 m/min; gas flow rate 5m^3/h (air), h = 10 mm; a) type of cut; b) type of cut in the profile; c) type of cut from above.

In case of increasing the thickness of the material, a higher laser radiation power requires. The maximum speed decreases sharply with increasing the thickness, because a continuously increasing power is required for cutting. The edge of the cut becomes darker with increasing the thickness and power of the radiation.

3. The conclusion

The wood and veneer are well cut to a thickness of 9-10 mm, depending on the type of wood. Pine, spruce, aspen, poplar are better than other types of wood suitable for laser cutting. Birch, beech or oak have much less suitable properties. The hardness and density of wood complicate the process of laser cutting, make it more time-consuming. The nature of cutting along and across of fibers is different. The general problem - twigs in the presence of excellent quality the cut it is very difficult to achieve. The cut edge is light brown to almost black, slightly charred. The cutting edges are darker and so the wood is thicker and harder.

Acknowledgments

This work was supported by the research grant of Kazan Federal University.

References

[1] Kobayashi A 1974 *Processing of plastics cutting* (Moscow: Engineering) 192 p
[2] State Industry Standard Plastic 12020-72 (ST SEV428-89). Methods for Determination of Resistance to Chemical Environments
[3] Granovsky G I 1982 *Processing of the results of experimental studies of metal cutting* (Moscow: Engineering) 112 p
[4] Zharin D E, Gumerov A F, Zagidullin A J, and Shafigullin L N 2012 The development of mathematical models of the cutting-purpose thermoset composites *Journal of International Scientific Publications* **6(3)** p 200-214
[5] Hemdi A and Taha 2007 Simplex method *Introduction to Operations Research – perations Research: An Introduction* (Moscow: Williams) p 95-141
[6] Ehrgott M and Gandibleux X 2004 Approximative Solution Methods for Multiobjective Combinatorial Optimization *Sociedad de Estadística e Investigación Operativa* **12(1)**
[7] Petrosyan L A, Zenkevich N A and Shevkoplyas E V 2012 Game Theory *St. Petersburg:BHV-Petersburg* p 432
[8] Trifonov A G 2000 Multi-criteria optimization. Optimization Toolbox 2.2 User's Guide. *http://matlab.exponenta.ru/optimiz/book_1/16.php*
[9] Khisamutdinov R M, Zvezdin V V, Saubanov Ruz R, Israfilov I H, Rakhimov R R, Spirin A A 2016 Study of processes of steels surfaces modification with highly concentrated energy flows *IOP Conf. Ser.: Mater. Sci. Eng.* Volume **669** Issue **1** Article number **012024**

[10] Israfilov I H, Saubanov R R, Rakhimov R R 2011 Prospective application of highly concentrated energy for the surface heat treatment products *Socio-economic and technical systems: research, design, optimization* № **1** p 25-30
[11] Grigoryants A G, Shiganov I N, Misyurov A I 2006 *Processes of laser processing* (Moskow: MGTU im. N. Uh. Bauman) 664 p
[12] Kostecki V V, Lissovsky A L 2013 Increasing the absorption capacity of the surface layer of metal during laser exposure by using absorbing coatings *Bulletin of Polotsk state University. Series: Industry. Applied science* No. **3**. p 97-101
[13] Stepanov A A 1987 *Cutting of High-Strength Polymer Composites* (Leningrad: Mashinostroenie) 176 p
[14] Stuchnyi BP 1974 *Cutting Plastics: A Handbook* (Moscow: Mashinostroenie) 144 p
[15] Mikul'skii V G 2000 *Structural Materials* (Moscow: Izdatelstvo ASV) 536 p
[16] Komar AG 1983 *Structural materials and Components* (Moscow: Vysshaya Shkola) p 487
[17] Yashcheritsyi P I, Fel'dshtein E E, Kornievich M A 2006 *Theory of Cutting* (Minsk: Novoe Znanie) 512 p
[18] Zharin D E, Yurasov S Y, Gumerov M I, Shafigullin L N 2009 Optimal machining of loosely filled polymer composites *Russian Engineering Research* **29** (**10**) pp 1007-1009
[19] Bashmakov A D, Gabdrakhmanov A T, Israfilov I H, Galiakbarov A T 2013 Electrostatic field as factor of increase of stability of quality of laser processing of metals *MNTK "IMTOM-2013"* Part **1** p 325-328
[20] Gabdrakhmanov A T, Israphilov I H, Galiakbarov A T 2017 Preparation of metal surfaces for application of functional coatings *IOP Conf. Series: Journal of Physics: Conf. Series.* Volume **789** Article number **012009**

IOP Conf. Series: Materials Science and Engineering **481** (2019) 012046 doi:10.1088/1757-899X/481/1/012046

Testing steelfibroconcrete centrifuged specimens of pipes used in drainage and water disposal systems at places of automobile and railway transport heavy traffic

M S Kuznetsov[1], A S Kuznetsov[2], A S Noskov[3] and V G Dubinina[1]

1Department of Technological Education, Ural State Technical University, 59 Krasnogvardeyskaya St., Nizhny Tagil, 622000, Russia
[2]The limited liability company "Proektsia", 4B Lenin St., Nizhny Tagil, 622000, Russia
[3]Department of Hydraulics, Ural State Technical University, 59 Krasnogvardeyskaya St., Nizhny Tagil, 622000, Russia

E-mail: kms.rf@rambler.ru

Abstract: Ferroconcrete pipes, used in drainage and water disposal systems of roads, works in conditions, in which the direction of the main tensile stresses is either unknown, or can change during exploitation. One of the ways of solving this problem is using of steel fibrous concrete, i.e. concrete on hydraulic tenacious substances with addition of reinforcing elements in the form of short lengths (fibers). Using of steel fibrous concrete allows to get the construction of low pressure centrifuged pipe for drainage and water disposal systems at places of automobile and railway transport heavy traffic with higher tensile and bending strength, crack resistance for the work under conditions when the direction of the main tensile stresses is unknown.

1. Introduction

Low tensile strength and as a result of it bad crack resistance are known to be characteristic features of reinforced concrete constructions.

Traditional strengthening concrete by the directional reinforcing is effective only in those cases when the direction of the main tensile stresses in constructions is known.

However, there are a number of structural elements working at perception of shock, dynamic and temperature actions, in which the direction of the main tensile stresses is either unknown, or can change during exploitation. Examples of such constructions are ferroconcrete pipes, used in drainage and water disposal systems at places of automobile and railway transport heavy traffic.

Not exact knowledge of stressed and deformed state of these constructions during their operation results in overconsumption of concrete and reinforcement, what finally makes manufacture of these constructions more difficult and costly [3].

2. Theoretical stresses in fibroconcrete

One of the ways of solving this problem is using of steel fibrous concrete, i.e. concrete on hydraulic tenacious substances with addition of reinforcing elements in the form of short lengths (fibers) [1].

Fiber adhesion with concrete occurs due to friction, adhesion and mechanical interlocking of the contact surfaces. The effect of each of these processes on the strength of fiber anchoring can be different depending on the type of fiber.

For example, for smooth fibers, adhesion and friction will play a large role, and the mechanical interlocking of fiber protrusions with the protrusions of concrete for shaped fibers. Fiber profiling enables increased load they can take depending on the profile parameters.

The length of the fiber affects the average tensile stress σ_B^{cp} which is produced in it, which determines the overall strength of the composite material.

At fracture of the composite material the fibers which happen to be in the area of the fracture, are oriented in different directions towards the fracture surface of the composite material. With some fibers, the fracture surface runs in the middle of the length, with most others it does between the middle and one of the ends. In the latter case, the tension in the fiber at the moment of fracture will depend on the distance between the surface of the composite material fracture and the end of the fiber closest to it. The equation of the composite materials mix with the tailored solid reinforcement is shown below:

$$\sigma_{\kappa} = \sigma_{м}(1 - V_a) + \sigma_a V_a$$

where $\sigma_{\kappa}, \sigma_a, \sigma_{м}$ – stresses in the composite material, reinforcement and the matrix correspondingly; V_a – the amount of reinforcement materials in the composite material.

In [2], it is proposed to reduce the fiber reinforcement to an equivalent unidirectional reinforcement by using a complex coefficient λ_{red}, equal to the multiplication of three coefficients that take into account: the chaotic nature of fiber orientation λ_p, the deviation of the force from the direction of the fibers λ_{or}, the anchoring of the fibers λ_{an}. Thus, the mix equation looks like below:

$$\sigma_{\kappa} = \sigma_{м}(1 - V_a) + \lambda_{red} \sigma_a V_a$$

After the fracture initiation, the strength of the composite material relies only on the strength of the fibers and the bonding strength of the fibers with the matrix and cannot exceed $\sigma_a V_a$. The ultimate strength of the composite material in this case will be equal to:

$$\sigma_{\kappa} = \lambda_{red} \sigma_a V_a$$

The theory of composites treats the composite material as a combination material (matrix and fiber), with individual components working elastically, and the non-slip interaction of these components with each other.

3. Problems of the theoretical approach

Based on these assumptions, the above analytical dependencies are derived. However, despite the apparent self-consistency of the equations which describe the properties of composite materials, the theory of composites, generally, works both ways.

First, although the computational theory of a composite material takes into account various coefficients which affect the accuracy of the calculation results, it nevertheless does not consider such a critical parameter as the steel fiber concrete mixing, placement and curing process. The experience of using steel fiber concrete shows that its properties can vary in a wide range and depend on the technology adopted by the manufacturer.

Second, the above theory does not take into account the geometric properties of the fiber, which determine the number of fibers per unit volume of concrete and the quantity of the fiber side face which is engaged with the matrix.

Third, it does not consider the cracking moment.

Therefore, the principle of the fiber performance at the stage of destruction is mis-addressed. In fact, if the fiber cuts across the crack, then the tangential stresses along the fiber are distributed differently from those in the intact concrete (Fig. 1).

In this case, the stress in the fiber σ_B^{II} directly within the crack is about 10 times that in the intact fiber (σ_B^{I}) [4]. Thus, the theory of composites is quite good at describing the behavior of the material at the elastic stage of work. However, due to the incorrect description of the mechanism of destruction, it is impossible to accurately predict neither strength nor deformation during fracture.

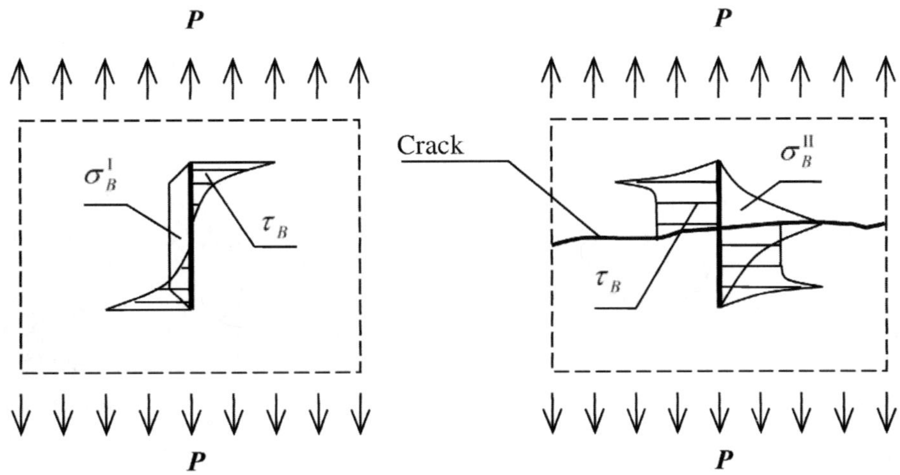

Figure 1. Distribution of normal and tangential stresses along the fiber length in the cracked and intact members.

Along with the above concept of calculating a composite material, there is the so-called "gap concept", which states that the strength and crack resistance of dispersed-reinforced concrete depend on the calculated distance between the centers of the fibers in the fibrous reinforcement.

This concept is based on the hypothesis that the geometric properties of the reinforcing elements (their diameter and the degree of dispersal) should be commensurable with "congenital lesions".

The "congenital lesions" here are both voids in the matrix and cracks initiated in the concrete body as it hardens, triggered by internal stresses, mainly in the "cement-aggregate" interface area.

This problem has been the subject of research by numerous authors. The authors have developed the techniques and proposed various analytical dependencies for calculating the distance between the fibers. We believe the calculation formula below [3] is suitable for practical purposes:

$$S = 12{,}5 d_f \sqrt{\frac{l_f}{\mu}},$$

where S is the calculated distance between the fibers, mm; d_f, l_f respectively the diameter and the length of the fiber, mm; μ reinforcement percent by volume, %.

Assuming that the fracture of the steel fiber reinforced concrete is induced by shearing of concrete between the fibers, the member's axial tension formula with due account for the fiber orientation is given in [3]:

$$R_{bt}^{f} = 0{,}375 R_{bt} l_f \sqrt{\mu} \frac{k_0}{d_f},$$

IOP Conf. Series: Materials Science and Engineering **481** (2019) 012046 doi:10.1088/1757-899X/481/1/012046

where k_0 is the coefficient of the fiber orientation.

The calculation, based on the gap concept, describes the physical processes of crack propagation and inhibition during the fracture of concrete. However, this technique does not take into account the work of the fibers after the formation of cracks. As shown by numerous tests [1, 4], the steel-fiber-concrete member won't run out of the carrying capacity after it has suffered from cracking. Steel fiber reinforced concrete member can sustain a significant extra load after the concrete has become ruptured all the way through. Therefore, the gap concept should be considered as a crack resistance criterion rather than that of the strength of steel fiber reinforced concrete.

4. Experiment results

Investigations revealed that the most characteristic features of steel fibrous concrete are strength, toughness and work of destruction at axis extension and bending. As for the last feature is concerned steel fibrous concrete may be 10-15 times better than usual concrete [2]. Besides, steel fibrous concrete pipes, used in drainage and water disposal systems at places of automobile and railway transport heavy traffic differ from the traditional ones in the following things: they have higher shock strength, crack and frost resistance, water tightness, cavity resistance, fatigue and high-temperature strength and failure toughness [2].

Comparing strength characteristics of steel fibrous concrete and reinforced concrete was held at testing pipe rings, made by centrifuging (Figure 1).

Figure 2. The specimen for testing strength and crack resistance.

Two series of ring pipes, containing three specimens in each series were made in typical shuttering at the joint-stock company "Reinforced Concrete" Ltd. in Nizhny Tagil. The rings were 1000 mm in outer diameter, 800mm in inner diameter and 1000 mm high. The rings of the first series (specimens R-1, R-2, R-3) were strengthened by spiral frames made of wire, 5mm in diameter class Vr-1 and made of heavy concrete M300. The rings of the second series (specimens Rf-1, Rf-2, Rf-3) were made of steel fibrous concrete, and there were no reinforcing frames.

Specimens Rf-1and Rf-2 contained 2 % dispersion reinforcement, and the specimen Rf-3 – 1.5%. In these specimens the volume of reinforcement was adopted a priori, with the aim of revealing the peculiarities of mixtures at concrete placement.

Motion-time study of the process of making rings of dispersion reinforced concrete showed that it differs by more adaptability to manufacture in comparison with the technology of making traditional rings. So, forming a dispersion reinforced ring took 6 -8 minutes less than making a traditional ring. It is explained by additional waste of time for installing a reinforcing frame and by the fact that an installed frame prevents from filling the mould with concrete. Visual examination of rings after steaming showed high quality of the surface of steel fibrous concrete specimens in contrast to typical ones having contraction cracks, cavities, and concrete flow. For estimating strength characteristics of products rings and control specimens (cubes with an edge 100 mm high) were formed of concrete of the same composition. They were subjected to heat-and-moisture treatment. Rings of both series had undergone testing by static short-term load. The results of the testing are given in the table 1.

Table 1. The results of the testing reinforced pipe rings.

№ of the ring	Percentage of dispersion reinforcement, %	Strength of concrete rings after steaming, MPa	Strength of concrete at the moment of testing, MPa	Cracking load, kN	Failure load, kN
R-1	0	28.0	36.4	75.0	80
R-2	0	28.0	36.4	56.0	62
R-3	0	28.0	36.4	60.0	68
≤-1	2	27.4	37.8	115.0	135
≤-2	2	28.3	37.8	135.0	150
≤-3	1.5	27.4	37.0	95.0	105

During the process of testing deformations and loads at which the first cracks appeared and ring destruction took place, were fixed. The analysis of testing rings shows that deformation of rings made of dispersion reinforced concrete on the average is 30% lower and cracking load is 70-80% higher than that of usual rings. It is due to higher steel fibrous concrete resistance to tension. The way of cracking is the same for all rings. However, if the destruction of usual rings takes place just after cracking, steel fibrous concrete rings withstand load which is 15-20 kN higher than cracking load. To our mind, it is due to higher adhesion between dispersion reinforcement with concrete matrix, which breaks down gradually.

The experience of the factory making pipe rings showed that the operations of reinforcing and concreting were joined together at steel fibrous concrete placement. It results in increasing labour productivity and quality of products.

Besides, revealed strength reserves of dispersion reinforced rings and preliminary calculations open possibilities of reducing their thickness. It will allow to reduce 25-30% concrete consumption taking into account the percentage of concentrating the mixture by dispersion reinforcement.

Conclusions

Thus, using of steel fibrous concrete allows to get the construction of low pressure centrifuged pipe for drainage and water disposal systems at places of automobile and railway transport heavy traffic with higher tensile and bending strength, crack resistance for the work under conditions when the direction of the main tensile stresses is unknown.

It is also necessary to consider the type of fiber reinforcement and the work of the fibers after the formation of cracks. Numerous tests shows that steel fiber reinforced concrete member can sustain a significant extra load after the concrete has become ruptured all the way through.

References

[1] Dubinina V, Kuznetsov M and Ilemkova N 2003 Analysis of possibility of using steel fibrous concrete for making pipes by different ways (Ekaterinburg: Ural State Technical University) p 71

[2] Mikhailov K 2001 *Reinforced concrete in XXI century: State and prospects of concrete and reinforced concrete development in Russia* (Moscow: Gothics) p 683

[3] Hans R 1999 Reinforcing and Prestressing Materials and Systems: mission and activities *Structural Concrete Journal of the fib* vol. **1** Number 4 (ISSN 1464-4177) p 34

[4] Kurbatov L 1978 Comparison bend tests of the rod and fiber reinforced concrete elements: Research and calculation of experimental fiber reinforced concrete structures (LenZNIIEP) p 60

[5] Rabinovich F 1989 *Fiber-reinforced Concretes* (Moscow: Stroyizdat) ISBN 5-274-00506-3 p 176

IOP Conf. Series: Materials Science and Engineering **481** (2019) 012046 doi:10.1088/1757-899X/481/1/012046

[6] Stepanova G 1974 Revisiting the Predicted Steel Fiber Reinforced Concrete Bend Tensile Strength: Research and Improvement of Reinforcement, Concrete and Reinforced Concrete Structures (Stepanov G: Volgograd) p 33

[7] Mangat P 1974 Compactilibity of Steel Fiber Reinforced Small Concrete Slabs vol. 8 (Concrete № 5) p 34

[8] Kuznetsov M 2005 Application of steel fiber reinforced concrete pipes in the drainage and waste water systems *Industrial and Civil Engineering* № **7** p 49

[9] Dubinina V 2005 Study of the rheological and physico-mechanical properties of the steel fiber reinforced concrete mixes (Krasnoyarsk: Youth and Science - the third millennium) p 78

[10] Dubinina V 2006 Calculation of steel fiber reinforced concrete members of the annular cross-section according to the Group 1 ultimate limit states (Penza: PGUAiS) p 29

IOP Conf. Series: Materials Science and Engineering **481** (2019) 012047 doi:10.1088/1757-899X/481/1/012047

Experience of using the ultrasonic low-frequency tomograph for inspection of reinforced concrete structures

M S Kuznetsov[1], O V Maltseva,[2] A S Noskov,[3] A S Kuznetsov[4]

[1]Department of Engineering, Nizhny Tagil technological Institute (branch) Ural Federal University, Russia
[2]Nondestructive Testing Laboratory, LLC “Proyektsia”, Russia
[3]Department of Hydraulics, Ural Federal University, Russia
[4]LLC "Proyektsia", Russia

Abstract. The article shows the hands-on examples of using the ultrasonic low-frequency tomograph type A1040MIRA which has been developed and created by the Research and Manufacturing company "Acoustic Control Systems" (LLC "ACS") for detection of internal flaws in concrete and reinforced concrete structures. The device is shown to be used for detecting the bodies of unrammed concrete (discontinuities of concrete) in the framework and measuring of its residual load-bearing capacity. Furthermore, the tomograph has been found useful to evaluate consolidation of concrete in hydraulic facilities. A1040MIRA resolves the issues of measuring the thickness of a concrete structure from either both sides if readily accessible or from just one side if the access is limited, or those of measuring a layer thickness in a sandwich structure. The article shows that for reliable "reading" of the images produced by the instrument the preliminary study of the reinforcement principle of a structure, of its geometry, the configuration of the reinforcements and the working environment is required.

1. Introduction

The detection of flaws both explicit and hidden is one of the main objectives during inspection of the reinforced concrete structures. They provide maximum data on the technical condition of the structure, its bearing capacity and durability.

The evidence from practice shows that the lack of high production standards, unskilled labor used for installation, a substandard quality control system or a general lack of any quality control system can produce significant flaws in the concrete body of both the load-bearing and the walling members during their casting.

The flaws can arise from disintegration of a concrete mix where inadequate ingredients are used, the mix is ill-proportioned or the batching time is failed, or the mix drop height is incorrect. Defects in the structure of concrete occur where the concrete mix is insufficiently consolidated or completely loose or where the construction joints are failed.

During operation of buildings and facilities made of the reinforced concrete structures the additional flaws may be triggered by corrosion of rebars, the effect of freeze-thaw temperatures, water proofing failures and other.

All these flaws show as cavities, cracks both on the surface of a concrete structure, and inside its body. And while the surface defects are visible and detected by the construction inspectors, the internal defects are difficult to reveal.

Defects of the structure directly affect the density and uniformity of concrete, and, consequently, its strength and durability.

At the end of the day the challenge is to identify the severity level of a flaw revealed in a concrete or reinforced concrete structure. Here are some of those:

- what are the size and the area of the unconsolidated portion of concrete in the section of the load bearing structure;
- how dense is the concrete in the hydraulic facilities;
- does the concrete structure comply with the design criteria in the case of a one-sided access with no access for hole boring.

2. Method of inspection of reinforced concrete structures

The above challenges can be resolved by using the ultrasonic low-frequency tomograph type A1040MIRA which has been developed and manufactured by the Research and Manufacturing company "Acoustic Control Systems" (LLC "ACS"). The device is intended for inspection of monolithic concrete and reinforced concrete structures in order to detect voids, ducts, rebars, foreign inclusions, segregations, cracks and other cavities, both unfilled and filled with liquid or solid material which differs from surrounding concrete in its physical and mechanical properties.

The device can scan concrete to the depth of up to 2,500 mm and up to 800 mm in case of reinforced concrete. Thickness of the structure can be measured in the range between 50 and 600 mm.

Multiple data from the inspected area is collected by a matrix antenna array of 48 low-frequency broadband transverse wave transducers with dry point movable contacts. The total area of the antenna array is 380x140 mm. The inspected structure and its sections are visualized via special software which uses a three-dimensional Cartesian coordinate system which is referenced to the position of the device on the structure, as in Figure 1.

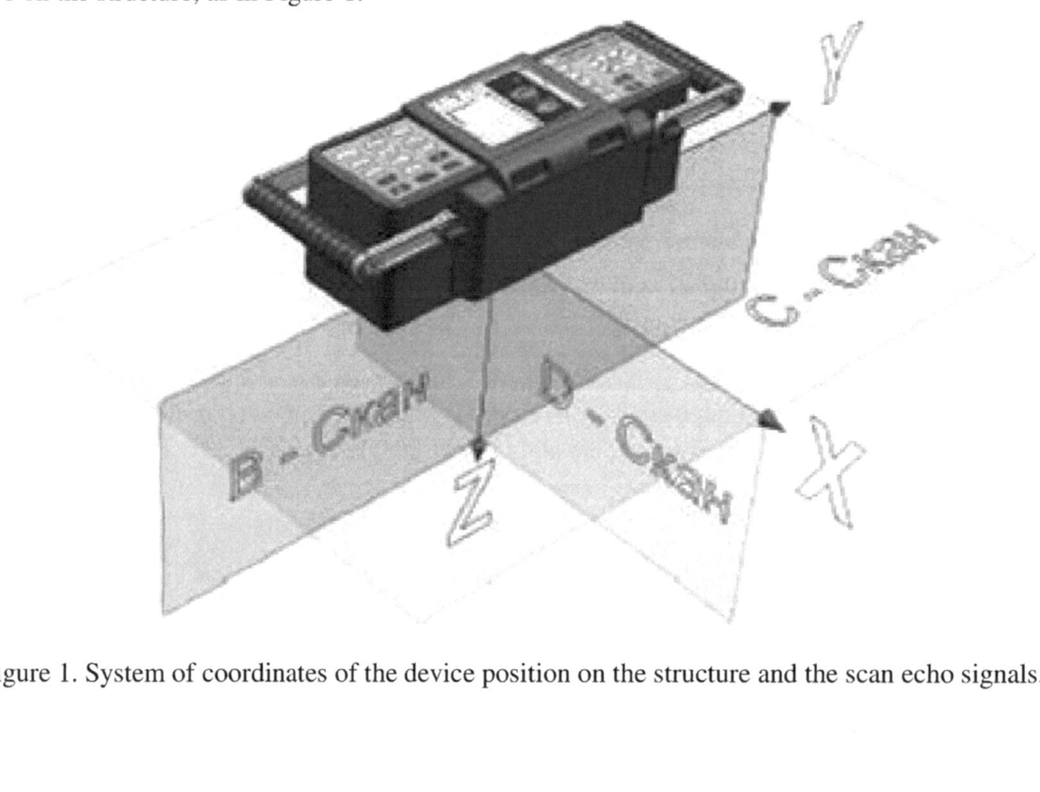

Figure 1. System of coordinates of the device position on the structure and the scan echo signals.

Signals from the antennas are processed by the built-in computer, displayed on a monitor and stored in the device's memory. The software offers reading of data and their displaying as images. Visual images of the sectional views of the inspected structure are produced where each point of the scanned visualized volume is color coded based upon its echoing property.

The device enables the UT inspection of the area in different modes. The "Overview" mode is intended for customizing the parameters of the device for the object under inspection: obtaining the image sharpness and reflecting the corresponding thickness of the structure, setting the average ultrasonic waves propagation velocity for fast search for defective areas.

The so-called B-Scan is produced on the screen of the device, which shows the section of the structure tested in the direction normal to the antenna surface (see Figure 1).

The "Map" mode is used for the areas where the flaws or their indications are detected. An array with certain spacing is applied to the surface of the tested area. The device collects the body of scans produced by a step-by-step motion of the antenna array. Eventually either a three-dimensional image or those of the scans of three planes, so-called image maps, are displayed via the IdealView software. Furthermore, you can view all the images of the tested area of the structure by moving the secant planes of the area.

See [1, 2, 3] for the real working principles of the tomograph as well as the process of imaging with various noises and images. The capabilities of the device in detecting large defects in concrete are shown, both in specially constructed trial blocks and in structures at existing facilities. Before starting to work with device you are recommended to acquire the experience of scanning the concrete trial blocks which contains various artificial defects. See the examples of the structure thickness measurements where the one-sided access is available. Furthermore, [2] shows various methods of non-destructive testing and the instruments for solving these problems. The conclusion is that the tomograph is the most informatory device for inspection of the reinforced concrete structures.

"Proektsiya" LLC has been using the ultrasonic low-frequency tomograph type A1040MIRA to measure the density of the structure (so-called continuity) of concrete in the erected reinforced concrete structures at various facilities inclusive of those hazardous and sophisticated.

A series of inspections have also been carried out to reveal internal defects and unconsolidated areas of concrete in building structures. Thicknesses of the concrete members such as the walls of tanks, tunnels and others have been measured many times.

When we just started to use the device we hoped to see the commonly understandable images similar to those we expect to see at the drawings, i.e. the structures with dimensions, with the rebar locations, with internal defects (cavities). And in three dimensions to top it off.

However, the very first test showed blurry “spot” patterns. We could hardly make out the images which belonged to the rebars and those of the concrete flaws and whether they were in there altogether since the designed locations of the rebars in the structure were unknown to us. Upon that we made sure to study the structure design first and that of the reinforcement before starting to use the device. Then the images became friendlier.

Please find below several examples. Scanning of the 250-mm floor. The reinforcement diagram and the antenna array application area are shown in Figure 2 below. The flooring is reinforced by two meshes, the bottom and the top, made of dia. 12-mm bars arranged in two orthogonally related directions with a 200-mm spacing. The protective mesh concrete layer thickness is 50 mm on both sides.

Figure 3 shows a B-scan of the flooring where some images are visible due to the difference in their color. The position of the images in the section of the structure can be judged by from the axes of the coordinates, X and Z. The horizontal axis X origin is at the center of the antenna array and rubs along the longitudinal side of the device. The vertical axis Z origin is on the surface of the structure to which the device is applied, and shows the location of the images along the thickness of the structure. The coordinates of the image of interest on the B-scan are identified by moving the horizontal and vertical cursor on the screen. Further, the A-scan on the instrument’s screen shows a graph of the images echoing property. The images which are lined up horizontally along the X axis, at a distance of about 50 mm from the structure's surface where the device is located, show the bottom reinforcement mesh of the flooring. The top reinforcement mesh, judging by the scan, is located about 150 mm from the floor surface. The ultrasonic waves are echoed from the bottom surface of the flooring at the interface between concrete and air (by reference to the device location), which is seen on the scan as a large image with Z coordinates equal to 250 mm, or the so-called backwall image. Its location confirms the thickness of the flooring being tested. The B-scan shows no images other than those which belong to the design elements. Thus, there are no structural flaws (discontinuities) of concrete on the scanned area of the floor structure detected.

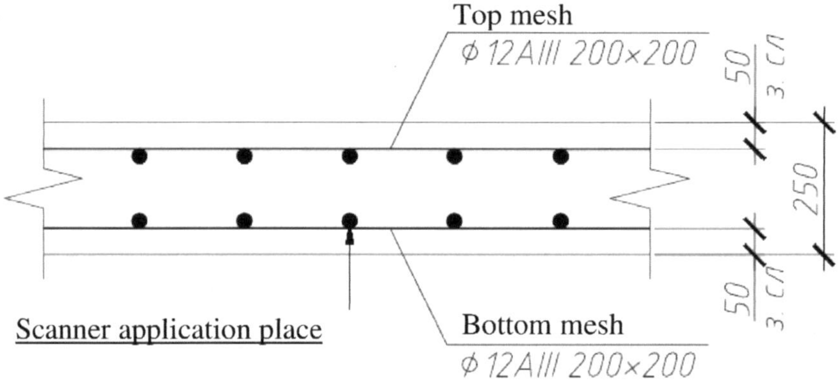

Figure 2. The floor reinforcement pattern and the device application place.

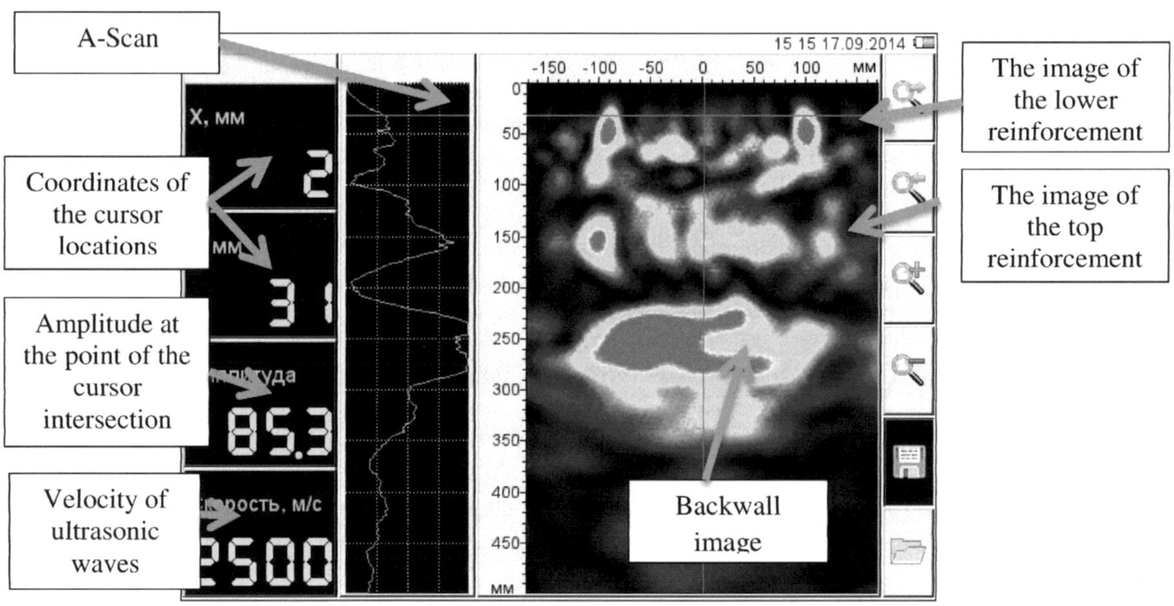

Figure 3. Scan of the 250-mm flooring with images of the nearby and distant rebars and the backwall image.

The scan shows no extraneous images, thus, there are no defects in the structure (discontinuities) of the concrete.

The scans below are made on the 300-mm wall of a tank. The wall reinforcement pattern and the tomograph's antenna array application diagram are shown at Fig. 4. The wall is reinforced by two meshes interconnected by transverse frames.

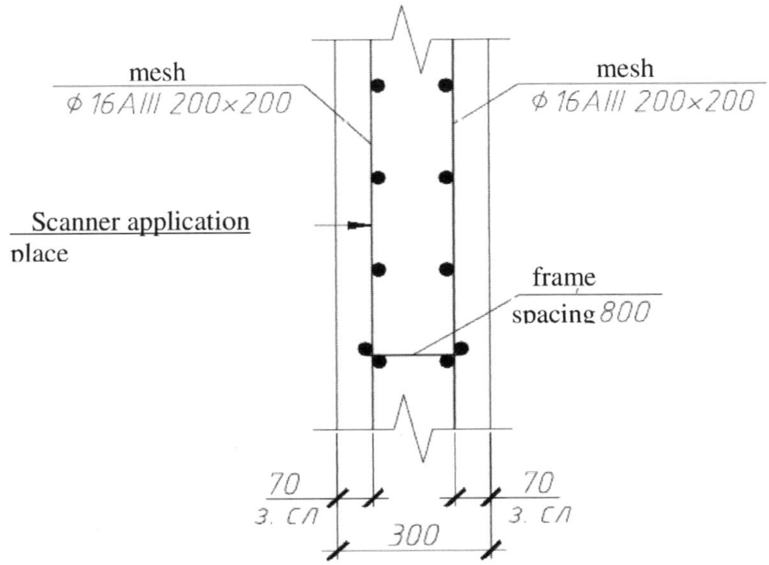

Figure 4. The tank wall reinforcement pattern and the tomograph's antenna array application diagram.

The tomography at Figure 5 clearly shows the images of two layers of reinforcement (two meshes) which are located at about 70 mm from the surfaces of the structure (protective layer of concrete). The backwall image of the structure is clearly visible at a distance of 300 mm. The interface between concrete and air shows the thickness of the structure. The scan shows no extraneous images, no structural flaws (discontinuities) of concrete in this area.

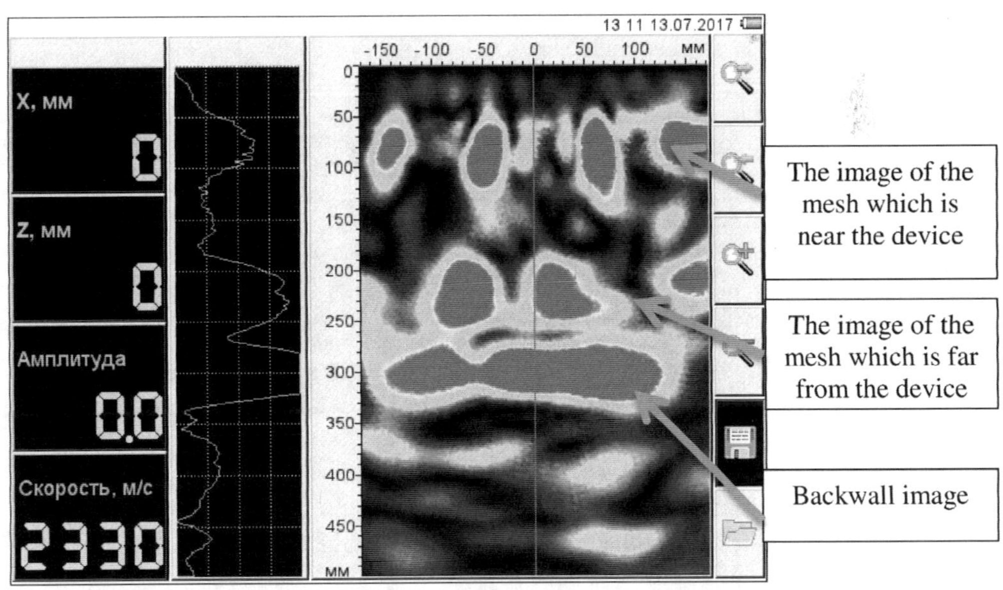

Figure 5. The tank wall scan with the reinforcement images and the backwall image.

A scan (Figure 6) with additional images lined up between two tiers of the mesh images was made at another area of the same tank wall. The transverse frame which connects the reinforcement meshes is obviously located here. A small number of additional images are seen at the location of the reinforcement. Obviously, the concrete is not dense enough in the area of the reinforcing cage. Therefore, the ultrasound velocity goes down.

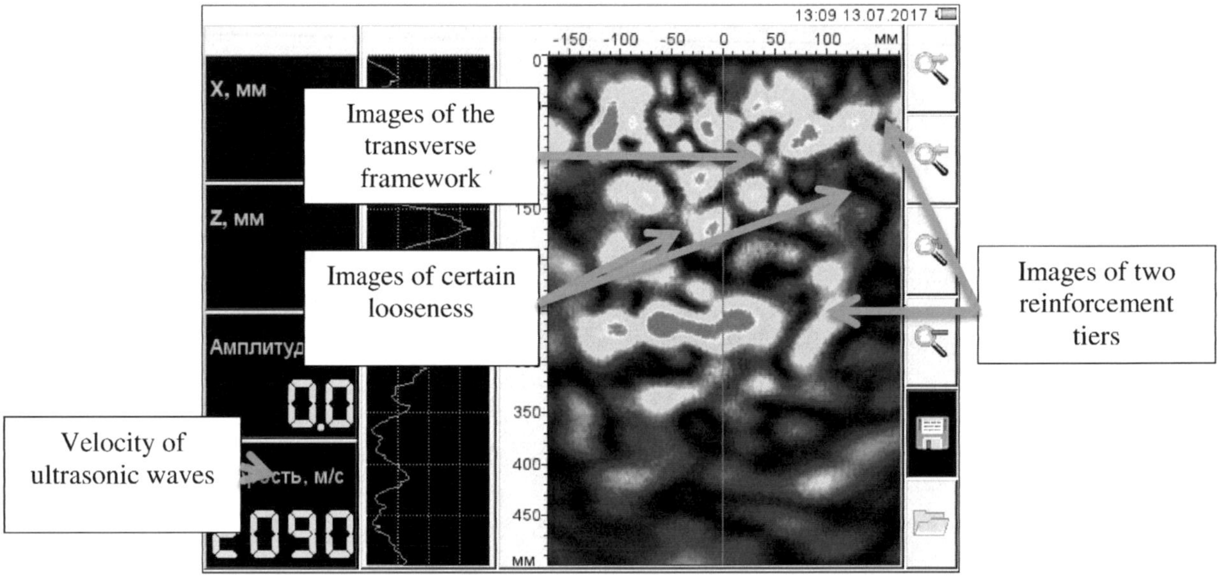

Figure 6. The tank wall scan with images of the transverse reinforcing cage between the near and distant meshes and the unexpected images of some loose concrete.

The same tank wall was further scanned in the loose concrete areas (Figure 7). The ultrasonic waves velocity here is lower than that in dense concrete. A large number of foreign images indicates a lack of consolidation of concrete to the depth equal to 250 mm of the structure thickness. The waves stop short of the backwall of the structure and are echoed from the indications. Consequently, the backwall image is not seen on the scan.

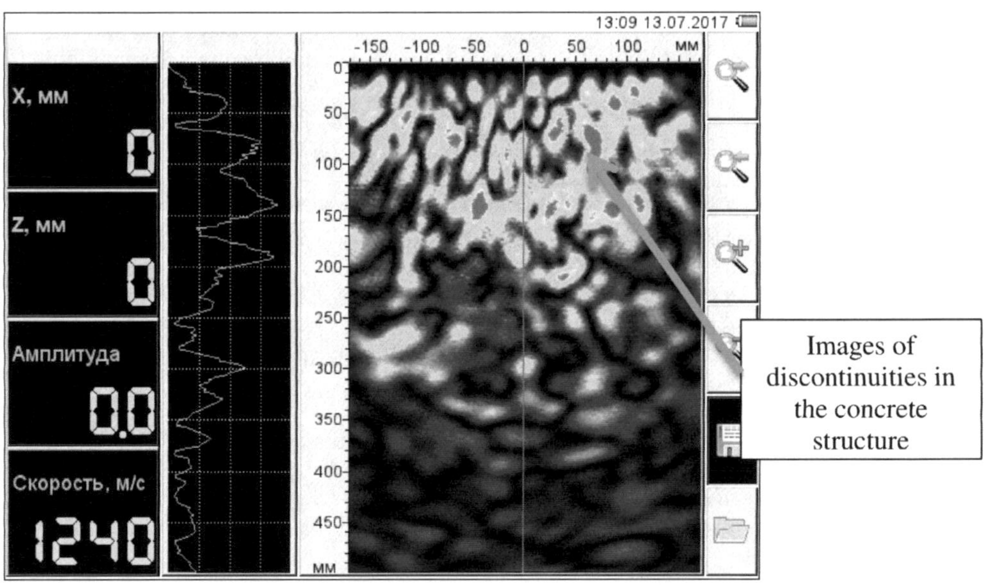

Figure 7. The scan of tank wall in the area of loose concrete.

Furthermore, the tomographic image mapping was performed for these areas of the tanks with loose concrete. The scans of the structure body in this area were collected for mapping by moving the device in 100-mm increments along the grid which has been preliminary made on the surface of the structure.

See Figure 8 for a screenshot of the plane scans of the tested area. The B-scan shows extra images undesigned between two reinforcement tiers, which may prove to be foreign matter or most probably cavities resulted from loose concrete. The C-and D-scans are shown along the secant planes of the body at the place where they run through the foreign images.

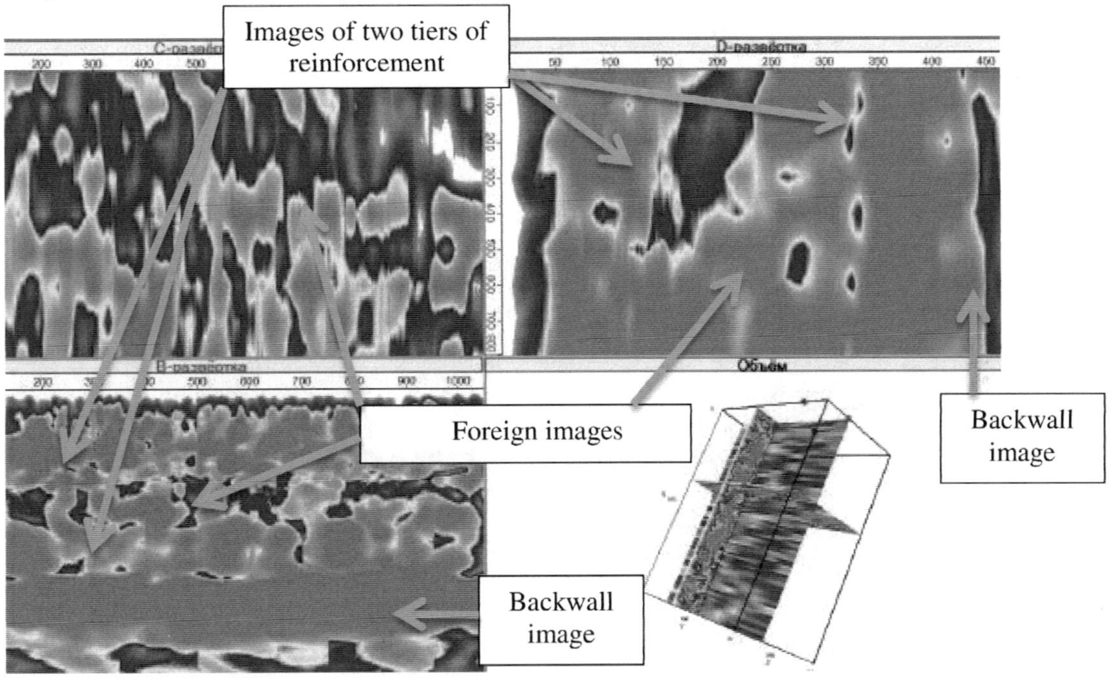

Figure 8. Screenshot of the plane scans at the tank wall area with echoes from the two tiers of reinforcement, the backwall and foreign images.

Built-in columns seen as wall piers were scanned in the same tank

Here the wall thickness is increased to 400 mm. The scan performed on a dense area of concrete is shown in Figure 9. It clearly shows the images of the reinforcement and the backwall image of the structure at a distance of 400 mm.

One of the piers showed a centerline defect seen as a long narrow cavity running deep into the body of the structure. At this area, the scan shown at Figure 10 has been made. As can be seen from the scan, the cavity runs deep inside and transforms into a large cavern which opens onto the opposite surface of the structure. The defect echoes the waves, the backwall image is missing.

From the scans it is clear that the images of the defects and those of the reinforcement look identical. A huge amount of scans of the structure has been collected to identify the typical patterns with the images of the reinforcement elements at their designed positions to understand the location of the images and detect their flaws. Any deviations, foreign images located in places where they are not expected to be, required interpretation and could be indications of the concrete structure discontinuities.

Operation of A1040MIRA requires a wide experience of UT scanning of various items. To acquire the experience the device developers and some scientific works recommend manufacturing and scanning of the trial concrete blocks with artificial reflectors (simulated flaws). However, this is labor-intensive and does not offer an accurate simulation of the defects encountered.

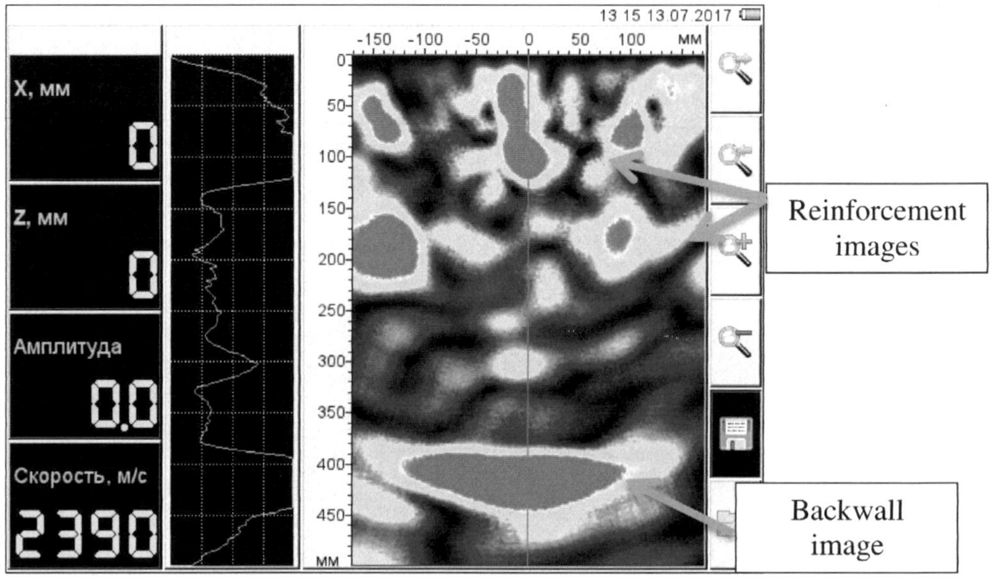

Figure 9. The tank wall scan at the pier location.

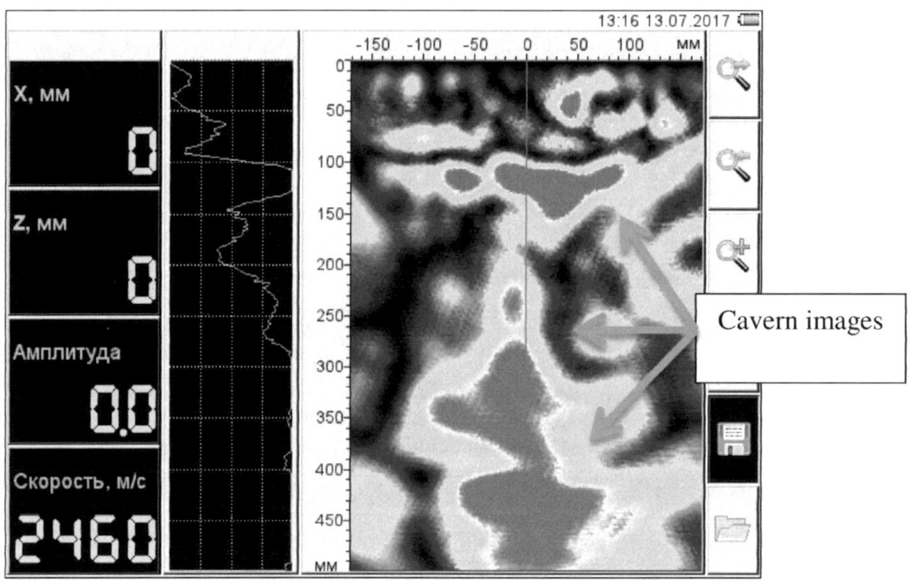

Figure 10. The tank wall image at the pier location with a large cavern image in the center of the pier.

To acquire the experience a 600-mm column was scanned in situ with a conduit in the center (figure 11). The scan clearly shows the images of the near reinforcement and those of the embedded part. The backwall image is missing since the waves won't go beyond the conduit and are reflected by it.

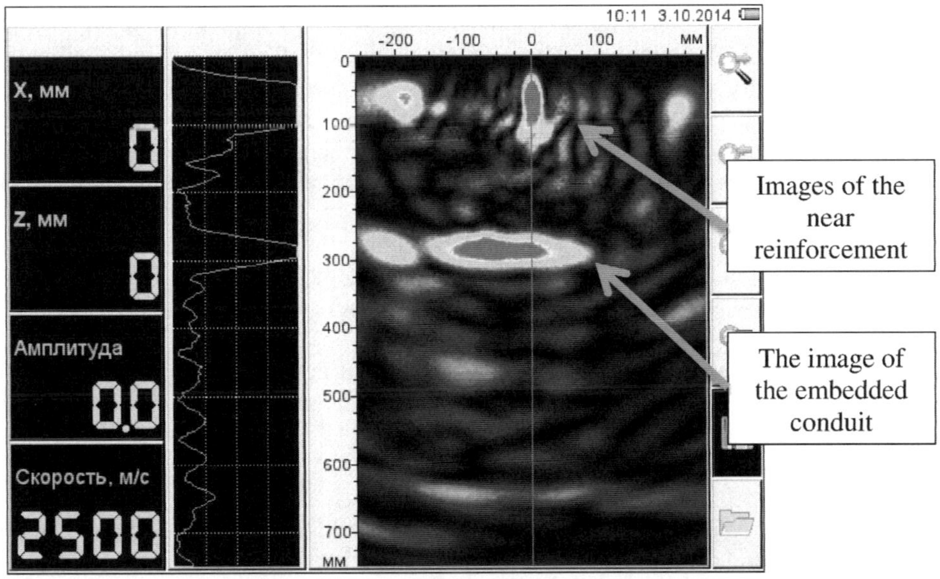

Figure 11. Scanning of the column with the embedded pipe at a distance of 300 mm from the surface.

Figure 12 shows a scan of the same column at the conduit free area. The images of the near reinforcement and the backwall image of the structure are clearly visible at a distance of 600 mm which is the section size of the column.

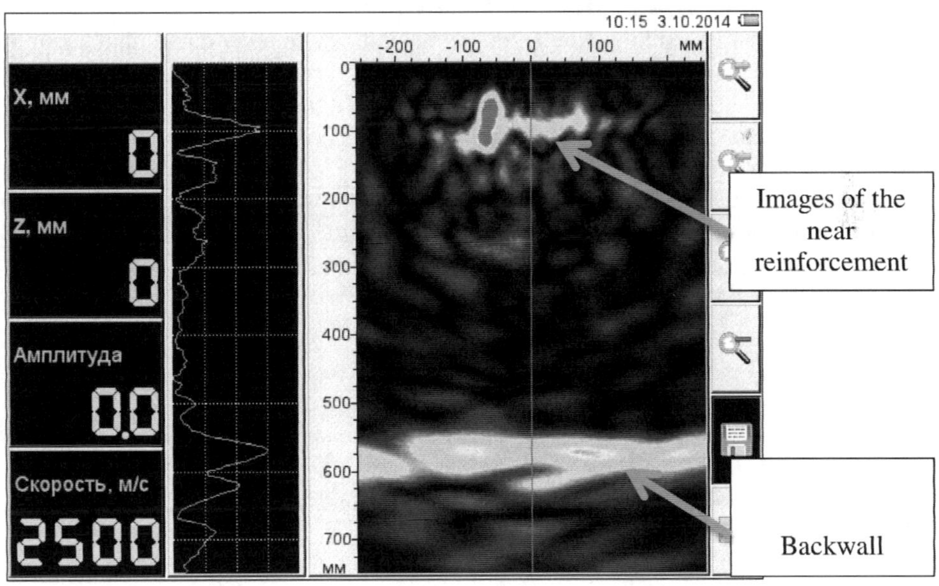

Figure 12. The column scan at the conduit free area.

Identification of defects in the body of concrete and reinforced concrete structures of buildings and constructions by A1040MIRA tomograph has prevented emergencies at hazardous facilities. Decisions on the further operation of defective structures were based on the size and location of the defects:

- “heal” the structure with special repair mixes as deep as the defect runs;
- reinforce structures with metal yokes and carbon reinforcement systems from outside;
- dismantle the structure.

3. Conclusions

Before A1040MIRA scanning of structures it is necessary to know the reinforcement principle of a structure, its geometry, the configuration of the reinforcements and the working environment. Information on the concrete placement procedure, construction joints, undesigned concrete placement interruptions, consolidation methods, the mix drop height, environment effecting the structure will explain the unexpected images on the scans. Just the reliable tomographic images will assess the effect of internal defects on the structure performance.

A1040MIRA tomograph is helpful in detecting the body of loose concrete (discontinuities in concrete) in the load bearing structure and identifying its residual bearing capacity.

A1040MIRA helped assess the degree of concrete consolidation in the structures of hydraulic facilities.

A1040MIRA helped resolve the issues of measuring the thickness of a concrete structure from either both sides if readily accessible and from just one side if the access is limited, or those of measuring a layer thickness in a sandwich structure.

References

[1] Samokrutov A 2012 Conclusion on the "A1040 MIRA" tomograph ultrasound examination for defection of flaw (Moscow: LLC "ACS-Service") p 52

[2] Kozlov V, Samokrutov A and Shevaldykin V 2002 The ultrasonic flaw detection of concrete by the echo method: status and prospects *In the world of nondestructive testing* № **2** p 6

[3] Samokrutov A and Shevaldykin V 2012 Opportunities of the ultrasonic tomography of concrete (Moscow: LLC "Acoustic control systems") UDC 620.179.16

[4] Samokrutov A, Shevaldykin V and Kozlov V 2003 Ultrasonic low-frequency piezoelectric transducers with dry point contact and their application for non-destructive testing *Inspection. Trouble shooting* № **2**(**16**) p 30

[5] Shvabovich K and Suvorov V 2012 Non-destructive testing and plotting of the backwall surface profile using the ultrasound tomography methods (Moscow: LLC "ACS-Service") p 13

[6] Shtengel V 2002 On the NDT methods and tools for inspection of the operated reinforced concrete structures *In the world of NDT* № **2**(**16**) p 12

[7] Samokrutov A, Kozlov V and Shevaldykin V 2006 Ultrasonic testing of concrete objects using dry acoustic contact. Methods, instruments and possibilities *The 5th Int. Conf. on Non-Destructive Testing and Technical Diagnostics in Industry* (Moscow) p 152

[8] Schabowicz K and Hoła J 2012 Nondestructive elastic-wave tests of foundation slab in office building *Materials Transactions* vol. **53** p 296

IOP Conf. Series: Materials Science and Engineering **481** (2019) 012048 doi:10.1088/1757-899X/481/1/012048

Modified fine-grained concretes based on highly filled self-compacting mixtures

T A Nizina[1], A S Balykov[1], D I Korovkin[1], V V Volodin[1]

[1]Department of Building Structures, Ogarev Mordovia State University, Russia

E-mail: artbalrun@yandex.ru.

Abstract. In this article the results of study of rheo-technological characteristics of cement-carbonate suspensions and physical and mechanical characteristics of cement composites from highly filled self-compacting mixtures are presented. The optimal dosage of polycarboxylate superplasticizer (Melflux 5581 F) and rheologically active mineral filler (microcalcite) in cement-mineral suspensions when developing self-compacting fine-grained concrete mixtures is defined. The water demand of fine natural quartz sands using the methodology of Skramtayev and Bazhenov is studied. The compositions of self-compacting fine-grained concrete mixtures of workability class SF1 with fine natural quartz sand are developed. High physical and mechanical characteristics of fine-grained concretes from highly filled self-compacting mixtures are fixed: density in normal humidity conditions at the age of 28 days up 2393 kg/m^3; bending and compressive strength at the age of 28 days up 14.9 and 115.8 MPa, respectively.

1. Introduction

One of the top-priority strategic trends in modern construction material science is the development of advanced energy-efficient modified cement concretes differing in a number of high physical and technical characteristics [1, 2, 3, 4, 5]. A common feature of modern modified cement composites is high strength (60÷100 MPa and higher), while other strength parameters increase when the content and size of the aggregate decrease reaching maximum when forming a high-strength fine-grained structure characterized by high homogeneity and no destruction hazard in the large aggregate/cement stone interface [6]. The efficiency of the concrete fine-grained structure increasing highly strong cement composites has been proved by the results of own studies [7, 8, 9, 10].

It should be noted that in the development of highly strong heavy and fine-grained concretes, a determinant role has been played by a scientific basis for modifying cement systems with chemical and active mineral additives, which have been formed as a result of multiple studies and proved in practice [11, 12, 13, 14, 15, 16, 17]. Using efficient individual additives (superplasticizing additives, based on polycarboxylate and polyglycol, active pozzolanic additives – microsilica, dehydrated kaolin and finely-dispersed ashes, etc.) and their systems has become a key to solving many process tasks of the concrete science, including manufacture of modified cement composites of increased strength from self-compacting mixtures (SCC) [6, 18, 19, 20].

The term "self-compacting concrete" (SCC, English, Selbstverdichtender Beton (SVB), German, Beton autoplacant (BAP), French) suggested in 1986 by the Japanese Professor Okamura [18] combines concrete mixtures with high workability (slump cone flow above 55-60 cm with a reduced cement-water ratio down to 0.35-0.4 and less) caused by high deformability of the suspension matrix along with its high resistance to segregation and lamination in displacement [6].

2. Formulation of the problem

A distinctive feature of the formulation of self-compacting heavy or fine-grained concretes with coarse sands is an increased content of rheologically active components, the amount of which in different concretes varies from 40 to 50% [18, 19, 20]. Increasing the share of water-dispersive suspension is achieved by introducing fine powders of rocks of sedimentary, volcanic and metamorphic origin in the amount of up to 100÷150% of the cement weight that are compatible with known plasticizers and are rheologically active in mixture with cement [21] (limestone, marble, dolomite, shredded quartz sand, etc.). However, the issues of manufacturing self-compacting fine-grained concrete mixtures using widely spread fine natural quartz sands with the fineness modulus below 2.0 are almost not investigated, especially in low-cement mixtures with the consumption of the binder up to 400 kg/m^3, self-compacting of which cannot be achieved with minimal water and no lamination for the defined filling extent of cement systems due to high inter-grain cavernosity and water demand of the given fine aggregates, as well as low efficiency of plasticizers in such mixtures irrespective of their dosage. The described complications require scientific and theoretical basis to be developed to increase the plasticizing and water-reducing efficiency of modern plasticizers in concrete mixtures based on fine water-demanding sands as well as recommendations for optimal dispersion and concentration of mineral aggregates in such systems to achieve their high mobility with minimal water of mixing.

3. Problem solution

The defined research aim has been reached by solving the following tasks:

- defining the optimal dosage of a polycarboxylate superplasticizer and rheologically active mineral filler in cement-mineral suspensions when developing self-compacting fine-grained concrete mixtures;
- studying the water demand of fine natural quartz sands using the methodology of Skramtayev and Bazhenov;
- developing compositions of self-compacting fine-grained concrete mixtures of workability class SF1 with fine natural quartz sand;
- studying the physical and mechanical characteristics of fine-grained concretes from self-compacting mixtures.

4. Results and discussion

4.1. Defining the optimal dosage of polycarboxylate superplasticizer and rheologically active mineral filler in cement-mineral suspensions

To prepare suspensions, Portland cement CEM I 32.5R (C) by Mordovtsement PJSC was used. The mineral part included a carbonate filler from KM100 microcalcite (MKM) by Polipark LLC with the dosage of 0÷300% of Portland cement weight (0÷75% of solid phase weight) with the variability pitch of 100%. Melflux 5581 F (SP) polycarboxylate superplasticizer by BASF Construction Solutions (Trostberg, Germany) was used as a plasticizer.

The study was carried out with the fixed water-solid ratio W/S = 0.15 with the varying factors being:

- ratio MKM /C, x_1 = 0÷3.0 relative units;
- ratio SP/(C + MKM), x_2 = 0÷1.5%.

The study results were used to develop an experimentally statistical (ES) model describing the changes in the flow diameter of cement-carbonate suspensions (Portland cement + microcalcite) [mm] from the Hegermann cone from the content of the varying factors x_1 and x_2. ES-model was set as polynome being indicated below

$$y = 288.9 + 52.9 \cdot x_1 + 64.8 \cdot x_2 + 19.8 \cdot x_1 \cdot x_2 - 51.33 \cdot x_1^2 - 89.3 \cdot x_2^2 + 37.97 \cdot x_1^3 + \\ + 32.34 \cdot x_2^3 - 59.06 \cdot x_1 \cdot x_2^2 - 15.19 \cdot x_1^2 \cdot x_2 + 31.64 \cdot x_1^2 \cdot x_2^2, \quad (1)$$

where x_1, x_2 are the values of variable factors.

Using the numerical values of the coefficients of the polynomial (1), isolines have been built, which reflect the changes in the flow diameter of cement-carbonate suspensions from the Hegermann cone depending on the content of microcalcite and Melflux 5581 F superplasticizer (Figure 1). It has been found that for the constant water/solid ratio of W/S = 0.15 relative units, an increase in the dosage of the superplasticizer and the mineral filler (microcalcite) causes a significant increase in the flow diameter of the cement-mineral suspension.

The analysis of the ES-model (1) and Figure 1 found the optimal levels of the varying factors that allow reaching the self-compacting of suspensions for the flow diameter from the Hegermann cone above 280 mm and the water/solid ratio of 0.15 relative units: the dosage of Melflux 5581 F superplasticizer is 0.5÷1.0% of the solid phase weight; the filling degree of suspension with microcalcite is at least 105% of the Portland cement weight. An increase in the plasticizer content to 1.5% of the Portland cement weight does not cause a significant growth of the rheo-technological parameters. Reaching the self-compacting of the dispersive system with reduced dosages of the plasticizer equal to 0.45÷0.6% of the solid phase weight is possible for the aggregate content above 180% of the Portland cement weight.

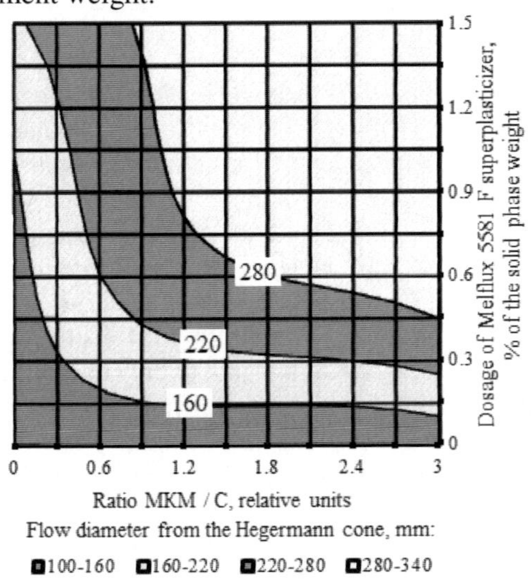

Figure 1. Changes in the flow diameter of cement-carbonate suspensions from the Hegermann cone depending on the superplasticizer dosage and microcalcite filling degree.

4.2. Studying the water demand of fine natural quartz sands

The water demand of fine and large natural quartz sands was defined upon the methodology of Skramtayev and Bazhenov [22] based on selecting the mixtures of the same mobility with comparative tests of the cement paste and mortar with a fine aggregate. To find the water demand of sand (W_s), the at the initial stage water content of the cement paste was found ($(W/C)_c$), for which it has the flow diameter from the Hegermann cone of about 170 mm, which corresponds to normal cement paste density. The water-cement ratio of the mortar was then found for the composition of sand/cement = 1/2 ($(W/C)_m$), for which it has the same flow diameter from the Hegermann cone (170 mm). The water demand of sand (%) was calculated under the formula:

$$W_s = \frac{(W/C)_m - (W/C)_c}{2} \cdot 100\%. \quad (2)$$

As a binder for the preparation of the cement paste and mortar, the above Portland cement CEM I 32.5R (C) was used with the normal cement paste density 26.5% and the water-cement ratio $(W/C)_c = 0.265$.

As a fine aggregate, the natural quartz sand of the Mordovia Novostepanovskiy pit was used with the coarseness modulus $M_{coar} = 1.6$ and the content of pulverulent and clay particles of 1.6% (Russian GOSTs 8735 and 8736). As a result of sand fractioning and selecting fraction below 0.63 mm, the coarseness modulus changes within $M_{coar} = 1.4 \div 1.6$. In the study, this type of a fine aggregate was used in two versions: as a selected fraction below 0.63 mm with $M_{coar} = 1.4$ and average grain size $d_{av} = 0.23$ mm (S1); in natural occurrence with the grain size below 5 mm, $d_{av} = 0.29$ mm and $M_{coar} = 1.6$ (S2).

The results of calculating the water demand of the studied types of sand are given in Table 1.

Table 1. Study results of water demand of applied sands.

Type of applied sand	Marking of sand	Dispersivity and coarseness modulus of sand	$(W/C)_m$, relative units	W_s, %
natural quartz sand of the Mordovia Novostepanovskiy pit	S1	fraction below 0.63 mm, $d_{av} = 0.23$ mm and $M_{coar} = 1.4$	0.430	8.3
	S2	natural occurrence with the grain size below 5 mm, $d_{av} = 0.29$ mm, $M_{coar} = 1.6$	0.413	7.4

4.3. Developing of compositions of self-compacted fine-grained concrete mixtures with fine natural quartz sand

The study results of the water demand of sands and rheological efficiency of formulations of cement-carbonate suspensions were used to develop self-compacting fine-grained concrete mixtures of workability class SF1 (the flow diameter from the Hegermann cone nevertheless 260÷280 mm, slump and flow of the standard Adams cone above 25÷25.5 cm and 500÷550 mm, respectively) with fine natural sand S1 (Table 1), consumption of Portland cement CEM I 32.5R (C) within 239÷739 kg/m^3 of the concrete mixture, MKM shares – 45÷321% of Portland cement weight, SP Melflux 5581 F and water contents – 0.7% and 0.202÷0.218 relative units of the mixture (C + MKM), respectively. These high-mobility mixtures are highly filled systems, characterized by an increased content of fine rheologically active components (C + MKM) and rheologically active suspension (C + MKM + mineral additive + water): 340÷360 and 580÷600 l/m^3 of the concrete mixture, respectively, ensuring a sufficient inter-layer between particles of fine sand.

To increase the dispersive phase of the binder, in addition to microcalcite, three types of mineral additives were used: condensed non-compacted microsilica MC-85 (MC) from Kuznetskiye Ferrosplavy JSC; highly active white metakaolin (HAWM) from Meta-D LLC; integral crystalline waterproofing additive Penetron Admix (Admix) for concrete mixture from Penetron Waterproofing Factory LLC.

4.4. Studying the physical and mechanical properties of fine-grained concretes from highly filled self-compacting mixtures

At the final stage, 7 compositions of highly filled self-compacting fine-grained concrete mixtures with fine natural quartz sand S1 were selected (Table 2) for experimental studies of the physical and mechanical properties of modified cement composites: density in normal humidity conditions at the age of 28 days, bending strength and compressive strength (cube strength) at the age of 1, 7 and 28 days. The research results are given in Table 2 and Figure 2.

Table 2. Compositions, physical and technical properties of modified fine-grained concretes from highly filled self-compacting mixtures with fine sand S1.

Components \ Compositions	1	2	3	4	5	6	7
Prescription and technological parameters							
MKM / C, relative units	0.45	0.45	0.45	0.45	1.11	1.63	3.21
C, kg/m^3	701÷739				239÷485		
S1 / C, relative units	1.50	1.50	1.50	1.50	2.23	2.92	4.68
Additive (A), % of weight (C + A)	-	10 (HAWM)	10 (Admix)	10 (MC)	10 (HAWM)	-	-
SP, % of weight (C + MKM)	0.7	0.7	0.7	0.7	0.7	0.7	0.7
W / C, relative units	0.29	0.30	0.32	0.32	0.44	0.57	0.89
Flow diameter from the Hegermann cone, mm	295	260	280	280	260	290	300
Physical and mechanical properties at the age of 28 days							
Density, kg/m^3	2393	2343	2320	2367	2320	2323	2344
Bending strength, MPa	12.3	12.8	10.6	14.9	9.9	7.4	5.3
Compressive strength, MPa	92.5	104.2	82.7	115.8	73.8	47.5	28.3

Experimental studies found high densities of cement composites made from highly filled self-compacting mixtures with sand S1 amounting to 2320÷2393 kg/m^3 (Table 2). Strength at the age of 28 days for the studied modified fine-grained concretes with the Portland cement consumption of 239÷739 kg/m^3 of the concrete mixture vary within a wide range: bending strength – 5.3÷14.9 MPa; compressive strength – 28.3÷115.8 MPa.

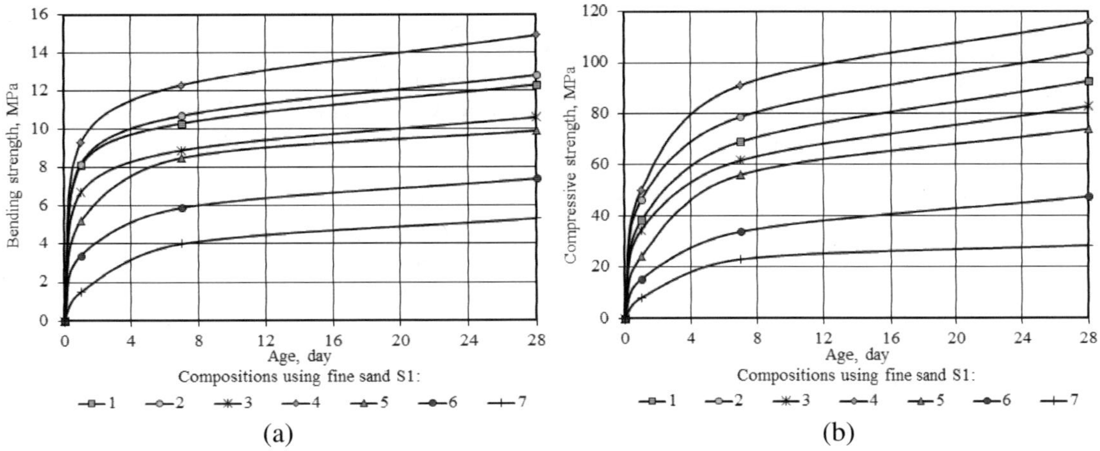

Figure 2. Kinetics of growth of bending (a) and compressive (b) strength of compositions of modified fine-grained concretes from highly filled self-compacting mixtures using fine sand S1.

The studied compositions of modified cement composites from highly filled self-compacting mixtures differ in high rates of strength gain: the bending strength at the age of 1 and 7 days, respectively, was 28÷66 and 75÷86%, and the compressive strength at the age of 1 and 7 days was 28÷44 and 71÷81% of the design strength at the age of 28 days (Figure 2).

The data analysis of Figure 2 and Table 2 found an opportunity to produce highly strong fine-grained concretes with no active mineral additives with the Portland cement consumption of 739 kg/m^3 of the concrete mixture: composition No. 1 with the ultimate bending and compressive strength at the age of 28 days: 92.5 and 12.3 MPa, respectively.

By studying groups of cement composites with increased consumption of Portland cement (701÷739 kg/m^3 of the concrete mixture), it has been found that introducing active mineral additives of condensed non-compacted microsilica (MC) and highly active white metakaolin (HAWM) into the formulation results in the increased strength of composites. In particular, the strength characteristics of fine-grained concretes with MC and HAWM of compositions Nos. 2 and 4 are higher than the similar indicators of the control composite No. 1 non-modified with mineral additives with almost the same consumptions of Portland cement: up to 21 and 25% for bending strength and compressive strength, respectively.

Introducing Admix (composition No. 3) into the formulation promotes some decrease in the strength of fine-grained concretes at the design age (28 days) as compared to composites of the control composition No. 1 with almost the same consumptions of Portland cement: by 14 and 11% for bending strength and compressive strength, respectively. This effect can be explained by a somewhat increased water demand of fine-grained concrete mixtures with this mineral additive (Table 2) and the specifics of the phase composition of composites produced on their basis.

5. Conclusion

The results of experimental studies were used to develop compositions of modified fine-grained concretes of compression strength classes from C16/20 to C80/95 made of highly filled self-compacting mixtures with the Portland cement consumption of 239÷739 kg/m^3 and the water content of 212÷226 l/m^3 of the concrete mixture, including high strength concretes of class C45/55 and higher with the Portland cement consumption of 485÷739 kg/m^3 of the concrete mixture, including carbonate filler, Melflux 5581 F superplasticizer, active mineral additives (if necessary), fine natural quartz sand with the coarseness modulus of 1.4. These cement composites are characterized by the following quality indicators: density in normal humidity conditions at the age of 28 days –2320÷2393 kg/m^3; bending and compressive strength at the age of 28 days – 5.3÷14.9 and 28.3÷115.8 MPa, respectively. Mass use of such materials forms pre-requisites to implement into the production complex resource-saving, energy-saving and more environmentally friendly technologies by reducing the buildup of material and energy-consuming production of Portland cement, reducing emissions of harmful gases to the atmosphere, decreasing the energy and fuel consumption for laying and forming concrete mixtures, reducing the volumes of automobile and railroad transportation of raw materials and replacement of expensive imported aggregates with cheap local sands.

Acknowledgments

The reported study was funded by RFBR according to the research project № 18-29-12036.

References

[1] Bazhenov Yu M, Demianova V I and Kalashnikov V I 2006 *Modified High-quality Concrete* (Moscow: Publishing house of the Association of Construction Universities) p 386

[2] Kalashnikov V I 2016 Evolution of composition development and change in strength of concrete. Concretes of the present and the future. Part 1. Changing the composition and strength of concrete *Build. Mater.* **1-2** 96 – 103.

[3] Caprielov S S, Sheinfeld A V and Kardumyan G S 2010 *New Modified Concrete* (Moscow: Navruz) p 258

[4] Aitcin P C 2004 *High Performance Concrete* (London: T&F e-Library) p 591

[5] Nawy E G 2001 *Fundamentals of High-performance Concrete* (New-York: J. Wiley and Sons) p 441

[6] Falikman V R 2011 New efficient high-performance concretes *Concrete And reinforced concrete Equip. Mater. Techn.* **2** 78 – 84.

[7] Nizita T A and Balykov A S 2016 Experimental-statistical models of properties of modified fiber-reinforced fine-grained concretes *Mag. of Civil Engin.* **2** 13 – 25

[8] Nizina T A, Ponomarev A N, Balykov A S, Volodin V V and Korovkin D I 2017 Fine-grained fibre concretes modified by complexed nanoadditives *Intern. Journal. of Nanotech.* **14** 665 – 679

[9] Nizina T A, Selyaev V P, Balykov A S, Volodin V V and Korovkin D I 2017 Optimization of compositions of multicomponent fine-grained fiber concretes modified on different scale levels *Nanotech. In Constr.: A Sci. Intern.-Journal. B* **9** 43 – 65

[10] Nizina T A, Balykov A S, Volodin V V and Korovkin D I 2017 Fiber fine-grained concretes with polyfunctional modifying additives *Mag. of Civil Engin.* **4** 73 – 83

[11] Usherov-Marshak A V 2006 Additives in concrete: progress and problems *Build. Mater.* **10** 8 – 12

[12] Falikman V R 2009 Polycarboxylate hyperplasticizers: yesterday, today, tomorrow *Pop. Concrete Sci.* **2(28)** 86 – 90

[13] Lashley L 2009 *The Compatibility and Performance of Cementitious Materials and Chemical Admixtures: Master of Applied Science Thesis* (Toronto: University of Toronto) p 185

[14] Selyaev V P, Nizina T A and Balbalin A V 2013 Multifunctional modifiers of cement composites based on mineral additives and polycarboxylate plasticizers *Bulletin of Volgograd State Architect. And Constr. University. Constr. And architect. B.* **31 (50)** 156 – 163.

[15] Kirsanova A A and Kramar L Ya 2013 Organometallic modifiers based on metakaolin for cement concretes *Build. Mater.* **10** 54 – 56.

[16] Kalashnikov V I, Moskvin R N, Belyakova E A, Belyakova V S and Petukhov A V 2014 High-dispersion fillers for powder-activated concrete of a new generation *Systems. Methods. Technologies* **2 (22)** 113 – 118.

[17] Tarakanov O V and Kalashnikov V I 2017 Prospects for the application of complex additives in new-generation concretes *Izvestiya KGASU* **1 (39)**. 223 – 229.

[18] Okamura H and Ouchi M 2003 Self-compacting concrete *Journal of Advanced Concrete Techn. A* **1** 5 – 15

[19] De Shutter G, Bartos P, Domone P and Gibbs J 2008 *Self-compacting Concrete* (New-York: Taylor & Francis Group) p 296

[20] EFNARC 2005 *The European Guidelines for Self-compacting Concrete. Specification, Production and Use* (UK: Farnham) p 68

[21] Kalashnikov V I 2008 Calculation of compositions of high-strength self-compacting concrete *Build. Materials* **10** 4 – 6

[22] Bazhenov Yu M 2003 *Technology of Concrete* (Moscow: Publishing House of the DIA) p 500

IOP Conf. Series: Materials Science and Engineering **481** (2019) 012049 doi:10.1088/1757-899X/481/1/012049

Fly Ash-based Geopolymer Binders: Perspective Materials for Sustainable Building

I K Domanskaya[1], A Yekimovskaya[1]

1Department of building materials, Ural Federal University, Mira Street 19, Ekaterinburg, 620002, Russia

E-mail: i.k.domanskaya@urfu.ru

Abstract. The possibility of obtaining a geopolymer binder on the basis of fly ash of the Reft thermal power station and fine-grained concrete on its basis is investigated. It is shown that the chemical-mineral composition and properties of this fly ash allow it to be used as an aluminosilicate component without additional grinding and introduction of mineral additives. In this case, the alkaline activator is an industrial sodium liquid glass diluted to a density of 1200 kg/m^3. The dependences of the strength of geopolymer binder and fine-grained concrete on its basis on the conditions of hardening, aggregate consumption and the sequence of mixing of components are established. Based on the fly-ash geopolimer and the screenings crushing granite dropout picked up the fine-grained concrete class B35.

Key words: geopolymer binder, fly ash, fine-grained concrete, alkaline activator, strength, water resistance, conditions of hardening

1. Introduction

Features of the life cycle of any building and/or structure are largely determined by the materials chosen for its construction. They affect the speed of construction and performance of the construction site, its maintainability and durability, the possibility and manufacturability of recycling after dismantling structures. Therefore, ensuring the construction of economically and environmentally efficient materials that meet the requirements of Sustainable development is an urgent task. The question of alternative binders that can compete with Portland cement, the production technology of which is resource-intensive and dangerous for the environment, is especially acute today [1-5]. Binders obtained by alkaline activation of mineral raw materials without the use of energy-intensive firing processes (characteristic of Portland cement technology) are considered as the most promising. There are several names of this class of binders: alkali-activated cements, slag-alkaline binders, alkaline cements and others; most often they are called "geopolymers". The validity of the use of this term is based on the fact that during their hardening the synthesis of high-molecular structures occurs by polymerization or polycondensation of monomer silicate and aluminate groups. The latter can be formed by the action of alkaline activators on the materials of aluminosilicate composition [6-8]. Aluminosilicate components are usually crushed rocks, mineral industrial waste or a combination thereof; solutions of sodium silicates, sodium or potassium hydroxides and other chemical compounds having an alkaline character are used as activators [9-18].

The literature describes numerous positive examples of the use of geopolymer binders in the composition of heavy and light concretes, mortars [19-21]. Despite this, to date, they have not been widely used in construction. One of the reasons is the complexity of the technology for producing geopolymer composites. It involves the combination of technology for the preparation of the binder and building material, based on it (for example, concrete). At the same time, the technology of obtaining the majority of composites using traditional binder analogues (cement, gypsum, etc.) provides for the use of the finished binder as a raw component [22]. Therefore, in each case for each type of aluminosilicate component, alkaline activator and type of reinforcing element (crushed stone, sand, fibrous materials, etc.) it is necessary to select the optimal technological parameters that provide the necessary properties of the building composite.

In this study, fly ash from the Reft thermal power plant (FARTP) located in the Sverdlovsk region was used as an aluminosilicate component. The chemical composition of ash is comparable with clays, and the phase composition is dominated by glass phase [23]. This suggests its high potential for the creation of alkali-activated binders. The aim of the study was to select the optimal technological parameters for the production of geopolymer ash binder and fine-grained concrete (FGC) based on it.

2. Initial materials and research methods characteristic

A representative sample of FARTP with the following characteristics: the specific surface 260 m^2/kg, bulk density 790 kg/m^3, true density 1900 kg/m^3 was used in this research. Chemical composition of FARTP, % (by weight): 58.70 SiO_2, 25.60 Al_2O_3, 7.21 Fe_2O_3, 5.59 CaO, 0.80 MgO, 2.10 loss on ignition. As an alkaline activator, industrial liquid sodium silicate glass with a density of 1450 kg/m^3, a silicate module 3.2 (manufactured by LLC «Himreaktivy», Yekaterinburg), and a water solution of sodium hydroxide were used. To obtain a liquid glass of the required density, it was mixed with distilled water. Density was controlled by the areometer with a range of measurement of 1000-1500 kg/m^3.

The geopolymer binder was prepared by thoroughly mixing FARTP with an alkaline solution. The resulting paste was formed in the form of samples of a cubic shape measuring 20 × 20 mm, which were hardened under normal conditions and under heat treatment conditions: in an autoclave (autoclave mode: 1 hour rise in temperature, 4 hours at 123 °C and 2.2 atm pressure, 1 hour temperature release), in a drying oven (drying for 8 hours at 100 °C), in the curing room (curing room mode: 1 hour temperature rise, 4 hours at 85 °C, 1 hour temperature release). For the production of FGC, the screenings crushing granite fraction of 1.25-2.5 mm with a bulk density of 1320 kg/m^3 was used. The content of pulverized and clay particles was 3 % by mass. 2 methods of preparation of molding mass were used. Method 1 (suspension): the beginning of fly ash was closed with sodium liquid glass, then mixed with granite screenings. Method 2 (dry mix): first, fly ash and granite screenings were mixed in a dry state, then the mixture was closed with sodium liquid glass (Fig.1). The ratio of the solid (ash-removal and screening) and liquid (sodium silicate solution) component did not change and was 0.48.

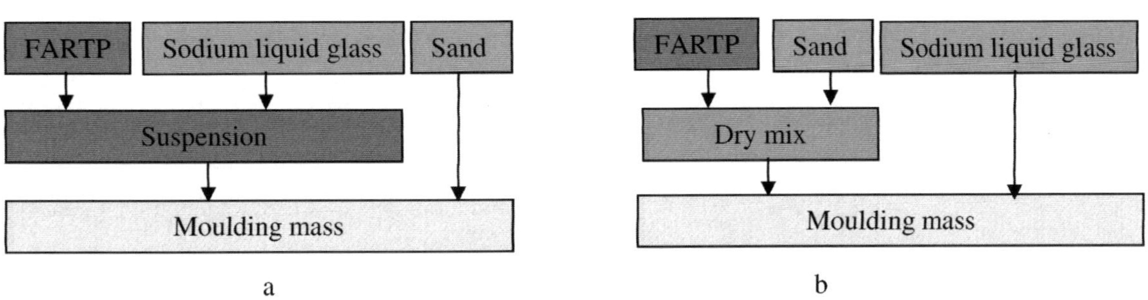

Figure 1. Methods of preparation of molding mass: a – method 1 (suspension); b – method 2 (dry mix)

The compression strength test was carried out on a PRG-1 benchtop press after a predetermined hardening time. To determine water absorption, the samples were placed in water for 24 hours, after measuring their mass. The water resistance of the hardened stone was evaluated by the softening coefficient:

$$Cs = R_{sat}/R_{dry}$$

where R_{sat} is the strength of a stone saturated with water; R_{dry} – the strength of the stone in a dry state.

3. Experiment results and discussion

The influence of some technological parameters (type and concentration of alkaline activator, hardening conditionson) the strength and water resistance of fly ash-based geopolymer binder and fine-grained concrete based on it was investigated.

Industrial liquid sodium silicate glass provides 5 times more high-strength artificial fly ash-stone than an aqueous solution of NaOH (Table 1).

Table 1. The influence of alkali activator on properties of fly ash geopolimer.

Alkaline activator	The ratio of the liquid and solid component (L/S)	Water Resistance (Cs)	Compressive strength, MPa, at the age of days (normal hardening conditions)	
			3	7
Liquid sodium silicate glass	0.48	0.57	5.5	15.1
Water solution of sodium hydroxide	0.48	0.30	1.1	2.9

The optimal density of the liquid glass solution in this case was 1200 kg/m^3 (Table 2).

Table 2. Effect of liquid glass density on compressive strength of fly ash geopolimer.

Density, kg/m^3	Compressive strength, MPa, at the age of days (normal hardening conditions)	
	3	7
1350	6.5	14.6
1300	6.9	15.4
1250	7.2	18.8
1200	7.5	22.1
1150	7.4	21.5

Heat treatment accelerates chemical reactions and improves the strength characteristics of the ash geopolymer. After it, when samples reached 1 and 3 days, they were tested for strength and water resistance. Regardless of the type of treatment, the softening coefficient of the ash geopolymer was 0.9. The maximum compressive strength had samples subjected to steaming, the minimum – after autoclaving heat treatment (Figure 2).

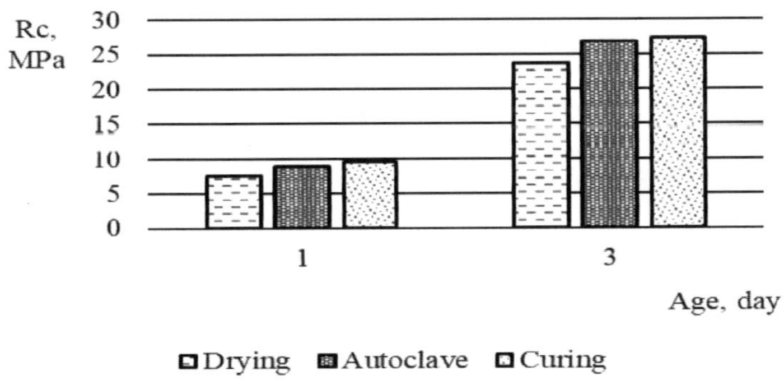

Figure 2. Influence of the type of heat treatment on the strength of the ash geopolymer.

It was found that the optimal ratio of the aluminosilicate component (fly ash) and the fine aggregate (granite screenings) was 2.0 (Figure 3).

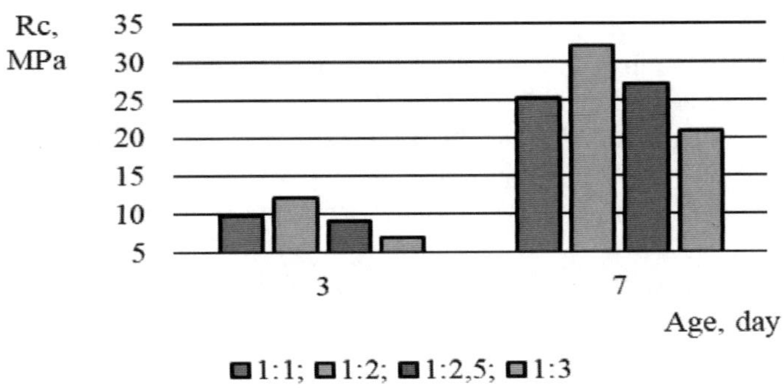

Figure 3. Effect of addition of fine aggregate on the strength of the geopolymeric composite.

When organizing the technological process of obtaining building composites based on ash geopolymers, it is important to observe the sequence of mixing components. it is advisable to pre-mix the fly ash with the reinforcing element, and then mixing them with an aqueous solution of an alkaline activator. The strength of FGC samples produced by method 2 (dry mixture) was 14 % higher than that obtained by method 1, all other things being equal.

As a result of the studies, a high water resistance fly ash-based geopolymer FGC was obtained, the strength of which corresponded to class B 35 (Table 3).

Table 3. est results of geopolymer fine-grained concrete.

Property	Value
Density, kg/m^3 (dried condition)	1870
Tensile strength, MPa, at the age of 28 days	4,8
Compressive strength, MPa, at the age of 28 days	47,3
Coefficient of softening (water resistance)	1

4. Conclusion

Studies have shown that it is possible to obtain geopolymer binder at the fly ash of the Reft TPP without additional grinding and introduction of mineral additives. In this case, the alkaline activator is an industrial sodium liquid glass diluted to a density of 1200 kg/m^3. The dependences of the strength of geopolymer binder and fine-grained concrete on its basis on the conditions of hardening, aggregate

IOP Conf. Series: Materials Science and Engineering **481** (2019) 012049 doi:10.1088/1757-899X/481/1/012049

consumption and the sequence of mixing of components are established. Based on the fly-ash geopolimer and the screenings crushing granite dropout picked up the fine-grained concrete class B35. After further studies of durability, this material can be recommended for implementation in production as cost-effective, environmentally friendly and promising to replace cement concrete of appropriate quality.

References

[1] Huntzingera D N and Eatmonb T D 2009 J of Cleaner Production 17 pp 668–75

[2] Gartner E M and MacPhee D E 2011 Cement and Concrete Research 41 (7) pp 736–49

[3] Khozin V G, Khokhryakov O V, Sibgatullin I R, Gizzatullin A R and Kharchenko I Y 2014 Carbonate cements low water demand – green alternative cement industry Russia Building materials 5 pp 76-82

[4] Tang S W, Yao Y, Andrade C and Li Z J 2015 Cement and Concrete Research 78 A pp 143–54

[5] Domanskaya I, Bednyagin S and Fisher H-B 2018 IOP Conf. Ser.: Earth Environ. Sci. 177 012029

[6] Loewenstein W 1954 The Distribution of Aluminum in the Tetrahedra of Silicates and Aluminates American Mineralogist 39 pp 92–96

[7] Khale D and Chaudhary R 2007 Mechanism of geopolymerization and factors influencing its development: a review J. of Materials Science 42 pp 729–46

[8] Rakhimov R Z, Rakhimova N R and Stoyanov O V 2014 Geopolymers Vestnik Kazanskogo tekhnologicheskogo universiteta 17 (23) pp 189–96

[9] Glukhovsky V D 1979 Alkaline and alkaline-earth hydraulic binders and concretes (Kyiv: Vyscha shkola) p 232

[10] Davidovits J 1993 Geopolymer Cements to minimize Carbon-Dioxide Greenhouse Warming Ceramic Transactions 37. Cement-Based Materials: Present, Future and Environmental Aspect. pp 165–182

[11] Barbosa V F F, MacKenzie K J D and Thaumaturgo C 2000 Synthesis and characterization of materials based on inorganic polymers of alumina and silica: sodium polysialate polymers Intern. Journal of Inorganic Materials 2 pp 309–17.

[12] Komnitsas K and Zaharaki D 2007 Geopolymerisation: a review and prospects for the minerals industry Miner. Eng. 20 pp 1261–77

[13] Rashad A M 2013 Alkali-activated metakaolin: A short guide for civil Engineer-An overview. Construction and Building Materials. https://doi.org/10.1016/j.conbuildmat.2012.12.030

[14] Robayo-Salazar R, Mejía-Arcila J, Mejía de Gutiérrez R and Martínez E 2018 Construction and Building Materials 176 pp 103–11

[15] Bernal S A, Provis J L, Fernández-Jiménez A, Krivenko P V, Kavalerova E, Palacios M and Shi C 2014 Binder chemistry – High-calcium alkali-activated materials. RILEM State-of-the-Art Reports, 13, pp 59–91

[16] Rodríguez E, Bernal S, Provis J, Paya J, Monzo J and Borrachero M 2013 Effect of nanosilica-based activators on the performance of an alkali-activated fly ash binder Cement and Concrete Composites 35 (1) pp 1-11

[17] Panizza M, Natali M, Garbin E, Tamburini S and Secco M 2018 Assessment of geopolymers with Construction and Demolition Waste (CDW) aggregates as a building material Construction and Building Materials 181 pp 119–33

[18] Marjanović N, M. Komljenović M, Baščarević Z, Nikolić V and Petrović R 2015 Physical-mechanical and microstructural properties of alkali-activated fly ash-blast furnace slag blends *Ceramics International* **41** pp 1421–35

[19] LI Z and LI S 2018 Carbonation resistance of fly ash and blast furnace slag based geopolymer concrete *Construction and Building Materials* **163** pp 668–80

[20] Kozhuhova N I, Zhernovski I V and Strokova V V 2012 Evaluation of biopositive geopolymeric binders on the basis of low calcium fly ash *Building Materials* **9** pp 84–86

[21] Kireety C H and Raghavarani Y 2016 Global Journal of Engineering Science and Researches, 3 (10) http://doi.org/10.5281/zenodo.163105
[22] Voitovich E V, Kozhukhova N I, Zhernovsky I V, Cherevatova A V and Niezvet D D 2013 The concept of quality control aluminosilicate binders degidrattingo hardening *Building Materials* **11** pp 68–70
[23] Potapov S O, Sviridova M N, Tanutrov I N and Toloknov D A 2016 Physico-chemical properties of fly ash from the combustion of coal Butlerov communications **45** (3) ROI: jbc-01/16-45-3-36

IOP Conf. Series: Materials Science and Engineering **481** (2019) 012050 doi:10.1088/1757-899X/481/1/012050

Composition and Properties of Non-Fired Fly Ash Gravel and Concrete on Its Basis

F L Kapustin[1], D V Kokorina[1], I V Fomina[1]

[1]Department of materials sciences in construction of the Institute of new materials and technologies, Ural Federal University, 19 Mira street, Yekaterinburg 620002, Russia

E-mail: f.l.kapustin@urfu.ru

Abstract. Non-fired fly ash gravel was obtained on the basis of fly ash of thermal power plant № 5 of JSC "SIBEKO", 15 % of Portland cement and the hardening accelerator. The gravel of 5-20 mm fraction and 6 months' air-moist curing is 900 density grade, P300 strength grade, frost-resistance grade – F35. The compositions of cement concretes of strength class from B10 to B25 were established and their physical and mechanical properties were determined using obtained non-fired fly ash gravel. They belong to the lightweight concrete according to the density (1817-1857 kg/m^3). The introduction of superplasticizer SP-1 reduces water demand of concrete mixture and increases strength of concrete on fly ash gravel.

1. Introduction

Construction of modern buildings and structures requires the use of concrete and reinforced concrete structures based on high-strength concrete with low density. To increase their strength highly active cements and hardening accelerators are used. These accelerators reduce water consumption for the preparation of concrete mixture and increase the density of concrete [1-3]. The decrease of concrete density is achieved mostly with use of lightweight aggregate and cement stone porosity. It is recommended to use keramzite, agloporite, fired fly ash and non-fired fly ash gravels [4-7] as aggregates in the composition of lightweight high-strength concrete. The physical and mechanical properties of aggregates mentioned above are presented in the table 1.

Table 1. Properties of artificial aggregates based on fly ash and slag wastes.

Parameter	Keramzite gravel	Fired fly ash gravel	Agloporite gravel	Non-fired fly ash gravel	
				Dense	Light-weight
Bulk density (kg/m^3)	250-800	450-900	450-900	800-950	450-600
Water absorption (mass %)	8-20	14-20	8-20	5-9	7-15
The strength when compressed in the cylinder (MPa)	0.5-6.5	1-5	0.5-3.5	5-8	1-3

Non-fired fly ash gravel (NFFYG) is an artificial porous aggregate obtained in the form of granules from a pre-moistened mixture of fly ash from the coal combustion and Portland cement followed by hardening during steaming or in natural conditions. For the NFFYG manufacturing Portland-slag cement and lime can also be used as binders, ash-slag mixture from the dumps of thermal power plants and ground metallurgical slag can be used as fillers and calcium chloride, sodium silicate, sodium sulfate, nitrite-nitrate calcium chloride and others can be used as cement hardening accelerators. In Russia about 30 million tons of ash and slag waste are generated annually during the burning of hard and brown coals at thermal and electric power plants, up to 40 % of their total volume is produced in Siberia.

These wastes pollute the environment and only less than 10 % of their current output are utilized, mainly in the production of construction materials. Power plant № 5 of JSC "SIBEKO" (Novosibirsk) more than 100 thousand tons of ash and slag are produced annually, then they are stored in dumps hydraulically. One of the most effective fields of application of ashes and slags can be production of artificial aggregates for concrete, in the concrete composition ashes occupy 70-85 % pf volume. Feasibility study of obtaining and using of high-strength NFFYG based on fly ash of Reftinskaya district power plant in concretes is considered in [8, 9].

2. Raw materials and methodology

To obtain NFFYG and concrete based on it the following materials were used:

- the fly ash from the combustion of stone coal at thermal power plant № 5 of JSC "SIBEKO";
- Portland cement CEM I 42,5 N, Russian Standard 31108-2016 [10] by JSC "Sukholozhskcement" (Sverdlovsk region);
- technical sodium sulfate (Na_2SO_4) A class, second grade according to Russian Standard 6318-77 [11].

The ash used in the work is acidic by chemical composition, it is fine by the grain composition (average particle diameter of 8.8 microns), it is characterized with a high specific surface which is 358 m^2/kg, has a true density of 2.36 g/cm^3 and it has no ability to self-hardening. The phase composition of ash is mainly represented by glass, quartz, mullite and magnetite minerals. According to the basic properties of ash meets the requirements of Russian Standard 25818-91 [12] (table 2), refers to the 3 type and can be used in the concrete production.

Table 2. Accordance of dispersion and chemical composition of fly ash of thermal power plant № 5 of JSC "SIBEKO" to requirements of Russian Standard 25818-91.

Parameter	Sieve residue № 008 (%)	Oxides content (mass %)						Weight loss on ignition (mass %)
		SiO_2	Na_2O	Fe_2O_3	MgO	SO_3	CaO	
Actual	19.40	55.70	1.29	5.38	2.23	0.27	4.93	3.25
Russian Standard requirements	≤ 20	≥ 45	≤ 3	≤ 10	Not standardized	≤ 3	≤ 10	≤ 10

NFFYG was obtained as follows: fly ash and Portland cement were mixed in a certain amount in a drum mixer, then it moistened with an water solution of sodium sulfate (calculated as 2 % Na_2SO_4 by cement weight) heated to a temperature of 80-90 °C; the mixture was granulated on a laboratory plate granulator with a plate diameter of 0.7 m and a rotation speed of 24 rpm; the obtained raw granules hardened in air-wet conditions for 6 months. Cement quantity was 15 % by weight of the ash-cement mixture, water for moisture was about 20 % and the duration of preparation and granulation of the mixture was 15 minutes.

Physical and mechanical properties of NFFYG were determined according to Russian Standard 9758-2012 [13] and compared with the requirements of Russian Standard 33928-2016 [14].

For establishing the concrete composition. NFFYG, Portland cement (CEM I 42,5 N), natural quartz sand and sand screenings from granite crushing (class 2, moderate) and superplasticizer SP-1 were used.

Selection of the composition of natural hardening concrete was performed by calculation and experimental method. After testing the aggregates it was determined:

- the optimal water consumption which provides the required workability of the concrete mixture and the cement-water ratio;
- cement consumption per 1 m^3 of the mixture which is providing the required strength of concrete in a given period under certain conditions of compaction and hardening;
- the consumption of aggregates per 1 m^3 of the placed and compacted mixture.

According to the results of determining the concrete strength the experimental composition of the mixture was refined [15].

3. Results and discussion

For the disposal and storage of ash and slag wastes so-called "dry" ash and slag removal technologies are developed and used at thermal power plants, for example, their granulation and storage in a stack [16, 17]. This technology of removal of granulated ash is close to the production of natural hardening NFFYG and can be implemented at the thermal power plants with the production of artificial aggregate for construction concretes. The NFFYG of 5-20 mm fraction, P300 strength grade and 900 density grade was used for preparation of concrete mix (table 3). The appearance of NFFYG is shown in figure 1.

Table 3. Comparative characteristics of NFFYG of fraction 5-20 mm.

Parameter	The test results after (days)		Requirements of Russian Standard 33928-2016
	28	180	
Bulk density (kg/m^3)	846	865	800-900
Average density (kg/m^3)	1532	1545	Not standardized
Grain composition:			
- content of grains of $\geq$ 5 mm (%)	99.7		85-100
- content of grains of $\geq$ 20 mm (%)	0.3		up to 10
- content of grains of $\geq$ 40 mm (%)	-		not allowed
Water absorption (mass %)	13.7	12.5	$\leq$ 20
Humidity (mass %)	11.1	1.4	$\leq$ 5
Frost resistance	F15	F35	$\geq$ F15
Content of split grains (%)	-	3	$\leq$ 15
Weight loss in silicate decomposition (mass %)	-	0	$\leq$ 5
Weight loss on ignition (mass %)	-	8.9	$\leq$ 5
The strength when compressed in the cylinder (MPa)	4.4	6.8	6.5-8

Concretes based on NFFYG are characterized by relatively low consumption of cement, its resistance to tension as well as the adhesion strength of gravel to the mixture is much higher compared to concretes based on other types of aggregates (heavy concrete, keramzite concrete, etc.), ranging from B25 class and above. NFFYG is an effective aggregate on the basis of which it is possible to produce lightweight structural concretes with compressive strength of 10-40 MPa and above, density of 1700-2100 kg/m^3 [18-21].

Concrete compositions with a design strength class of B10 and B25 were established on the basis of NFFYG from fly ash of thermal power plant № 5 of JSC "SIBEKO". Concrete mixtures had a PK2 brand of mobility. The actual consumption of materials of the selected concrete is presented in table 4, their physical and mechanical properties is presented table 5. Concretes are lightweight with density from 1817 to 1857 kg/m^3. Concrete based on sand screenings from granite crushing as fine aggregate (composition 3) has a compressive strength after 28 days of normal hardening by 25 % more in comparison with strength of concrete based on natural quartz sand (composition 1) and respectively belongs to the B10 strength class (table 5). The concrete of 2 and 4 compositions with the compressive strength to 36.7 and 17.8 MPa belong to classes B25 and B12,5 respectively. It was found that the introduction of a superplasticizer SP-1 in the amount of 1.7 kg/m^3 in the concrete mix the water consumption decreased by 8.7 % and the concrete strength increased from 14.9 to 17.8 MPa.

Figure 1. Non-fired fly ash gravel appearance.

Table 4. Component composition of concrete mix.

Composition number	Material consumption (kg/m^3)						Density of concrete mix (kg/m^3)	Design strength class
	Cement	NFFYG	Natural sand	Sand screenings	Water	SP-1		
1	177	708	840	-	236	-	1961	B10
2	367	716	664	-	250	-	1997	B25
3	178	704	-	859	249	-	1990	B10
4	179	708	-	865	230	1.7	1984	B10

Table 5. Physical and mechanical properties of concrete.

Composition number according to table 4	Dry concrete density (kg/m^3)	Compression strength (MPa)		Actual strength class of concrete
		after curing	after 28 days	
1	1843	7.9	11.9	B10
2	1857	24.0	36.7	B25
3	1817	7.8	14.9	B10
4	1834	12.2	17.8	B12.5

4. Conclusion

On the basis of fly ash of thermal power plant № 5 of JSC "SIBEKO", 15 % of Portland cement and 2 % of sodium sulfate as the hardening accelerator the 5-20 mm fraction NFFYG was obtained which after 6 months of air-moist curing had the 900 density brand, P300 strength brand and F35 frost-resistance brand. Lightweight concrete compositions of strength classes from B10 to B25 were established using the obtained gravel.

The work was supported by Act 211 Government of the Russian Federation, contract № 02.A03.21.0006.

5. References

[1] Bazhenov Yu M 2007 Concrete Technology (Moscow: ASV Publishing)
[2] Bazhenov Yu M, Demyanova V S and Kalashnikov V I 2006 Modified High-Quality Concretes (Moscow: ASV Publishing)
[3] Batrakov V G 1998 Modified Concretes Theory and Practice (Moscow: Technoproject)
[4] Ivanov I A 1993 Light Concretes based on Artificial Porous Aggregates (Moscow: Stroyizdat)
[5] Gorlov Yu P, Vasilkov S G, Onatsky S P and Elinson M P 1987 Artificial Porous Aggregates and Lightweight Concretes based on Them (Moscow: Stroyizdat)
[6] Bazhenov Yu M, Danilovich I Yu and Vysotskaya O B 1980 Non-fired fly ash gravel is a new effective aggregate for concretes *Construction Mater.* **8** 19-20
[7] Kapustin F L and Ryzhkova I V 2012 Non-fired fly ash gravel is a new effective aggregate for construction concretes *Construction Mater.* **8** 57-60
[8] Kapustin F L and Fomina I V 2014 Generation of a light aggregate based on fly ash of Reftinskaya urban district power plant for structural concretes *Ecology and industry of Russia* **8** 32-34
[9] Kapustin F L and Fomina I V 2013 Generation and use of a non-fired fly ash gravel for structural concretes *Promising Mater.* **10** 76-78
[10] Russian Standard 31108-2016: Cements for General construction Specifications (Moscow: Standardinform)
[11] Russian Standard 6318-77: The technical sodium sulfate. Specifications (Moscow: Standards publishing)
[12] Russian Standard 25818-91: Fly ash from thermal power plants for concretes Specifications (Moscow: Standards publishing)
[13] Russian Standard 9758-2012: The porous inorganic aggregates for construction works Technical Test methods (Moscow: Standards publishing)
[14] Russian Standard 33928-2016: The artificial porous aggregates on the basis of ashes and slags from thermal power plants Specifications (Moscow: Standardinform)
[15] Russian Standard 27006-86: Concretes. Rules of selection of the composition (Moscow: Standardinform)
[16] Vishnya B L, Ufimtsev V M and Kapustin F L 2006 A Promising Technology of Removal, Storage and Use of Ashes and Slags from Thermal Power Plants (Ekaterinburg: USTU-UPI Publishing)
[17] Vishnya B L, Ufimtsev V M and Sirota Yu Ya 1994 Ashes and slags removal and storage: options and prospects for development *Hydraulic Eng.* **11** 24-28
[18] Volzhensky A B, Gladkikh K V and Yudina A M 1970 Non-fired artificial aggregates for lightweight concretes *Construction Mater.* **7** 16-17
[19] Kudryavtsev A A and Romanov Yu M 1983 Concretes based on coarse aggregates made of thermal power plants ash *Concr. Reinf. Concr.* **4** 19-20
[20] Kudryavtsev A A and Murzabekov E R 1986 Long-term strength of concrete based on non-fired fly ash gravel *Concr. Reinf. Concr.* **11** 29-30
[21] Urkhanova L A and Efremenko A S 2011 Structural lightweight concretes based on non-fired porous aggregates vol 1(48) (Irkutsk: Irkutsk state technical University) pp 100-103

IOP Conf. Series: Materials Science and Engineering **481** (2019) 012051 doi:10.1088/1757-899X/481/1/012051

The analysis of technological parameters of precast-monolithic system with permanent formwork walls

S A Serbin[1], P O Dedyukhin[2], N I Fomin[3]

[1]Department of civil engineering, Ural Federal University, 17 Mira Street, Yekaterinburg, 620002, Russia
[2]Department of civil engineering, Ural Federal University, 17 Mira Street, Yekaterinburg, 620002, Russia
[3]Department of civil engineering, Ural Federal University, 17 Mira Street, Yekaterinburg, 620002, Russia

E-mail: Andrey-sergey.serbin@yandex.ru

Abstract. This article presents a brief description of the research, which consists of determine the optimal parameters of the bay of concreting permanent formwork walls with a spatial reinforcing cage for the construction of precast-monolithic reinforced concrete structures in the construction of civil, public and industrial buildings. The results are presented in the form of graphs, by means of them it is possible according to the specified parameters of permanent formwork, such as the step of the spatial reinforcing cage, the thickness of the monolithic core and the total thickness of the formwork, to determine the recommended technological parameters of construction work, which will not only avoid the destruction of permanent formwork in the process of concrete work in the construction of precast-monolithic reinforced concrete structures in the construction of civil, public and industrial buildings, but also to maximize their manufacturability.

1. Introduction

Construction companies using precast-monolithic structures as wall elements have repeatedly faced the problem of destruction of the panels outer shell during concreting.

The aim of the research is to determine the optimal parameters of the bay of concreting permanent formwork walls with spatial reinforcing cage for the construction of precast-monolithic reinforced concrete structures in the construction of civil, public and industrial buildings and development of technological recommendations for the production of concrete works.

When we had studied native and foreign experience on this topic [6–10] it was determined that the calculation part of the research is to determine the maximum values of the concreting speed (m/h), at which the prefabricated elements of the permanent formwork of the walls will be able to accept technological loads from both the longitudinal forces and the bending moment without destruction. To do this, consider the section of the prefabricated part of the wall element of size bxb (size b varies depending on the types and sizes of the selected reinforcing cage [5]. Note that the calculated area is selected in such a way that there is no reinforcing bars (Fig. 1).

IOP Conf. Series: Materials Science and Engineering **481** (2019) 012051 doi:10.1088/1757-899X/481/1/012051

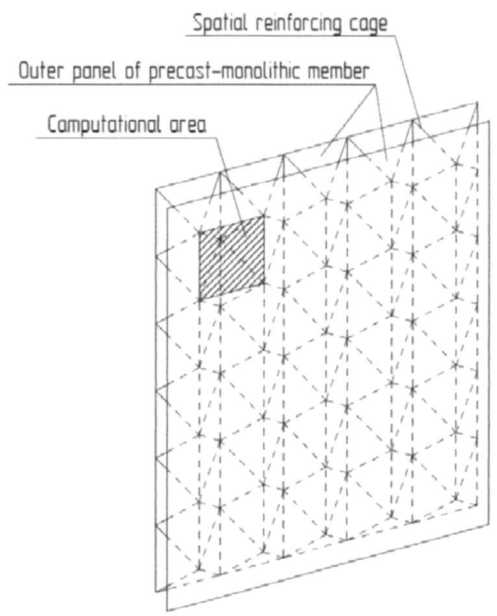

Figure 1. Computational model.

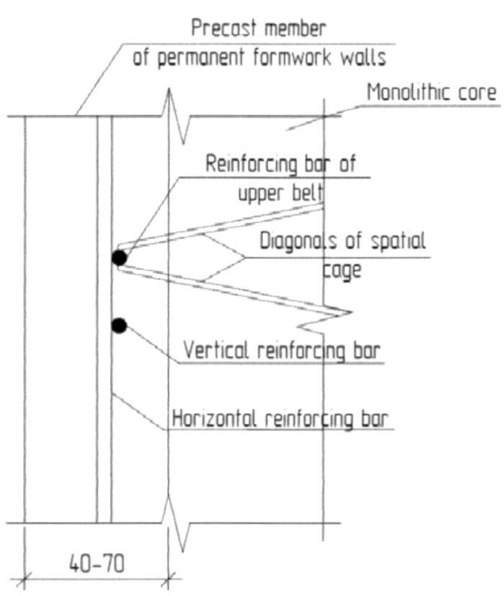

Figure 2. Cross-section of the precast member of permanent formwork with a monolithic core.

2. Calculation part

According to the loads (from the pressure of the concrete mixture and from the effects of wind) acting on the calculated area, we obtain two types of destructive effects:

1. The maximum values of the longitudinal force in the area of the cargo area $A_{ar} = b \cdot b$ (Fig. 1);

2. Bending moment in the aisle and on the bearer of the selected area.

To begin with, we started toanalyze the first condition: the value of the distributed load from external forces should be less than the holding force distributed over the cargo area and resulting from the welded connection of diagonals with the lower and upper belt and the length of the anchoring of the diagonals of the spatial frame into the body of the assembled part of the permanent wall formwork [7]:

$$R_{br} \leq R_h, \tag{1}$$

where R_{br} – breaking pressure on the section of the cargo area $A_{c.a.} = H \cdot B$, kPa; H, B –step of reinforcing cage in the longitudinal and transverse directions; R_h – maximum holding pressure that can be sensed by the cargo area, kPa.

The breaking pressure is determined by the formula:

$$R_{br} = P_{c.m.} + P_w, \tag{2}$$

where $P_{c.m.}$ – maximum lateral pressure of concrete mixture, kPa; P_w –the wind pressure occurring at a height of 80 m (the height is conventionally accepted as the highest point of the element location), kPa.

The maximum lateral pressure of the concrete mixture is determined by the formula for calculating the lateral pressure of the concrete mixture when compacted by vibrators on the ADJ. T [1]:

$$P_{c.m.} = g \cdot (0{,}27 \cdot V + 0{,}78) \cdot K_1 \cdot K_2, \tag{3}$$

where g – volume weight of concrete mix, kg/m^3; V – concrete casting rate (rate of filling of the formwork height), m/h; K_1 – coefficient taking into account the influence of mobility (stiffness) of the concrete mixture; for the worst case, at OK more than 8 cm. K_2 – coefficient taking into account the influence of temperature of concrete mix; $K_2 = 1{,}15$ – for the lowest possible concreting temperature of 5-10°C.

Wind pressure is determined by the formula11.1 [2]:

$$P_w = w_m + w_p, \tag{4}$$

where w_m – average component of the main wind load, kPa; w_p – pulsation component of wind load, kPa.

Determination of the standard value of the average component of the main wind load:

$$w_m = w_0 \cdot k(z_e) \cdot c, \tag{5}$$

where w_0 –standard value of wind pressure, determined by the table. 10.1 [2], kPa; $k(z_e)$ – coefficient taking into account the change in wind pressure for height, determined by table. 11.2 [2], which is taken equal to 1.45 for 80 m; c – the aerodynamic coefficient is determined by ADJ. B.1.1 [2], which is taken equal to 1.8 m for territories of type B.

Determination of the pulsation component of the main wind load:

$$w_p = w_m \cdot \zeta(z_e) \cdot v, \tag{6}$$

where $\zeta(z_e)$ – the coefficient of wind pressure pulsation, taken according to table. 11.4 [2], numerically equal to 0.7 for a height of 80 m; v – the coefficient of spatial correlation of wind pressure pulsations, determined by table. 11.6 [2], which is taken equal to 0.73 for 80 m.

The holding pressure is determined by the formula:

$$R_h = \frac{F_h}{A_{ar}}, \tag{7}$$

where F_h – the total holding force, kN;

The total holding force is determined by the formula:

$$F_h = F_j + F_{an}, \tag{8}$$

IOP Conf. Series: Materials Science and Engineering **481** (2019) 012051 doi:10.1088/1757-899X/481/1/012051

where F_j – strength in shear of a welded joint, determined by tables A.1-A.3 [3], which is takenequal to 6 kN for the lower belt at a diagonal Ø 5 mm; F_{an} – the resistance of the adhesion reinforcing bar with concreteon the anchorage length

The adhesion resistance of the reinforcement with concrete on the anchorage length is determined by the formula:

$$F_{an} = R_{bond} \cdot u_s \cdot l_{an}, \tag{9}$$

where R_{bond} – the design resistance of reinforcing bar adhesion with concrete, taken uniformly distributed along the length of the anchorage, determined by the formula 10.2 [4], kN; u_s – the entire cross-section of anchorage reinforcing bar, m^2; l_{an} –anchorage length of reinforcing bar, m.

$$R_{bond} = \eta_1 \cdot \eta_2 \cdot R_{bt}, \tag{10}$$

where η_1 – coefficient, taking into account the influence of the type of surface reinforcing bar, taken equal to 2.5 for high-rolled reinforcement; η_2 – coefficient, takingintoaccountthe influence of diameter size of reinforcing bar, which is taken equal to 1,0 for $d \le 32$ mm; R_{bt} – the design resistance of concrete to axial stretching, as defined in the table. 6.8 [4], which is taken equal to 1.05 MPa for concrete class B25.

The essence of the second condition is that the bending moment arising from the action of the external load in the aisle or on the bearer was less than the limiting bending moment, which can be perceived by the cross section of the element of the precast member of the permanent wall formwork, p.7.1.12 [4]:

$$M \le M_{ult}, \tag{11}$$

where M – the bending moment from the external loads, kNm; M_{ult} –the ultimate bending moment, kNm.

The value of the limiting bending moment is determined by the formula 7.9 [4]:

$$M_{ult} = R_{bt} \cdot W, \tag{12}$$

where W – resistance moment of the element cross section theoutermost tensile fiber;

The moment of resistance of the cross section of the element for the outermost stretched fiber of rectangular elements is determined by the formula 7.10 [4]:

$$W = \frac{b \cdot h^2}{6}, \tag{13}$$

The bending moment of the external load is determined by the formula 14:

$$M = k_1 \cdot R_{br}, \tag{14}$$

where k_1 –coefficient taking into account the nature of the action of the bending moment (at bearer or in the aisle), determined in accordance with table 14.31 [5], which is taken equal to the aisle – 0,0179 in the bearer – 0,0417.

As a result of mathematical calculations using the presented formulas and values, taking into account the requirements for wall elements and the specification of reinforcing cages presented in [3], the following data were obtained, which are presented in Fig. 3, Fig. 4, Fig. 5, Fig. 6. The values are given for the concrete pump rated capacity of 164 m^3/h, with an increase in its performance requires recalculation.

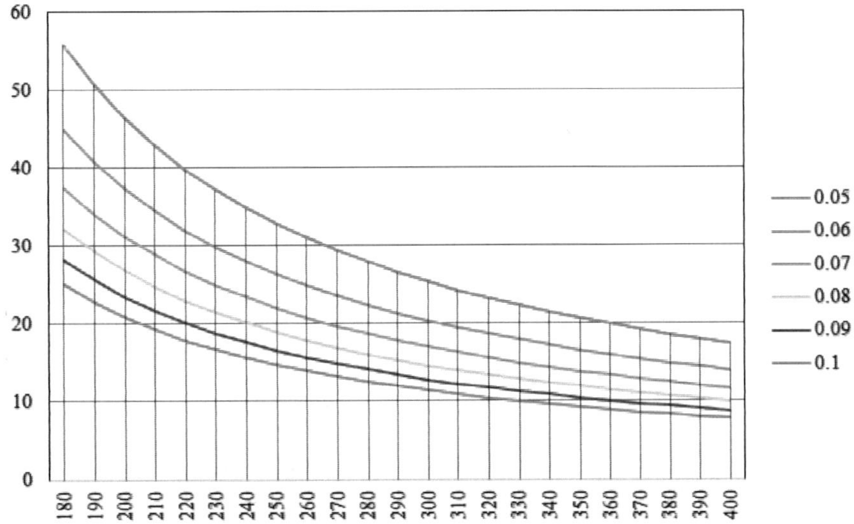

Figure 3. The dependence of the minimum total bay length on the thickness of the wall element with the thickness of the wall precast member h=0.04 m. In the figure on the abscissa axis shows the total thickness of the wall element, mm, and on the ordinate axis–the total length.

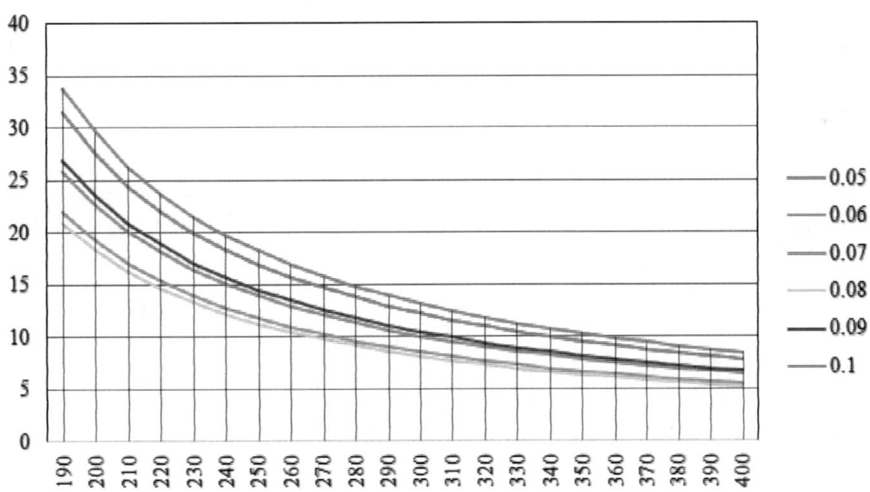

Figure 4. The dependence of the minimum total bay length on the thickness of the wall element with the thickness of the wall precast member h=0.05 m. In the figure on the abscissa axis shows the total thickness of the wall element, mm, and on the ordinate axis–the total length.

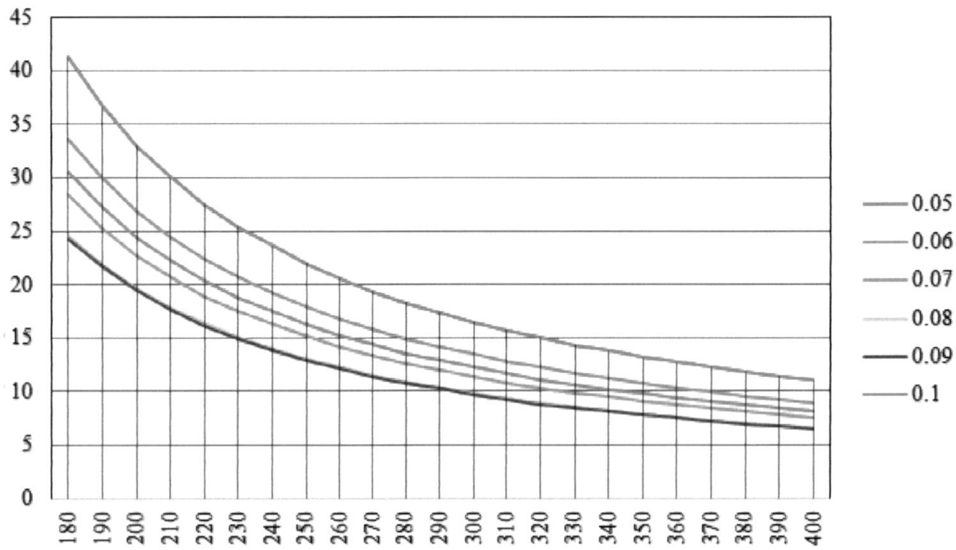

Figure 5. The dependence of the minimum total bay length on the thickness of the wall element with the thickness of the wall precast member h=0.06 m. In the figure on the abscissa axis shows the total thickness of the wall element, mm, and on the ordinate axis–the total length.

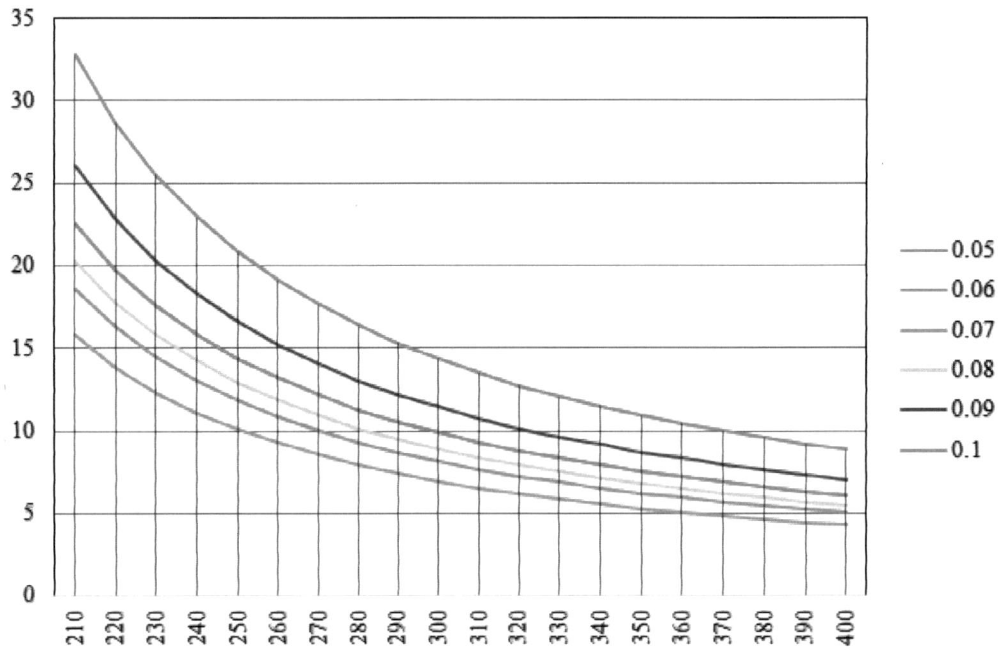

Figure 6. The dependence of the minimum total bay length on the thickness of the wall element with the thickness of the wall precast member h=0.07 m. In the figure on the abscissa axis shows the total thickness of the wall element, mm, and on the ordinate axis–the total length.

To make it more simple algorithm is made on DRAKON language. It shows the actions sequence without any detailed description (Figure 7).

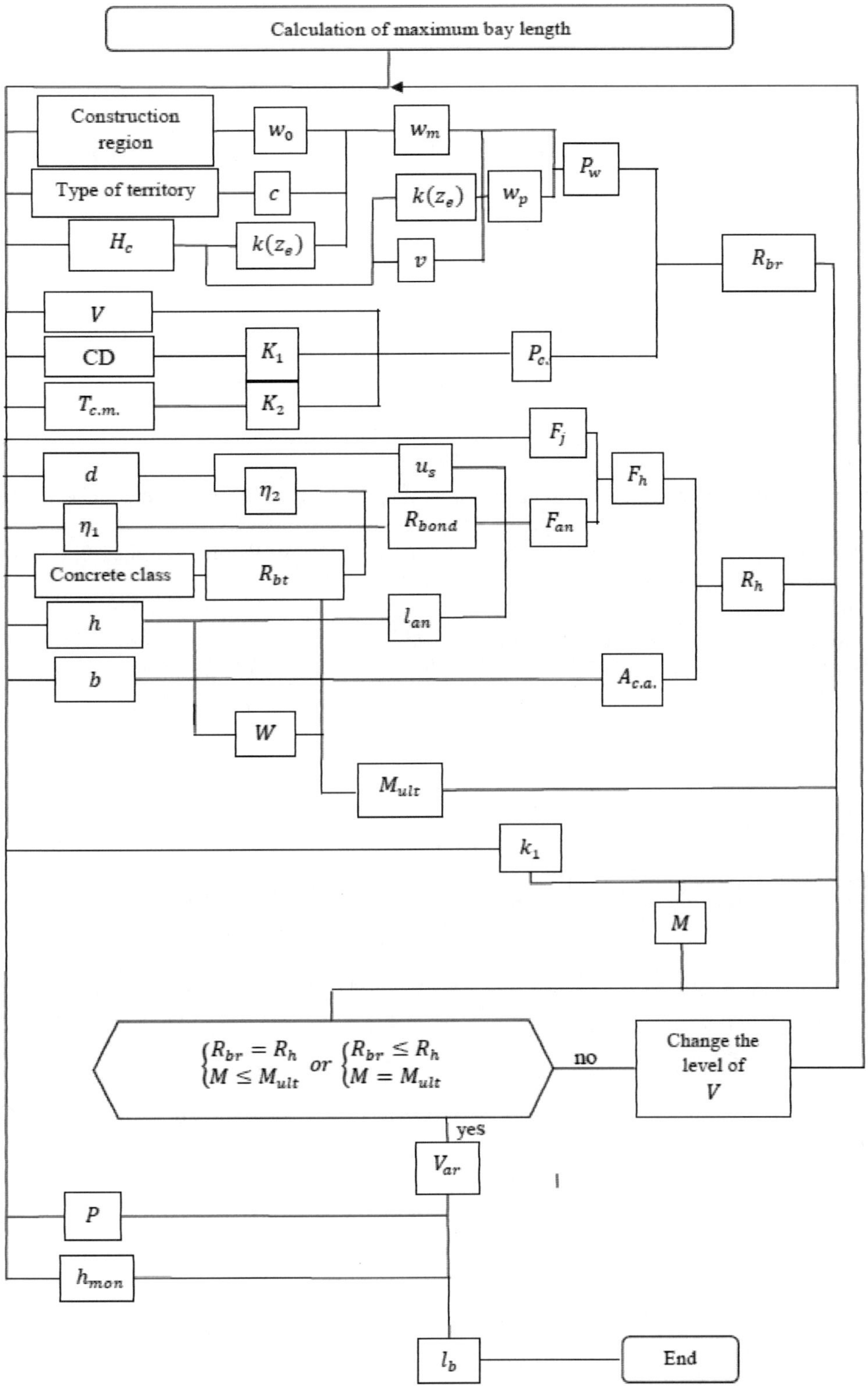

Figure7. Algorithm for calculation ofprecast-monolithic permanent formwork walls.

IOP Conf. Series: Materials Science and Engineering **481** (2019) 012051 doi:10.1088/1757-899X/481/1/012051

3. Conclusion

The graphs show the maximum bay length of the wall elements, concreted to the full height.

Note that the graphs also show the purposefulness of the possible choice of a particular step of the reinforcing cage at a given thickness of the precastmember. It makes sense to mention that the total thickness of the wall element is the sum of the thickness of the outer panels of the prefabricated element of a fixed wall panel, as well as a monolithic reinforced concrete part that fills the empty space between them, which clearly follows from Fig. 1.

References

[1] SR 70.13330.2012 Set of rules Carrier and filler constructions.
[2] SR 20.13330.2016 Set of rules Stress and loads.
[3] OS 2.6.15-2011 Elements of precast concrete walls and slabs with spatial reinforcement cage
[4] SR 63.13330.2012 Set of rules Concrete and reinforced concrete elements
[5] Zoteeva E 2017 Precast-monolithic building system: native construction experience *AlleyaNauki* pp 15-7
[6] Zoteeva E 2017 New solution for permanent walls formwork in underground floor *Molodoy Ucheniy* pp 17-20
[7] Fomin N 2017 The research of the technology of precast-monolithic permanent walls formwork *Vestnik grajdanskih injenerov* pp 131-6
[8] Fomin N and Zoteeva E 2018 Stay-in-place formwork for walls of reinforced concrete *Russian Patent №178930blank №12* 2p
[9] Fomin N, Berngardt K and Shavrin D 2013 Stay-in-place formwork for walls of reinforced concrete *Russian Patent №135671 blank №35* 2p
[10] Fomin N and Shavrin N 2014 Stay-in-place formwork for walls of reinforced concrete *Russian Patent №145947 blank №27* 2p

IOP Conf. Series: Materials Science and Engineering **481** (2019) 012052 doi:10.1088/1757-899X/481/1/012052

New Design and Technological Solutions for Ensuring Technological Reliability of Erecting Pre-Cast with Cast-in-Place Slabs in Civil Buildings

N I Fomin[1], K V Berngardt[1], A V Vorobiyov[1]

[1]Ural Federal University, Institute of Civil Engineering and Architecture, Academic Department of Industrial, Civil Engineering and Real Estate Expertise, 19 ulitsa Mira, Ekaterinburg, 620002, Russia

E-mail: ni.fomin@urfu.ru, k.v.berngardt@urfu.ru, av.vorobyev@urfu.ru

Abstract. The study of erecting cast-in-place and hybrid pre-cast/cast-in-place civil buildings in the Middle Urals revealed the need to develop new design and technological solutions aimed at ensuring high technological reliability of construction processes being implemented, as well as at enhancing service properties of slabs. Description is made of a set of new design and technological solutions developed with the use of a Russian heuristic formalized structural synthesis method called a Theory of Inventive Problem Solving (hereinafter referred to as TIPS).

1. Introduction

The study of how slab cast over pre-cast joists are erected in civil buildings at the construction sites in Ekaterinburg and other cities of the Middle Urals revealed that the technological reliability of three processes performed with the help of conventional methods and standard outfit [1] is insufficient. The existing low level of technological reliability often leads to functional defects, and, as a consequence, to deterioration of service properties of a slab structure. Thus, new design and technological solutions are necessary for ensuring high level of technological reliability of construction procedures. The development of new technical solutions in construction is related to the solution of a conventional inventive problem, and the correct answer to it depends on many factors. A fair number of inventive techniques with varying degrees of effectiveness [2, 3] have been developed to solve such problems, the theory of inventive problem solving (TIPS) being the best of them [4].

2. TIPS, a heuristic synthesis method, as an effective methodological tool for solving technical problems in construction

A review of a number of literature sources on the theory of inventive problem solving [4, 5, 6] showed the adaptability of this heuristic formalized structural synthesis method not only for solving practical problems in many industries, but also for training technical experts in universities around the world.

The method was developed by Genrikh Saulovich Altshuller, a Russian scientist and inventor. The purpose of this method is to solve an inventive problem by means of consistently revealing system contradictions (administrative, technical and physical ones) and eliminating them using laws, patterns, and trends in the development of technical systems. To resolve the said contradictions, G.S. Altshuller and other researchers developed other methods based on the assumption that in most cases the contradictions are resolved by similar methods.

The detailed procedure of solving an inventive problem of practically any level of complexity is presented in the respective algorithm for solving inventive problems (hereinafter referred to as ASIP), the latest version of it consisting of about 85 actions (steps).

The main advantage is that it is the most methodically researched heuristic synthesis method based on the fundamental laws of technical systems' development. The TIPS methodical toolkit allows to solve inventive problems of practically any level. Though the method has all the benefits mentioned above, it also has one major drawback – it is rather complicated for practical use because of high requirements to the inventor's education. The ASIP states that an expert, who uses this method effectively, i.e. who is capable of solving complex inventive problems (those of the 3rd and the 4th levels, according to TIPS classification) should possess a university level of knowledge in physics and in that field of technology, to which the inventive task to be solved is related. Using standardized techniques for eliminating technical contradictions will be enough for solving relatively simple problems. The algorithm of using such techniques is so simple that after a short training even schoolchildren of 5th-9th forms can learn how to use it.

The solutions described below are based on the following ASIP principles: revealing administrative contradictions, revealing technical contradictions. The revealed technical contradictions were eliminated using a standard table containing 40 techniques. No determination of physical contradictions was provided for in the development of the solutions described.

3. Design and technological solutions for the arrangement of cast-in-place part of a slab cast over pre-cast joists

3.1. Element for linking the reinforcement bars

The reinforcement bars perpendicular to one another need to be linked in order to arrange the cast-in-place part of a floor slab without welding. The most common way of linking such bars is to tie a place of connection with a thin wire with the help of hand-guided tools. This rather painstaking technique contains a set of technological operations and is not always convenient to use, especially for the lower row of bars in the floor slab. Bar benders often use this technique at the construction site, and the connections thus made are faulty, and thinner than it is required by the design. This leads to the reduction in stiffness of reinforcement cage and, as a consequence, to the reduction of load-bearing capacity of the cast-in-place part of a slab. We used the recommendations listed in [7, 8] to make the patent search of well-known elements for linking the reinforcement bars, and found one constructive solution patented in China [9]. Representing three wire hooks connected with one another, this solution has sufficient technological effectiveness but does not provide for reliable fixing of reinforcement bars due to insufficient length of those parts of the hooks that touch the surface of bars to be connected. We have developed a new deign solution of a linking element to achieve technological and reliable connection.

The element for linking reinforcement bars, perpendicular to one another, is made of spring steel wire and consists of three hooks. Two lower hooks (the second and the third ones) lie in mutually parallel planes and are symmetrical with respect to the upper hook (the first one) located in the plane formed by arc-shaped curves. To link the reinforcement bars, the upper bar is passed through the upper hook, while the lower hooks are put on on the lower bar. Thus, the process of linking the bars involves just two simple technological operations. The shape of the plane, where the upper hook is situated, allows to not only increase the length of those parts of the first hook that touch the upper bar, but also to increase the length of those parts of the lower hooks that are in contact with the upper part of the lower bar. Figure 1 below presents the prototype and the developed design solution of the element in schematic drawings of parts of the hooks touching the bars.

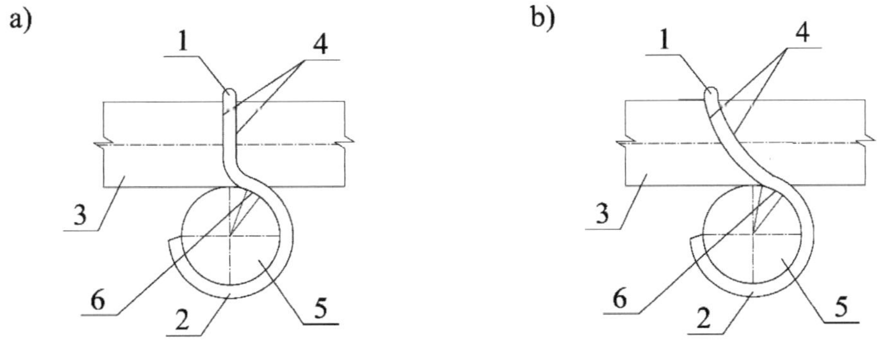

Figure 1. Schematic drawing of parts of the element's hooks for linking the reinforcement bars: a) for the prototype (patent CN No 102691381); b) for the developed solution; 1 – the first hook; 2 – the second hook (the third hook); 3 – upper bar; 4 – fixing length of the upper bar; 5 – lower bar; 6 – fixing length of the upper part of the lower bar.

An application was made on the basis of the developed solution, and a Russian Federation utility model patent RU No 180555 was obtained.

3.2. Method of shaping the edges of a slab's cast-in-place part and a device for its implementation

We also revealed another process with low technological reliability – a process of shaping the edges of a slab's cast-in-place part.

Nowadays the technology of shaping the edges of the slab's cast-in-place part implies the installation of standard end beams on the standard or non-removable formwork of a slab, these elements being fastened against displacement during concrete casting with AV standard brackets (for PERI formwork) or universal fireproof concrete shut-off valves (for DOKA formwork) etc. End beams and brackets are dismantled after concrete casting is completed. The described method implies shaping the inner surface of multiple-use removable end beams by the edge of the cast-in-place part of the slab. This method has low technological reliability because the installation of an end beam is highly painstaking, and because the dismantling has to be performed with high technical discipline, (which requires high quality of works). Our study has shown that both the quality of edge surface of the slab's cast-in-place part, and its service properties largely depend on the wear and tear of end beams, and the weaker is the technical discipline during their dismantling, the more intensive is their wear and tear. Besides, in this method end beams do not allow to obtain a decorative surface of the edge without carrying out additional site works.

We have developed a new method of shaping the edge of the slab's cast-in-place part, and a design element for carrying out this process, which not only ensures the increased technological reliability of shaping the edge, but also allows to obtain the decorative surface of it without any additional works performed at the construction site.

The method involves the following sequence of technological operations. In accordance with the regulations, provision is made for the arrangement of the inverted-U-shaped reinforcement bars (stirrups) on the end sections of the cast-in-place slab located along its edge. These bars are fastened to the main reinforcement (reinforcement cage) of the slab. Next, tying rods oriented along the slab's edges are fastened to the vertical part of the stirrups. Before that the tying rods are passed through the loop outlets of the developed design element (Fig. 2a). The factory-made pre-cast design elements (Fig. 2b, Fig. 2c) are installed along the slab edge on its removable or non-removable formwork. The concrete casting of the slab's cast-in-place part is carried out after installing the said design elements.

The design element for implementing the proposed technique is made in the form of a pre-cast thin-walled reinforced concrete panel and contains loop outlets located on the inner side of it. The inner side of the panel is rough for its adhesion with the concrete of the slab's cast-in-place part, whereas its external side is smooth.

IOP Conf. Series: Materials Science and Engineering **481** (2019) 012052 doi:10.1088/1757-899X/481/1/012052

In order to obtain the decorative edge surface of the slab's cast-in-place part, the external side of the panel can be made with brickwork rustication or in the colours of the façade.

In accordance with the recommendations listed in [10], Figure 2 presents the graphic explanation of the developed method in isometric views.

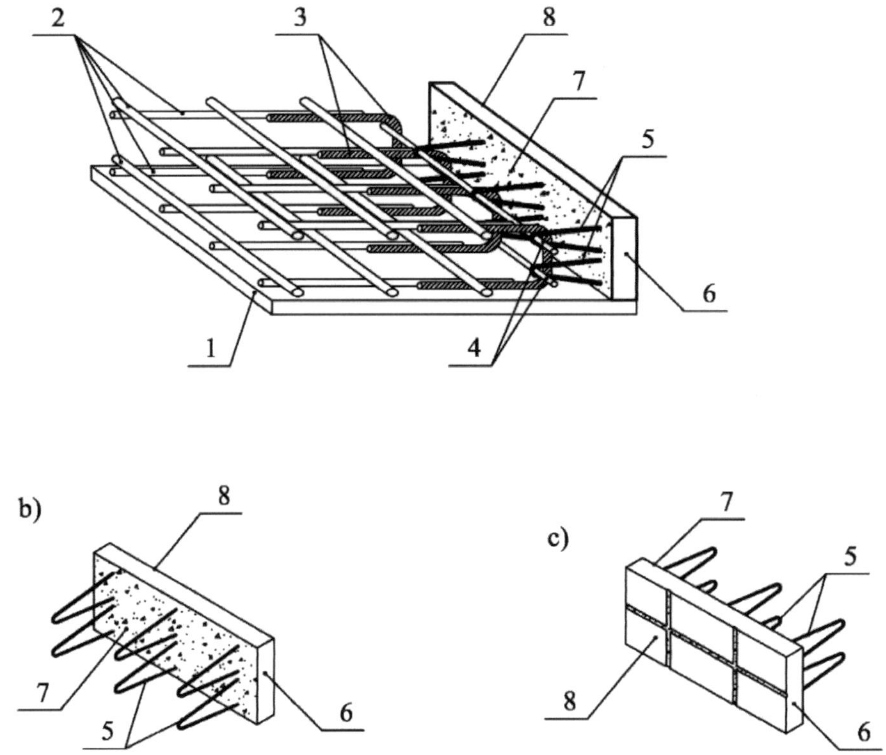

Figure 2. Method of shaping the edges of a slab's cast-in-place part: a) slab fragment before concrete placing; b) design element (inside view); c) design element (outside decorative view); 1 – formwork of a slab's cast-in-place part (removable or non-removable); 2 – reinforcement bars of the end sections of a slab's cast-in-place part; 3 – inverted U-shaped stirrups; 4 – tying rods; 5 – loop outlets of the design element; 7 – inner side of a design element panel; 8 – external side of a design element panel.

An application was made on the basis of the developed solution, and a Russian Federation patent on invention RU No 2658687 was obtained.

4. Technological solution for assembling the pre-cast part of a slab cast over pre-cast joists

The method used for assembling bents for the arrangement of slab cast over pre-cast joists of the building implies the installation of the bent on the mounting tables attached to the built-up columns, and on the intermediate supports in the form of telescopic jack columns of the collapsible forms of the slab. The bent should be subject to height and horizontal alignment by means of assembly tools and then installed on the beam of the slab. In case pre-cast with cast-in-place bents are present in the slab, the cast-in-place part of the bent should also be arranged.

It was established that this method for installing the bent has low technological reliability due to the fact that: height alignment of the bent is very painstaking when using telescopic jack columns of the standard formwork of the slab, particularly in case of uneven or inclined base, on which the jack columns are resting;horizontal alignment of the bent is very painstaking, too, and there is no opportunity for technological and reliable horizontal fixation of the bent when the slabs are installed thereon, and the cast-in-place part of the bent is being arranged.

The latter is a particularly important prerequisite for high-quality erection in case of using a pre-cast bent with the reinforcement bars being pre-stressed and the bent is subject to one-sided loading with pre-cast slabs.

We have developed a new way of erecting a pre-cast part of the bent and an erection device for it. Their application will allow to improve the technological reliability of erection.

The method involves the installation of the bent on mounting tables fastened to the columns. At least two erection devices are installed vertically before installing the bent, and screw-jacks are used to align their vertical position together with the height position of their U-shaped trays, so that the bent should rest on the lower part of these U-shaped trays while being in the design position. After installing the bent on mounting tables and fixtures, it shall be aligned horizontally using the screws located on the U-shaped trays, and fixed until the completion of subsequent works related to installation of slabs on the bent and arrangement of the cast-in-place part thereof.

An erection device has been developed in order to implement the proposed method (Fig. 3). The device consists of two columns linked with one another, an U-shaped tray with side parts equipped with screws able to move horizontally being rigidly fastened to the upper part of the columns. The fixing plate is rigidly fastened to each screw, and the lower part of each column is equipped with a screw jack and a supporting plate.

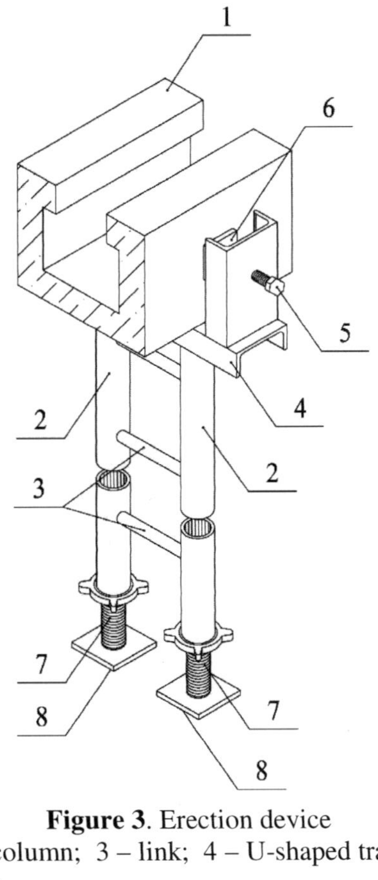

Figure 3. Erection device
1 – pre-cast part of the bent; 2 – column; 3 – link; 4 – U-shaped tray; 5 – screw; 6 – fixing plate; 7 – screw-jack; 8 – supporting plate.

An application for a patent of Russian Federation No. 2018127991, which is currently undergoing expert review, had been drawn up on the basis of the developed solution.

5. Conclusion

The existing level of technological reliability of the erection of slab cast over pre-cast joists does not provide for the required service properties of load-bearing structures of the buildings. The use of modern inventive techniques, in particular, of the theory of inventive problem solving, allows to obtain new patentable design and technological solutions, which, if applied, improve technological reliability of construction operations and reduce the possibility of construction defects. The following things have been developed according to the results of the study: an element for linking the reinforcement bars; a method for shaping the edges of the slab's cast-in-place part and the device for carrying it out; a method for erecting a pre-cast part of the bent and an erection device therefor. The novelty and the industrial applicability of the developed solutions have been confirmed with the Russian patents.

References

[1] Peurifoy R and Oberlender G 2010 *Formwork for Concrete Structures* (New York: McGraw-Hill Education) 544 p ISBN 978-0-07-163917-0
[2] Arciszewski T 2016 *Inventive Engineering: Knowledge and Skills for Creative Engineers* (Boca Raton: CRS Press) 417 p. ISBN 978-1-498-71124-1
[3] Chakrabarti A 2002 *Engineering Design Synthesis: Understanding, Approaches and Tools* (London: Springer) 330 p ISBN 978-1-849-96876-8
[4] Chechurin L 2016 *Research and Practice of the Theory of Inventive Problem Solving (TRIZ): Linking Creativity, Engineering and Innovation* (New York: Springer) 281 p ISBN 978-3-319-31782-3
[5] Hua Z, Yang J and Coulibaly S (2006) Integration TRIZ with Problem-solving Tools: A Literature Review from 1995 to 2006 1 (1-2) *International Journal of Business Innovation and Research* pp 111 – 128
[6] Gadd K 2011 *TRIZ for Engineers: Enabling Inventive Problem Solving* (Chichester: John Wiley & Sons) 504 p ISBN 978-0-470-74188-7
[7] Kennedy J and Watkins W 2012 *How to Invent and Protect Your Invention: A Guide to Patents for Scientists and Engineers* (New Jersey: John Willey & Sons) 226 p ISBN 978-1-118-36937-1
[8] Hitchcock D 2013 *Patent Searching Made Easy: How to do Patent Searches Online and in the Library* (Berkeley: NOLO) 272 p ISBN 978-1-413-31872-2
[9] Yu Jin, Chen Ronglin, Cai Yanyan Patent CN No 102691 A IPC E04C 5/16, 5/18, 5/20 Simple fixing device for staggered tube rod pieces
[10] Pressman D 2015 *How to Make Patent Drawings* (Berkeley: NOLO) 256 p ISBN 978-1-413-32156-2

IOP Conf. Series: Materials Science and Engineering **481** (2019) 012053 doi:10.1088/1757-899X/481/1/012053

Structure of a lesson on the development of the students' creative personality in the CFCT-TRIZ pedagogical system

Vyacheslav Kim[1], Marina Kim[2], Michael Kim[3]

[1]Voronezh State Technical University, Voronezh, Russia

[2] Voronezh State Technical University, Voronezh, Russia

[3]Sber-tech, Moscow, Russia

e-mail: vyachhkim@yandex.ru

Abstract. The article describes the theory and practice of introducing the system of creative education CFCT-TRIZ into the Voronezh State Technical University.The system was developed by Professor, Doctor of pedagogical sciences Zinovkina M.M. We give an example of teaching the substantive part of the course for the students of the construction specialty. The article will be of interest to the specialists working in the field of creative pedagogy.

Key words: CFCT-TRIZ, creative pedagogy, creativity, contradiction, invention.

1. Introduction

Modern educational standards make high demands on graduates of higher educational institutions. They should have a whole range of competencies that help them improve and develop their intellectual potential throughout their lives, find non-standard solutions to emerging problems, adapt to new situations, analyze their capabilities and gain new experience, work in a scientific team, generate new ideas, and so on. The analysis of the stated requirements and the pedagogical situation in the education showed that the best result can be given by a multilevel continuous creative education. The system of the continuous formation of creative thinking (CFCT-TRIZ), developed under the supervision of Professor M. Zinovkina, is aimed at achieving this goal [1, 2]. This innovative pedagogical system covers all stages of the formation and development of a person from childhood to old age. As professors of higher education, we are more interested in the strategy of the continuous formation of creative engineering thinking (CFCET) in higher education. CFCET proposes to conduct an integrated cycle of courses on the methodology of technical engineering creativity in the university. The basis of such courses is TRIZ (in English the name is typically rendered as "the theory of inventive problem solving", or TIPS), created by the Soviet scientist and inventor G.S. Altshuller [3, 4, 5, 6, 7] and developed by his students [8, 9, 10, 11].The theory of Altshuller has received worldwide recognition and is widely used as a tool for solving creative problems[12, 13, 14,1 5, 16, 17, 18].

2. The structure of a creative lesson on the system of CFCT-TRIZ

The implementation of innovative pedagogical technologies in the multi-level system of CFCT-TRIZ means changing the structure of lessons and their original content.

IOP Conf. Series: Materials Science and Engineering **481** (2019) 012053 doi:10.1088/1757-899X/481/1/012053

Let us consider how this system is implemented in a particular lesson at the Construction Faculty of VSTU [19, 20, 21]. As a basis for the development of the methods of creative lessons with first-year students, we took the structure of a double creative lesson proposed by M.M. Zinovkina [2].

Block 1 (*motivation*) represents a specially selected system of original objects that can cause surprise to a student. This block ensures the student's motivation for the lessons and develops their curiosity. In this block, as a rule, the students were invited to guess from the photo or video of unusual objects what they were meant for.

Blocks 2 and **6** (*content part*) include the program material of the educational course and provide the formation of systemic thinking and the development of creative abilities.

Block 3 (*psychological unloading*) represents a system of tasks for psychological relief. For this purpose, the students were shown some exerciseson the harmonization of cerebral hemispheres development.

Block 4 (*brain-teaser*) represents a system of increasingly complicated brain-teasers, embodied in real objects, in the design of which an original, witty idea is realized. For this, the Tangram game was used, which allows one to assemble complex plot constructions from simple geometric figures, moving from relatively simple to more and more complex.

Block 5 (*intellectual warm-up*) represents a system of increasingly complex tasks aimed at developing the motivation, logical thinking and creative abilities of students. In this block, the students were offered to solve logical problems, allowing them to develop their ability to reflect.

Block 7 (*summary*) provides feedback from the students about the lesson and provides a qualitative and emotional assessment of the lesson by the students. Then some parts of the lesson were corrected, based on the results of the feedback.

3. Example of teaching the content part of the course for students of VSTU

Let us consider the methodology of teaching how to work with methods of resolving contradictions using intellectual tools for the development of creative thinking by examining the creative task.

The creative task: it is necessary to build a bridge with the span of more than 1 km across the strait so that ocean liners can pass under it.

3.1. Let us analyze the creative task

In order to build a stationary bridge over a water obstacle, the design of a beam bridge is usually used as a classical solution. The main bearing elements of a beam bridge are beams or girders. They work in bending and transmit the load to the piers installed on the base of the bridge. The beams and girders of a bridge are assembled into larger parts called spans. Each span is supported by piers. Spans can be split and continuous, and also cantilever. Each split span rests on two piers at the edges, and continuous spans can be supported by three or more piers. If the span of a bridge is small, then you can use the construction of a movable bridge. However, there is no such design of a movable or beam bridge, so that ocean liners can pass under it. What should one do?

To solve the creative problem, we will apply methods of activating creative thinking [3, 4]. The thing that lies at the base of every creative inventive task is a contradiction. The intensification and deepening of the contradiction is the main way to solve the problem: from the contradiction lying on the surface (administrative contradiction) to the deep contradiction present inside the system (technical contradiction) and,in the end, the ultimate development of the contradiction (physical contradiction), when some part of the technical system is immediately in two mutually exclusive states.

Let us define an administrative contradiction. An **administrative contradiction** is a contradiction between a need and a possibility of its satisfaction. It is often set by some administration or a customer and is formulated in the form: "It is necessary to do this and that, and how – it is not known."

Let us formulate the administrative contradiction in this problem: it is necessary to build a bridge with the span more than 1 km over the strait so that ocean liners can pass under it, but no one knows how to do this.

Let us now try to deepen the administrative contradiction in order to find its causes and move on to the technical contradiction and the model of the problem.

A technical contradiction is a contradiction between certain parts, qualities or parameters of a system. A technical contradiction arises with the improvement of some parts of the system due to unac ceptable deterioration of the others. It represents the cause of the administrative contradiction and deepens it.

As a rule, improving some characteristics of an object, we sharply worsen the others. Usually, we have to find a compromise, that is, to sacrifice something.

A technical contradiction arises as a result of a disproportion in the development of different parts or parameters of a system. With significant quantitative changes in one part of the system and a sharp "lagging behind" of its other parts, situations arise when quantitative changes of one of the system's sides conflict with the others. The solution of such a contradiction often requires a qualitative change in the technical system itself. This is the law of the transition of quantitative changes to qualitative changes. Generating a creative problem model, we can facilitate the formulation of a technical contradiction.

A **creative problem model** is a mental, conventional scheme of the problem, reflecting the structure of the conflict in a system. One of the elements of a **conflicting pair** is the main object of consideration, and it is called a **product** or an object, and the second element is a **tool**. A **product** is an element of the technical system thatneeds to be processed (manufactured, moved, change, improve) according to the problem requirements. A **tool** is an element that directly interacts with the product.

In this problem, the conflicting pair is the span length of the bridge and the bridge pier. **The product** here is the span length of the bridge, and **the tool** is the bridge pier. The longer the span of the bridge, the more piers are required, which makes it complicated for ocean liners to passunder the bridge.

Let us aggravate the conflict and formulate the creative problem model as the technical contradiction: if the length of the bridge span will be enough to let ocean liners pass under the bridge, the problem is solved, but the bearing capacity of the span without bridge piersis not provided, which does not allow one to create such a span.

3.2. Analysis of the creative problem model

Let us define an **operational zone**, that is, a space within which the conflict specified in the problem model arises. In our case, the operational zoneis the span of the bridge with piers.

Let us now define **substance-field resources** of the system, the environment, and the product in question. To do this, we will perform a su-field analysis of the problem.

As is known [3, 4, 5], to improve the efficiency of technical systems, their structure should be specific. A model of such a structure is called a **su-field model**.

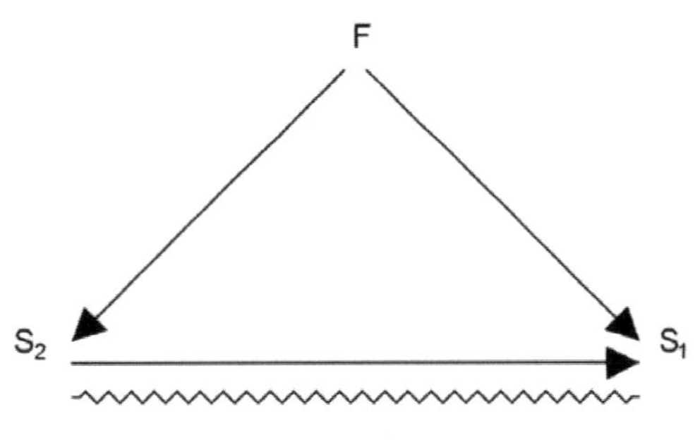

Figure 1. Schematic representation of the su-field model.

A **su-field model** is a minimally controlled technical system consisting of two interacting objects and the energy of their interaction. The interacting objects are conditionally called **substances** and are denoted $\mathbf{S_1}$ and $\mathbf{S_2}$, and the interaction energy is called a **field** is denoted **F**.

The term "**Su-field**" is derived from the words "Substance" and "Field".Typically, $\mathbf{S_1}$ is a product, $\mathbf{S_2}$ is a tool that "processes" the product $\mathbf{S_1}$, and **F** is a field. Let us construct the su-field model: we have $\mathbf{S_1}$ – the length of the span (product), $\mathbf{S_2}$ – the bridge pier (the tool "processing" the product $\mathbf{S_1}$), and **F** is the gravitational field. The su-field model is represented by the following scheme (Fig. 1).

Here the wavy line denotes a harmful interaction, which, according to the conditions of the creative task, must be eliminated.

3.3. Definition of an ideal final result and the physical contradiction

An ideal technical system is a system that does not exist, but its functions are fulfilled, i.e. the goals are achieved without the means. **An ideal final result** is an ideal to which one should strive for to solve a problem. The proximity of the solution to the ideal one determines the level and quality of the solution.

Let us formulate the ideal final result for our case: the span of the bridge by its own allows itself to be longer than 1 km without loss of bearing capacity and without the installation of intermediate bridge piers.

The analysis of the formulation of the ideal final result allows us to proceed to the formulation of a physical contradiction.

A physical contradiction is the presentation of diametrically opposite demands to the properties in a certain part of the technical system, that is, some part of it has to be in two mutually exclusive states.

The stage of revealing a physical contradiction represents an exact post-setting of the problem.

Let us formulate the physical contradiction for our creative problem: the span of the bridge has to be longer than 1 km without loss of bearing capacity and without the installation of intermediate bridge piers, and should not be longer than 1 km without loss of bearing capacity and without the installation of intermediate bridge piers.

The solution of the creative problem consists in resolving the physical contradiction.

Let us use the typical methods of eliminating physical contradictions created by Altshuller G.S. and collected in the information fund [3, 4, 5]. The information fund includes standards for solving inventive problems, technological effects, techniques for eliminating contradictions, ways of applying nature resources and technologies.

Firstly, we use the rule for the destruction of a su-field model, which says that we should destroy unnecessary or harmful su-field model by introducing extraneous substance $\mathbf{S_3}$ (Fig. 2).

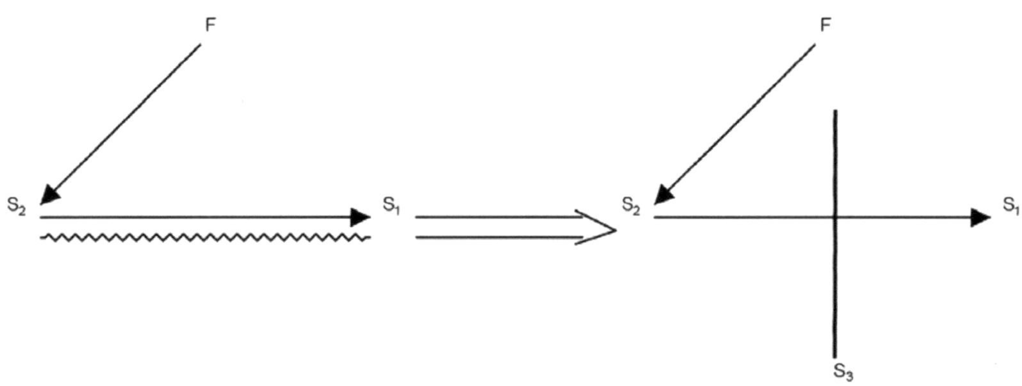

Figure 2. Elimination of harmful connections between the substances in the su-field model

IOP Conf. Series: Materials Science and Engineering **481** (2019) 012053 doi:10.1088/1757-899X/481/1/012053

Secondly, we use a certain **X**-element, which takes on the functions of the element $\mathbf{S_3}$.

Thirdly, we use the Altshuller table and analyze what needs to be changed by the requirements of the problem and what impermissibly worsens. The analysis of the table makes it possible to use the following set of typical methods for resolving technical contradictions [3, 4, 5]: the division principle, the taking out principle, the anti-weight (balance preserving) principle, the inversion (upside down) principle, the dynamics principle, the "blessing in disguise" (convert harm into benefit) principle, the intermediate means or "fitting" principle, theself-servicing principle.

The logic of solving creative problems according to TRIZ sharply narrows the field of searches and purposefully leads to the solution of the problem. Basing on the analysis of the methods above, we cometothe solution in the form of a hanging or suspension (cable truss) bridge design.

Hanging bridges, or suspension bridges, are installed where it is impossible or impractical to install bridges with intermediate piers. Bearing elements of such bridges are pylons and connecting cables. Pylons are installed on opposite banks or slopes, and between them, the main cables are stretched, which extend to the ground and are fixed there. Vertical cables, chains or beams supporting the canvas of the bridge are attached to them.

Cable truss bridges are a kind of suspension bridges. The difference is that the cables connect the canvas of the bridge directly to the pylons or to one, central pylon, and not to the horizontal cables. This construction makes the bridge more rigid and stable in comparison with the classic suspension bridges.

The construction of one of the world's largest cable truss bridges was completed in the Far East of Russia in the spring of 2012. The new bridge goesover the Bosporus East Passage and connects the mainland with the Russky Island. This is the first bridge of such dimensions and design in the territory of Russia. It can rightly be called a unique achievement of Russian engineers. It has the longest cable-stay span in the world (1104 m), the longest suspension cables (580 m), its pylons reach the height of 320 m. The total length of the structure is 3100 m, and the height of the main canvas is 70 m above the water, which allows even the most cumbersome ocean liners to pass under it (Fig. 3).

Figure 3. General view of the bridge to the island Russky
Source: website https://bigpicture.ru

References

[1] Zinovkina M M 2002 Multilevel continuous creative education *Collection of Sci. Papers of MSIU* (Moscow: Moscow State Industrial University Press) p 310

[2] Zinovkina M M 2008 *CFCT-TRIZ: the Creative Education of the 21st Century. Theory and Practice* (Moscow: Moscow State Industrial University Press) p 306
[3] Altshuller G 1979 *Creativity as an Exact Science. Theory of Inventive Problem Solving* (Moscow: Sovetskoe Radio) p 184
[4] Altshuller G and Selyutsky A B 1980 *Wings for Icarus: How to Solve Inventive Problems* (Petrozavodsk: Kareliya) p 224
[5] Altshuller G 2003 *To Find the Idea. Introduction to the Theory of Inventive Problem Solving* (Petrozavodsk: Skandinaviya) p 186
[6] Altshuller G 1996 *And Suddenly the Inventor Appeared: TRIZ, the Theory of Inventive Problem Solving* (Technical Innovation Center, Inc.) p 171
[7] Altshuller G 1999 *The Innovation Algorithm: TRIZ, Systematic Innovation, and Technicalcreativity* (Technical Innovation Center, Inc.) p 312
[8] Petrov V and Zlotina E 2002 *Structural substance-field analysis* (Tel Aviv) p 74
[9] Litvin S 2013 Types of psychological inertia. Further development of su-field analysis. Development of inventive thinking *Collection of Sci. Papers. TRIZ Developers Summit Library* **5** 261
[10] Orloff M A 2006 *Inventive Thinking through TRIZ* (Springer Berlin-Heidelberg) p 352
[11] Orloff M A 2016 *Abc-triz: Introduction to Creative Design Thinking with Modern TRIZ Modeling* (Springer) p 516
[12] Cohen Y H and Reich Y 2016 *Biomimetic Design Method for Innovation and Sustainability* (New York: Springer) p 258
[13] Fey V and Rivin E 2005 *Innovation on Demand: New Product Development Using TRIZ* (Cambridge University Press) p 256
[14] Haines-Gadd L 2016 *TRIZ for Dummies* (John Wiley & Sons) p 292
[15] Harrington J H 2017 *Lean TRIZ. How to Dramatically Reduce Product-Development Costs with this Innovative Problem-Solving Tool* (CRC Press) p 233
[16] Kim S-K Chai K-H Tan K-C Siy L A B and Shie A-J 2012 *Innovative Service Design Workbook. Five Step Process and TRIZ Tools for Service Designers* (Bookboon) p 87
[17] Rantanen K et al. 2017 *Simplified TRIZ: New Problem-Solving Applications for Technical and Business Professionals* (Productivity Press) p 304
[18] Silverstein D DeCarlo N and Slocum M 2007 *Insourcing Innovation: How to Achieve Competitive Excellence Using TRIZ* (Auerbach Publications) p 304
[19] Kim V 2013 The structure of a creative lecture-practical lesson on the development of the creative personality of undergraduates in the pedagogical system CFCT-TRIZ. *Concept: Sci.-Methodological Electronic J.* **5** 37-41
[20] Kim M 2017 The structure of a creative lesson on the development to the creative personality of the students of constructional speciality using the pedagogical system CFCT-TRIZ *Concept: Sci.-Methodological Electronic J.* **16**
[21] Kim M and Kim V 2017 The application of methods of scientific creativity in the teaching of students of construction specialties *XVII Int. Conf. on Sci. and Tech. Russia – Korea – CIS* (Yuzhno-Sakhalinsk) pp 99-106.

IOP Conf. Series: Materials Science and Engineering **481** (2019) 012054 doi:10.1088/1757-899X/481/1/012054

Training Business Simulation Game «Commercial Property Management in a Competitive Environment»

Salamatina A S[1] Andronova A A[1] Alekseeva I E[1]

[1]Department of Construction Engineering and Materials Science, Civil Engineering Faculty, Perm National Research Polytechnic University, Komsomolsky av., 29, Perm, 614990, Russian Federation

E-mail: salamatina@cems.pstu.ru, andronova@cems.pstu.ru, alekseeva@cems.pstu.ru

Abstract: The article describes the training business simulation game "Commercial property management in a competitive environment" and a special software module. The suggested business simulation game and the created computer simulator are used for training the students, who study in the field "Expertise and management of real estate property". Several examples of game scenarios and results are given to demonstrate their application in the training and educational process. The game simulator allows students to obtain the following knowledge: the commercial property management task is cognitively complex; the optimal allocation of costs for the development and promotion of commercial property depends on the objective function, which can be profit or index of profitability; for shopping centers, the optimization problem has a sustainable solution to the opponent's actions, but such solutions are unstable to the behavior of customers and their preferences; in the case where students select the minimum costs of development and promotion of shopping centers, the profit becomes negative, it explains the necessity of development and promotion of commercial property; in another case – students select the maximum cost, the effect of an "excessive utility" arises. The main result of the game-like training is that students acquire new decision-making skills in the tasks of commercial property management in a competitive environment.

1. Introduction

Nowadays, simulation modelling is widely used in the research of various phenomena and processes in construction industry. It is due to the fact that mathematical modelling is, perhaps, the only method of research because of the high cost and duration of natural experiments. Information about simulation modelling in construction industry is presented in the literature review [1]. Simulation modelling can be used for following tasks: Inventory management [2], Strategy [3], Project management [4, 5] and etc.

Another method of modelling is game simulation in the form of business games. In these games, employees of organizations or invited experts carry out certain roles, corresponding to their operational activities. In relation to construction industry, there are works [6, 7] using the simulation gaming technique. The aim of such games is developing of effective strategies for interaction between project participants, the design and construction processes of buildings and structures. In addition, this type of games is often used as an interactive method of training and education, as well as for the certification of existing employees, testing the knowledge and skills of applicants for any positions. Different scenarios can be created in a business game. All scenarios can reflect to some extent the changes in the real world. However, despite numerous works in the field of mathematical and game modelling in the construction, the task of existing buildings and structures management were practically not considered, with the exception of few works, for example [8].

It is important that life cycle of constructed buildings is measured by decades and in some cases they are hundred years, so the actual problem is the effective management of real estate objects. A particular task of this problem is teaching and training of students and new specialists. This research, which deals with the case of commercial property management in competition conditions on an example of trade centers, is devoted to this particular task.

The task of managing shopping centers has a high degree of uncertainty that is connected with the difficulty of predicting the results of managerial and entrepreneurial activity of the management agents, increasing the number of heterogeneous economic agents and carrying out their entrepreneurial activities autonomously [8].

A key source of uncertainty is the human factor and it is required the research of the strategic behavior of the management agents of shopping centers in competition conditions.

2. Training software module

For teaching and training it is offered to use software module based on MS Excel, which is shown in the Figure 1 created by authors of work [9]. The mathematical model used in this software module is described in detail in works [9] and [10].

In the suggested business simulation game we have two competitive shopping centers (SC) ($j \in J$, $J = \{1, 2\}$), each of them has three pedestrian-transport zones and their intersection forms ten sectors (K). The task of SC management is formulated as a conditional optimization problem, where profit (1) or index of profitability (2) can be used as an objective function (OF). Constrained optimization problem have inequality constraints to the costs of SC development and promotion: a set of acceptable values $c_y \in C_y$, $y \in Y$, where Y – set of managed parameters.

$$OF = P_j(c_{jy}; c_{-jy}) = \mu \cdot Ar \cdot n_j\left(c_{jy}; c_{-jy}\right) - \sum_{y\in Y} c_{jy} - TFC_j, \tag{1}$$

$$OF = RR_j\left(c_{jy}; c_{-jy}\right) = \frac{\mu \cdot Ar \cdot n_j\left(c_{jy}; c_{-jy}\right)}{\sum_{y\in Y} c_{jy} + TFC_j} - 1, \tag{2}$$

TFC_j – total fixed costs of the the j-th SC Manager; μ – customer conversion ratio (the percentage of visitors who purchase something); Ar – the average amount of purchases that visitors make; c_{jy}– costs of managing j-th SC for the development of managed parameters of the object; n_j – the number of attracted visitors in the j-th SC. The last one is determined by the equation:

$$n_j = \sum_{k=1}^{10} N_k \cdot \frac{\prod_{y\in Y} Q_j(c_y)^{q_y} \cdot \prod_{ny\in NY} Q_j(c_{ny})^{q_{ny}} \cdot T_k^{-\lambda_k}}{\sum_{j=1,2}\prod_{y\in Y} Q_j(c_y)^{q_y} \cdot \prod_{ny\in NY} Q_j(c_{ny})^{q_{ny}} \cdot T_k^{-\lambda_k}}, \tag{3}$$

$Q_j(c_y)$ – quality of managed parameters of object j; $Q_j(c_{ny})$ – quality of unmanaged parameters of object j; T_k – correspondence time to j-th SC of visitors to k-th sector; λ_k – parameter that affects the importance of correspondence time for visitors; N_k – number of inhabitants in sector k, $k=\{1,\ldots,10\}$; ny – parameters not controlled at the tactical and operational level; q_y, q_{ny} – power parameters. Since the number of visitors is constantly ($n_1 + n_2$ = const), and by changing the quality of SC, managers actually pull visitors from each other.

Students have role of the competitive shopping center Manager. The purpose of every player is to maximize personal profit or index of profitability in managing the costs of developing their SC. It is worth noting that the optimal allocation of costs depends on the objective function, and may be different.

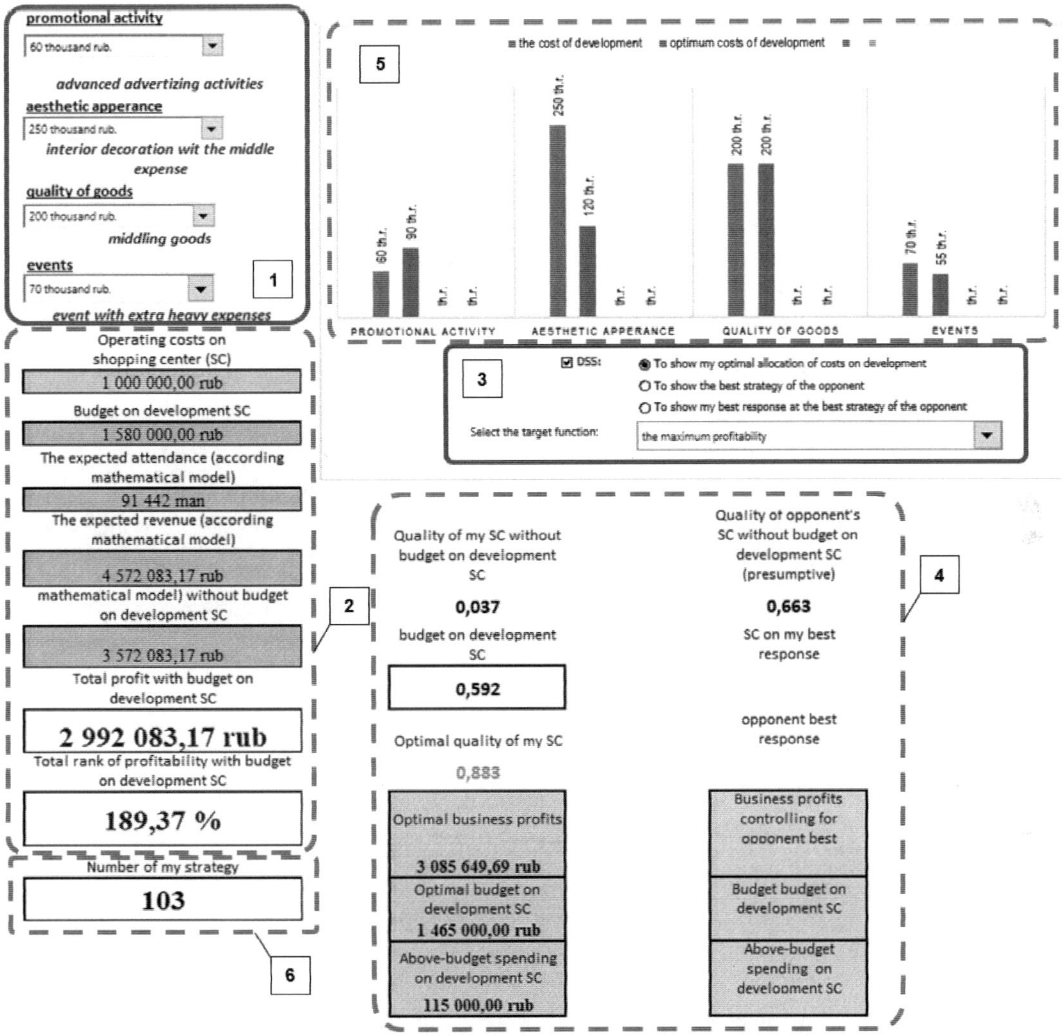

Figure 1. The display form of the player – student in role of Manager SC.

Note for Figure 1:

– changeable parameters

– automatically calculated parameters

1	– options costs of management the four criteria realized in the form of the dropping-out lists
2	– numerical data expressing the quantitative parameters of SC
3	– strategy performance dashboard
4	– quantitative characteristics of indicators of quality of SC
5	– histogram of the current allocation of costs on advance of the player and also the realized Decision Support System (DSS) with a possibility of the choice of criterion function for player (profit or profitability)
6	– automatically defined number of strategy for player at the current allocation of costs on SC development

IOP Conf. Series: Materials Science and Engineering **481** (2019) 012054 doi:10.1088/1757-899X/481/1/012054

The rules of the business game provide for the possibility of scenario modeling, during which the player searches for the optimal solution (or optimal solution in his opinion, this case is called acceptable pseudo-optimal [11] solution) on the allocation of costs for the management of the SC (see Figure 7, area 1). When searching, the changes of the resulting parameters are analyzed (see Figure 7, areas 2 and 4).

When a player makes a decision on the allocation of costs for the SC development, he can use the Decision Support System (DSS) (see Figure 7, area 3, "DSS" checkbox) to compare player's solution with the optimal one. During the training session, the teacher is carried out so that the DSS is turned on by the player only after the decision he made.

When connecting the DSS, players can use the next options by choosing relevant ratio button.

«To show my optimal allocation of costs on development» (it is calculated without taking into account the actions of the opponent, with the settings of the game as a competitor); in this case, the optimal allocation is described by the following equation:

$$c^*_{1m} = c^*_{1m}\Big|_{Q_{2_0}} = \arg\max_{c_{1m}\in C_m}\left(OF\left(c_{1m} \Bigg| \prod_{m\in M} Q_2(C_m)^{q_m} \cdot Q_{o2} = Q_{2_0} \right)\right), \quad (4)$$

Q_{2_0} – this is the quality of the opponent's SC given by the game conditions.

«To show the best strategy of the opponent» (is determined under the assumption that the strategy of the player will correspond to the optimal allocation, see the previous strategy; this mode is designed to see what the quality of opponent' SC is beneficial to him); the best strategy of the opponent is described as follows:

$$c^*_{2m} = \arg\max_{c_{2m}\in C_m}\left(OF\left(c_{2m} \Big| c_{1m} = c^*_{1m}\Big|_{Q_{2_0}} \right)\right). \quad (5)$$

«To show my best response at the best strategy of the opponent» (determined by the assumption that the opponent action according to his best answer; this mode is implemented as double best response); the double best response is described as:

$$DBR_1 = c^*_{1m}\Big|_{c^*_{2m}} = \arg\max_{c_{1m}\in C_m}\left(OF\left(c_{1m} \Big| c_{2m} = c^*_{2m} \right)\right). \quad (6)$$

3. Scenarios of the business game

To get the best result when conducting business games it is necessary to conduct a preparatory stage. As a preparatory stage, you should use familiarization with the rules and goals of the game, the objective functions and the functional dependence of the values of the objective functions from the decisions made by a player. Also, to reduce the time of familiarization with conditions of the simulation game for participants, it is necessary to present clear and understandable guidance to it.

Information can be presented in the form of an introductory course of lectures, methodological material, as well as necessary literature.

In the preparatory stage it is necessary to inform:

- basic concepts and laws of functioning of the commercial property market;
- distinctive features of the commercial property market;
- basic provisions of management theory used in the management of commercial property;
- a shopping and entertainment complex as an object of commercial property;
- factors, affecting on consumer demand and supply;
- termsof profit, profitability and optimality.

As a result of conducting of business game, the participant should gain knowledge of methodical bases, master the skill of the search of the optimal strategy of behavior in competition conditions in the commercial property management.

Consider the basic scenarios of a business game in the process of conducting a training lesson with using a software simulator. The teacher sets the task for all students to apply the minimum cost strategy in all parameters. As a result, all students see that the shopping center is receiving losses (Figure 2).

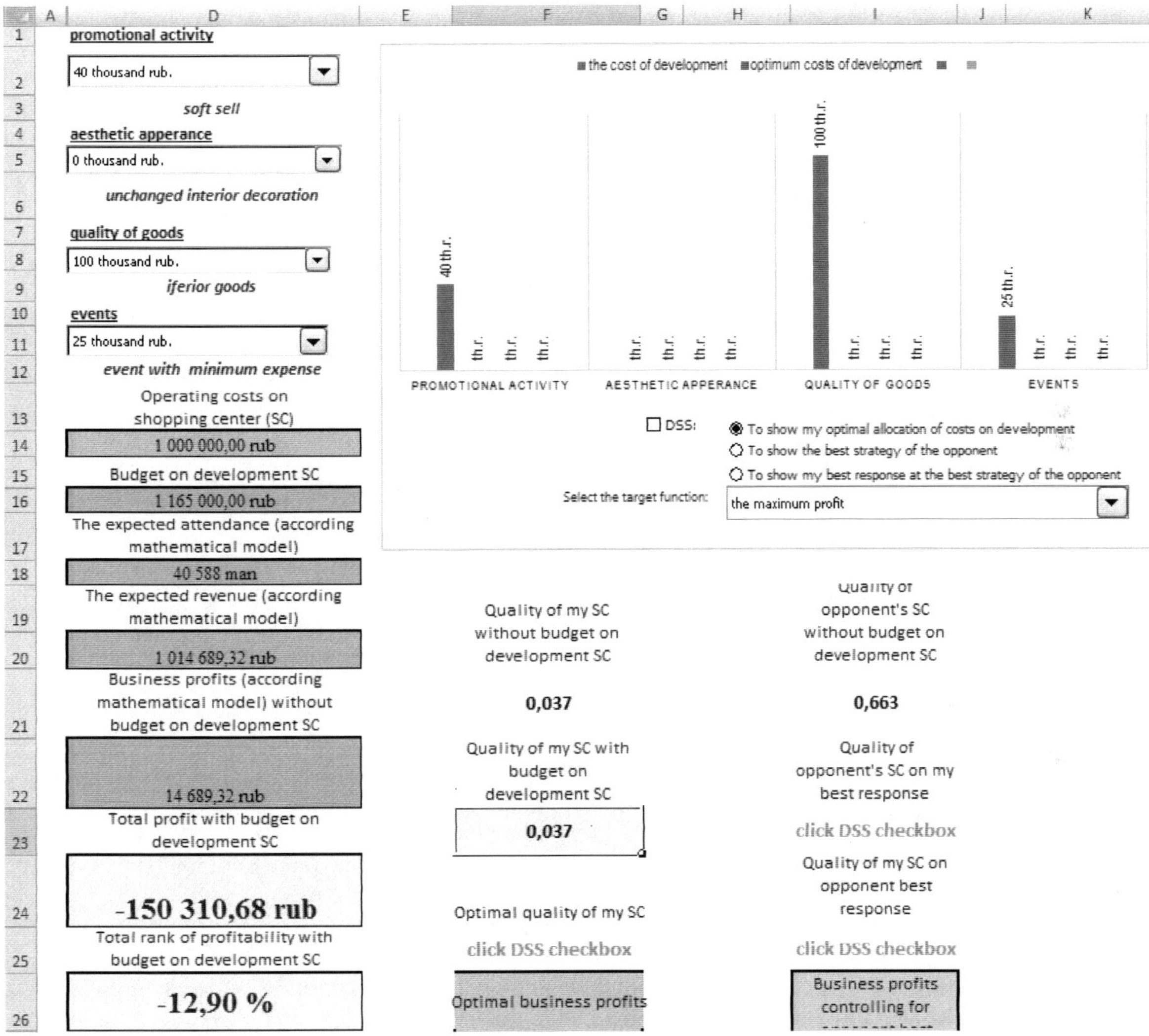

Figure 2. The display form of the player with data at minimal costs.

In the next stage of the business game, the teacher sets the task for students to apply the maximum spending strategy for all parameters. As a result, all students see that the shopping center is making a profit (Figure 3).

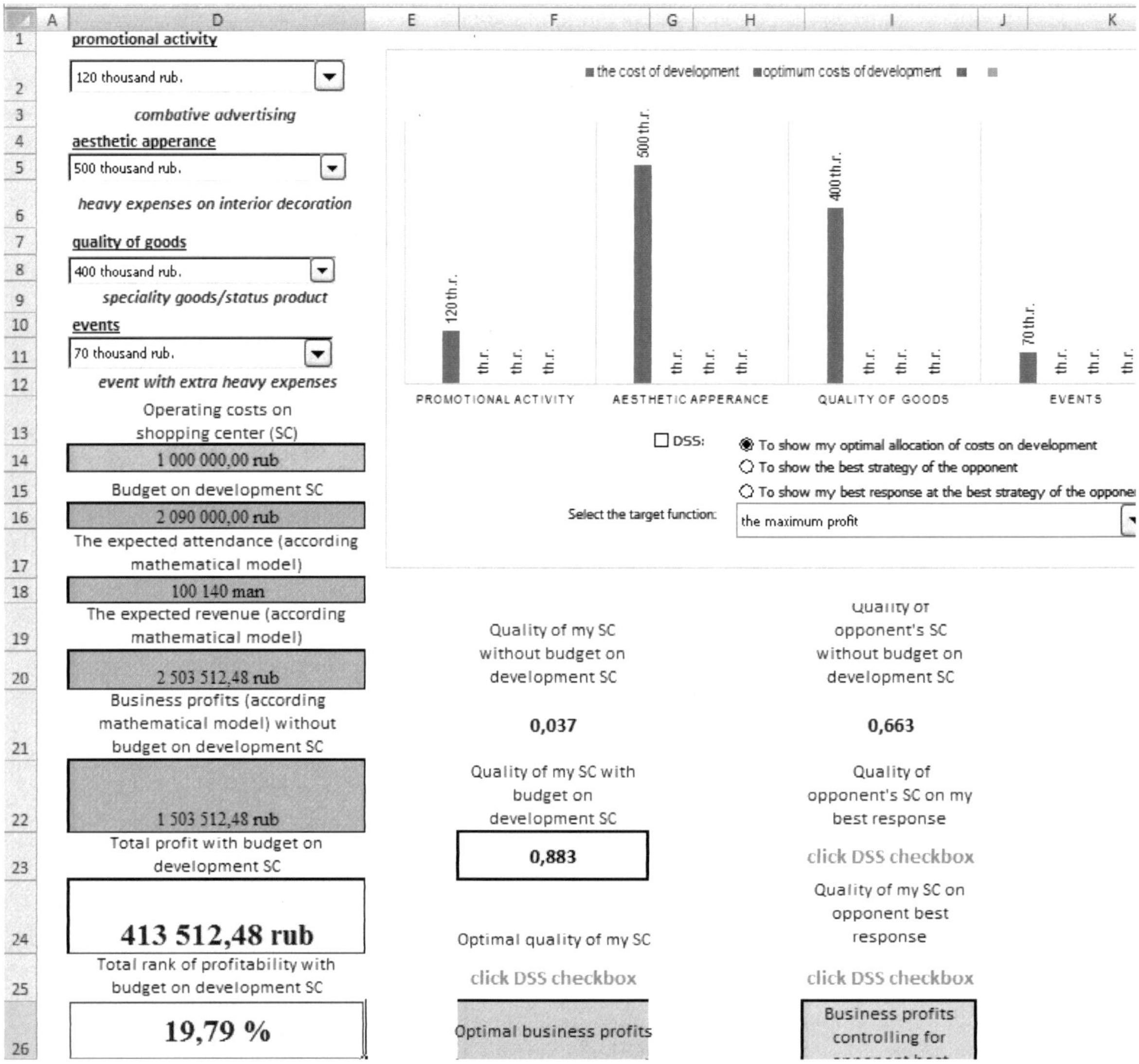

Figure 3. The display form of the player with data at maximum cost.

However, students have to answer following questions: Is this the maximum profit? Is it possible to incur less management costs without losing profit? To get a clear answer to this question, the teacher suggests that students to change the quality expenses and specify instead of 400 thousand – 200. As a result, students see that when saving money for management and development, we even get additional profit (Figure 4).

This phenomenon is called "excessive utility", and then the teacher can elaborate on this concept and phenomenon, for example, give examples from the real practice of managing commercial property.

Further, the teacher offers students to try other combinations of possible controls on their own and gives time to find the best strategy.

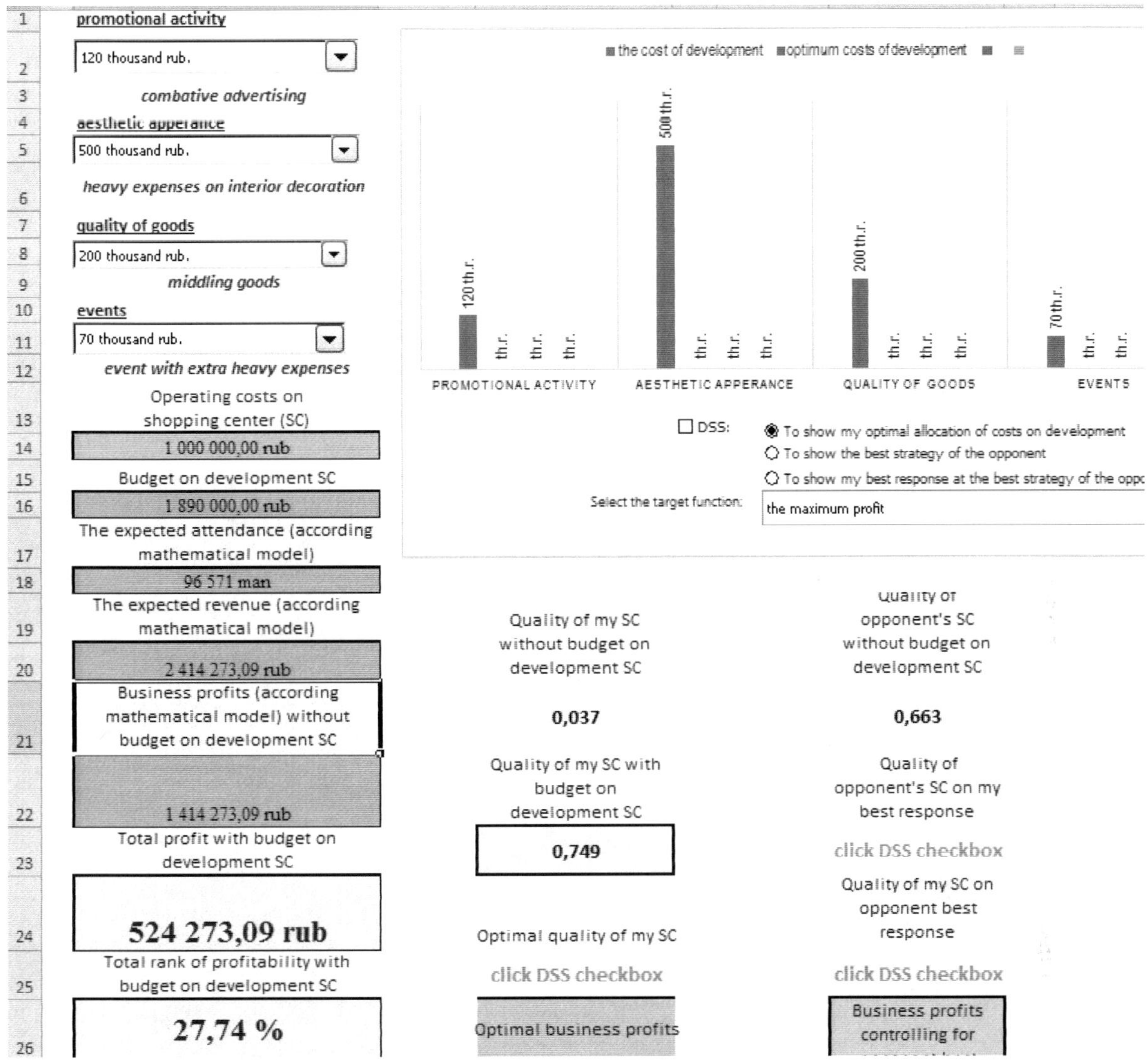

Figure 4. The display form of the player after adjusting costs.

After the students have coped, the teacher allows include the decision support system and compare the found strategy with the optimal one.

In order to clarify the meaning of the concept of optimality, the teacher talks about the task of optimization, the concept of the objective function, and draws attention to the fact that in the game played by students, profit is used as the objective function. The teacher proposes to change the objective function and see if the optimal strategy changes. To do this, it is sufficient to enable the DSS mode (Figure 5).

In order to show that the result of the management company's activities depends on environmental factors, the teacher asks all students to change the value of the two parameters in the game settings: average check and conversion rate. The average check at the beginning of the game was 500 rubles and was changed to 1000 rubles. The conversion ratio was 0.05 and became 0.10.

It turns out that the optimal strategy depends on the external environment (Figure 6).

Using the computer simulator, the teacher can show how the result of the management company's activities depends on the actions of competitors. To do this, you can change the perceived quality of the competitor (Figure 7), specifically, to assume that the competitor will maximize his quality to unity. In this case, students will see immediately that their profit has decreased (compare Figures 6 and 7).

For the same purpose, students can be asked to exchange messages about what quality they get with their optimal strategy. And together try to find the optimal strategy.

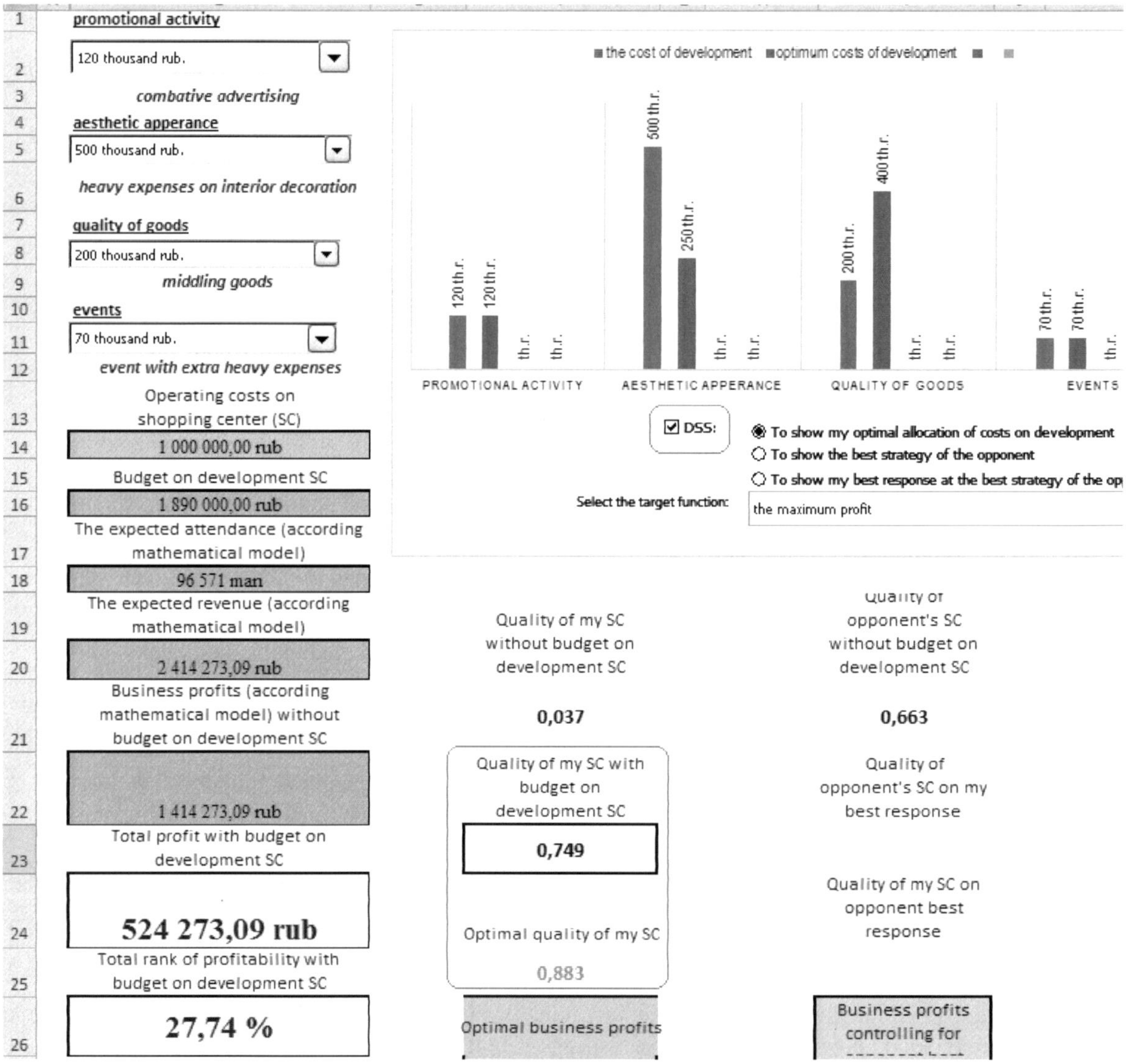

Figure 5. The display form of the player with turn on DSS checkbox.

All these actions must be performed with the DSS mode off. After making several moves, students can make certain of their profit decreases when the competitor develops his object, but their optimal strategy does not change. Thus, students get acquainted with the term of a sustainable solution, to which the teacher should emphasize and explain that strategies in the simulator are resistant to the actions of opponents, but are not resistant to changes in the level of consumption.

Business game and computer simulator of the subject area allows demonstrate the following effects:

1. The task of optimization is cognitively complex, because in the simple case – when the management strategy of the shopping center depends on four controlled parameters and, in this case, the number of strategies of each manager is 256, with 5 controlled factors, the number of strategies increases to 1024, with 6 to 4096 and etc.

2. The task of optimization depends on the objective function, in which quality profit or index of profitability can be used.

3. In spite of the fact that the optimization problem is cognitively complex, students can quickly come to a sustainable solution (stable to the changing situation, depending on the opponent's behavior).

4. In the case when the students select the maximum values of all four controlled parameters, the effect of "excessive utility" arises. In this case, there is an overspending for promotion.

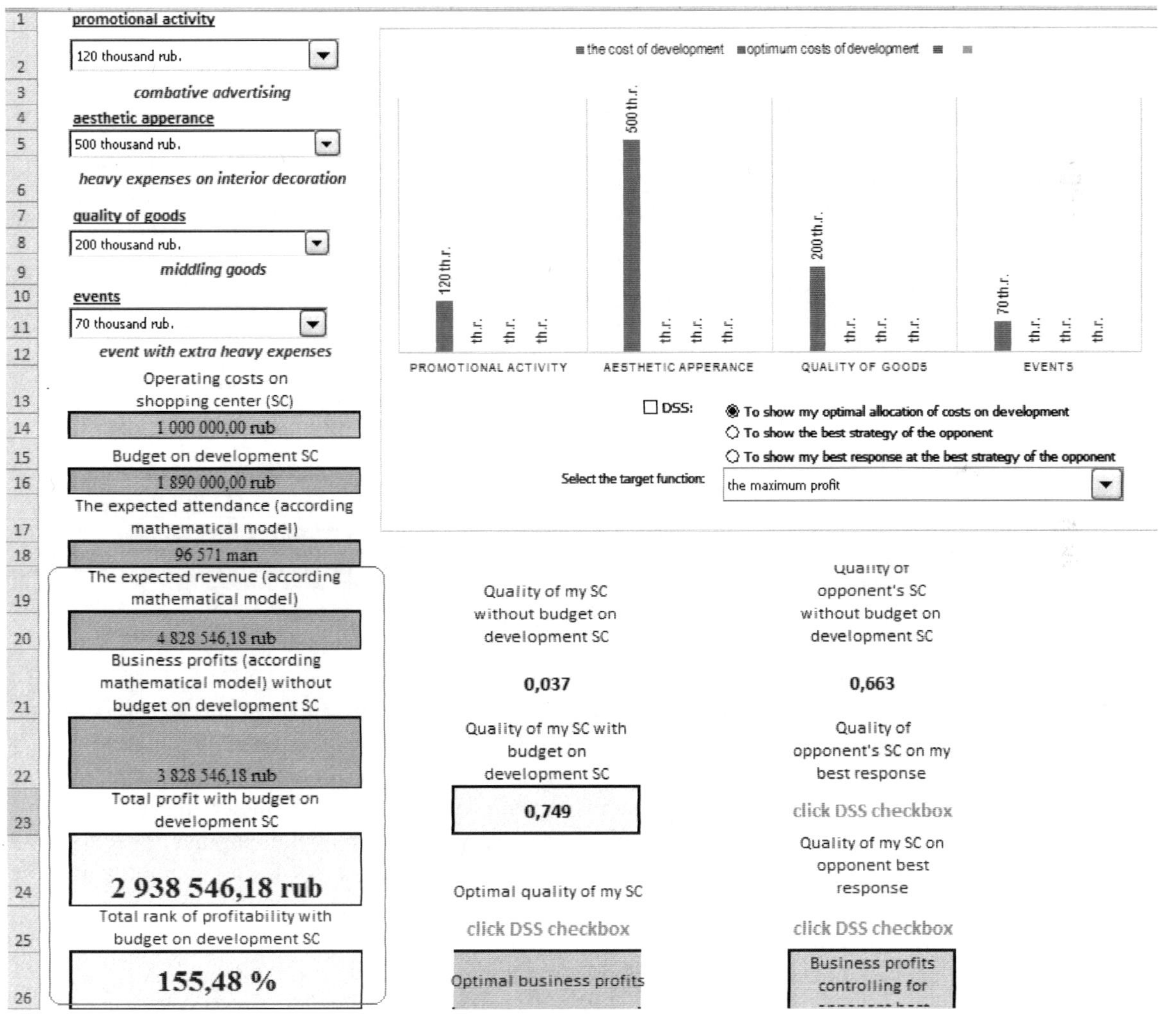

Figure 6. The display form of the player with changed game settings.

5. In the case where the students select the minimum values of all four controlled parameters, the profit becomes negative, which explains the need for the development and promotion of shopping and entertainment complexes.

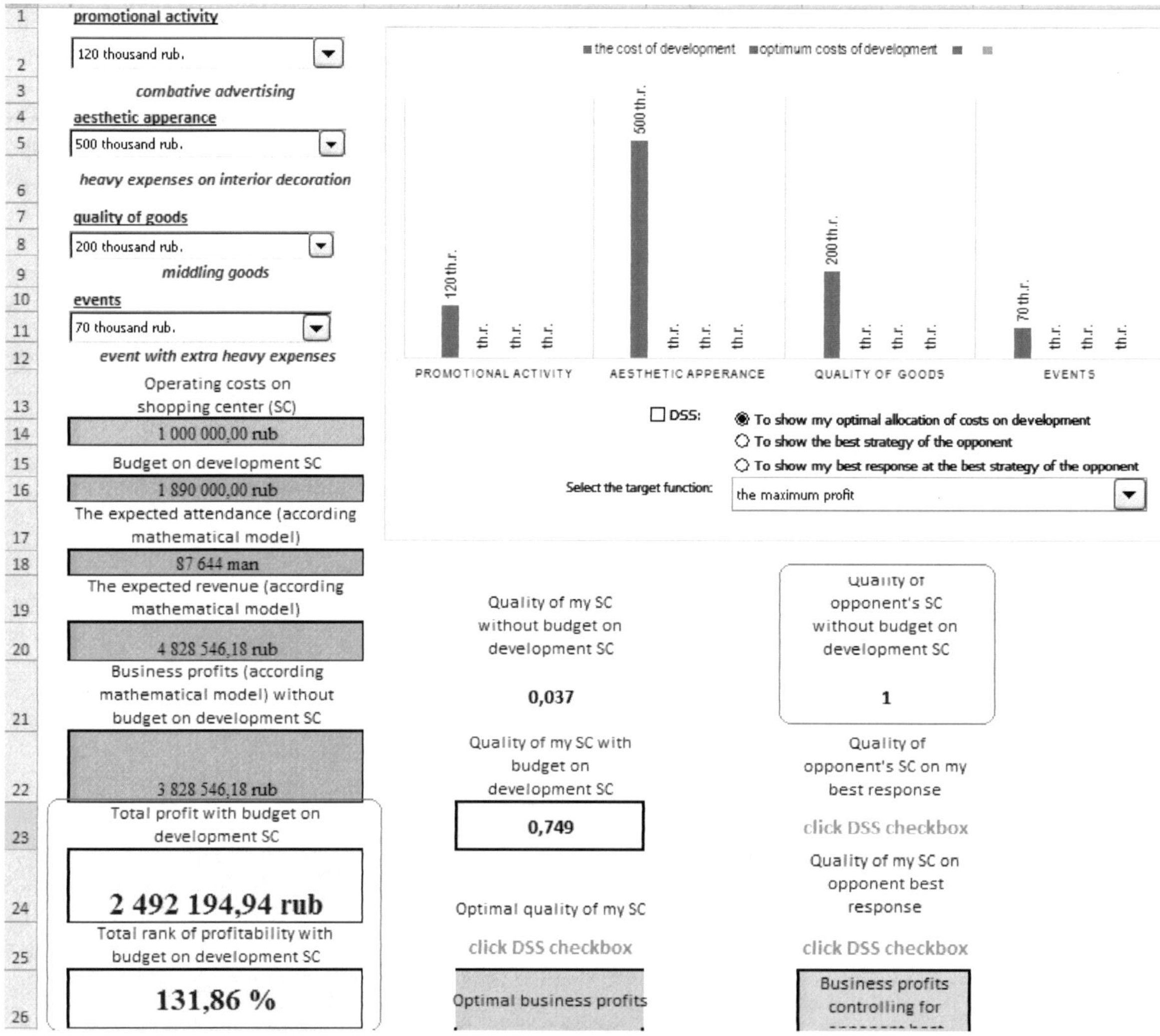

Figure 7. The display form of the player in case of changing quality of opponent's SC.

6. It is necessary to develop those parameters that have a greater impact on profit, that is, they are important for the consumer (the aesthetic appearance and quality of the products).

The business game shows that in the performance of the normative function, that is, the determination of the necessary actions, only quantitative methods can give an answer to the question - how much it is necessary to spend resources on the effective promotion of the commercial real estate object. Qualitative methods can give an answer to the question of what to spend, but cannot determine the size of the required resources.

References

[1] Jahangirian M, Eldabi T, Naseer A, Stergioulas L K and Young T 2010 Simulation in manufacturing and business: A review *European Journal of Operational Research* **203(1)** pp 1-13

[2] Polat G and Arditi D 2005 The JIT materials management system in developing countries *Construction Management and Economics* **23(7)** pp 697–712

[3] Lee S H, Pena-Mora F and Park M 2006 Dynamic planning and control methodology for strategic and operational construction project management *Automation in Construction* **15(1)** pp 84–97

[4] Love P E D, Manual P and Li H 1999 *Construction Management and Economics* Determining the causal structure of rework influences in construction **17(4)** pp 505–517

[5] Sawhney A and Mund A 1999 Hierarchical and modular modeling of structural steel erection process using Petri nets *Civil Engineering and Environmental Systems* **17(1)** pp 63–88

[6] Nassar K 2002 Simulation gaming in construction: ER the equipment replacement game *Journal of Construction Education* **7(1)** pp 16–30

[7] Zulch G and Fischer J 2003 The benefits of using a market share model in a simulation aided planning game *Production Planning and Control* **14(2)** pp 146–154

[8] Spirina V S 2017 Proof of the necessity of counting strategic behaviour of tenants at the management of commercial real estate *Proceedings of Universities. Investment. Construction. Real estate* **7(3)** pp 42–52

[9] Alekseev A O, Spirina V S and Korgin N A 2016 Commercial Real Estate Management Technology Taking Into Account Consumer Preferences *Large-Scale Systems Control* **62** pp 124-168

[10] Alekseev A O, Spirina V S and Andronova A A 2018 Theoretical and experimental study of the strategic behavior of the managers of the of shopping centers *Applied Mathematics and Control Sciences* **1** pp 91-107

[11] Burkov V N, Goubko M, Korgin N and Novikov D 2015 *Introduction to Theory of Control in Organizations* (Boca Raton: CRC Press) 346 p

The Assortment Approach to the Selection of Building Materials for the Construction of Real Estate

D N Krivogina, N I Safonov, V A Kharitonov [1]

[1]Department of Civil Engineering and Material, Perm National Research Polytechnic University, 29, Komsomolsky prospekt, Perm, 614990, Russia;

E-mail: darya.krivogina@gmail.com, saf.nikita@gmail.com, cems@pstu.ru

Abstract. During the construction of real estate, the choice of building materials for the manufacture of building structures is carried out according to the regulatory requirements developed on the principle of unification (uniformity of technical characteristics). These requirements should be opposed to the optimization of building structures in accordance with their functional purpose and operating conditions in the building to reduce the growth of costs of resources of the construction industry. The normative approach is proposed to be combined with a more economical assortment approach. This approach taking into accounting of preferences of subjects of management of building materials and construction, reflecting the interests of the subsystems of consumer and producer of building materials and constructions. The optimization procedure is proposed to be carried out on the basis of the development of technical task. In the presented model describing a set of alternatives using the proposed algorithms search for the optimal control of the production of building materials in the solution of multicriteria optimization problems is the search for an alternative with the maximum comprehensive assessment is carried out.

1. Introduction

Today during constructing of projects of real estate the choice of building materials, which are used for manufacture of concrete building structures is carried out according to regulatory requirements $\sigma_{regul.}$. These requirements are developed according to the principle of unification (uniformity of technical characteristics).

$$\sigma_{regul.} = \bigcap_{m=1}^{M} \sigma^{m}_{regul.} \quad (1)$$

where $\sigma^{m}_{regul.}$, $m \in \overline{1,M}$ are many options for managing the characteristics of building materials, in accordance with regulatory requirements (Figure 1).

The field of regulatory requirements includes a limited number of recipes of the full range of options for the manufacture of building materials. It is expediently to counteract regulatory requirements with the optimization of building structures. It will allow to reduce growth of costs of resources of construction industry. It is natural that alternatives with optimum sets of values of characteristics for specific conditions of operation and functional purpose can be outside the field of regulatory requirements. At the same time the principle of unification is shown in the form of more economic assortment approach. In this research, unit of the assortment of building constructions is an enough exact compliance of characteristics of building materials to the set conditions of their operation and to functional purpose in real estate objects [1, 2].

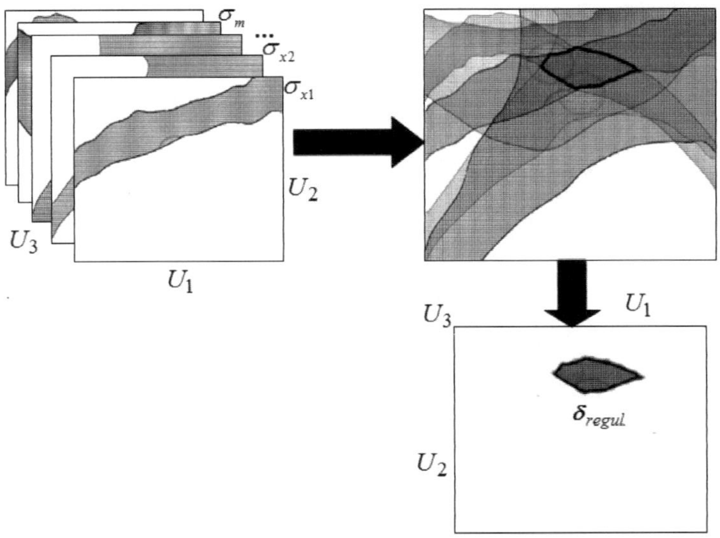

Figure 1. Formation of field of compliance to regulatory requirements of full set of characteristics of building material.

It has been developed the concept of optimization of release of the assortment of building materials for account and the description of features of assortment approach by the system analysis and the subject focused management methods. [3]. The concept includes design of model of the description of a set of feasible controls of production of construction materials, a stage of drawing up the technical specification and the optimization procedure.

2. The model of the description of a set of admissible controls by assortment of building materials

For creation of model of the description of managements of the assortment of concrete building materials were chosen following control's factors: U_1 – the water cement relation (W/c); U_2 – a ratio between large and small filler, sand and crushed stone the maximum fineness of 20 mm (r); U_3 – the content of the superplasticizing PFM-NLK additive increasing frost resistance, % of the mass of cement. The chosen factors, their codes and values of the top, average and lower level are given in Table 1

Table 1. Chosen factors, their codes and values of the top, average and lower level.

Code	Value	Researching factors		
		U_1 (W/c)	U_2 (r)	U_3 (D)
Top level	+1	0.7	0.8	0.6
Average level	0	0.6	0.65	0.5
Lower level	-1	0.5	0.5	0.4

The regression equations for such characteristics (2)–(6) as x_1 – durability at compression, x_2 – frost resistance, x_3 – watertightness, x_4 – density, x_5 – concrete workability have been received as a result of carrying out an orthogonal experiment [4] and processing of the obtained data, including checks of reproducibility of experiences and assessment of the significance of coefficients.

$$x_1 = 133.54 + 71.03U_1 - 272.5U_2 - 110.122U_3 - 133.93U_1^2 + 197.19U_2^2 + 106.33U_3^2 + 12.74U_1U_2 + 62.06U_1U_3 - 18.72U_2U_3 \quad (2)$$

$$x_2 = 33.86 - 618.65U_1 - 63.19U_2 + 1764.12U_3 + 1127.57U_1^2 + 855.9U_2^2 - 10219.91U_3^2 - 1171.61U_1U_2 - 1334.26U_1U_3 - 388.89U_2U_3 \quad (3)$$

$$x_3 = 47.23 - 92.72U_1 - 19.31U_2 - 0.36U_3 + 60.66U_1^2 + 5.1U_2^2 + 5.9U_3^2 + 14.07U_1U_2 - 5.19U_1U_3 + 0.83U_2U_3 \quad (4)$$

$$x_4 = 2679.66 + 1053.08U_1 - 1093.6U_2 - 22.5U_3 - 1188.48U_1^2 + 962.14U_2^2 + 10.65U_3^2 - 740.74U_1U_2 + 19.44U_1U_3 - 21.29U_2U_3 \quad (5)$$

$$x_5 = 7.93 - 23.99U_1 + 7.5U_2 + 1.62U_3 + 11.44U_1^2 - 11.52U_2^2 - 1.9U_3^2 + 7.53U_1U_2 + 1.57U_1U_3 + 3.89U_2U_3 \quad (6)$$

The list of characteristics of materials is considered by consumer. The main characteristics of materials for producer are costs of primary components at a mixing process and production costs. These characteristics depend on a ratio of primary components and are defined at the enterprise depending on technological process.

Two sets of matrixes-arrays are composed on base on the regression equations obtained using the software product Decon BM (Figure.2)

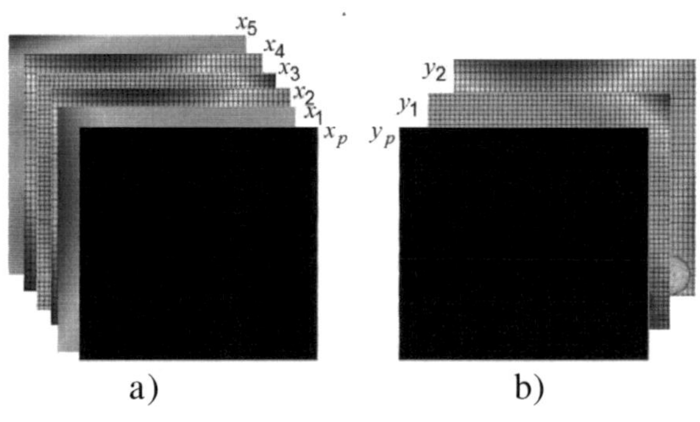

Figure 2. Matrixes-arrays of characteristic of finished products from concrete, considered from the position of the consumer (a) and the producer (b).

Material's characteristics in matrixes-arrays of producer and consumer have differences in dimension in arguments. This difference is an important obstacle in complex estimation of variants of manufacturing of building materials. Using subject-oriented management for solving this problem is required bringing material's characteristics to one universal qualitative scale, where the abscissa axis is the physical values of the finished product's quality, the ordinate axis is the qualimetric scale (Figure 3).

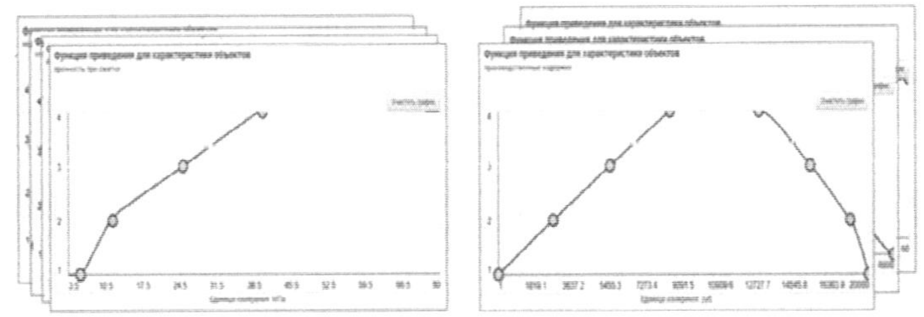

a) b)

Figure 3. The reduced functions of the characteristics of the finished product, constructed on the basis of the preferences of the subsystems of the consumer (a) and the producer (b).

One of the most significant characteristics for both subjects of management the assortment of building materials and structures is their selling price [5]. To determine the price, it is necessary to develop a mechanism for subject-oriented pricing based on modeling of preferences of stakeholders in relation to the expected rate of return of the producer and product quality indicators by the consumer [6].

Pricing for variants of building material's production can be presented as a composition of models of stakeholders' preference, that reproduces the process of functioning of market mechanism and it has a goal to establish the demand and supply functions, also the contract price in equilibrium point (Figure 4). Modeling preferences is offered to realize in accordance with dimension of phase space F.

In the framework of the offered approach it is required to change interpretation of functions of supply and demand from "quantity of the acquired (made) product" to "satisfaction degree" from the chosen variant of material's production by the producer and consumer. The required result of the "fair" not manipulated pricing of an alternative of building material is determined by parameters of an equilibrium condition of the market that is formed in intersection point of curves of supply and demand in a subspace where satisfaction degrees from the transaction get the most equal values.

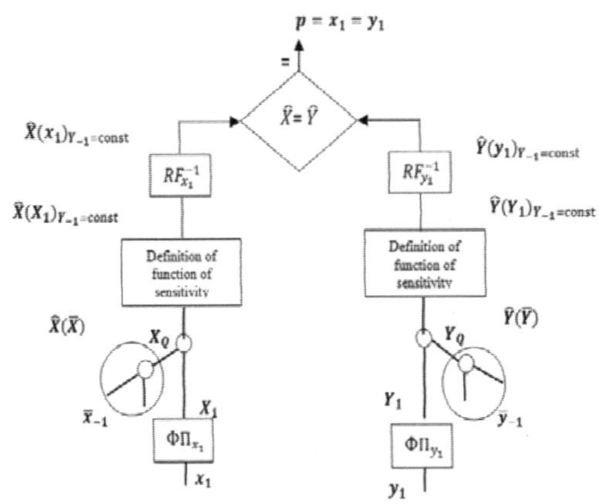

Figure 4. The mechanism of subject-oriented pricing.

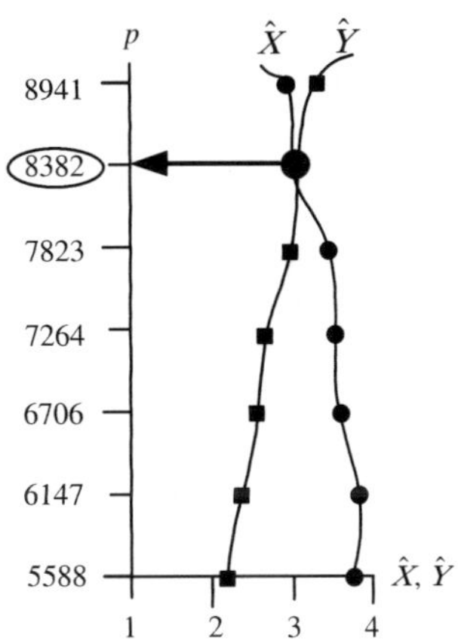

Figure 5. Determination of the agreed price.

The procedure of pricing is connected with using the phase (price P) and qualimetric (levels $\hat{X}$, $\hat{Y}$ of satisfaction of stakeholders) scales of the model of option of production building materials for fixed quality levels $X^*_{Q_x}, Y^*_{Q_y}$ and ends with occurrence of an event $\hat{X}(p^*)=\hat{Y}(p^*)$ with the result in the

IOP Conf. Series: Materials Science and Engineering **481** (2019) 012055 doi:10.1088/1757-899X/481/1/012055

form of an agreed price p^* for the alternative (Figure 5). With increasing validity of pricing's models within the framework of this mechanism, it is advisable to use the procedures for finding agreed solutions that are based on the development of the principles of active non-manipulative expertise and the generalized median.

The pricing process is held by experts one by one in the whole of set of alternatives of building material's production. At the end we have the model of description of set of feasible alternatives of release of building materials, which we will use for optimization in accordance with technical specification on production of assortment of building materials.

3. Technical specification for the production of assortment units of building constructions

As an example, consider the technical specification for the assortment of concrete slabs. On the basis of the information about the functional purpose and conditions of operation of the real estate object coming from the external environment, the designer representing the consumer's interests carries out the procedure for formalizing the requirements for the constructions [7], which can be presented in the following form (Table 2):

Table 2. Operational requirements for building structures.

№	Operational impacts	The unit of measurement	
Assortment unit №1			
1	Static and dynamic loads	16.7	kN/cm^2
2	Noises and sounds (percussion and air)	<49	Decibel
3	Temperature drops	≥ F200	cycles
Assortment unit №2			
1	Static and dynamic loads	14.8	kN/cm^2
2	Noises and sounds (percussion and air)	<32	Decibel
3	Temperature drops	≥ F130	cycles

After the procedure of formalizing the requirements for the characteristics of finished products it is carried out the procedure of transfer of requirements for the characteristics of building materials and setting multicriteria optimization tasks of building materials [8].

Table 3. Requirements for the technical characteristics of building materials.

x / №	Strength, megapascal	Density, kg/m^3	Frost resistance, cycle	Watertightness
		Assortment unit №1		
1	$x_1 \to \max$ $x_1 \geq 30$	$x_2 \to \min$ $x_2 < 2400$	-	$x_4 \to \max$ $x_4 > 2$
2	$x_1 \to \max$ $x_1 > 10$	$x_2 \to \min$ $x_2 < 2400$	-	$x_4 \to \max$ $x_4 > 4$
3	$x_1 \to \max$ $x_1 > 10$	$x_2 \to \min$ $x_2 < 2400$	$x_3 \to \max$ $x_3 > 200$	$x_4 \to \max$ $x_4 \geq 4$

Requirements for the characteristics of materials	$x_1^{max} \geq 30$	$x_2 < 2400$	$x_3 > 200$	$x_4^{max} \geq 4$;
	Assortment unit № 2			
1	$x_1 \to max$ $x_1 \geq 25$	$x_2 \to min$ $x_2 \langle 2400$	-	$x_4 \to min$ $x_4 \leq 6$
2	$x_1 \to max$ $x_1 > 10$	$x_2 \to min$ $x_2 < 2400$	-	$x4 \to min$ $x_4 \leq 6$
3	$x_1 \to max$ $x_1 > 10$	$x_2 \to min$ $x_2 < 2400$	$x_3 \to max$ $x_3 > 130$	$x_4 \to min$ $x_4 \leq 6$
Requirements for the characteristics of materials	$x_1^{max} \geq 25$	$x_2 < 2400$	$x_3 > 130$	$x_4^{max} \leq 6$

The problem of optimization of the most significant characteristics of building materials, according to the reviewed example, is presented the next formula:

$$\hat{X}_{1,4} = (K_1 X_1 + K_2 X_4 + K_3 X_p) \to max \tag{7}$$

Where X_1, X_4 – technical characteristics of building materials; X_p – price per unit of product, rub.; K_1, K_2, K_3 – weighted coefficients of building material's characteristics.
The number of assortment units of building structures is determined by the number of setting of multicriteria tasks of optimization of characteristics of building materials.

4. Algorithms of search of optimal variants of production of assortment units of concrete building material

The development of this algorithm is based on the tool of complex estimation based on hierarchical linear bundles - "Jobs-Decon" [9].

The work of the algorithm begins with the identification of system limitations to the characteristics of the material (Figure 6). Using standard programs, a cyclic procedure for truncating all infeasible regions of matrix- arrays is performed, corresponding to the number of given limitations and creating a new array with feasible alternatives to realization. After that, a model of consumer preferences is determined through a phased implementation of procedures including the construction of reduced function, weighted coefficients, ranking of priority material characteristics and calculation of integrated assessment for all feasible alternatives. As a result of applying the algorithm, we obtain one or a set of several alternatives having maximal complex estimation. In the case of obtaining several alternatives having the same complex estimation, the incorrect choice is eliminated by involving the manufacturer in the solution of the problem and choosing the most economically and technically advantageous variant of manufacturing the building material, or, the choice problem is solved without taking into account the manufacturer's preferences, the best result is assigned automatically on the basis of maximizing or minimizing the addressing of cells.

As part of the choice of assortment of slabs, we obtain the following model of consumer's preferences (8)

$$\hat{X} = 0.43 * X_1 + 0.18 * X_2 + 0.38 * X_p \tag{8}$$

where X_1 – compressive strength, MPa, X_2 – water resistance, coefficient; X_p – price per unit of output, rubles; $K_1 = 0.43$, $K_2 = 0.18$, $K_3 = 0.38$ – values of weighted coefficients of material characteristics.

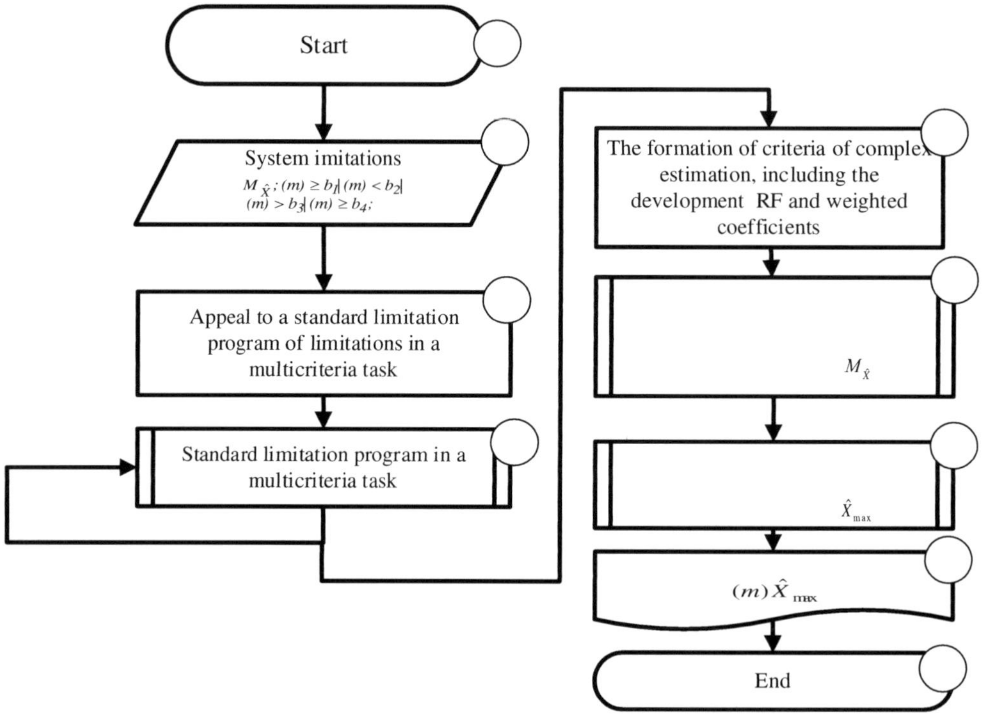

Figure 6. The flowchart of an algorithm of search of optimal controls of release of building materials in the context of solving the multicriteria optimization problem.

On the basis of this model of preferences, a procedure of complex estimation of feasible alternatives of building materials was carried out, among which the most complex estimation of has an alternative under No. 3 (Figure 7). Since we have a single alternative with the maximum value of a complex estimation it is not necessary to involve the manufacturer in the solution of the problem of choice. Otherwise, it is necessary to build a model of the manufacturer's preferences and to implement the repeated complex estimation with feasible alternatives.

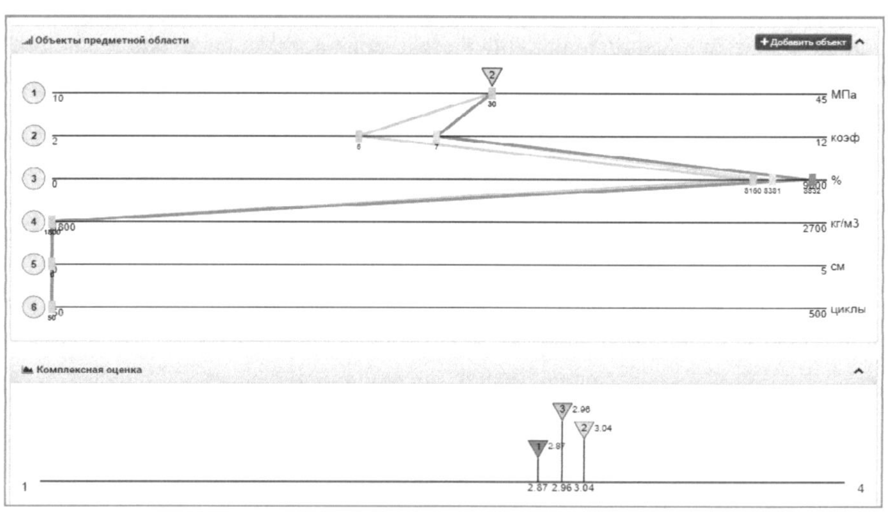

Figure. 7. Complex estimation of alternatives by the consumer.

IOP Conf. Series: Materials Science and Engineering **481** (2019) 012055 doi:10.1088/1757-899X/481/1/012055

The final stage is the procedure for selecting assortment units for all the tasks of multicriteria optimization of the release of building materials.

There was an efficiency evaluation of selected alternatives of the release of building materials at the end of the optimization procedure. It is proved that in the conducted experiment the proposed algorithms ensured the obtaining of optimal concrete compositions, intended for the production of two assortment units of slabs, that allowed to obtain on average an economic benefit of 453 rubles per unit of production in the processes of justifying the optimal properties of building materials.

5. Conclusion

As a result of the presented assortment approach, it becomes a possibility to take into account the functional purpose and operating conditions of structural elements in the real estate object, which contributes to the rational use of the primary components and the achievement of the optimal set of operational characteristics of building structures in buildings.

References

[1] Bol'shakov V I 2006 *Fundamentals of the theory and methodology of multiparameter composition of concretes* (Dnepropetrovsk: Pridneprovskaya State Academy of Civil Engineering and Architecture Press) p 360

[2] Krivogina D N 2018 Optimization of production of assortment of building materials on the basis of methods of system engineering *Applied Mathematics and Management Issues* vol **2** (Perm: Perm National Research Polytechnic University) pp 173–214

[3] Krivogina D N and Haritonov V A 2018 The concept of subject-oriented optimization of production processes for the assortment of building materials *The bulletin of IZHGTU. M.T. Kalashnikova* vol **21** (Izhevsk: Izhevsk State Technical University. M.T. Kalashnikov) pp 167–172

[4] Borovikov V 2003 *STATISTICA.The art of data analysis on a computer: for professionals* (Saint-Petersburg: Piter)

[5] Muhina M K 2000 Studying the lifestyle of consumers and segmenting the market on the basis of psychographic types *Marketing in Russia and Abroad* vol **3** ed E P Golubkov (Moscow: Finpress) Available at: http://www.mavriz.ru/articles/2000/3/262.html

[6] Haritonov V A and Gejhman L K and Krivogina D N 2017 Mechanisms of subject-oriented pricing in the tasks of managing venture projects (Perm University Herald. ECONOMY vol **12**) ed T V Mirolyubova (Perm: Perm State National Research University) pp 61–77

[7] Setkov V I and Serbin E P 2007 *Building construction* (Moscow: Infra)

[8] Polyak B T 1983 *Introduction to Optimization* (Moscow: Nauka)

[9] Haritonov V A and Safonov N I and Krivogina D N 2017 Development of subjectivity in tasks of automation and management of technological processes *Information and mathematical technologies in science and management* vol **6** (Irkutsk: Federal State Budgetary Institute of Science L.A. Melentieva Siberian Branch of the Russian Academy of Sciences) pp 79–87

IOP Conf. Series: Materials Science and Engineering **481** (2019) 012056 doi:10.1088/1757-899X/481/1/012056

Enhancement of residential construction of urban agglomerations based on the modern financial mechanisms

V Ya Mishchenko[1], O K Meschcherjakova[2], M A Meschcherjakova[3]

[1]Department of Technology, Organization of Construction, Examination and Management of the real estate, Voronezh State Technical University, 20-letija Oktjabrja, 84, Voronezh, 394006, Russia
[2]Department of Technology, Organization of Construction, Examination and Management of the real estate, Voronezh State Technical University, 20-letija Oktjabrja, 84, Voronezh, 394006, Russia
[3]Department of Technology, Organization of Construction, Examination and Management of the real estate, Voronezh State Technical University, 20-letija Oktjabrja, 84, Voronezh, 394006, Russia

E-mail: masha0207@mail.ru

Abstract. Primarily, new approaches to residential construction are not exclusively applied to the development of the new territories, but also to the renovation of the old ones. Paper reviews the situation of residential development using the principle of 'agglomeration'. The paper outlines perspective directions for the development of urban agglomerations based on the appropriate practices of the Voronezh Region. The comprehensive solution of the housing problem, which would consider the balanced development of the territories, is one of the utmost issues along this way. Authors consider solving this task to be the most effective solution, which would combine the interests of the state financing and private investors where the public-private partnership mechanism is employed. To this end, within the framework of the agglomeration process, an algorithm for housing development is proposed based on the sample territory of the city of Voronezh. This algorithm contributes to the efficient allocation of resources, financial interest, and responsibility of the parties, minimizing the costs of the process participants.

1. Introduction

Currently, we observe the consistent evolution of the cities, which in its turn leads to the improvement of the urban environment under the influence of innovative development processes, the concentration of production centers, and merging of labor and cultural ties between the cities and their surroundings [1, 2].

As a result, urban agglomerations evolve. Agglomerations, which not simply allow synchronizing the development of the settlements and improving the efficiency of spending public resources but also contributing to the improvement of the quality of life, which in its turn is an ultimate state priority.

2. Relevance

Considering the growing demand for the development of the new territories and the renovation of existing social spaces, the main task of urban development policy today is the development and implementation of the new approaches that promote the qualified utilization of the residential development areas.

However, designing and implementing specialized targeted programs aimed at developing an affordable housing market in each metropolitan area require balanced financing. That is, in addition to state participation, it is necessary to attract private investors, who are in great demand in current socio-economic conditions of particular regions and the country as a whole. The study and analysis of the development of territories within the framework of agglomerations are scientifically topical. This topic was reviewed and analyzed by various authors, in particular, E. Hoover, M. Rouget, A.A. Kruber, G.M. Lappo et al. It has to be mentioned that the features of the basic approaches and definitions of the term "urban agglomeration" have already been reviewed in the by M.G. Dikansky, P.I. Dubrovina, V.N. Petrova, V.A. Skutina and others.

3. Task

Nevertheless, despite the manifested scientific interest, as well as the enormous contribution of both domestic and foreign researchers to the study of this concept and its practical application, there remain a number of issues that have to be considered. Particularly, certain issues have not been adequately reflected. Such issue as the possibility of using the mechanism of public-private partnership as a model of financing and development of social space within the urban agglomerations, which facilitate the country's policy at the new level, but also helps to achieve a number of "social effects".

The paper reviews the possibility of financing of the aforementioned process of agglomeration of regional territories based on the Voronezh region example. Voronezh agglomeration is the largest monocentric agglomeration in the Central Federal District of Russia, with a population of about 1.3 million people.

Having analyzed the current situation for the prospective development of the Voronezh city agglomeration for 2018-2020, it is possible to outline its directions: a comprehensive solution of the housing problem, taking into account the balanced development of the territories; improvement of the ecological situation; overcoming the impending transport collapse.

Public sector resources do not appear to be sufficient to resolve such problems, therefore, it is obvious that business structures should be engaged in the process to achieve these goals.

4. Theoretical part

Cooperation between the state and the private sector have always existed in one form or another, however, the degree of private partner involvement was usually determined by the state and was often insignificant.

Currently, there are numerous different definitions for this kind of cooperation nonetheless the public-private partnership (PPP) mechanism is the most promising option, which implies close and fruitful interaction in socially significant industries - both on the part of the state and the business side [3, 4,5]. Considering the long studying experience of the PPP mechanism, it could be affirmed that most of the developed countries today are more adapted to its practical application than the Russian Federation. Russia, on the whole, has some positive dynamics whilst using the PPP mechanism, however, various barriers are still in place [6, 7]. Analysis of the formation and development of the PPPs in the regions showed that PPP mechanisms are widely applied in the 'subjects' of the Russian Federation [8, 9]. Among the afore subjects the Voronezh region was one of the first in Russia to adopt the law on public-private partnerships in November 2011 [10, 11].

The social space of urban agglomeration is a set of the society's places of residence and different forms of labor application, accompanied by the social, cultural and social purpose accessories. Those get localized based on the vital functions of society's livelihoods [12]. The set of residential areas constitutes a 'housing stock'. Thus, forming social space as one of the most important components of urban agglomeration development is primarily affected by the housing policy. Modern conditions affected the general economic model of the country and led to the transition of housing policy to the new principles, which would consider solving the afore problems in their entire complexity.

This has required the introduction of new approaches, which would provide for the conditions for the development of specially targeted programs for the development of affordable housing market in each specific agglomeration, taking into account the prevailing level of housing prices and the population's solvency potential [13, 14, 15].

Development and implementation of the state policy aimed at solving the housing problem and achieving the set goals are currently being reflected by the "Housing" federal program [16].

Analyzing several current 'subprograms' of the more general federal program, typical for the Voronezh region, it became obvious that the achievement of a result that positively affects the quality of life is only possible if the competent integrated approaches are applied to the development of territories, which presume the PPPs as a mechanism for balanced financing for its successful implementation [17].

5. Practical part

The proposed pattern for the adoption of the PPP mechanism within the framework of housing construction development based on the example of the Voronezh region is presented in Figure. 1.

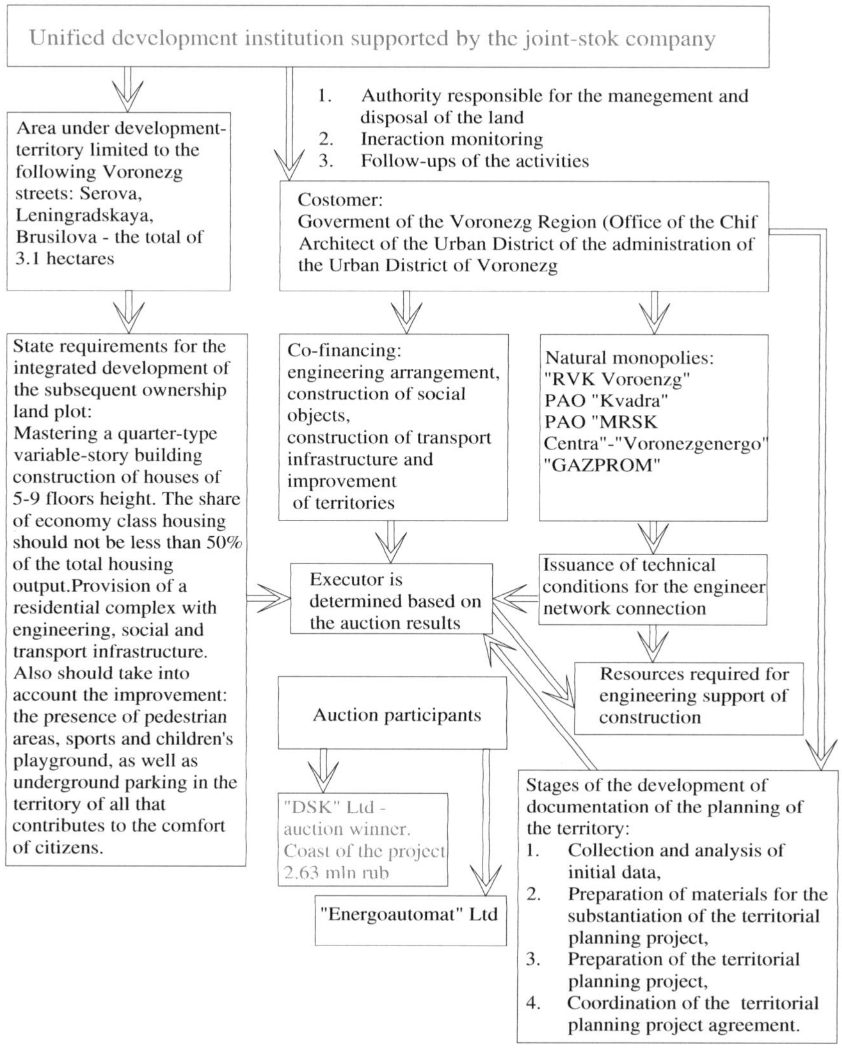

Figure 1. The process of the territory agglomeration limited by Serova, Leningradskaya and Brusilova streets of the city of Voronezh (housing construction with the PPP mechanism applied).

The left bank of the Voronezh river district was chosen as a 'sample territory', bounded by Serova - Leningradskaya - Brusilova streets, with a total area of 3.1 hectares [18]. The houses located in this micro-district are the subjects to demolition within the framework of the municipal program called "Demolition and reconstruction of dilapidated multi-family housing stock". The Unified Development Institute owns the site under consideration in the housing sector, which has been operating since 2014. It interacts with the Customer - a legal state person, who will co-finance engineering development, construction of social facilities, construction of transport infrastructure and landscaping .

Given the scope of the program, it is obvious that the private investor is involved. In this case, the 'Customer' (represented by the Government of the Voronezh Region) holds an auction. Following this auction, the second side of the project financing is determined. In our case, it will be LLC DSK-1, with a proposal of 2.63 million rubles based on the requirements for integrated development of the land plot, the Contractor must submit a project for the development of the territory under consideration. Based on the need for resources for engineering maintenance of the building, the relationship between the Contractor and natural monopolies is traced.

It has to be mentioned as well that the Customer prepares and approves investment programs for the integrated development of engineering infrastructure for Natural Monopolies [19, 20].

From the economic point of view, such pattern promotes a clear coordination of actions, as well as a competent allocation of resources, financial interest, and responsibility of the parties, thus, allowing to minimize the costs of the process participants.

At the same time, it also has to be noted that the social effect – namely - the implemented project within the framework of the PPP mechanism -contributes to the city image as a whole and also raises the quality of life of the citizens, since the balanced construction activities allow to improve the entire territory of the city.

6. Conclusion

Given the above, it can be argued that the mechanism of public-private partnership in the nearest future will be further developed and that it has practical application in the development of the social space of urban agglomerations, which in its turn will affect the integrated development of both new and existing territories. Such prospects certainly do contribute to the strengthening of economic indicators and the country's entrance to the new economic level, but also to ensure the quality of life of its citizens.

References

[1] Percik E N 1999 *Cities of the world: Geography of world urbanization* (Moscow: International Relations) p.382

[2] *Meeting of the Presidium of the State Council 2011 On Enhancing the Role of Regions in Modernizing the Russian Economy Voronezh* (Electronic resource access date 20.09.2018) URL:http://state.kremlin.ru/state_council/13477

[3] Meshcheryakova O K and Stepantsoova S A 2012 Analysis of the practical application of the mechanism of public-private partnership at the regional level *Construction: Science and Education* **2** 3

[4] Meshcheryakova O K, Meshcheryakova M A and Stepantsova S A 2017 Development of built-up areas as a reserve for sustainable development of the city *Modern trends in the construction and operation of real estate objects* (Voronezh:VSTU) pp.134-137

[5] Khrustalev B B and Ermolaeva A V 2016 Peculiarities of the market and internal strategies of functioning and development of the enterprise in modern production conditions *Competitiveness in the global world: economy, science, technology* **8-3(22)** 192-194

[6] Varnavsky V G 2005 *Public-private partnership: Forms, projects, risks* (Moscow: Science)

[7] *Public-private partnership in modern Russia* (Electronic resource access date 20.09.2018) URL:https://studbooks.net/2424985/ekonomika/mehanizm_razvitiya_regionov_rossii

[8] *Structure, goals, and objectives of the regional PPP center of the Voronezh region 2018* (The regional center of public-private partnership of the Voronezh region) URL: http://invest-in-voronezh.ru/eng/index.php?option=com_content&view=article&id=145&Itemid=137
[9] Gorchakova E V Stages of development of public-private partnership in Russia *Problems of theory and practice of entrepreneurship* (Electronic resource access date 20.09.2018) URL:https://www.docme.ru/doc/1508748/e-tapy-razvitiya-gosudarstvenno-chastnogo-partnerstva-v-rossii
[10] Fedorova E A, Dovzgenko S E and Gurina V A 2013 Forming an effective mechanism of public-private partnership in the Russian Federation Electronic resource *Finance and credit* **16(544)** 13-23
[11] Regions 2018 *PPP Development Center* (Electronic resource access date 20.09.2018) URL: http://pppcenter.ru/ru/regions
[12] Sirotkina NV 2017 Modeling of strategic parameters of social and economic development of urban agglomeration *Region: Systems, Economics, Management* **4(39)** 226-234
[13] Meshcheryakova O K, Maiburova E S and Fursov K A 2017 Development of agglomeration processes by the example of the urban district of Voronezh *Modern trends in the construction and operation of real estate objects* (Voronezh:VSTU) pp.94-97
[14] Grabovyi P G, Lunyakov M A and Kamenetsky M I 2017 Features and problems of project financing in construction *Real estate: economics, management* **4** 6-14
[15] Belokrylova O 2006 Interaction of business and government in the process of forming a strategy for the economic development of regions *Problems of Forming State Policies in Russia* (Moscow: Scientific expert) pp.563-572
[16] Meshcheryakova M A, Chesnokova E A and Batova A V 2016 Analysis of financing the reproduction of housing facilities *Modern problems and prospects for the development of construction, the operation of real estate* (Voronezh:VSTU) pp.226-233
[17] Sviridova S V, Rassokha E V and Samodurova S A 2016 Formation of the system of innovative development of the region *Modern economy: problems and solutions* **11(83)** 151-163
[18] *Decision on the development of the built-up area, bordered by Serova - Leningradskaya - Brusilova streets in the city district of Voronezh* (Electronic resource access date 20.09.2018) URL:http://docs.cntd.ru/document/440510097
[19] Zagidullina G M and Sobolev E A 2014 Technological ways, their role, and importance in the development of Russia's innovative economy *Proc. of the Kazan State University of Architecture and Civil Engineering* **4(30)** 348-355
[20] Ganzha I V and Nabiev R A 2013 Public-private partnership as an instrument for realizing the interaction of the state and investors in the investment process *Vestnik of ASTU Series: Economics* **3** 9-20 (Electronic resource access date 20.09.2018) URL: http://vestnik.astu.org/Content/UserImages/file/economics_2013_3/01.pdf

Ways to Improve the Safety of Built-Up and Adjacent Areas

A E Sheveleva[1], N R Stepanova[1], P V Bochkov[1]

[1]Department of Economy and Management in Construction and Realty Market, Ural Federal University, Mira, 19, Yekaterinburg, 620002, Russia

E-mail: anchytka_Sh@mail.ru, n.r.stepanova66@gmail.com, bochkov05@mail.ru

Abstract. This paper deals with the issues of technospheric and environmental impact on the environment by artificially created objects, in particular, shopping centers. Commercial real estate is one of the investment-attractive sectors of commercial real estate. The article presents the classification of retail real estate, especially the development of retail real estate and analysis of shopping centers within the city of Yekaterinburg. Environmental safety is paramount and substantiated within any real estate design, including the surrounding areas. The paper also discusses the threat from man-made accidents in built-up areas. The classification of types of accidents is given, their causes and subsequent consequences are analyzed. Possible methods of prevention of technosphere risks in built-up areas, as well as measures of various directions for the prevention of emergency situations are considered in the form of, highlighting the environmental issues in university programmes, and improving the safety management system.

1. Introduction

At the present stage of rapid growth of cities and construction of new real estate – residential complexes, shopping and business centers, public places, the problem of ensuring the safety of built-up areas becomes urgent.

This is due to the increase in the number of accidents that occur for various reasons. In this regard, it is necessary to identify new ways to prevent or minimise the risks within built-up areas.

High rates of growth of production potential, new technologies of construction production have technogenic influence on all main components of environment. The impact of this growth increases with the size of the scale of construction and improved productivity of construction technologies.

Modern construction production creates a certain technological system that changes during technological processes over time, resulting in a destructive impact on biological and environmental systems.

Commercial real estate is one of the investment-attractive sectors of commercial real estate.

The aim of the study was to determine what combination of approaches and methods will be most effective from the point of view of safety [1].

2. Research methodology

2.1. Classification of commercial real estate

Shopping centers today are designed to fulfil the need for a popular place to visit and to provide a source of leisure. This is evidenced by the wide variety of types and forms of shopping centers.

Let us turn to the different classifications of retail real estate in Russia and in the world. Here is an example of the following classifications:

- retail real estate in the version of G. M. Sternik [2] (chief analyst of the Russian Guild of realtors);
- classification of Urban Land Institute (USA);
- classification of the international Council of shopping centers International Council of Shopping Centres;
- classification Of the Russian Council Of shopping Centers.

According to the classification of commercial real estate in the version of G. M. Sternik, there are the following types: the type of building and space-planning solution-built-attached premises, detached buildings, shopping malls; the type and nuances of the device; the range of goods sold and the trading area with its size; commodity specialization [2].

American Institute Urban Land Institute includes in the classification of district shopping center; district shopping center; district shopping center; super-circle shopping center; regional shopping center; super-regional center; specialized shopping centers [3].

The classification of the international Council of Shopping centers International Council of shopping Centres Arrd assigns to each Shopping center, one of eight classes: "the Store next door", "Regional", "Moda Center", "Power center", "Regional" and "Superregional", "Shop manufacturer", "Specialized" [1].

The classification of shopping centers developed by the Russian Council Of shopping Centers includes the following parameters: attractiveness for tenants, technological parameters, attractiveness for investors, compliance of the quality assessment of the object with the expectations of visitors [4].

2.2. Features of retail real estate development and analysis of shopping centers in Yekaterinburg

Development of retail real estate has its own characteristics. In particular, it is the need for an integrated and detailed approach to this issue, which covers options for financing the construction of commercial real estate, the development of commercial concepts and tasks for architectural design, etc., despite the large volume of elements of the development of commercial real estate, one of the main issues are security issues [5].

Analysis of shopping centers (shopping centers) of the city of Yekaterinburg showed that the city is fully saturated with these objects. In a detailed description of the most well-known existing shopping centers with an indication of the range of services provided are presented in the Table 1.

Table 1. The most well-known existing shopping centers in Yekaterinburg with an indication of the range of services [4].

Name	Address	Range of services
KIT	Yekaterinburg Amudsena street, 65	Shops brands of clothing, accessories and shoes. On the territory of the shopping center is a fitness center and children's playroom
Parakhod	Yekaterinburg Amudsena street, 62	Shoe center. In the same building are: group of Companie, shoe house
Yekaterininskiy	Yekaterinburg Sherbakova street, 4	Hypermarket of food products, audio-video and home appliances salons, furniture, clothes and footwear, perfumes and cosmetics, food court, children's entertainment complex, cinema center, conference halls
Dirizhabl	Yekaterinburg Akademika shvartca street, 17	Shops: clothes, shoes, sports goods, leather goods, equipment for outdoor activities, perfumes. Places for recreation, large-scale events. Food-court. Theatre ticket offices, air and railway ticket offices

Megapolis	Yekaterinburg 8 Marta street, 149	Supermarket, children's Playground, cinema, food court, cafe, restaurant, bar, services: pharmacy, travel Agency, theater box office, pet store, banks, ATMs, manicure Studio
MEGA	Yekaterinburg Metallurgov street, 87	Hypermarkets, cheap clothing stores and clothing stores, cosmetics and perfumes. Beauty salons, banks, studios, cafes and restaurants
Raduga-Park	Yekaterinburg Repina street, 94	Hypermarket, children's entertainment center, cinema, food court, pharmacy, home life, manicure Studio, children's hairdresser
Alatyr	Yekaterinburg Malysheva street, 5	Shops, cafes and restaurants, food court: entertainment, fitness club
Grinvich	Yekaterinburg 8 Marta street, 46	Many shops the goods for the house and family. Modern multiplex, many restaurants
Fan fan	Yekaterinburg Yasnaya street, 2	Shops, cafes and restaurants, entertainment: cinema, zoo, karaoke, bowling, car wash

2.3. Structure of consumer expenditures of the population of Sverdlovsk region and their analysis of the areas in shopping centers of Yekaterinburg
The market potential of the Sverdlovsk region-consumer spending. Consumers spend 37.31% on food products and 37.13% on non-food products, which is reflected in the Table 2.

Table 2. Structure of consumer spending in Sverdlovsk region at the end of 2017 [6].

Indicator	Value, %
Food products	37,31
Non-food items	37,13
Services	25,56
Total goods	74,44
Total goods and services	100

The population prefers food products (37.31%) and non-food products (37.13%).

According to the Ural chamber of real estate, in the shopping center of Yekaterinburg can be observed a slow reduction in vacant space (Figure 1)

According to Figure 2, vacant areas in the 3rd quarter of 2017 amounted to super-regional and regional shopping centers – 11.5%; District shopping centers – 8.9%; District shopping centers – 8.0%.

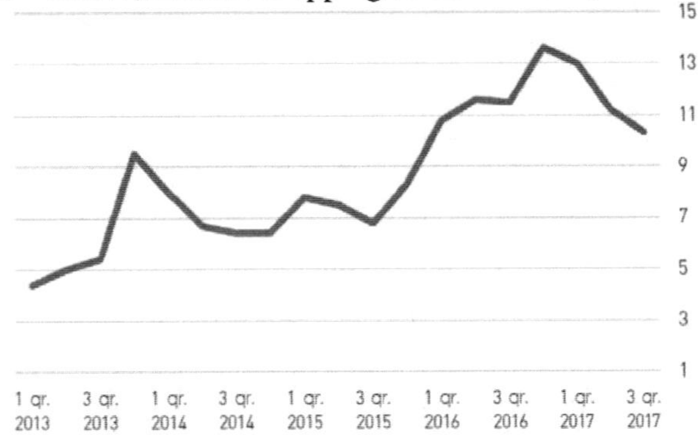

Figure 1. The share of empty space per annum
in the market of shopping centers of Yekaterinburg [5] (Source: upn.ru).

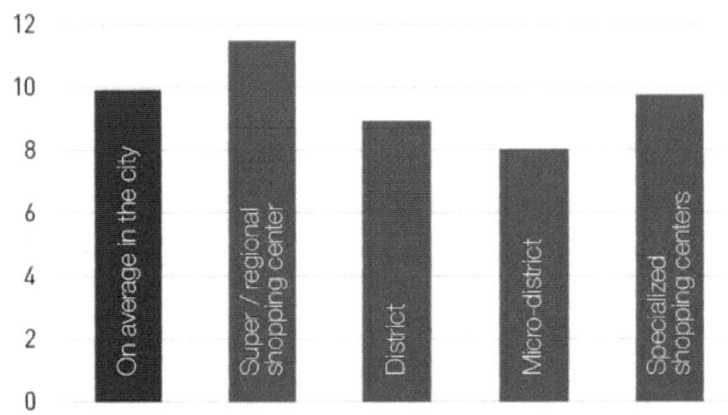

Figure 2. Vacant space in Yekaterinburg shopping centers in 2017, % [5] (Source: upn.ru).

Experts of the Ural chamber of real estate noted that the rates of commercial real estate do not generally have a downward trend (Figure 3).

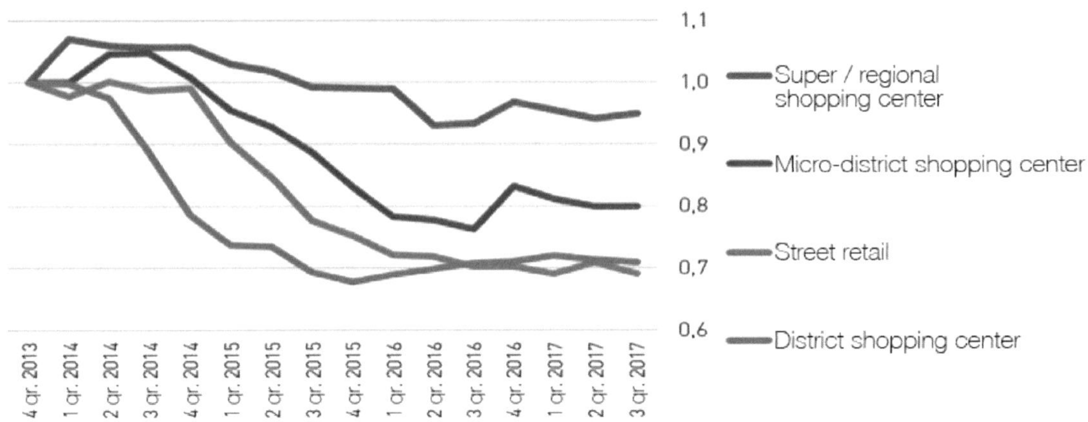

Figure 3. Rental rates index in the commercial real estate market of Yekaterinburg, units [5,6].

As the monitoring showed, the degree of saturation of shopping centers of the city of Yekaterinburg is quite high. In this regard, there is a need to ensure the safety of the environment.

3. Ensuring environmental safety

To ensure the safety of the environment, namely, these industrial commercial real estate and adjacent areas, it is necessary to develop a plan to prevent risks in built-up areas.

3.1. Risks and Classification of possible accidents in built-up areas

The principle of metrological support [7] in technosphere safety is considered by Akhobadze G N. Continuous monitoring of the technosphere taking into account metrological principles is expected to produce an efficient forecast for the technosphere development and make appropriate decisions [7].

Smirnova E considers the demonstration of threats related to the uncontrolled, the dangerous and economically expensive progress of society as they indicated the need for environmental and technospheric security [8].

Technosphere risk, described as the complex index of the reliability of the elements of the technosphere, characterizes the probability of emergency situations in the operation of machines, mechanisms, technological processes in the construction and operation of buildings and structures [9].

Experts cite a range of classifications of such accidents built-up areas as shown in Table 3.

Table 3. Classification of possible accidents of built-up areas [10].

Type of accident	Causes of occurrence	Aftermath
Fire	Careless handling of fire; explosions of hazardous components; arson; failure to comply with security measures; failure to observe safety precautions when using the equipment; environmental factors; failure of equipment; a consequence of the explosion	Damage to human health; territories, buildings, structures or their destruction; environment; property; economic losses; threats to human life
Bang	Careless handling of hazardous substances; failure to comply with safety measures; improper installation of equipment; failure to observe precautions when using equipment; failure of equipment	Damage to human health; territories, buildings, structures or their destruction; environment; property; economic losses; threats to human life
Sudden building collapse	Errors in the design of the building; deviation from the project in the conduct of construction works; violation of the rules of installation, the commissioning of the building or its individual parts with major deficiencies; rules of operation of the building; due to natural or man-made emergency	Building failure; occurrence of fires; destruction of utilities and energy networks; formation of blockages; injury and death of people
Utility system crashes	Outdated material and technical base; natural disasters; lack of financial resources	Ecological catastrophes; disruption of work to provide the population with resources; damage to territories, buildings; economic costs; injury and loss of life

3.2. Geological environment and risks

Various dangerous phenomena are not always possible to predict, however, modern knowledge, technology, and access to data can significantly reduce the possible risks. First of all, the risk refers to the natural-technosphere processes that have a connection with human activities.

In the presented connection, it makes sense to move to an advanced strategy of security in the construction and reconstruction of urban areas, focused on urban planning, taking into account the risks. At different stages of design and planning work should be carried out to analyse not only socio-economic, architectural and planning, commercial and other factors, but also the state of the geological environment.

Important elements of urban development that guarantee the safety of the environment are engineering-geological zoning, monitoring and forecasting of dangerous phenomena [11].

3.3. Ways to prevent technosphere risk in built-up areas

We will present ways to prevent man-made risks of built-up areas in order to ensure the safety of users of buildings and structures:

1. Engineering protection – a complex of structures aimed at protecting people, buildings, structures, or territory, from the impact of dangerous natural processes of man-made impact, threats of a terrorist nature, as well as to prevent the consequences of impacts.
2. Organizational, technical, regime and operational measures to avoid fires.
3. Timely examination of the condition of the building or structure for cracks and damage to the supporting structures.
4. Improvement of fire protection systems and strict observance of fire safety rulés in the process of construction and operation of buildings and structures.
5. Using the results of emergency forecasts to improve security systems.
6. Improvement of methods of preparation of life support systems of buildings for operation in the cold period [1].

4. Results and suggestions

4.1. Measures on prevention of emergency situations

Select the number of measures on prevention of emergency situations, reduction of their scope in case of occurrence and potential losses (Table 4) [12].

Table 4. Measures focus on the prevention of accidents.

Measure	Principles on the use
Social advertising	Tools to attract public attention to the problem of man-made disasters in built-up areas
The control of the management companies	Periodic inspection of buildings for signs of risk of accidents, monitoring of repairs
Internet portals, mobile applications	Development and implementation of channels to inform about accidents

4.2. Training in universities in the direction of "Technosphere safety"

Fire prevention and control, ensuring the safety of civil and industrial facilities requires highly qualified specialists.

As an option for the development of this issue, it is necessary to improve training in universities in the direction of "Technosphere safety", including the profile of "Fire safety" (Table 5) [1].

Table 5. Improvement of training at universities in the direction of "Technosphere safety".

Indicator	Position
Area of competence	Ensuring fire safety of industrial, public and residential buildings and structures; natural areas; minimization of technosphere impact on the natural environment; preservation of human life and health, through the use of modern technical means, methods of control and forecasting
Objects	Human and hazards associated with human activities; environmental hazards associated with human activities; environmental hazards associated with natural hazards; methods and means of assessing hazards, risk; methods and means of protecting humans and the environment from hazards; rules of regulation of hazards; methods of human rescue
Types of professional activity	Design and engineering; service and operational

IOP Conf. Series: Materials Science and Engineering **481** (2019) 012057 doi:10.1088/1757-899X/481/1/012057

It is also necessary to create special documents for the creation and implementation of the technological process in the field of environmental and technosphere safety, which aim to carry solutions to create conditions for the conservation of natural resources and environmental safety; improving ways to improve environmental safety in the field of industrial facilities, in particular shopping centers [12].

4.3. Improvement of safety management system

It is also necessary to create special documents for the creation and implementation of the technological process in the field of environmental and technosphere safety, which aim to carry solutions to create conditions for the conservation of natural resources and environmental safety; improving ways to improve environmental safety in the field of industrial facilities, in particular shopping centers.

The management system of the Federal Executive Supervisory authority-the Ministry of the Russian Federation for civil defense, emergencies and elimination of consequences of natural disasters – a set of management bodies, forces and means aimed at solving problems in the field of civil defense, protection of the population and territories from natural and man-made emergencies, fire safety and human safety at water bodies [11].

It should be noted the 2030 plan and the development of a mechanism for improving the system of emergency management in Russia.

The order of 23.01.2017 № 21 approved the action Plan of emergency situations of Russia for the implementation of the fundamentals of the state policy of the Russian Federation in the field of civil defense for the period up to 2030. In the field of improving the system of civil defense management, warning systems and informing the population about the dangers of peace and wartime, a number of measures have been reflected [9].

Taking into account the forms of achieving the Plan to improve the management system 2030, attention is focused on the creation of "a single technological chain of protection of the population and territories: monitoring – assessment – response – interaction – effective assistance", which includes the improvement of the Supervisory authority by optimizing the risk-oriented approach; training of qualified personnel to identify risks; technical equipment to identify risks [12, 10].

Also note that in order to avoid human and material losses, it is necessary to draw public attention to the described problem, to control the functioning of built-up areas, to adhere to precautions on a permanent basis. Conducting fact checks after the tragedies will not be an effective measure.

References

[1] Karavaeva N 2016 The fundamentals of real estate development *The organization of the development process* [In Russ] (Yekaterinburg Ural Federal University) p 10-13

[2] Sternik G M 2002 *Typology of retail real estate* [In Russ] (Tutorial Moscow) p 320

[3] Ozerov E S 2007 *Property management Classification Urban Land Institute USA* (Encyclopedia Bonnier Biznes Press Sankt-Peterburg) p 342

[4] Shopping centers of the Ykaterinburg city 2017 http://www.malls.ru/city/ekaterinburg.shtml

[5] Khor'kov M 2017 Report of the Head of the Analytical Department of the Ural Chamber of Real Estate Deputy Head of the Consulting Committee of the Russian Guild of Realtors (Ekaterinburg: Christmas Summit) p 5-11

[6] Report on the Social and Economic Situation of Russia for January-March 2017 Rosstat Information on the Social and Economic Situation of Russia Federal Service for State Statistics [In Russ]

[7] Akhobadze G N 2017 Metrological Support in Technosphere Safety *IOP Conference Series Materials Science and Engineering* p 262(1)

[8] Smirnova E 2018 *Control capability of environmental safety in the context of green construction paradigm* (Espacios) p 39(22)

[9] Order of the Ministry of Emergency Situations of Russia from 23 01 2017 No 21 On approval of the plan of measures of Russia to implement the fundamentals of the state policy of the Russian Federation in the field of civil defense for the period until 2030 [In Russ]
[10] Prevention and liquidation of emergency situations 2017 http://www.obzh.ru/pre/2-3.html
[11] Meeting of the Government of the Russian Federation On the establishment of crisis management centers [In Russ] 2012 (Moscow) № 35
[12] Bezdenezhnykh V M, Protasov K A 2016 Features of the risk-oriented approach in the management of small and medium-sized businesses [In Russ] (Scientific Records of Young Researchers) p 35-40

IOP Conf. Series: Materials Science and Engineering **481** (2019) 012058 doi:10.1088/1757-899X/481/1/012058

Sustainability analysis of the management strategy in a shopping centers

A O Alekseev[1], V S Spirina[1], A A Andronova[1]

[1]Department of Construction Engineering and Materials Science, Civil Engineering Faculty, Perm National Research Polytechnic University, Komsomolsky av., 29, Perm, 614990, Russian

E-mail: spirina@cems.pstu.ru; alekseev@cems.pstu.ru

Abstract. The managers' strategic behaviour of shopping centers in a competitive environment is discussed. Sensitivity analysis was made by using the methods of mathematical and experimental game theory. The stability of the strategy to the opponents' actions was investigated. Also, we investigated the stability of the strategies to the environment, expressed in the level of consumption. The level of consumption in our study is characterized by: the share of visitors, making purchases in the shopping centre (the conversion rate); the average amount of purchases that make visitors shopping centre (size of the average check). These parameters describe changes of the environment. All cases of the shopping centre management strategy stability are illustrated. The novelty is that at allocation costs of the Manager for the development and promotion of shopping centres, the strategies of managers are resistant to each other's actions, in contrast to changes in the external environment is shown.

1. Introduction

The task of management of shopping centers (SC) [6] is one of behavioral control in organization problems [1-5]. The competitive and implementation of SC managed autonomously requires a study of the strategic behavior of SC managers. This study makes is possible with using game theory and experimental game methods. The game-theoretical formulation of the competitive SC management problem corresponds to the zero-sum game, because the total number of visitors is constant, in other words, managers entice visitors from each other by changing the quality of SC.

Mathematically possible to determine the theoretically predicted behavior of the SC managers is showed in the theoretical section of this article. An experimental study of this problem in the form of a business simulation game will make it possible to find out – to which strategy the real people will come. Statement of experimental research of the shopping centers' managers' strategic behavior is described in experimental section. In particular, layout of a software module based on MS Excel used in preliminary experiments is shown. Results of preliminary experiments of strategic behavior managers of SC allow to found two aspects: with the apparent cognitive complexity of the problem, players were able to find the optimal distribution without using the decision support system in a fairly short period of time, in other words, players came to a theoretically predicted strategy; the optimal strategy, in turn, was resistant to the actions of the opponent, and therefore the players did not change their strategies, which led to a very rapid completion of the experiments.

2. An example of a decision theoretic game problem with two players engaged in the management of a competing SC

2.1. *Raw data for solving the problem of competing SC management*

The internal parameters characterizing the shopping center for the consumers are include: x_1 – square, x_2 – range of goods, x_3 – transport availability, x_4 – aesthetic parameter, x_5 – promotions and discounts, x_6 – quality of goods, x_7 – availability of brands and x_8 – events (here are the results obtained by sociological (marketing) survey of the SC in the Perm city[1]). The sociological (marketing) survey was interviewing of visitors (consumers) of shopping centers and main question was "what parameters are more important for you when you are visiting shopping centers?" Consumers (about 600 peoples) didn't have answers options and that so they answered about parameters for themselves. All answers were aggregate to 8 quality parameters because answers often repeated. For this study used this four parameters: m = 4, 6, 7, 8 as those which we will operate, and other remained parameters: o = 1, 2, 3, 5.

To predict the attendance of the commercial real estate object, the modified [6] Huff model [7-9] (1) can be used, which allows to estimate the consumer attractiveness of the commercial property:

$$A_{ij} = \frac{\{Q_j\}}{\{T_{ij}\}^{\lambda_k}}, \tag{1}$$

i – the number of the buyer (the i-th consumer means the consumer located in the point of i); j – number of the commercial property; A_{ij} – the attractiveness of the j-th property for the i-th consumer; Q_j – the quality of the property; T_{ij} – time spent by the i-th consumer on the road to the j-th property; $\lambda\epsilon[0; 1]$ – parameter reflecting the effect of different types of objects on the perceived time spent (this parameter is empirically); {} – the numerical value of the parameter. Quality in this study calculate by qualimetric model:

$$Q = \prod Q_m^{q_m} \cdot Q_o, \tag{2}$$

Q_m – quality of using parameters, q_m – importance of using parameters and Q_o – quality of others internal parameters with taking into account their importance. These parameters are constant and don't depend on costs in development. In this study Q_o have next values: Q_{o1} = 0.885; Q_{o2} = 0.803. These ones were calculated based on results of sociological (marketing) survey (table 1).

The time of correspondence (T) of consumers from the place of residence to the commercial object has a significant impact on the assessment of consumer attractiveness of shopping centers. The so-called pedestrian-transport zones [6] and corresponding parameters λ are formed to commercial real estate objects: for the first zone (from 45 to 80 min walk) – λ = 0; for the second zone (from 80 to 160 min walk) – λ = 0.5; for the third zone (over 160 min walk) – λ = 1. At the intersection of these zones, depending on the location of the studied commercial real estate objects, it is possible to allocate several sectors K (Fig. 1), each one are differently affects for consumers' time spent to property.

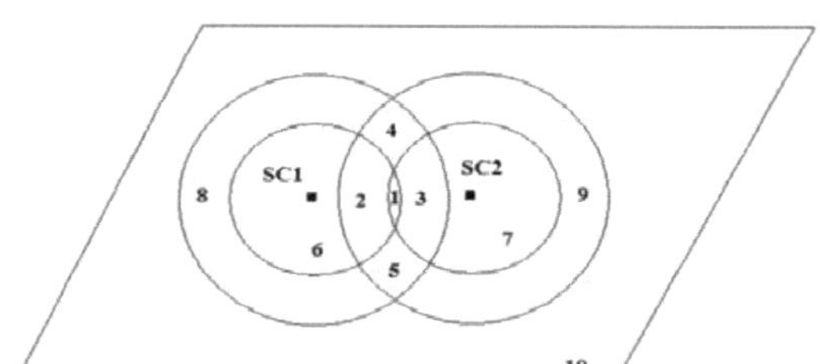

Figure 1. Allocation of sectors on the example of two shopping centers (SC).

[1] Advantages the Results presented in tablet form. Retried from https://goo.gl/enasqb (creation date: 18.09.2013)

Having calculated attractiveness of SCs according the modified Huff model (1), we can determine the probability of a potential consumer's choice of a trade object (3):

$$P_{ij} = \frac{A_{ij}}{\sum_{j=1}^{n} A_{ij}}, \tag{3}$$

and we can calculate the number of expected visitors (n) in the shopping center, knowing the number of residents in each sector k:

$$n = \sum_{k=1}^{K} \left(P_{ij}^{k} \cdot N_k \right), \tag{4}$$

P_{ij}^{k} – the probability of visiting the i-th visitor from the k-th sector of the j-th property; N_k – number of residents of k-th sector; K – number of sectors for the example with two shopping centers: $K=10$ (see figure 1).

Let there be two competing SC ($j \in J$, $J = \overline{1,2}$), each of them has three pedestrian-transport zones and their intersection forms ten sectors (K). The task of SC management is formulated as a constrained optimization problem, where profit (5) or index of profitability (6) can be used as a objective function (OF). Constrained optimization problem have inequality constraints to the costs of SC development and promotion. These costs are members of set of feasible values $c_m \in C_m$, $m \in M$, where M – set of using parameters.

$$OF_1 = P_j(c_{jm}; c_{-jm}) = \mu \cdot Ar \cdot n_j\left(c_{jm}; c_{-jm}\right) - \sum_{m \in M} c_{jm} - TFC_j, \tag{5}$$

$$OF_2 = RR_j\left(c_{jm}; c_{-jm}\right) = \frac{\mu \cdot Ar \cdot n_j\left(c_{jm}; c_{-jm}\right)}{\sum_{m \in M} c_{jm} + TFC_j} - 1, \tag{6}$$

where TFC_j – total fixed costs of the Manager of the j-th SC; μ – customer conversion ratio (the percentage of visitors who purchase something); Ar – the average amount of purchases that visitors make; c_{jm} – costs of managing j-th SC for the development of managed parameters of the object; n_j – the number of attracted visitors in the j-th SC, determined by the equation:

$$n_j = \sum_{k=1}^{10} N_k \cdot \frac{\prod_{m \in M} Q_j(c_y)^{q_y} \cdot Q_o \cdot T_k^{-\lambda_k}}{\sum_{j=1,2} \prod_{m \in M} Q_j(c_y)^{q_y} \cdot Q_o \cdot T_k^{-\lambda_k}}, \tag{7}$$

where $Q_j(c_m)$ – quality of using parameters of object j; Q_o – quality of other parameters of object j; T_k – correspondence time to i-th SC of visitors to k-th sector; λ_k – parameter that affects the importance of correspondence time for visitors; N_k – amount of residents in sector k, $k=\{1,\ldots,10\}$; q_m – importance of m-th parameter, m = 4, 6, 7, 8. The game-theoretical formulation of the competitive SC management problem corresponds to the zero-sum game, because the total number of visitors is constant $n_1+n_2=const$, in other words, managers entice visitors from each other by changing the quality of SC.

The solution of the optimization problem is the cost distribution for SC management, defined by the equation:

$$c_{jm}^{*} = \arg\max_{c_m \in C_m} \left(OF\left(c_{jm}, c_{-jm}\right) \right), \tag{8}$$

which corresponds to the game-theoretic formulation of the control problem, taking into account that the result of the Manager's activity depends on the actions of the opponent, and vice versa. The game situation is denoted by $a -j$ according to the theory of games [10, 11]. Accordingly, in the equation (8) $c{-}j_m$ is the opponent's costs to develop his SC.

The data presented below are the initial ones, at which the authors conducted theoretical and experimental research of strategic behavior of SC managers.

Table 1. Evaluation of quality parameters SC and values their importance, compiled by the authors based on the results of a survey of visitors SC, and brought to a scale [0; 1].

i	Parameter	Q_1	Q_2	q_i
1	Area	0.947	0.740	0.12
2	Range of goods	0.853	0.648	0.15
3	Transport accessibility	0.851	0.887	0.15
4	Aesthetic parameter	0.861	0.842	0.12
5	Promotions, discounts	0.660	0.516	0.11
6	Quality of goods	0.789	0.759	0.15
7	Availability of brands	0.884	0.748	0.12
8	Events, concerts	0.681	0.605	0.08

This survey was conducted among visitors of SC in the Perm city. They were offered to estimate SC in quality parameters (in a scale from 1 to 10 where 1 – poor quality, 10 – excellent quality) and to estimate (on a scale from 1 to 10 where 1 – it isn't important at all, 10 – it is very important) as far as him this parameter at the choice of this or that SC is important. Average values of parameters were brought to a scale [0; 1] for using the qualimetric model (2).

The next initial parameters (Table 2) need for using equations (1) and (4).

Table 2. The values of the parameter λ, the time correspondence to the SC and the number of residents, depending on the sectors

Initial parameters	Sectors (k) 1	2	3	4	5	6	7	8	9	10
Value of parameter λ:	λ1	λ2	λ3	λ4	λ5	λ6	λ7	λ8	λ9	λ10
for SC1	0	0	0.5	0.5	0.5	0	1	0.5	1	1
for SC2	0	0.5	0	0.5	0.5	1	0	1	0.5	1
Nk	5182	8292	8292	6215	6215	20729	20729	31094	31094	50000
T1	45	45	80	80	80	45	160	80	160	160
T2	45	80	45	80	80	160	45	160	80	160

In the discrete case, the total number of control strategies will be determined by the equation:

$$U = \prod_{m=1}^{M} \overline{u}_m, \tag{9}$$

m is the ordinal number of the parameter that can be controlled by the SC Manager; M – total number of parameters to be managed; $u_m = \{1, 2, \ldots, \overline{u}_m\}$, $\overline{u}_m$ – the maximum number of options for changes to the parameter m, $m = 4, 6, 7, 8$.

The first assumption is sets of the costs for the development of individual SC parameters m have values presented in Table 3.

Table 3. Example of cost distribution options to manage the four criteria

Variants of distribution of management costs	advertising $m=1$	aesthetic appearance $m=4$	quality of goods $m=6$	events $m=8$
$u=1$	40	0	100	25
$u=2$	60	120	200	40
$u=3$	90	250	300	55
$u=4$	120	500	400	70

Consider an example where the strategy of SC managers depends on four cost options for changing four using parameters. In this case (∀m: $\overline{u_m} = \overline{u} = 4$) the number of strategies of each manager will be determined by the equation:

$$U = \overline{u}^{M} = 4^4 = 256. \tag{10}$$

The next assumption was made in this study – the dependence between quality of the each managed parameter and costs to its development is linear. The quality values belongs to interval from 0 to 1, where 0 – bad quality, 1 – excellent quality. The costs to development the used parameter m belongs to interval from $\underline{c}_m$ to $\overline{c}_m$. That is why the suggested cost function determined as the ratio of the difference between the current cost value and the minimum to the cost range. However, according to equation (2) the quality of shopping center calculated as product of quality values of the used parameters and the quality value of other parameters. So if one of the managed parameters quality values will be zero, then quality of shopping center value will be zero too. For exclude such a situation the cost function equation is logic and determine as:

$$Q(c_m) = \begin{cases} \dfrac{c_m - \underline{c}_m}{\underline{c}_m - \overline{c}_m}, & c_m \in C_m \\ 0.001, & if\ c_m = \underline{c}_m \end{cases} \tag{11}$$

$\underline{c}_m$ – the minimum value of C_m, and $\overline{c}_m$ – the maximum value.

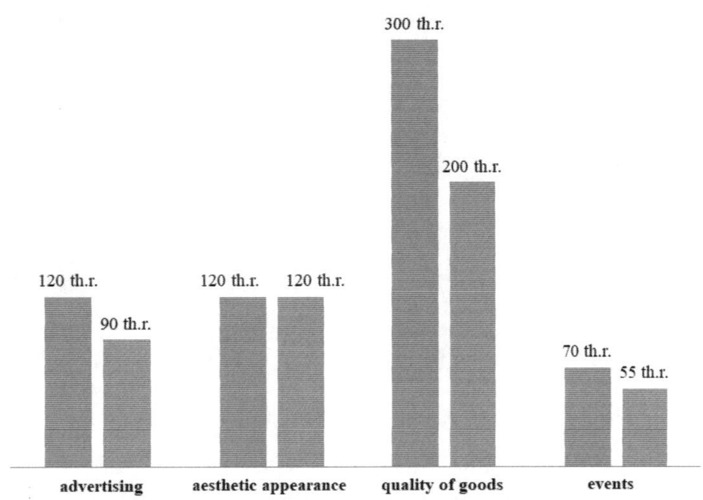

Figure 2. The optimal strategy for the first player.

Note for figure 2: at μ=0.1 the purchase is made every tenth visitor SC; AR – 0.5 th.r.; competitor's quality SC (Q) – 0.663, in relative scale [0;1].

According to the equation (8), the optimal solution to the management problem of the SC is to allocate the management budget to change the used parameters, which provides the maximum of their personal profit [6] or index of profitability. The optimal strategy of the SC Manager by profit maximization criterion (orange columns) and index of profitability maximization (gray columns) is shown below (Figure 2).

It is worth noting that these solutions are obtained with a known strategy of the opponent and a given level of consumption. The level of consumption is characterized by: the share of visitors, making purchases in the SC (the conversion rate - μ); the average amount of purchases that make visitors SC (size of the average check – AR). These parameters describe changes of the environment. Examples of the dependence of the optimal strategy of the SC's managers on the changing environment are given below (Fig. 3, 4).

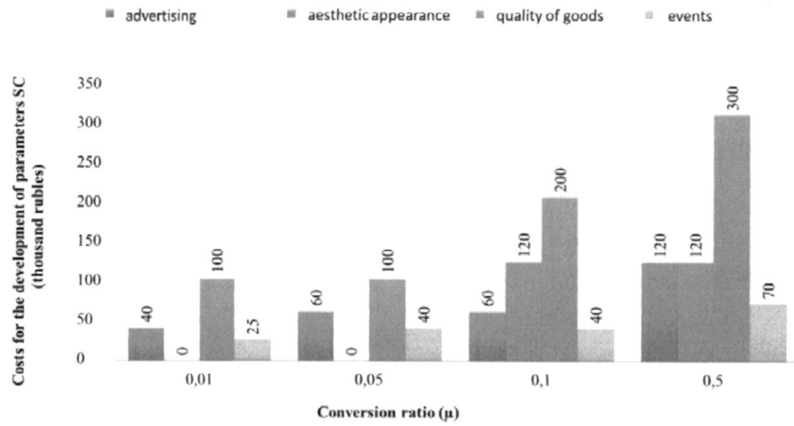

Figure 3. The optimal strategy distribution of development costs SC when the target profit function and the average bill is equal to 0.1 thousand of rubles (th.r.) at different conversion ratios.

Note for figure 3: at μ=0.01, the purchase is made by every hundredth visitor of SC; at μ=0.05, the purchase is made by every twentieth visitor of SC; at μ=0.1, the purchase is made by every tenth visitor of SC; at μ=0.5, the purchase is made by every second visitor of SC.

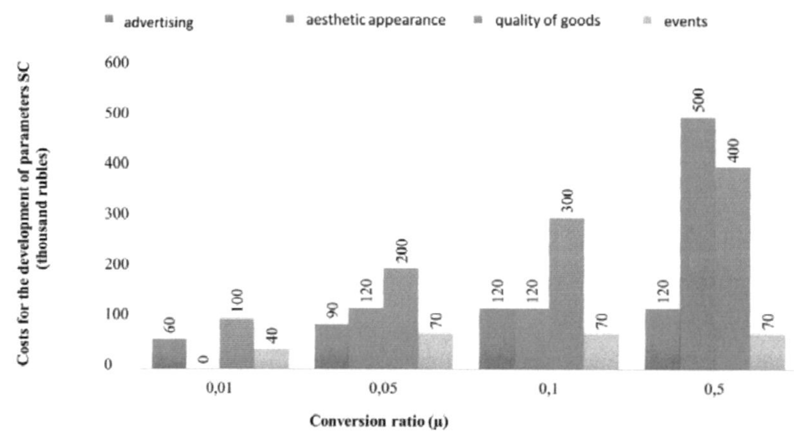

Figure 4. The optimal strategy distribution of development costs SC when the target profit function and the average bill is equal to 0.5 thousand of rubles (th.r.) at different conversion ratios.

Note for figure 4: at μ=0.01, the purchase is made by every hundredth visitor of SC; at μ=0.05, the purchase is made by every twentieth visitor of SC; at μ=0.1, the purchase is made by every tenth visitor of SC; at μ=0.5, the purchase is made by every second visitor of SC.

Allocations costs for SC development show that, when the level of consumption becomes costs of the development and promotion of SC becomes too, which is almost obvious. However, the task of searching for the optimal allocation of costs at the SC's management is not a trivial task and its solution is quite time-consuming. This fact makes necessary the management decision support system (DSS). It should also be noted that strategy is the most sensitive to changes in the external environment. For the purposes of this study, devoted precisely to the study of strategic behavior of competitive managers of SC, environment is fixed.

2.2. *An example of solving a game-theoretic problem with two players managing the competitive SC*

In this paper, we illustrate an example of the solution of the game-theoretic problem in the simplest formulations, where the participants in the game are two players that manage the competitive SCs and their strategies are determined by the costs of their development.

In a situation where the player does not know the opponent's strategy, it is suggested to search for strategies according to the concept of the maxmin (or also called maximum guaranteed result (MGR)). This situation is typical for players with the first rank of strategic reflection. The second rank of reflection means that the player calculates his Best Response (BR), knowing the opponent's strategy. The third rank of reflection means that the player uses the strategy corresponding to Double Best Response (DBR), with the known opponent's Best Response [11]. Comparative analysis strategies obtained based on these concepts is presented in Table 4. All solutions in table are present by number of player's strategy. It one is determined according to the equation:

$$1\cdot\left(u_{m=8}-1\right)+4\cdot\left(u_{m=6}-1\right)+16\cdot\left(u_{m=4}-1\right)+64\cdot\left(u_{m=7}-1\right). \quad (12)$$

Table 4. Matrix of strategies of players in the formulation of a game with two managers when using profit as an objective function.

Strategies		2nd player		
		MGR2	BR2	DBR2
1st player	*MGR1*	151 (MGR1) 151 (MGR2)	151 (MGR1) 151 (BR2(MGR1))	151 (MGR1) 151 (DBR 2(BR1(MGR2)))
	BR1	151 (BR1(MGR2)) 151 (MGR2)	151 (BR1(MGR2)) 151 (BR2(MGR1))	151 (BR1(MGR2)) 151 (DBR2(BR1(MGR2)))
	DBR1	151 (DBR1(BR2(MGR1))) 151 (MGR2)	151 (DBR1(BR2(MGR1))) 151 (BR2(MGR1))	151 (DBR1(BR2(MGR1))) 151 (DBR2(BR1(MGR2)))

The solution game-theoretic formulation of two SC players-managers shows that with an increase in rank of reflection, the players share the same strategies (Table 4). It is important to note that the solution depends on the level of consumption, which in this study deliberately not investigated.

In Table 4 the player's choice of strategy 87 corresponds to the following costs: for advertisement 60 th.r., for aesthetic appearance – 120 th.r., for the quality of goods – 200 th.r. and for holding events – 55 th.r.; strategy 151 corresponds to: an increase in advertising costs – 90 th.r., costs for aesthetic appearance – 120 th.r., costs for the quality of goods – 200 th.r. and increase the cost of events up to 70 th.r.

Thus mathematically possible to determine the theoretically predicted behavior of the SC managers and becomes relevant experimental study of strategic behavior through a series of business games. An experimental study of the above problem in the form of a business simulation game will make it possible to find out – to which strategy the people will come.

According to the authors of the experiment, these switches are necessary to visualize the optimal allocation of costs for the development of SC, as well as comparing the strategies of players at different ranks of reflection (see table 4). When switching, the histogram visualizes the solution for the corresponding mode, and in the area 4 the optimal values of the quality of the player and his opponent are displayed. Also in the layout there is a switch showing the optimal allocation of the objective function OF (profit maximization (5) or index of profitability (6)) selected by the rules of experiment.

At the end of the step, each player tells the experimenter the number of the strategy he selected and the value of the quality of his SC in the implementation of this strategy. At the beginning of the next step, each player replaces the quality of the opponent on the value that reached the opponent on the last step.

The end of the game is the situation when a certain number of steps is end in the game or the players do not change their strategies for the second consecutive step. The latter also takes into account the situation in which players would be beneficial (advisable) to change their strategy. This will show the difference between theoretically predicted and actual people behavior.

Before starting the experiment, players are informed of all the theoretical information about the essence of the game and methods of solving the SC management problem. The experimenter must be convinced that each player correctly understood the rules of the game and his role, through feedback.

Preliminary experiments were conduct in the Center of business simulation game of Perm national research polytechnic university (PNRPU). Participants in this study were master students of PNRPU, exactly, students of 1-st and 2-nd courses of master programs “Technologies of real estate property management” and “Risk management in civil engineering”. During one experiment session students were divided in 4 groups, each of one included 2 students. Total amount of participants’ token part in the preliminary experiments is 32 students. One of the purposes of the preliminary experiments was testing software module and that’s why the experiment didn’t incentivize.

In despite of preliminary character of the experiments two aspects were found: players were able to find the optimal allocation without using the decision support system in a fairly short period of time (see Figure 5 and Table 5), in other words, players came to a theoretically predicted strategy; the optimal strategy, in turn, was resistant to the actions of the opponent, and therefore the players did not change their strategies, which led to a very rapid completion of the experiments.

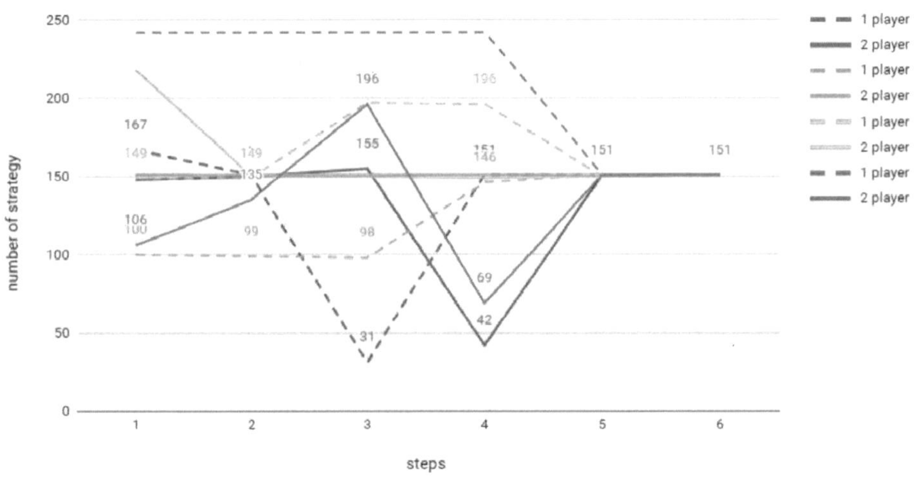

Figure 5. Results of a preliminary experiment.

Note for figure 5: In the figure preliminary results of a game of 4 groups of students (on 2 players in everyone) are presented. In other groups, similar results have turned out.

The resistance of the optimal strategy is explained by the experimental assumptions, including the values of used parameters of the costs to development and promotion SC (see table 3) and about the type of cost functions (see equation (11)). As it turned out, the costs are too large a change step and when the opponent tries to attract additional consumers to itself, the player's costs for their preservation in its SC exceed the expected economic effect, determined by the product of the average check (*AR*) on the number of visitors (n_j) and the conversion rate (μ). This fact make necessary for additional theoretical study of the SC managment problem with different cost functions.

Table 5. Matrix of strategies of players in the formulation of a game with two managers when using profit as an objective function.

Groups	Players	Steps *1*	*2*	*3*	*4*	*5*	*6*
1-st group	*1-st player*	167	151	31	151	151	151
	2-nd player	148	150	155	42	151	151
2-nd group	*1-st player*	100	99	98	146	151	151
	2-nd player	151	151	151	151	151	151
3-rd group	*1-st player*	149	149	197	196	150	151
	2-nd player	218	150	150	149	150	151
4-th group	*1-st player*	242	242	242	242	151	151
	2-nd player	106	135	196	69	151	151

2.3. Change of functions of costs for development and promotion of controlled parameters of SC in a computational experiment

To conduct the experiment in the mathematical model, the ranges of cost functions were increased (table 6). The conversion ratio is changed by 0.05, the average check is assumed to be 500 th.r.

Table 6. Example of allocation of increased cost options to manage four criteria.

Variants of distribution of management costs	Advertising m=7	Aesthetic appearance m=4	Quality of goods m=6	Events m=8
u=1	100	100	100	100
u=2	200	200	200	200
u=3	300	300	300	300
u=4	400	400	400	400

By increasing the range of costly functions, a stable strategy of player behavior and optimal resource allocation was revealed (number of optimal strategy – 0, in which m=4, m=6, m=7, m=8 correspond to the level of costs u=1). The resulting optimal strategy can be interpreted as follows: at the maximum cost functions should adhere to the minimum costs in the development and promotion of the SC.

The next stage was to reduce the range of cost functions (table 7). The conversion rate and the average check are the same.

Table 7. An example of the distribution of reduced variants of the cost of four criteria.

Variants of distribution of management costs	Advertising m=7	Aesthetic appearance m=4	Quality of goods m=6	Events m=8
u=1	5	0	3	8
u=2	10	10	6	16
u=3	15	20	9	24
u=4	20	30	12	32

When reducing the range of cost functions, it was also revealed a stable strategy of player behavior and optimal allocation of costs (number of optimal strategy – 255, in which m=4, m=6, m=7, m=8 correspond to the level of costs u=4). The resulting optimal strategy can be interpreted as follows: with minimal cost functions should adhere to the maximum cost in the development and promotion of the SC.

The last stage was the replacement of the quality evaluation function from geometric (2) to arithmetic in the mathematical model, with the cost functions adopted in section 2.2 (see table 3). As a result, the optimal strategy (without reflection) was for the first player – 5, for the second (with the first rank of reflection) – 196, for the first (with the second rank of reflection) – 151. As it turned out, the strategies of the players are resistant to the actions of opponents and, accordingly, to the quality of competing SCs, that is, the resistance of strategies does not depend on the function describing the quality of the object.

With the costs that were taken in the computational experiment, it turned out that there is a allocation that shows their optimal value. We can assume that the adopted objective function, the model adopted to assess the consumer attractiveness and quality of the SC, adopted in the experiment, the cost functions are adequate. Thus, it can be said that if in real practice of management, the quality of SC, which affects the choice of the consumer object to visit, would be described using a geometric weighted model, the strategy of the Manager would be resistant to the action of his opponent, but not resistant to changes in the environment, in the form of conversion ratio and average check. Previously the authors [6] showed on the example of SC of Perm city that the geometric model gives more accurate results of forecasting the consumer flow to real data than other recommended weighted models.

3. Conclusion

Sensitivity analysis showed resistance of SC management strategy to the opponent's behaviour, but sensitivity to environmental changes and consumer behaviour. The strategy of the players is resistant to changes in the cost functions (decrease or increase values) and to change the function describing the quality of the used parameters of the SC. The result of this study is: at the maximum cost functions should adhere to the minimum costs in the development and promotion of the SC; with minimal cost functions, adhere to the maximum cost in the development and promotion of the SC. As it turned out, the strategies of the players are resistant to the actions of opponents and, accordingly, to the quality of competitive SC, that is, the stability of the object. Thus, it can be said that if in real practice of management, the quality of the SC, which would be described using a geometric weighted model, of his opponent, but not of change in the environment, in the form of conversion.

Acknowledgments

The reported study was funded by RFBR according to the research project № 17-07-01550.

References

[1] Novikov D A 2013 *Theory of Control in Organizations* (New York: Nova Science Publishers) 341 p
[2] Burkov V N, Goubko M V, Korgin N A and Novikov D A 2015 *Introduction to Theory of Control in Organizations* (Boca Raton: CRC Press) 128 p
[3] Novikov D A 2013 Control Methodology (New York: Nova Science Publishers) 76 p
[4] Novikov D A and Rusjaeva E U 2012 Foundations of Control Methodology *Advances in Systems Science and Application* **12** pp 33–52
[5] Novikov D A 2016 *Cybernetics: from Past to Future* (Heidelberg: Springer) 107 p
[6] Alekseev A O, Spirina V S and Korgin N A 2016 Commercial Real Estate Management Technology Taking Into Account Consumer Preferences *Large-Scale Systems Control* **62** pp 124–168

IOP Conf. Series: Materials Science and Engineering **481** (2019) 012058 doi:10.1088/1757-899X/481/1/012058

[7] Huff D L 1963 A Probabilistic Analysis of Shopping Center Trade Areas *Land Economics* **39**(**1**) pp 81–90

[8] Huff D L 2003 Parameter Estimation in the Huff Model (ArcUser Post at October-December) URL: http://www.esri.com/news/arcuser/1003/files/huff.pdf

[9] Huff D L and Black W C 1997 The Huff model in retrospect *Applied Geographic Studies* **1**(**2**) pp 83–93

[10] Burkov V N, Goubko M V, Kondrat'ev V V, Korgin N A and Novikov D A 2013 *Mechanism Design and Management: Mathematical Methods for Smart Organizations* (New York: Nova Publishers) 198 p

[11] Goubko M V and Novikov D A 2008 Game-theoretical Models and Methods of the Organizational Systems Control Theory *Proc. Int. Conf. on Game Theory and Management* (SPb.: Graduate School of Management SPbU) p 140

IOP Conf. Series: Materials Science and Engineering **481** (2019) 012059 doi:10.1088/1757-899X/481/1/012059

Technologies of Development Decisions Making in Residential Civil Engineering

A O Alekseev[1], K S Koskova[1], E R Galiaskarov[1]

[1]Department of Construction Engineering and Materials Science, Civil Engineering Faculty, Perm National Research Polytechnic University, Komsomolsky av., 29, Perm, 614990, Russia

E-mail: koskova@cems.pstu.ru; alekseev@cems.pstu.ru

Abstract. Intelligent technologies of grounding the development decisions for conceptual design of residential properties are suggested. The concept of the real estate property determines a set of properties that should be possessed by property, e.g., location, infrastructure, engineering equipment, decoration, layout, etc. The problem of conceptual design of residential real estate properties is considered taking into account the consumer appeal and end-user preferences. Are methods of modelling consumer preferences, allowing to form the shape of the future building, satisfying most of the needs and desires of the real estate market? The technology is represented on the example of a ten-story residential single-entry house, where the factors influencing the choice of consumers use location, structural features, technical equipment and decoration. The competitive advantage of the conceptual design is the direct economic effect that an investor and/or developer receives as a result of saving money for constructing and promoting an item on the market. An indirect, but no less significant, economic effect is the reduction of losses arising from the disruption of the sales plan. The demand for finished premises is provided due to the account of real consumer preferences and opportunities of market participants.

1. Introduction

Recently, activities in the field of real estate construction are generally called development [1], and the construction is understood as either the industry or directly the process of creating real estate. In this paper, we will talk about strategic decisions made at the pre-investment stage by developer's activity participants – the developers. Since the concept of strategic decisions is quite general, to specify the decisions on which property to construct, we will call development decisions.

Designing and construction of residential property should be carried out considering consumer preferences [2] since the selection of housing is made on the basis of self-perception by the consumer of the housing quality properties. Therefore, at the pre-investment stage, even before designing a real estate item, it is worth considering the concept [1, 2] of the real estate property and taking into account the factors that influence the consumer choice.

The authors believe that the conceptual design [2, 3] of real estate properties is currently in demand and understood as front end engineering design, including marketing analysis [4-6] of the real estate market and identification of consumer preferences [7-11], as a result of which such a concept (appearance) of the future building or complex of buildings and structures, including external and internal infrastructure of the property that meets most of the needs and desires of end users.

Traditional design practices for the development of the project concept [12] are based on brainstorming methods that can take into account the opinions of the project team members and only indirectly – the real preferences of end users. Modern methods of mathematical modeling consumer preferences [13, 14] are provided to shape the future building, which meets most of the needs and desires of real estate market participants.

2. Methodology

If we consider real estate under construction as a multiparameter object described with the vector of properties (location, number of floors, finish, materials, area, etc.), then integrated estimation mechanisms can be used, for example, based on objectives (criteria) trees and binary convolution matrices [2, 15]. With the help of these or other mechanisms, it is possible to formalize the procedure for selecting by consumers the real estate items on the market and their utility function.

The matrix procedure of integrated assessment is based on reducing all the essential characteristics that influence the managerial decision making to a single scale of integrated assessment (to the criterial space with metrics). In the criterial space, the heterogeneous characteristics can be compared at a qualitative level. The convolution of qualitative criteria is performed by the formation of binary convolution matrices that describe the ratio of the preferences carrier to the pair of parameters to be convoluted. A complete convolution describing the functional relationship between the integrated assessment and a set of input criteria is described by a composition of binary packages [5].

In order to convolute heterogeneous parameters, it is necessary to reduce them to a single qualitative scale. For the purposes of this study, a four-point scale {1, 2, 3, 4} was used, each value of which has a qualitative interpretation: 1 - low attractiveness; 2 - average attractiveness; 3 - high attractiveness; 4 - very high attractiveness. A fragment of the descriptive scales of the factors that affect consumers when purchasing housing is given in Table 1.

Table 1. Fragment of the descriptive scales of the factors that affect consumers when purchasing housing.

Interpretation	Estimate			
	low attractiveness	**average attractiveness**	**high attractiveness**	**very high attractiveness**
Design features of the house	IV group of solidity	III group of solidity	II group of solidity	I group of solidity
House infrastructure	non-off-the-grid house, outmoded equipment	non-off-the-grid house, modern equipment	partially off-the-grid house, modern equipment	off-the-grid house, modern equipment
Improvement	shell condition	ordinary finish	improved finish	Design project / complete renovation
Location	suburb	average distant area	Areas adjacent to the center	downtown

To formalize the model of consumer preferences, the authors carried out a marketing research in the form of a social survey. Participants in the real estate market (potential buyers – preferences carriers) were asked to compare different combinations of factors and determine which combinations are attractive for them. To do this, during the survey respondents were asked the question "Is this combination of properties (factors) of the real estate item attractive for you?", which required an unambiguous answer "yes" or "no." When processing personal details, the answers "yes" were replaced with 1, "no" – with 0. A fragment of the results of the consumer survey on the attractiveness of the building is presented in Table 2.

Table 2. Fragment of the results of the consumer survey on the attractiveness of the building.

Left – Right	**«Building attractiveness» – «Improvement»**		**Respondent 1**	**Respondent 2**	**…**	**Portion of affirmative answers, %**	**Attractiveness assessment**
	Factor 1	**Factor 2**					
1-2	shell condition	partially off-the-grid house - II group of solidity	1	0		50	2
1-3	shell condition	partially off-the-grid house - I group of solidity	0	1		60	2
1-4	shell condition	off-the-grid house - II group of solidity	1	1		70	2
2-1	ordinary finish	modern equipment - I group of solidity	1	1		60	2
2-2	ordinary finish	partially off-the-grid house - II group of solidity	0	1		60	2
2-3	ordinary finish	partially off-the-grid house - I group of solidity	1	1		80	3
2-4	ordinary finish	off-the-grid house - II group of solidity	1	1		80	3
3-1	improved finish	modern equipment - I group of solidity	0	0		60	2
3-2	improved finish	partially off-the-grid house - II group of solidity	0	0		60	2
3-3	improved finish	partially off-the-grid house - I group of solidity	0	1		80	3
3-4	improved finish	off-the-grid house - II group of solidity	1	1		90	4

To determine the degree of attractiveness of combinations, the following rule was adopted:

- if less than 50% of the respondents answered in the affirmative, this combination of properties was accepted as low attractive (estimate - 1; interpretation - almost no one will buy);
- if 50% to 70% of the respondents answered in the affirmative, this combination of properties was accepted as average attractive (estimate - 2, interpretation - some will buy);
- if 70% to 90% of the respondents answered in the affirmative, this combination of properties was accepted as high attractive (estimate - 3, interpretation - many will buy);
- if over 90% of respondents answered in the affirmative, this combination of properties was accepted as very high attractive (estimate - 4, interpretation - almost everyone will buy);

The marketing research allowed to formalize possible combinations of different qualitative states of factors in the form of a matrix mechanism of integrated assessment (Fig. 1), which is a "compact" form of recording possible combinations of apartment house properties (in the example with 4 factors and 4 possible states of each factor, the full set of object properties combinations represents $4^4 = 256$ variants):

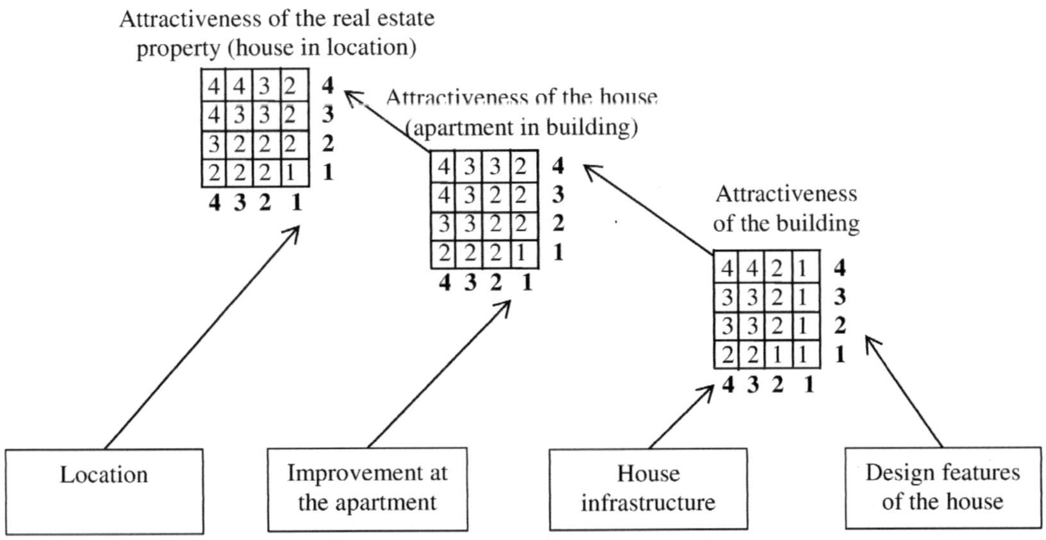

Figure 1. Fragment of a chain of possible variants for apartment house construction

To determine the financially cost-based functions that specify the cost of providing the quality of the factors to be considered, estimates were made on the basis of uniform norms and quotations in case of a ten-story residential single-entry house with the following technical and economic indicators:

- building area: 737 m2;
- number of storeys: 10 storeys;
- house dimensions: 41x16m;
- gross building area: 6394 m2;
- total number of flats (11 on the story) – 110 flats;
- single-room – 40 flats;
- two-room – 50 flats;
- three-room – 20 flats;
- total area of flats – 4450 m2;
- living area of flats – 2711 m2;
- estimated approximate number of persons: 230 persons;
- size of a land plot for this 10-story house: 6290 m2.

The results of averaged estimates (cost-based functions) are given in Table 3.

Table 3. Averaged summary estimate of cost-based functions (in mio. rub.)

Interpretation	**Estimate**			
	low attractiveness	**average attractiveness**	**high attractiveness**	**very high attractiveness**
Design features of the house	32.4	33.2	3.2	53.5
House infrastructure	8.6	9.2	8.3	14.3
Improvement	0	3.8	4.8	8.2
Location	3.1	3.6	6.3	5.6

3. Results

Normally, using the matrix mechanism of integrated assessment (see Fig. 1), it is possible to solve the inverse problem of integrated assessment - to search for a variant that provides the required value of the integrated assessment. In this paper, there is a variant of residential house construction with a given level of consumer appeal, for example, an estimate of 3, which corresponds to the fact that 70% to 90% of respondents will be satisfied with such housing. Using the method of constructing a network of strain variants [16], it is possible to determine such combinations of factors to be considered, among which there is an optimal variant in terms of the prime cost criterion. Knowing the financially cost-based functions (see Table 3), each variant can determine the total cost of providing such a set of properties (Table 4).

Table 4. Cost-based function of finite sets of variants (mln. rub.).

Attractivenes s estimation	Variant	Location	Design features of the house	Infrastructur e	Improv ement	Total value, mio rub.
3	1	2 (3.6)	4 (53.5)	2 (9.2)	2 (3.8)	70.1
	2	2 (3.6)	2 (33.2)	3 (8.3)	2 (3.8)	67.8
	3	2 (3.6)	3 (32.4)	1 (8.6)	4 (8.2)	52.8
	4	2 (3.6)	2 (33.2)	2 (9.2)	4 (8.2)	54.2
	5	3 (6.3)	3 (32.4)	1 (8.6)	1 (0)	47.3
	6	3 (6.3)	2 (33.2)	2 (9.2)	2 (3.8)	52.5
	7	3 (6.3)	1 (32.4)	1 (8.6)	2 (3.8)	51.1

So, we found an optimal development solution for the construction of a ten-story apartment house in terms of the prime cost, which is satisfied by the majority of requests of potential consumers (see Table 4). This solution determines the set of properties that an item should possess; from now, such a set of properties will be called the concept or appearance of the property. The concept of a real estate item is actually a technical design assignment and a prototype of the sketch.

Table 5. Fragment of the concept (appearance) of real estate property.

Criteria	Features	Attractiveness estimation	Cost, mio.rub.
Geographical location	areas adjacent to the center (3)	3 – high attractiveness (many will buy)	6.3
Design features	II group of solidity (3)		32.4
House infrastructure	non-off-the-grid house, modern equipment (1)		8.6
Improvement	shell condition (1)		0
	Fragment Total value*:		47.3

Note: the table shows the total cost of providing the required attractiveness of the property only for the factors to be considered, so the table does not include the costs of connection to engineering networks, improvement, etc.

4. Discussions

The problem of conceptual design of residential real estate properties is considered taking into account the consumer appeal and end-user preferences. The competitive advantage of the conceptual design is the direct economic effect that an investor and/or developer receives as a result of saving money for constructing and promoting an item on the market. An indirect, but no less significant, economic effect is the reduction of losses arising from the disruption of the sales plan. The demand for finished premises is provided due to the account of real consumer preferences and opportunities of market participants.

Acknowledgments

The reported study was funded by RFBR according to the research project № 17-07-01550.

References

[1] Peiser R B and Hamilton D 2012 *Professional Real Estate Development: The ULI Guide to the Business 3rd edition* (Urban Land Institute) p 400
[2] Alekseev A O and Galiaskarov E R 2018 Statement of the problem of conceptual designing of residential real estate taking into account consumer preferences *Proceedings of Universities Investment Construction Real estate* **8(3)** pp 11-25
[3] Teslinov A G 2009 *Concept design of complex decisions* (Saint Petersburg: Piter) p 288
[4] Christensen C M, Levitt T and Kotler Ph 2013 *HBR's 10 Must Reads on Strategic Marketing* (Harvard Business Review Press) p 224
[5] Kotler Ph, Kartajaya H and Setiawan I 2010 *Marketing 3.0: From Products to Customers to the Human Spirit* (New York: John Wiley & Sons Limited) p 208
[6] Prahalad C K, Ramaswamy V 2004 *The Future of Competition: Co-Creating Unique Value with Customers* (Harvard Business School Press) p 256
[7] Llinares C and Page A F 2011 Kano's model in Kansei Engineering to evaluate subjective real estate consumer preferences *International Journal of Industrial Ergonomics* **41(3)** pp 233–246
[8] Gibler K and Nelson S 2003 Consumer Behavior Applications to Real Estate Education *Journal of Real Estate Practice and Education* **6(1)** pp 63–83
[9] Black R, Brown G, Diaz J, Gibler K and Grissom T 203 Behavioral Research in Real Estate: A Search for the Boundaries *Journal of Real Estate Practice and Education* **6(1)** pp 85–112
[10] Lundgren B 2013 Customer-perceived Value in Residential Developments. *International Real Estate Review* **16(1)** pp 1–27
[11] Opoku R A and Abdul-Muhmin A G 2010 Housing preferences and attribute importance among low-income consumers in Saudi Arabia. *Habitat International* **34(2)** pp 219–227
[12] Sternik G M and Sternik S G *Real Estate Market Analysis for Professionals* (Moscow: Publisher Economic) p 605
[13] Alekseev A O 2015 The agent-based concept of multiple factor risks modeling in multi agent systems *Upravlenie ekonomicheskimi sistemami* **76(4)** p 11
[14] Alekseev A O and Alekseeva I E 2015 Mathematical modelling of economic agent's preferences *Upravlenie ekonomicheskimi sistemami* **76(4)** p 18
[15] Kharitonov V A, Alekseev A O, Gureev K A, Lykov M V, Shaidulin R F and Bukalova A Yu 2013 *Intelligent technologies of the property management*, ed A O Alekseev (Perm: Publisher of Perm national research polytechnic university) p 170
[16] Burkov V N, Novikov D A and Shchepkin A V 2008 *Control mechanisms of ecology-economics systems*, ed S N Vasilyev (Moscow: Izdatel'stvo fiziko-matematicheskoj literatury) p 244

IOP Conf. Series: Materials Science and Engineering **481** (2019) 012060 doi:10.1088/1757-899X/481/1/012060

Challenges of Entering International Markets for Russian Construction Companies

A Platonov[1], V Larionova[1], Yu Davy[1]

[1]Department of Economy and Management in Construction and Realty Market, Ural Federal University, Mira, 19, Yekaterinburg, 620002, Russia

E-mail: v.a.larionova@urfu.ru

Abstract: The paper reviews key challenges faced by Russian construction companies while entering the international markets. International construction company ratings demonstrate that the amount of work performed by Russian companies is several times less than the ones performed by international contractors. At the same time the world level companies successfully work overseas. Instead of waiting for beneficial investment climate changes, Russian companies must shift the focus from national market towards foreign expansion and extending operations abroad. That would allow extension of foreign operations and positive changes in the system of values and mentality of Russian contractors.

Keywords: global investment and construction sphere, investment attractiveness, investment and construction expansion, organization behavior.

1. Introduction

International construction market has great potential for Russian construction companies. Analysis of TOP-200 world construction companies (Table 1) performing construction overseas [1, 2], and a similar rating of largest Russian construction companies developed by the Top-RF.ru agency using 2014-2015 data (Table 2) [3] demonstrated that it is impossible to compare Russian and international companies by the amount of revenue Russian. For example, top four Russian companies demonstrated revenue within the range of 260-105 bln roubles according to Forbes, while top four companies from TOP-200 list demonstrated overseas revenue within the range from 2.6 to 1.2 trillion roubles.

Table 1. TOP-10 of the Construction companies providing international construction services (2013 data).

Rating	Name and location	2013 revenue, mln USD
1	Grupo ACS, Spain	44053.80
2	HOCHTIEF AG, Essen, Germany	34845.0
3	Bechtel, San-Francisco, CA, USA	23637.0

Rating	Name and location	2013 revenue, mln USD
4	VINCI, Rueil-Malmaison, France	20292.6
5	Fluor Corp., Irving, TX, USA	16784.3
6	Strabag SE, Vienne, Austria	15392.0
7	BUIG, Paris, France	14789.0
8	Saipem, San-Donato Milanese, Italy	12137.6
9	Skanska AB, Stockholm, Sweden	14141.1
10	Communications Constructions Group Ltd, Beijing, China	13162.50
11	Velestroy, Moscow, Russia	1070.0

Table 2. Largest construction companies of Russia: 2016 (Forbes).

	Position in Forbes rating	Name and Headquarters
1	21	Stroygazmontazh (Moscow)
2	46	Mostotrest (Moscow)
3	79	LSR Group (St. Petersburg)
4	82	Renaissance-Construction (St. Petersburg)
5	96	Velestroy (Moscow)
6	112	Globalstroy -Engineering (Moscow)
7	122	Morton (Moscow)
8	123	FSK Leader (Moscow)
9	124	Stroygazconsulting (Moscow)
10	136	PIK Group (Moscow)

Leading Russian companies are “Stroygazconsulting”, “Stroygazmontazh”, “Mostotrest” and “Tamir”. Leading foreign construction companies are “Grypo ACS” (Madrid, Spain, “HOCHTIFE AG” (Essen, Germany), “Bechtel” (San-Francisco, CA, USA) and “VINCI” (Rueil-Malmaison, France).

TOP-200 rating included large number of construction companies from different countries (table 3). Rating analysis demonstrates that Spain, US, China, Italy, South Korea, Japan and Turkey occupy more than 70% of the world construction market. Other countries are also trying to keep their positions in the rating.

Four last lines of the rating for Russian Companies working in Russia (Forbes rating 156-192) were occupied by ARKS Group (34.1 bln rubles), Velestroy (332 bln rubles), Inteco (30.0 bln rubles), and Capital Group (27 bln. rubles).

Table 3. Results of analyzing TOP-20 of international construction companies (2013 data).

	Country	Number of companies in the rating	Range of positions in the ratinge		Country	Number of companies in the rating	Range of positions in the ratinge
1	Spain	13	1-159	16	Turkey	27	52-200
2	Germany	5	2-106	17	Portugal	2	59-125
3	USA	17	3-192	18	Belgium	1	69
4	France	4	4-165	19	Ireland	1	80
5	Austria	2	6-72	20	Lebanon	2	99-144
6	Sweden	2	8-32	21	Canada	1	100
7	China	47	9-196	22	Egypt	1	105
8	Italy	15	11-199	23	Israel	1	109
9	Brazil	4	12-141	24	Greece	3	24-156
10	South Korea	12	13-189	25	Iran	2	118-197
11	Netherlands	3	18-67	26	UAE	2	119-161
12	Great Britain	3	21-193	27	Mexico	1	122
13	Japan	13	27-182	28	Taiwan	1	131
14	Austria	4	33-190	29	Denmark	1	151
15	India	4	47-173	30	Thailand	1	186

Four last lines of the global rating (lines 197-200) were occupied by "SUNIR Co" (Tehran, Iran), TACA Construction Ink (Istanbul, Turkey), Impresa Construzioni Giuseppe Maltauro SpA (Italy), and "Metag Inssaat Ticaret AS" (Ankara, Turkey) with annual revenue from 15.6 to 14.3 bln. rubles. In general the spread of revenue for construction companies is quite substantial and is close to 2.6 trillion rubles.

According to rating data most of the revenue of the construction companies was obtained from the operation overseas. At the same time these companies also implement local projects with the quantities similar to international projects, but they display only local performance.

This difference is probably due to expansionism of international construction companies, and local orientation of the Russian ones.

There are the following questions – why did it happen, what are the issues, and what can be done to fix the situation?

2. Expansion of Russian construction companies in the world market as a new paradigm

There are external reasons – political, financial and economic reasons that cut expansion ambitions of Russian constructing companies. Expansion was also limited by the trends in construction related the way construction resources are allocated across the country.

In order to counterbalance this international market trends it is important to form expansionist mindset of Russian construction companies' top management and staff. Internal investment-related problems of construction companies can be resolved only through competition on international markets. The amounts of work required for development of construction companies and regional economy can be located both within the country and outside of it. All it takes- is intent and qualification to bid against the international players in their field of operations, because today technologies, equipment, materials, prefabricated elements and components are available on the market, along with generally accepted international project management standards that are used in construction in Russia and other countries.

It is now necessary to combine increasing the investment attractiveness of local markets with the trend of expanding operations abroad to the attractive markets. For example, TOP-10 of the attractive countries for direct investments include countries that are actively working on foreign construction markets, namely, US, China, Canada, Germany, Great Britain, Japan, Australia, France, India, and Singapore [4].

If almost all of the construction companies are working abroad, the Russian companies shall also move in this direction.

In order to do that there are changes to be made in two aspects – both internally by changing the company operations model, and externally by considering specifics of competition and investment environment on global markets.

In terms of internal changes, it is necessary to rearrange operation strategy and management values, rules of organizational behavior, and existing approaches to values within the companies towards the creative organizational and economic expansion. The companies will also need to change mentality of the workforce in general, of the personnel groups, and specialists.

3. Main problems of the construction services export and their solutions

Russia has examples of using successful expansion approaches to solving industry level or national level problems. That includes space exploration, higher education institutions attempting to enter international education markets, construction of nuclear power plants, railroads, bridges, and GLONASS nodes abroad.

Construction companies are to start that work on their own, here and now. The main strategy of changing management and workforce mentality is the combination of nonstandard creative approaches, unchaining and support of creative thinking with strict production discipline, and real devotion to profession on all levels of management and operations.

It is possible to implement these changes by basing upon primary (family, colleagues, and friends), and secondary (non-commercial organizations, trade unions, and religious bodies) referent groups that really affect operations of construction companies. Habermas [5] stated that the combination of social relations for the person (family, relatives, and friends) with the systematic world of business and anonymous relations occurring on the markets, in the society and its institutions, creates internal models of interacting and interconnecting stakeholder mentalities within each company. All construction company operation actors affect each other, intentionally, or without such an intention, and finally affect mentality of a worker – a person acting as a member of a family, and a circle of peers.

Worker personality and its formation within a company are generally affected by owners, other stakeholders, and staff members of different levels that act upon the person within the system of worker's business relations. The system shall be transparent, and implemented through organization ethics and values. At the same time social relations of a person with family, friends, and colleagues can affect personal values and behavior either positively, or in a negativistic, and opportunistic way [6].

The correspondence of the values system with external and internal environment can и described by the efficiency of company operations. Acceptance of the values by the staff allows formation of the new company mission meanings, and new approaches to resolution of actual problems. In general that would signify transition from organizational ideology to strategy, and efficient management procedures.

It is known that organizational theory is based upon the hypothesis of correct relationships and positive motivations of the people. The real company forms a spectrum of interrelations, and multivectored motivation motives, and values aimed at same and different directions.

Owners, stakeholders, top managers, management, and workers shall determine the corresponding formats of interactions, and determine motivations for achieving stable development of the company and solving the appearing problems.

Some reasons for unsuccessful company adaptation to the occurring changes in competition include outdated value-based approaches, rules and assumptions acting as a base for organization, management approach that does not fit the strategy and external challenges, lack of high-order goals that unite the personnel, failure to resolve current issues, and skills of the staff unsuitable for the selected development strategy.

The necessity to move away from technocratic approaches to management is evident. All experience from USSR and Russia demonstrates that without some invisible virtual delta of spiritual culture all material resources of the company supported by its assets, history record, technologies, machinery, and equipment do not allow company to compete internationally.

Spiritual culture includes the values, social norms, meanings, skills and knowledge that are usually ignored by the management. Also the national economy displays the trend of recreating unfavorable conditions for modern social and economic systems.

Speaking about the external expansion vector for construction companies, it is necessary to note the following. Crisis will affect operations of construction companies for years to come. That would definitely cause the decreased in amount of capital investments and construction activities on a local market, increase in amount s of unfinished construction [7, 8], and bankruptcy of construction companies [9].

In order to overcome the trend it is necessary to increase level of efforts in the area of foreign expansion for companies and corporations [10].

One of the strategic directions of expansion is the development of regional and national mechanisms for supporting export of construction services. International experience shows successful private companies and non-commercial partnerships that support construction companies in terms of financing and expanding construction services export by assisting in establishing contacts with foreign partners. It is necessary to form a system of tools (foundations, programs, and financing lines) that would provide targeted financial and administrative support for Russian companies participating in international construction projects and acting as general contractors abroad. Support mechanisms must be created with involvement of specialized companies, and use public funds or funds of public-to-private partnerships. Specialized organizations must create favorable conditions for national construction companies on foreign construction markets, and assure compliance with international construction market requirements.

For general contract market the requirements can include the following:

- Financial stability and ability to take responsibility for project risks;
- Use of up-to-date construction technologies;
- Internal design force aimed at providing financial, technological, technical and schedule optimization for investment and construction activities;
- Qualified and skilled workforce possessing experience and equipment required to implement construction projects;
- Access to international financing
- International marketing.

For Russian companies the solution may be in the plane of creating joint ventures with international construction companies to take part in international bidding procedures. In case the international partner has requisites necessary for bidding, the joint venture will also be allowed to participate. In case of success, and successful project execution, the experience and financial results will be credited to the Russian company too. Later on the national company can take part in international tenders, possessing required experience and money. It is important that tender procedures and bidding requirements are quite uniform, so after gaining experience in a joint venture regional construction companies can take part in international tenders on their own.

IOP Conf. Series: Materials Science and Engineering **481** (2019) 012060 doi:10.1088/1757-899X/481/1/012060

4. Conclusion

Thus, it is possible to formulate the following goals and priority directions of operations for construction companies in crisis conditions – expansion to international markets, and corresponding changes of management models and methods.

Investment attractiveness of a territory and construction industry has an infinite time lag until the current collisions will somehow resolve. At the same time expansion is the action that depends upon its initiator. Attractiveness and expansion do not oppose, but supplement each other.

Acknowledgments

The authors would like to express their deepest gratitude to the Russian Foundation for Basic Research (RFBR) for the support of the research within the project No. 17-22-07001 "The complex algorithm of culture-based regeneration of minor industrial cities in the context of agglomeration processes in Russia and Europe".

References

[1] TOP-200 of the largest construction companies for 2014, available at http://total-rating.ru/459-krupneyshie-stroitelnye-kompanii-mira.html Febr. 2017

[2] World's largest construction companies for 2015, available at http://total-rating.ru/459-krupneyshie-stroitelnye-kompanii-mira.html, Oct. 2018

[3] Russia's largest construction companies of 2014, available at http://7idey.ru/samye-krupnye-stroitelnye-kompanii-rossii/, Febr. 2017

[4] TOP-10 of the most attractive countries for direct investments, available at https://prian.ru/news/32631.html, Febr. 2017

[5] Habermas J Die Zukunft der menschlichen Natur. Auf dem Weg zu einer liberalen Eugenik, available at http://onscene.ru/Die-Zukunft-der-menschlichen-Natur--auf-dem-Weg-zu-einer-liberalen-Eugenik-J%C3%BCrgen-Habermas/2/cghicdh, Febr. 2017

[6] Belkin V and Bezobrazova Yu 2011 *Socio-economic foundations of opportunism employees of Russian companies* (Ekaterinburg: Institut ekonomiki UrO RAN) p 160

[7] Kashkinbaev T 2018 *Development of construction business under conditions of activating commissioning of incomplete construction objects* PhD Thesis (Moscow: Moscow State University of Civil Engineering) p 178

[8] Nikishina O B and Nikishina O V 2015 Unfinished Building: Causes of Occurrence and Possible Ways of Solution *News of universities. Investments. Building. Real Estate* **4 (15)** (Irkutsk: Irkutsk National Research Technical University) pp 83-91

[9] Federal Service of National Statistics Data for 2016 and 2017 available at http://www.gks.ru/wps/wcm/connect/rosstat_main/rosstat/ru/statistics/enterprise/building/, Oct. 2018

[10] Asaul A 2001 *Corporate structures in regional construction complex* ed A N Asaul and A V Batrak (Moscow: ASV, Saint Petersburg: Saint-Petersburg Architecture and Construction University) pp 168

[11] Platonov A and Larionova V 2017 Investment and Operational Expansion Strategy for National Construction Contractors *Herald of Saint-Petersburg State Economic University Faculty of Management* **1-2** (Saint-Petersburg: Saint-Petersburg State Economic University) pp 225–31